THE LIBRARY
ST. MARY'S COLLEGE OF MARYLAND
ST. MARY'S CITY, MARYLAND 20686

D1787017

Chimeras in Developmental Biology

Two-month-old quail↔chick "spinal cord chimera"

Chimeras in Developmental Biology

EDITED BY

Nicole Le Douarin

Institut d'Embryologie du CNRS et du Collège de France
Nogent-sur-Marne, France

Anne McLaren

MRC Mammalian Development Unit
University College London
London, England

1984

ACADEMIC PRESS
(Harcourt Brace Jovanovich, Publishers)
London Orlando San Diego New York
Toronto Montreal Sydney Tokyo

COPYRIGHT © 1984, BY ACADEMIC PRESS, INC. (LONDON) LTD.
ALL RIGHTS RESERVED.
NO PART OF THS PUBLICATION MAY BE REPRODUCED OR
TRANSMITTED IN ANY FORM OR BY ANY MEANS, ELECTRONIC
OR MECHANICAL, INCLUDING PHOTOCOPY, RECORDING, OR ANY
INFORMATION STORAGE AND RETRIEVAL SYSTEM, WITHOUT
PERMISSION IN WRITING FROM THE PUBLISHER.

ACADEMIC PRESS, INC. (LONDON) LTD.
24-28 Oval Road,
London NW1 7DX

United States Edition published by
ACADEMIC PRESS, INC.
Orlando, Florida 32887

LIBRARY OF CONGRESS CATALOG CARD NUMBER: 83-83403

ISBN 0-12-440580-0

PRINTED IN THE UNITED STATES OF AMERICA

84 85 86 87   9 8 7 6 5 4 3 2 1

Contents

Contributors xi
Preface xiii

PART 1 Technical Aspects

I
Making Chimeras
V. E. Papaioannou and F. Dieterlen-Lièvre

    I. Introduction   3
   II. Avian Chimeras   5
  III. Mammalian Chimeras   18
  IV. Conclusions   32
       References   34

II
Cell Markers
John D. West

    I. Introduction   39
   II. Applied Markers   40
  III. Genetic Markers   42
      References   63

PART 2 Chimeras in Cell Lineage Studies

III
Early Development of Birds
L. Vakaet

    I. General Introduction   71

II. Somatic Cell Lineage Studies	72
III. Chimeras in the Study of Embryonic Induction	81
References	87

IV
Somatic Cell Lineages in Mammalian Chimeras
J. ROSSANT

I. Introduction	89
II. Allocation of Cells to Distinct Lineages	92
III. Restriction of Cells to Different Lineages	101
IV. Conclusions	106
References	107

V
Germ Cell Lineages
ANNE MCLAREN

I. Germ Cell Lineages	111
II. Conclusions	126
References	127

PART 3 Hemopoietic and Immune System

VI
Blood in Chimeras
FRANCOISE DIETERLEN-LIÈVRE

I. Introduction	133
II. Avian Chimeras	134
III. Amphibian Chimeras	152
IV. Mammalian Chimeras	155
V. Conclusion	161
References	161

VII
Human Chimeras
PATRICIA TIPPETT

I. Introduction	165
II. Twin Chimeras	166
III. Dispermic Chimeras	169
IV. A Fertile XX/XY Dispermic Chimera	171
V. Twins with an Unusual XX/XY Chimerism	172

VI.	Human Chimera Disclosed by Her Progeny	174
VII.	Chimeras of Unestablished Type	175
VIII.	XX/XY Mosaicism without Other Evidence of Chimerism	176
IX.	Other Conditions Simulating Chimerism	176
X.	Summary	177
	References	177

VIII
Primary Lymphoid Organ Ontogeny in Birds
NICOLE M. LE DOUARIN, FRANCINE V. JOTEREAU,
ELISABETH HOUSSAINT, AND JEAN-PAUL THIERY

I.	Introduction	179
II.	Thymic Histogenesis	181
III.	Ontogeny of the Bursa of Fabricius	203
IV.	Concluding Remarks	212
	References	214

IX
The Use of Chimeric Mice in Immunology: How Does the Immune System Know Self and Non-Self?
TAKESHI MATSUNAGA

I.	Introduction—A Problem of Recognition of Self and Non-Self by the Immune System	217
II.	Various Chimeric Mice Used for Experiment	219
III.	Allogeneic MHC Tolerance in Chimeras and Some Other Problems	221
IV.	MHC Restriction, Immune Response (*Ir*) Gene, and Cell Interaction	226
	References	235

PART 4   Muscle and Skeleton

X
The Use of Chimeras in Analyses of Craniofacial Development
DREW M. NODEN

I.	Introduction	241
II.	Defining Embryonic Origins of Cephalic Tissues	244
III.	Experimental Analyses of Craniofacial Development	261
IV.	Extrapolation to Mammals	269
V.	Methodological Considerations	271
VI.	Summary and Perspectives	275
	References	277

XI
Muscle and Skeleton of Limbs and Body Wall
MADELEINE GUMPEL-PINOT

I. Introduction	281
II. Birds	282
III. Mammals	302
IV. Concluding Remarks	307
References	308

PART 5 Nervous System

XII
Chimeras in the Study of the Peripheral Nervous System of Birds
NICOLE M. LE DOUARIN, M. A. TEILLET, AND J. FONTAINE-PERUS

I. Neural Crest and Placodal Origin of the PNS	318
II. Distribution of Developing Potentialities along the Neural Crest	339
III. Development of Peptidergic Neurons in Chimeras	341
IV. All Embryonic PNS Ganglia Contain Greater Developmental Potentialities (Both Quantitatively and Qualitatively) Than Are Expressed in Normal Development	344
V. Quail↔Chick "Spinal Cord Chimeras" Can Hatch and Survive: Observations and Perspectives	348
References	349

XIII
Ontogeny and Genetics of the Mammalian Nervous System
RICHARD J. MULLEN

I. The Normal Cerebellum	355
II. Cerebellar Mutants	359
III. Other Regions of the Nervous System	365
IV. Conclusion	367
References	367

XIV
Mouse Chimeras in the Study of Behavior
MURIEL N. NESBITT

I. The Background	369
II. The Problem	372
III. How Can Chimeras Provide Answers?	373

IV. Methodology	374
V. Results	375
VI. Significance of the Data	377
References	379

PART 6 Urogenital System

XV
Chimeras and Sexual Differentiation
ANNE MCLAREN

I. Hydra	382
II. Nemertine Worms	383
III. Amphibia and Birds	386
IV. Mammals	387
V. Conclusions	397
References	397

XVI
Chimeric Tissue Combinations in the Analysis of Developmental Mechanisms in the Embryonic Kidney
LAURI SAXÉN

I. The Problems	401
II. Transmission and Spread of the Inductive Signal	402
III. Exclusion of an Assimilatory Mode of Induction	403
IV. Demonstration of the Migratory Capacity of Induced Cells	404
V. Origin of the Glomerular Endothelium	405
VI. Origin of the Glomerular Basement Membrane	407
VII. Comment	407
References	408

XVII
Teratocarcinoma Chimeras and Gene Expression
C. L. STEWART

I. Introduction	409
II. The Isolation of Teratocarcinomas and the Formation of Teratocarcinoma Chimeras	411
III. *In Vitro* Derived Mutants of EC Cells	412
IV. Chromosome-Mediated Gene Transfer into EC Cells	414
V. Viral- and DNA-Mediated Transformation of EC Cells	415
VI. Summary and Conclusions	422
References	424

PART 7 Perspectives

XVIII
Mammalian Chimeras—Future Perspectives
R. L. Gardner

I. Introduction	431
II. Size Regulation in Chimeric Embryos	432
III. Refinements in Chimera Production	433
IV. Cell Markers	436
V. Cell Lineage	438
VI. Tissue Growth	439
VII. Mutant Genes	440
VIII. Concluding Remarks	441
References	441

XIX
Quail–Chick Chimeras: Concluding Remarks and Perspectives
Nicole M. Le Douarin

Text	445
References	446

Index 447

Contributors

Numbers in parentheses indicate the pages on which the authors' contributions begin.

F. Dieterlen-Lièvre (3, 133), Institut d'Embryologie du CNRS et du Collège de France, F-94130 Nogent-sur-Marne, France

J. Fontaine-Perus (318), Institut d'Embryologie du CNRS et du Collège de France, F-94130 Nogent-sur-Marne, France

R. L. Gardner (431), Sir William Dunn School of Pathology, University of Oxford, Oxford OX1 3RE, England

Madeleine Gumpel-Pinot (281), Laboratoire de Neurochimie, INSERM, L'Hôpital de la Salpêtrière, 75651 Paris, France

Elisabeth Houssaint (179), Faculté des Sciences, INSERM, F-44035 Nantes, France

Francine V. Jotereau (179), Faculté des Sciences, INSERM, F-44035 Nantes, France

Nicole M. Le Douarin (179, 318, 445), Institut d'Embryologie du CNRS et du Collège de France, F-94130 Nogent-sur-Marne, France

Takeshi Matsunaga (217), Department of General and Oncologic Surgery and Division of Biology, City of Hope National Medical Center, Duarte, California 91010

Anne McLaren (111, 381), MRC Mammalian Development Unit, University College London, London NW1 2HE, England

Richard J. Mullen (353), Department of Anatomy, University of Utah School of Medicine, Salt Lake City, Utah 84132

Muriel N. Nesbitt (369), Department of Biology, University of California at San Diego, La Jolla, California 92037

Drew M. Noden (241), Department of Anatomy, New York State College of Veterinary Medicine, Cornell University, Ithaca, New York 14853

V. E. Papaioannou (3), Department of Pathology, Tufts University School of Medicine, Boston, Massachusetts 02111

J. Rossant (89), Department of Biological Sciences, Brock University, St. Catherines, Ontario L2S 3A1, Canada

Lauri Saxén (401), Department of Pathology, University of Helsinki, SF-00290 Helsinki, Finland

C. L. Stewart* (409), Heinrich-Pette-Institut für Experimentelle Virologie und Immunologie an der, Universität Hamburg, 2000 Hamburg, Federal Republic of Germany

M. A. Teillet (318), Institut d'Embryologie du CNRS et du Collège de France, F-94130 Nogent-sur-Marne, France

Jean-Paul Thiery (179), Institut d'Embryologie du CNRS et du Collège de France, F-94130 Nogent-sur-Marne, France

Patricia Tippett (165), MRC Blood Group Unit, Wolfson House, University College London, London NW1 2HE, England

L. Vakaet (71), Laboratorium voor Anatomie, Rijksuniversiteit Gent, B-9000 Gent, Belgium

John D. West (39), Radiobiology Unit, Medical Research Council, Harwell, Didcot, Oxon OX11 ORD, England

---

*Present address: EMBL Laboratory, 6900 Heidelberg, Federal Republic of Germany.

Preface

The construction of chimeras—composite animals containing genetically different cell populations derived from more than one zygote—has been a longstanding method in embryology. The recognition of the exact fate of the blastomeres in some species and the demonstration that embryonic cells do not develop independently but actively interact with each other have been decisive landmarks in unravelling the enormous complexity of developmental processes. This progress in our knowledge was made possible as a result of imaginative experimental procedures by which cells of one species were grafted into another, using recognizable markers to allow their identification. Was not the role of the "primary organizer" in the amphibian embryo demonstrated by Spemann through grafting experiments between *Triturus cristatus* and *Triturus taenatius?*

Following these pioneering studies, the use of chimeras in developmental biology expanded tremendously once it was demonstrated that a single "tetraparental" mouse could be constructed by the aggregation of two embryos. A new era in the analysis of mammalian development arose from the initial experiments of Tarkowski and Mintz. The flux of information has been so overwhelming that 6 years after Anne McLaren's comprehensive review on mammalian chimeras, it was felt useful to assemble the most prominent data on the subject. In addition, making avian chimeras by combining quail and chick cells has been fertile area of developmental biology over the past 10 years; that is why this volume, devoted essentially to chimeras in mammals and birds, has been devised.

The technical aspects of constructing chimeras and how to analyze them by cell markers are treated in the first two chapters. Cell lineage studies in early development are reviewed next. The largest part of the book is devoted to the considerable advances that have been brought about by the use of chimeras in the fields of differentiation and organogenesis. So rich is the range of problems that can be addressed with chimera experimental paradigms, that they have a say even in the area of behaviour. The last chapter of the book deals with chimeras that involve teratocarcinoma cells.

A critical reflexion on the perspectives that the use of chimeras offers developmental biology closes the book but decidedly not the research in the field. The pace is such that what was, at the time of writing, the state of the art concerning the contribution of chimeras to knowledge, will, when this book appears, most probably have been overtaken by the progress resulting from a combination of chimera systems with the most recent methods of molecular biology.

1

Technical Aspects

CHAPTER I

Making Chimeras

V. E. PAPAIOANNOU
F. DIETERLEN-LIÈVRE

Department of Pathology
Tufts University School of Medicine
Boston, Massachusetts

Institut d'Embryologie du CNRS et
du Collège de France
Nogent-sur-Marne, France

I.	Introduction	3
II.	Avian Chimeras	5
	A. Germ Layer Chimeras	7
	B. Neural Tube Chimeras	8
	C. Hemopoietic Chimeras	10
III.	Mammalian Chimeras	18
	A. General Considerations in Making Mammalian Chimeras	18
	B. Making Chimeras before Implantation	21
	C. Making Chimeras after Implantation	31
IV.	Conclusions	32
	References	34

I. INTRODUCTION

Many vertebrate embryos are highly regulative in their early development and are capable of withstanding a surprising amount of experimental abuse and perturbation. In amphibians, birds, and mammals in particular, this characteristic has allowed experimenters to manipulate the cellular composition of embryos so that animals can be produced that are composed of cell populations derived from two or more separate embryos. Although a variety of names have been applied to these individuals, such as allophenics, mosaics, and tetraparentals, they are commonly called chimeras in recognition of their composite nature. The term chimera, as used in the present work, thus describes any animal that is composed of different cell populations that have been derived from more than one fertilized egg (McLaren, 1976).

Although chimeras sometimes arise spontaneously during gestation, it is the deliberate experimental production of chimeras that has, for the past 70–80 years, provided biologists with experimental material to pose and investigate a range of questions important in developmental biology: What is the fate and potential of particular embryonic cells? What is the nature of determination? Can differentiation be reversed? How is gene expression controlled? How do cells interact during development?

In the early part of this century, many of the fundamental concepts of embryology and developmental biology were, at least in part, formulated on the basis of results of cell or tissue transplants between two separate embryos, usually different species of amphibian. Spemann (1962) discussed his own and other workers' experiments with such transplants and described their use in conjunction with other types of embryonic manipulation in testing ideas on induction, cell fate, cell potency, and determination in amphibian embryos. Even before the turn of the century, embryologists made use of contrasting colors of different amphibian species to follow cells after interspecific tissue transplantation, that is, after the production of chimeric embryos. Chimeras were used to delineate the potencies of different areas of the amphibian gastrula and to map out regional determination both within and between germ layers of the gastrula. Induction of the lens by the optic cup was studied using chimeric eyes. The inducing potential of the optic cup and the capacity of various tissues to respond to the inducer was analyzed with respect to the age of the embryo, using tissue transplants between species. Even the control of fine structural differences in lenses of different species was investigated with tissue transplants between species, the results indicating that the inducing stimulus determines that a lens should be formed but does not dictate its fine structure.

Chimeras were also used in experiments where embryonic regulation was less complete. The organizing power of the dorsal lip of the blastopore was investigated by interspecific tissue transplanatation of this area. Mangold (reviewed by Spemann, 1962) determined that the organizing force of this region was functional regardless of the material in its field, since whole secondary embryos that formed under the influence of a transplanted "organizer" were chimeric even within tissues.

Technical aspects of making such amphibian chimeras are straightforward thanks to the independence of the embryos from their parents and their relatively simple requirements during embryogenesis. They are easily accessible to the experimenter and easy to maintain, large enough to manipulate without sophisticated equipment, receptive to foreign tissue, even across species barriers, and heal rapidly following manipulation. There are cellular pigmentation differences that serve as easily identified markers for cells of different species. Not all of

I. Making Chimeras

these qualities are shared by avian and mammalian embryos, however, so it is not surprising that there has been a long gap between the pioneering experimental chimera work in amphibians and that in the higher vertebrates. Nonetheless, similar problems in the developmental biology of birds and mammals are now being approached through the production and study of chimeric embryos.

This chapter will concentrate on the technical aspects of making chimeras in birds and mammals. In general, progress in two areas has allowed the successful exploitation of these experimental procedures in recent decades. One is the considerable advance in culture techniques, particularly for the growth and maintenance of mammalian embryos. In order for experimental manipulations to be performed, it is necessary to disturb the intimate maternal–fetal relationship for at least a brief period during embryogenesis. Until the requirements of embryos were known and could be met during the period of experimentation, production of chimeras was impossible. Now it is a routine matter to remove preimplantation mammalian embryos from the maternal reproductive tract, manipulate them, culture them for one or more days, and return them to a foster mother for further development. Even the postimplantation stages of development are becoming accessible for experimentation as postimplantation embryos are now being grown for longer and longer periods outside the mother's body. Reimplantation of these embryos has not yet been achieved, but some short-term experiments with chimeric, postimplantation embryos have been possible.

Hand in hand with improvements in culture methods has come the use of new, innovative marker systems for the recognition of cell types within chimeric animals. Pigmentation differences, such as those used in amphibian chimeras, are not present in the embryonic cells of mammals or birds and differences in hair, skin, and feather pigmentation are of limited use in adult chimeras because of limited tissue distribution. A variety of other markers have been developed in recent years for use both between and within species. These include characteristics such as enzyme differences, differences in nucleolar morphology, and nuclear DNA differences. The type and scope of analysis of developmental problems that can be approached with chimeras has been enormously expanded by the use of these markers. They will be discussed in more detail in Chapter II.

II. AVIAN CHIMERAS

Bird embryos have several advantages over other vertebrate embryos, which have made some interesting approaches feasible. They are accessible to experimentation *in situ* during organogenesis, whereas this period is less easily approached in mammalian embryos that are implanted in the uterine wall at that time. Another advantage of the avian embryo is its relatively large size and the ease with which the different rudiments can be delineated. It is thus possible to remove and replace them with exquisite precision. Several markers are available, one of which is particularly simple to use, clear to identify, and endowed with great power of resolution, i.e, the quail-chick nuclear marker, devised by N. Le Douarin (1969) (see Chapter XII). With the advent of this marker, heterospecific avian chimeras have been extensively exploited to investigate ontogenetic mechanisms. Such studies provide guidelines, which may thereafter be confirmed in homospecific avian chimeras or extended to mammalian embryos. Markers used to construct homospecific chimeras are the classic tritiated thymidine labeling, sex chromosomes, molecular variants of enzymes and, more recently, histocompatibility antigens.

Organogenesis is indeed the most favorable time for experimentation on developing birds. During segmentation and gastrulation, the blastodisc is difficult to visualize and very thin, the tissues are fragile, and the adhesive properties such that foreign tissues do not implant very successfully. Thus, experiments on these stages are usually carried out *in vitro*. From about 30 hours of incubation, it becomes possible to experiment *in ovo*. After 4 days, vascularization becomes very dense and hemorrhages are difficult to avoid.

Avian chimeric organs or chimeric embryos have been especially useful to study the derivatives of the neural crest, the development of the hemopoietic system, and the differentiation of the excretory system. Chimeric hemopoietic organs may be obtained by grafting either on the chorioallantoic membrane or in the somatopleura. Chimeric embryos are constructed by replacing an area of the blastodisc or a rudiment by an equivalent area or rudiment of another embryo.

The preparation of the incubated egg prior to any type of operation is always the same. Its aim is (1) to locate the shell window precisely above the blastodisc or embryo; (2) to avoid damaging the blastodisc, the embryo, or the extraembryonic vessels when opening the shell; and (3) to prevent the growing embryo from being injured on the rim of the shell window. Earlier than the third day of incubation it is not possible to detect the embryo and surrounding vessels by observation over a candle. Thus the eggs should be incubated in a fixed position at least 24 hours before opening, allowing the blastodisc to settle at the uppermost surface of the yolk. The eggs can be set either with their long axis horizontal or standing on their pointed end. In the first case, the yolk and supported blastodisc should be dropped away from the shell, before proceeding further. To do this, a small hole is bored with blunt forceps at the pointed end of the egg and 2.5 ml of albumen are withdrawn using a syringe. The hole is sealed with scotch tape. A round window is then cut out above the embryo. If the eggs are incubated standing on their pointed end, the opening will be made above the air chamber; in this case it is not necessary to withdraw any albumen. The shell membrane is split into two sheets in this region. The sheet just over the embryo must be torn apart before the embryo becomes accessible. "Horizontal" and "vertical" methods each have their advantages; the latter disturbs as little as possible the

I. Making Chimeras

relationships which will be established between the chorioallantoic membrane and the shell; this method is preferred when the operated embryos are to be incubated for as long as possible and when hatching is eventually desirable. On the other hand, the embryo is more accessible in the "horizontal" method, so that it is more convenient for delicate operations.

A. Germ Layer Chimeras

An attempt was made by Marzullo in 1970 to produce avian chimeras similar to tetraparental or allophenic mice. A cell suspension from unincubated blastodiscs was slipped between the vitelline membrane and the surface of another blastodisc. The donor blastodisc was from a pigmented strain and the host from a white Leghorn chicken. Three embryos with some pigmented feathers were obtained out of 19 that survived to day 15 of incubation. This experiment has not been repeated since, thus it is difficult to judge its rate of success or its usefulness.

On the other hand, germ layer chimeras have been used to study gastrulation (Vakaet, 1974; Fontaine and Le Douarin, 1977). As mentioned above, these experiments were done *in vitro*. Blastoderms incubated for 5 to 8 hours were dissociated mechanically into hypoblast and epiblast. The layers were then exchanged between quail and chick, and the recombinants were cultured epiblast side down for 24 to 50 hours according to New's (1955) culture technique. At slightly later stages [head process to head fold (Hamburger and Hamilton, 1951)] pieces of the area pellucida were dissociated by trypsin treatment (0.1% in Tyrode solution minus calcium and magnesium) into ectoderm and endoderm plus mesoderm. Recombined layers were cultivated for 12 hours on semisolid

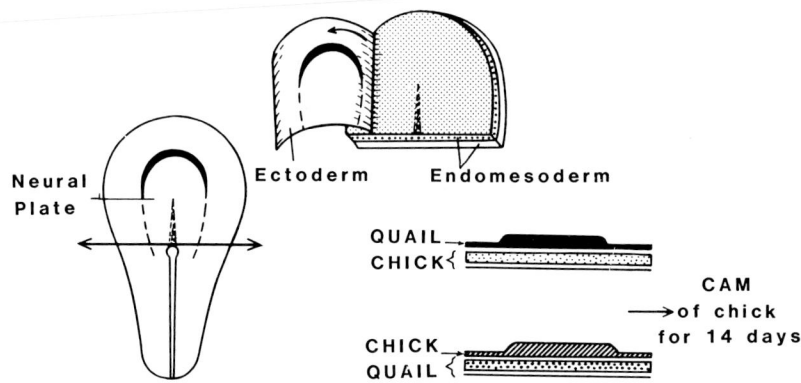

Fig. 1. Scheme of the dissociation-recombination experiment yielding a germ layer chimera. CAM, chorioallantoic membrane.

medium (Wolff and Haffen, 1952) to ensure their association (Fig. 1), then grafted onto the chorioallantoic membrane of chick hosts.

B. Neural Tube Chimeras

A transitory structure, the neural crest, develops at the apex of the neural tube as it closes. Its cells very soon move out again, giving rise to a variety of structures throughout the body. To follow the fate of these cells, a segment of the neural tube bearing the neural crest cells is replaced by a homologous structure provided with a marker, for instance, quail cells or cells labeled with tritiated thymidine. The operation involves three successive steps:

1. NEURAL TUBE EXCISION IN THE RECIPIENT EMBRYO (Fig. 2a)

The neural tube is excised bilaterally during the second day of incubation. When the brain level is involved, it is possible to excise only one side (Nakamura and Le Lièvre, 1982). Using dissecting needles that have been sharpened into fine knives, incisions are cut on each side of the neural rudiment to separate the neural tube and crest from dorsal ectoderm and somitic or cephalic mesoderm. The neural tube is then separated from the underlying notocord and discarded.

The developmental stage is chosen according to the level where the neural crest is to be replaced, since the neural crest develops progressively in a cephalocaudal direction, as the neural tube closes; soon after these structures

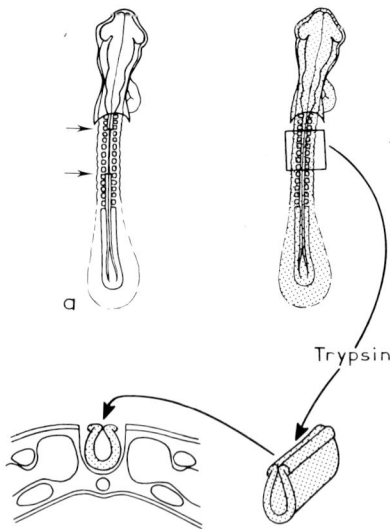

Fig. 2. Scheme of isotopic neural tube grafting. In (a) the naked notochord appears as a black line between the somites, where the tube has been removed in the recipient embryo (arrows).

I. Making Chimeras

have formed, the crest cells begin migrating. Thus the tube must be excised immediately after its closure.

2. ISOLATION OF A SEGMENT OF THE NEURAL RUDIMENT FROM THE DONOR EMBRYO (Fig. 2b)

A donor is selected at a developmental stage that may be identical to, or different from, that of the recipient. This depends on whether the graft is to be isotopic, in which case it will also be isochronic, or heterotopic; in the latter instance the grafted rudiment must be taken from a level where the neural crest has just formed and has not yet begun migrating; thus it will also be heterochronic (Fig. 3). A transverse section of the embryo, including precisely the neural segment wanted, is taken and trypsinized for 10 to 15 minutes. Trypsin (Difco) is diluted to a final 1% concentration in Tyrode's solution minus calcium and magnesium and is used at 4°C. The neural rudiment is finally separated mechanically from surrounding tissue with microscalpels. After being rinsed quickly in complete Tyrode solution to which a few drops of horse serum may be added, the rudiment is transferred onto the donor embryo. Thus treated the rudiment is purely ectodermal and uncontaminated by either mesodermal or endodermal cells.

3. GRAFTING THE NEURAL TUBE (Fig. 2c)

The isolated neural segment is deposited on the recipient embryo in the cavity resulting from the excision of the neural rudiment. The original anteroposterior orientation of the rudiment is maintained.

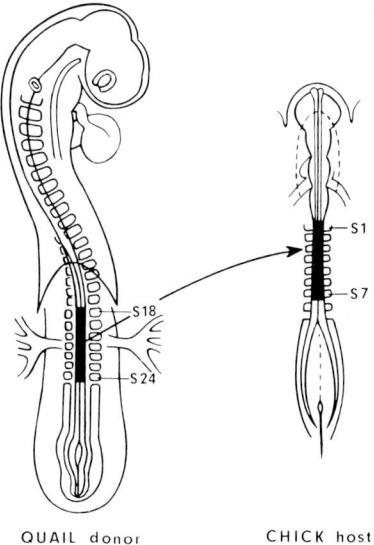

QUAIL donor   CHICK host

Fig. 3. Scheme of heterotopic, heterochronic neural tube grafting.

Usually the quail embryo is chosen as a donor, first, because the operation is carried out with more ease in the larger chick egg, and, second, because it is easier to identify one isolated quail cell among chick cells than the reverse. On the other hand, the reverse graft is feasible, and it is often necessary to verify results using both combinations.

C. Hemopoietic Chimeras

1. CHORIOALLANTOIC GRAFTS

The chorioallantoic membrane (CAM) of the chick embryo is an excellent culture environment for many viruses, microorganisms, and normal or tumorous tissues. When avian hemopoietic organ rudiments are grafted onto it, they are colonized by blood-borne extrinsic stem cells and thus become chimeric. The advantage of this type of grafting is its ease and rapidity. Indeed it suffices to deposit the tissue or rudiment in an area devoid of large vessels on the CAM of 6–10 day embryos. The tissue embeds into the CAM and becomes vascularized. If only the stromal frame of the rudiment is grafted, the hemopoietic cells will be provided by the host. Alternatively, it is also possible to deposit a suspension of hemopoietic cells on the CAM or to inject the cells into a vein of the CAM at 13 days of incubation. To that end, the egg is candled, branching vessels are identified and marked on the shell. A triangle is cut from the shell around the branching, using a circular saw and maintaining the shell membrane undamaged. A drop of paraffin oil is applied to the window, making it transparent. The vessels become visible. They remain adherent to the shell membrane and the needle may be inserted tangentially into the direction of the branching. This procedure is mainly used to study the influence of adult lymphoid cells on the avian embryo (graft versus host reaction).

2. SOMATOPLEURAL, COELOMIC, AND INTRAMESENTERIC GRAFTS

These three types of graft differ in the stage at which the recipient embryo is taken; the nature of the grafted tissue also determines the implantation site, so that the technique chosen depends upon the grafted material and the problem under study. The procedures are very similar.

Somatopleural grafting has been used extensively by Le Douarin's group to investigate the colonization schedule of the thymus and bursa of Fabricius (Fig. 4) (see Chapter VIII). When the host has about 30 pairs of somites, i.e., 2 to 3 days of incubation, the vitelline membrane, still present in these young embryos, is torn apart using watchmaker's forceps, and the amnion is split away. The graft, marked with a few particles of carbon black, is deposited near the grafting site. The somatopleura, i.e., the body wall constituted by ectoderm and meso-

I. Making Chimeras

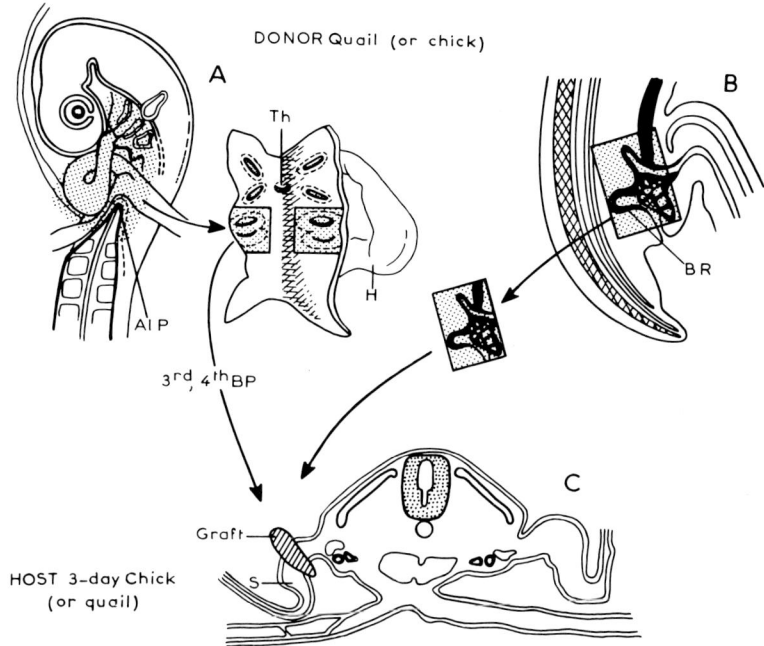

Fig. 4. Scheme of somatopleural grafting of thymus or bursa of Fabricius. AIP, anterior intestinal portal; BP, branchial pouch; H, heart; S, somatopleura; Th, thyroid. (A) 3-day donor for the floor of the pharynx; (B) 6-day donor for the bursa rudiment (BR); (C) transverse section in trunk of 2-day recipient showing position of the graft.

derm, is cleaved and the graft is inserted into the cleft, so that it remains wedged into the opening. The size of the cleft has to be adapted to that of the graft.

For a coelomic graft, the embryo is older (3 to 4 days) and a larger cleft is made, so that the graft falls into the coelom. Hemorrhages are easily avoided by taking care not to hit large vessels. Grafts may be implanted in a cephalic or caudal position in relationship to the omphalomesenteric veins. After 24 hours they become vascularized. They can be identified and recovered at any time through the carbon marks. It is useful to retrieve the grafts with the surrounding host tissues into which they have become incorporated. For coelomic grafting, the implant may also be inserted through the umbilicus between $3\frac{1}{2}$ and 5 days of incubation (Dossel, 1954). Vascularization in this site is, however, very dense and impedes good development.

Coelomic grafts usually attach to homologous tissues; for instance, a grafted hepatic rudiment will usually be found on, or even within, the host liver. When purely mesenchymal tissues are grafted, they will embed within the dorsal mesentery of the host. This is the case for the wall of the dorsal aorta that, grafted in this location, gives rise to blood islands.

3. PARABIONTS

Chimerism of the hemopoietic system occurs to a high degree when two embryos are joined through vascular connections. The technique was first devised in order to join two chick embryos (Hasek, 1953; Moore and Owen, 1965). Eggs incubated for 6 days are candled, and the most vascular region of the chorioallantoic membrane is identified. A small hole is then made through the shell into the air space, and a second triangular opening is cut through the shell over the most vascular region of the CAM; penetration of the shell membrane should be carefully avoided in order not to damage the underlying CAM. A drop of saline is placed on the shell membrane, which is then torn open. Suction is applied to the air space, causing collapse of the CAM and creation of an artificial air space at the tear in the shell membrane. The holes in the two eggs treated in this way can then be enlarged into circular openings 0.1–1 cm in diameter. They are positioned facing each other, and the two eggs are sealed together in that arrangement with paraffin wax around the holes. Vascular parabiosis develops rapidly between the two contacting CAMs. If the eggs are incubated in a fixed

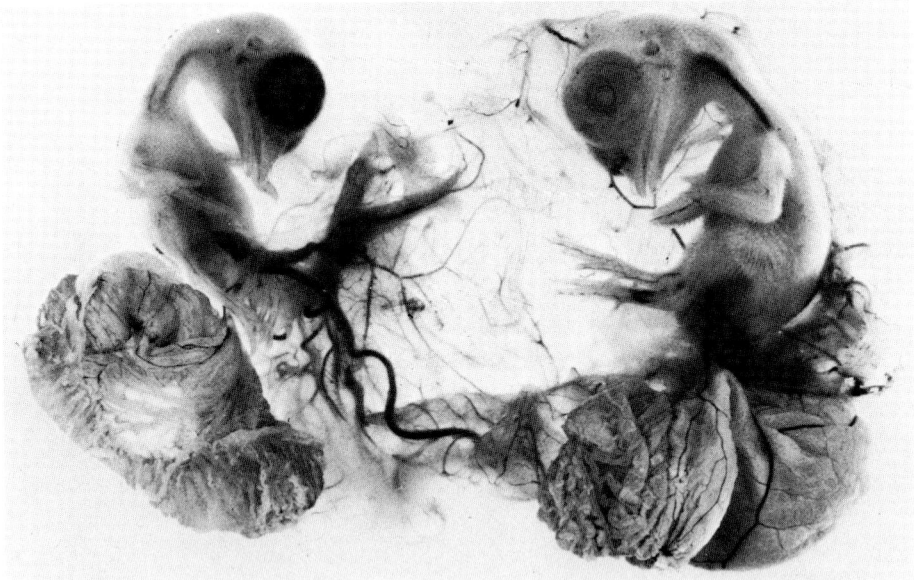

Fig. 5. Parabiosed quail embryos of opposite sexes. The embryos are from a strain carrying a sex-linked recessive albino mutation. Parents are ♀ Z^+/W and ♂ Z^{alb}/Z^{alb}. The F_1 progeny used for the experiments are albino ♀ Z^{alb}/W and pigmented ♂ Z^{alb}/Z^+. The embryo to the right can be identified as female by its unpigmented eye and the one to the left as male by its dark eye. The picture also shows that chorioallantoic vessels are anastomosed while yolk sac circulations remain separate. Original photograph provided by N. Le Dourain and M. Coltey.

position and joined at 4–5 days of incubation through the sites where the embryos are developing, anastomosis develops between the yolk sac vessels at an earlier stage than in the first technique (Moore and Owen, 1967).

A very simple method has been developed more recently to achieve parabiosis between chick and quail (C. Martin, unpublished) [Fig. 6 (part 5)] or between two quail embryos (Le Douarin et al., 1983). For the first combination, 5 ml of albumin are taken from an unincubated chick egg, and the whole content of a quail egg is poured into the chick shell next to the chick yolk. Vascular anastomoses form when the two embryos develop. Alternately the chick shell is completely emptied through a wide circular opening and used as a culture dish, into which the whole content of two unincubated quail eggs is transferred. Yolk transfer can be accomplished until 2 days on incubation. Later, when the vitelline membrane has ruptured, the yolks become too fragile to tolerate the manipulation. The establishment of vascular anastomoses has been carefully checked through injection of india ink in the blood of one of the parabionts. It appears that chorioallantoic vessels join during the eighth day of incubation but that yolk sac circulations never connect in these associated quails (Fig. 5).

4. YOLK SAC CHIMERAS [Figs. 6 (parts 1 and 2) and 7]

This technique, devised by Claude Martin (1972), yields embryos that receive primitive erythrocytes and eventually other hemopoietic cells from the associated yolk sac. Exchanges of cells also occur in the reverse direction. The operation is done during the second day of incubation. It can be carried out on embryos ranging from 8 to 22 pairs of somites. The blastodiscs giving the two components of the association are matched for stage. When quail and chick are involved, the chick eggs are incubated 6 hours earlier than the quail to achieve synchrony. The operation can only be performed in the chick egg, i.e., on the chick yolk sac.

The quail blastoderms, from which the embryonic area will be obtained, are taken out in Tyrode's solution and the vitelline membrane is removed. The central area of the blastoderm is trimmed leaving a small margin around the head and somites. This margin will be resected as the central area is seamed onto the extraembryonic area.

The recipient chick blastodisc may be rendered more visible by depositing a few particles of neutral red on the vitelline membrane. As neutral red diffuses, it will stain the blastodisc evenly. The trimmed donor embryo is transplanted by means of a wide-mouthed pipette onto the recipient blastoderm and positioned correctly with respect to the germ layers as well as with respect to the cephalocaudal axis, side by side with the original embryo. Only then is the vitelline membrane of the host blastoderm torn apart and the embryo excised. The transplanted embryo is moved above the cavity, and the edges of the two partners are seamed together by resecting their margins simultaneously with Paschew scissors. The egg is then sealed with tape and reincubated. In some 10%

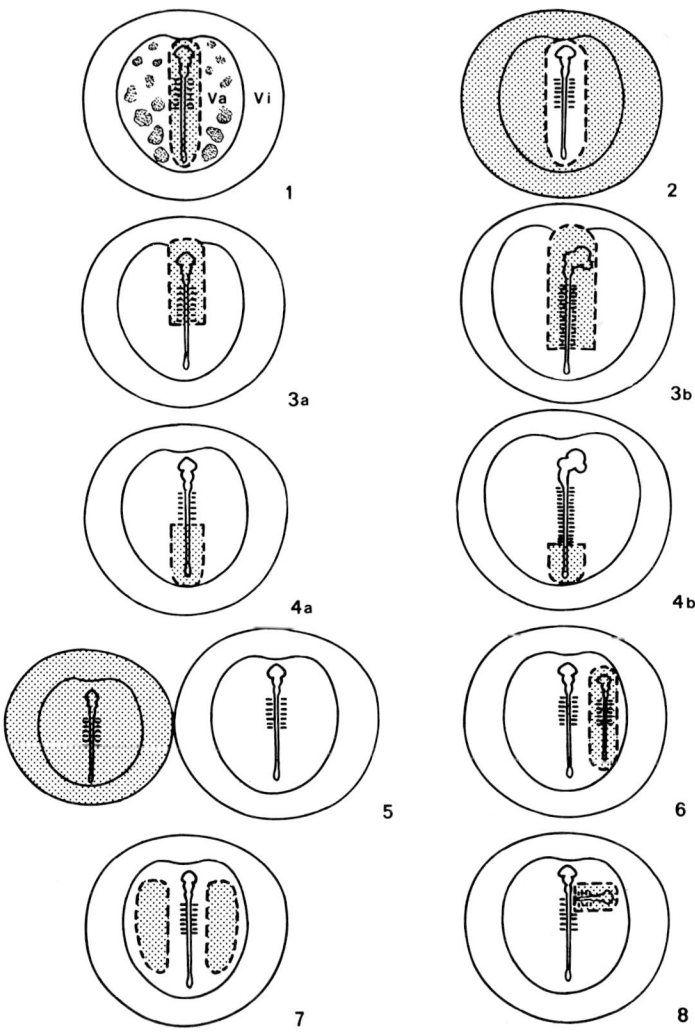

Fig. 6. Avian mosaics which can be obtained by exchanging homologous territories between two blastodiscs. (1), (2) Yolk chimeras; (3), (4) complementary chimeras; (5) parabionts; (6) supplementary chimera; (7) chimeric yolk sac; (8) additional head and neck graft.

of operated embryos, circulation will form across the seam and the chimaeras will develop according to a pattern very similar to the one that is normal for quail embryos. However, the weight of the quail embryo is statistically increased at all ages of incubation. Furthermore the retraction of the yolk sac, which is under thyroid control, proceeds more rapidly than in a control chick embryo.

This type of chimera has also been made between two chick partners from inbred lines differing by their genetic sex (Lassila *et al.*, 1978), by their pre-

I. Making Chimeras

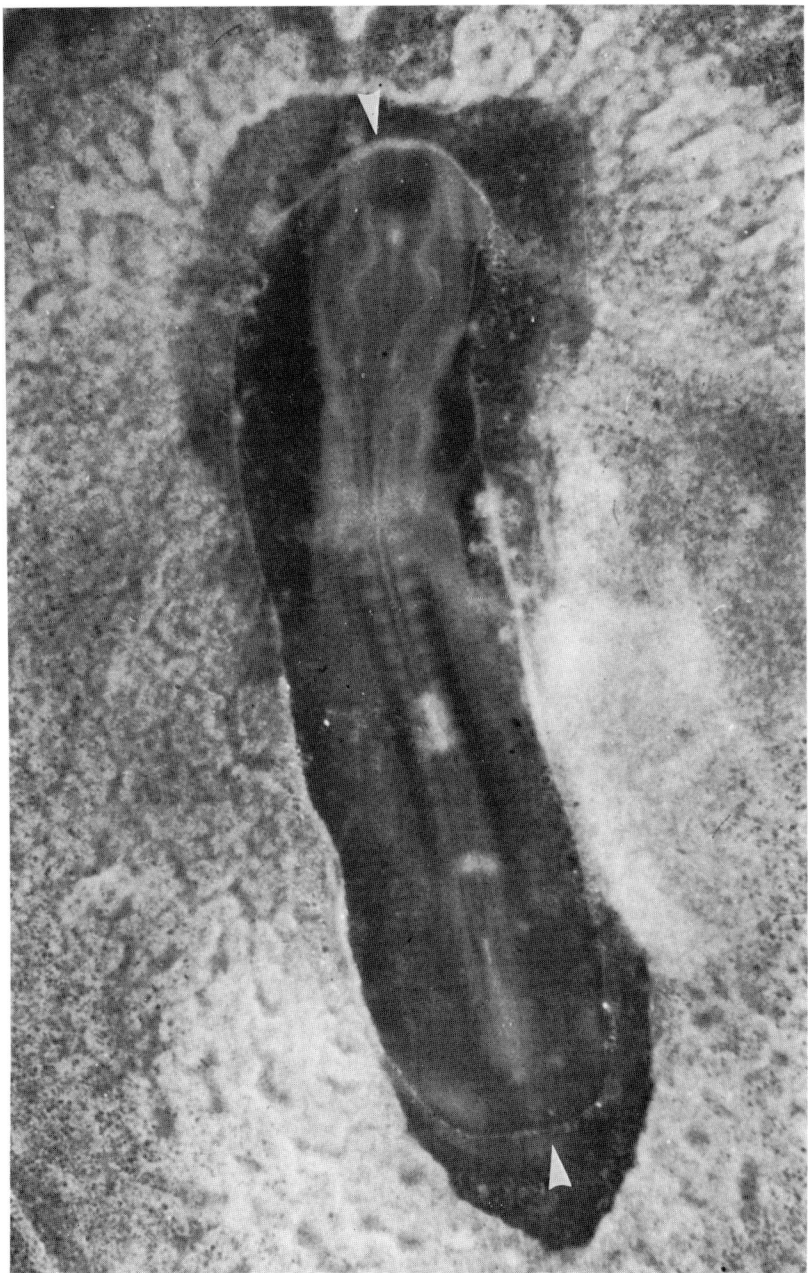

Fig. 7. Dorsal view of a quail↔chick yolk sac chimera. The white line (arrows) which closely surrounds the embryonic body is the suture between the chick extraembryonic area and the quail embryo.

sumptive immunoglobin allotypes (Martin et al., 1979) or by their major histocompatibility antigen (Lassila et al., 1982). Such chick↔chick chimeras are able to hatch and grow to adulthood.

5. MOSAIC EMBRYOS

Rather than grafting the whole embryonic area, it is possible to graft parts of it [Fig. 6 (parts 3 and 4)]. This operation first involves constructing *in vitro* the central mosaic area. The two embryos are explanted with a little margin. They are placed back to back and head to tail and cut simultaneously perpendicular to the neural axis and posterior to the somites. Cutting ensures adherence of the complementary superposed anterior end of one embryo and posterior end of the other; the latter is then flapped back into normal position. A skillful operator will thus obtain two mirror-image chimeras. The suture is always done behind the last somite, so that if different levels of suture are desired, different stages of development will be chosen.

In a second step the composite embryo is grafted onto a recipient yolk sac, following an identical procedure to that used to obtain yolk sac chimeras. The resulting chimera is usually a well-formed embryo, with good coaptation of the foreign anterior and posterior components (Martin et al., 1980). When wild-type quail and normal white Leghorn are thus associated, the borderline between the two parts appears as a more or less straight demarcation between pigmented and nonpigmented areas (Fig. 8). The borderline in internal organs, however, is not sharp, owing to the morphogenetic movements that occur after the operation. Actually the sorting out capacities of embryonic cells are demonstrated in spectacular fashion in these chimeras. Caudad movements of the pharyngeal endoderm that grows into more posterior mesodermal territories provokes the formation of complex chimeric organs. The pancreas, for instance, is usually constituted by mesenchyme belonging entirely to the "posterior" species, while the endoderm is partly "anterior," partly "posterior." The anatomy of the organ is always normal. The mesonephros is also chimeric, when the suture is done behind the level of somite 10. The development of that organ involves caudad progression of the Wolffian duct that induces nephric tubules into the renal blastema, along its pathway (Martin, 1976). In such chimeras, the derivatives of the Wolffian duct, i.e., some mesonephric collecting tubules and the ureteric bud, are of the anterior species, while blastema derivatives, namely, the main part of mesonephric tubules, are of the posterior species.

This type of mosaic embryo has been used to study the development of the kidney (Martin, 1976), the development of the hemopoietic system (Martin et al., 1980), and recently the interplay between primordial germ cells (PGC) and gonadal stoma in the determination of meiosis (Hajji et al., 1983). During day 2 of incubation, the PGC are located in an extraembryonic anterior crescent; they move into the blood when circulation sets in and become trapped in the attractive

1. Making Chimeras

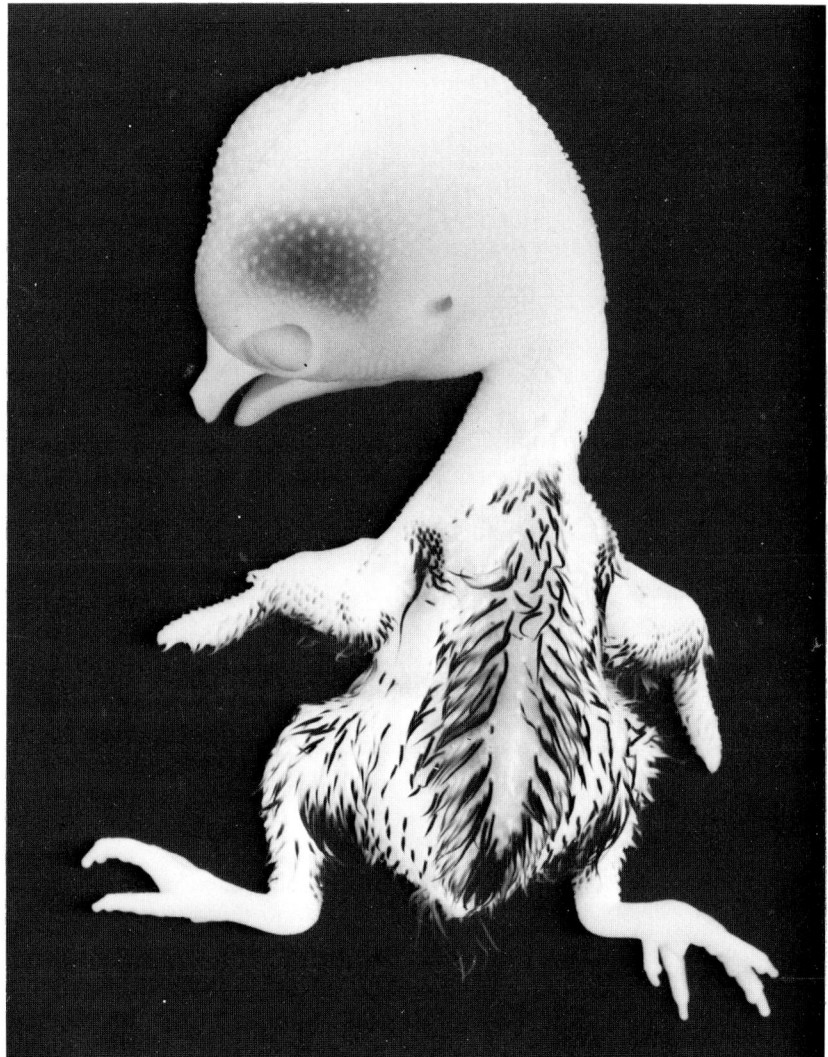

Fig. 8. Complementary chimera with chick head and neck on quail (pigmented) body. Thirteen days of incubation.

germinal epithelium (Simon, 1960; Dubois, 1968). Yolk sac chimeras thus have chimeric gonads. Actually their PGC population is not entirely derived from the anterior crescent (Hajji *et al.*, 1983) but is often "contaminated" by PGC from the embryonic area. To minimize this contamination, chimeras were made according to pattern 4a in Fig. 6 in which the grafted area is limited to the posterior embryonic territory containing the presumptive gonadal stromas. Mosaic embryos in this study were constructed from quail and chick strains carrying a sex-linked albino mutation (Fig. 5). Thus it was possible to recognize stromal or gonadal cells through the quail-chick nuclear difference, and to identify the genetic sex of each type of cell through the pigmentation (albino or wild type) of the anterior and posterior component that yielded, respectively, PGC and gonadal stroma.

6. OTHER CHIMERIC SCHEMES [Fig. 6 (parts 6–8)]

From the descriptions above, it is manifest that the avian blastoderm tolerates large foreign insertions (C. Martin, unpublished; Dieterlen-Lièvre et al., 1979). All the variants in Fig. 6 have been tested. The graft of extraembryonic areas [Fig. 6 (part 7)] is one of the simplest. On the other hand, grafting an additional embryo Fig. 6 (part 6), or part of one [Fig. 6 (part 8)] in the vicinity of the native embryo meets with exceptionally high mortality. In Fig. 6 (part 6), two embryos develop side by side on the same yolk sac, and it is possible that asynchronous heart beats are not compatible with a well-balanced circulation.

III. MAMMALIAN CHIMERAS

A. General Considerations in Making Mammalian Chimeras

1. DIFFERENT SPECIES

By far the most popular choice of mammal for making chimeras for developmental studies has been the common house or laboratory mouse, *Mus musculus*. The advantages this species offers are abundant: A vast amount of information is available on the biology of the mouse, including its reproductive biology and embryology. Mice are easily available in a variety of genetically well-characterized strains, which provide scope for using genetic markers. Gestation time of mice is short, and they breed round the year, providing a continuous supply of embryos. Because the embryos have so frequently been used in experimental work (Papaioannou, 1982a), the requirements for their maintenance in culture have been well worked out, and embryo culture is reliable and reproducible. Also, the techniques for transfer of preimplantation embryos into foster mothers have become routine in many laboratories.

This is not to say, however, that other species have not been used or that others species might not prove more valuable in chimera studies to answer particular questions. Information gained from mice can be generalized only to the extent that developmental concepts formulated from that information can be generalized. Specific data from other species will be helpful in understanding species differences and in formulating a comprehensive picture of mammalian develop-

I. Making Chimeras

mental biology. In some cases, interspecific chimeras have been useful, particularly for the exploitation of certain marker systems not available within a species. The techniques that have been worked out to make *Mus musculus* chimeras have been adapted to make chimeras of several other laboratory and farm animals. It appears that the limiting factors are not the actual manipulations on the embryos but rather the embryo culture and transplant techniques and an understanding of the animals' reproductive biology.

Rat chimeras have been produced by the aggregation of early preimplantation embryos followed by retransplantation into foster mothers (Mayer and Fritz, 1974; Mullen and LaVail, 1976). This technique involves the removal of the nonliving zona pellucida, the primary egg envelope, and thus cannot be extended to use with rabbits, since rabbit embryos fail to implant if the zona pellucida (albumina) is removed. This problem has been overcome in rabbits by the use of a different technique, also developed for mouse embryos: transplantation of cells by injection into an intact, zona-contained embryo. A number of live-born rabbit chimeras have thus been produced (Babinet and Bordenave, 1980; Gardner and Munro, 1974; Moustafa, 1974). Extensive work has not been reported in other species, although sheep chimeras have been successfully produced by blastomere transplantation (Tucker *et al.*, 1974), as well by aggregation combined with the agar-embedding technique devised by Willadsen (1979).

More effort has been applied to working out techniques to produce interspecific chimeras between, for example, the rat and mouse or between different species of *Mus*. Compatibility in the development of early stages of bank vole and mouse and of rat and mouse was demonstrated by the successful formation of interspecific chimeric blastocysts following aggregation of morulae (Mulnard, 1973; Mystkowska, 1975; Stern 1973; Zeilmaker, 1973). The rat↔mouse combination has been used to investigate cell lineage and determination using species-specific cell markers (Rossant, 1976; Gardner and Johnson, 1973, 1975). Although use of the injection technique to introduce rat cells into mouse embryos, followed by transfer to a mouse foster mother, allowed implantation and continued intrauterine development (despite considerable antigenic differences), the only proven chimeras that survived to term were runted and died shortly after birth (Gardner and Johnson, 1975), possibly due to the sensitization of the mother to rat antigens.

A more successful interspecific chimeric combination has been between the closely related species, *Mus musculus* and *Mus caroli*. The possibility of obtaining healthy, fertile adult chimeras (Rossant and Chapman, 1983; Rossant and Frels, 1980) combined with a marker system to identify individual cells in histological sections (Rossant *et al.*, 1983; Siracusa *et al.*, 1983) opens many avenues for research in different areas of developmental and reproductive biology (Rossant *et al.*, 1982). Chimeras between sheep and goats have also been produced and have proved fertile.

2. STAGE OF DEVELOPMENT USED FOR MAKING CHIMERAS

a. After Birth. By definition any individual with an organ, tissue, or skin transplant from another individual could be considered a chimera. This type of transplant, however, is normally done in adults, particularly in humans for therapeutic reasons, and has little bearing on developmental biology. An exception to this is the production of so-called radiation chimeras in different species. In making these chimeras, the power of an animal to respond to foreign antigens is first destroyed by radiation and then the spleen, thymus, and bone marrow are repopulated by the transplantation of foreign cells (Micklem and Loutit, 1966). A similar type of chimera can be produced if an animal is rendered tolerant of foreign cells before birth (Billingham et al., 1956).

Radiation chimeras have been used in investigations of erythropoietic, granulopoietic, and thrombopoietic stem cell populations in various grafted tissues. The survival of these chimeras and the permanence of the grafted cell population depends not only on an adequate number of stem cells in the graft, but also on the antigenic relationship between donor and host and the amount of radiation damage done to the host. Techniques for producing this type of chimera will not be further discussed here.

b. During Gestation. The chimeras that have been of the greatest interest to developmental biologists are those produced during different stages of embryogenesis. Although the stage at which the tissues of two embryos are associated will to some extent dictate the type of question that can be approached, chimeras made with preimplantation stage embryos have found wide application. Technically, this is the most convenient time for manipulation of mammalian embryos. During the first 3 to 4 days of gestation in the mouse, the embryos have not yet formed an attachment in the mother's reproductive tract and can thus be explanted, manipulated, and then transplanted to a suitable foster mother's reproductive tract to continue their development. Depending on the nature of the experiment, the chimeras may then be left to develop to term or they may be analyzed at any stage during their gestation. Alternatively, the experimental protocol may be such that more immediate analysis is required, in which case the chimeric embryos may be maintained *in vitro* for short periods.

Greater technical difficulties are encountered in attempting to make chimeras after implantation has occurred, although some success has been reported (Beddington, 1981, 1982; Fleischman and Mintz, 1979; Gootwine et al., 1982; Weissman et al., 1977, 1978). Since the fetuses cannot be returned to a uterus once they have been removed, tissue must either be grafted to fetuses *in utero* to make the chimeras or the entire experiment, including analysis, must be done *in vitro* following explantation of the fetuses.

I. Making Chimeras

B. Making Chimeras before Implantation

1. MORULA AGGREGATION

It is possible to aggregate two or more cleavage stage embryos together, making a composite morula that will develop into a chimeric animal when transplanted to a foster mother. Methods for the aggregation of cleaving mouse embryos were first described by Tarkowski (1961) and Mintz (1962a, 1964) and have since been adapted in various ways and used in many laboratories. In brief, cleaving eggs are removed from the oviducts of mated females, the zonae pellucidae are removed, and two (or more) suitably marked embryos from different mothers are placed together until the blastomeres stick to one another (Fig. 9). The pair or group of embryos will then aggregate to form a single morula that,

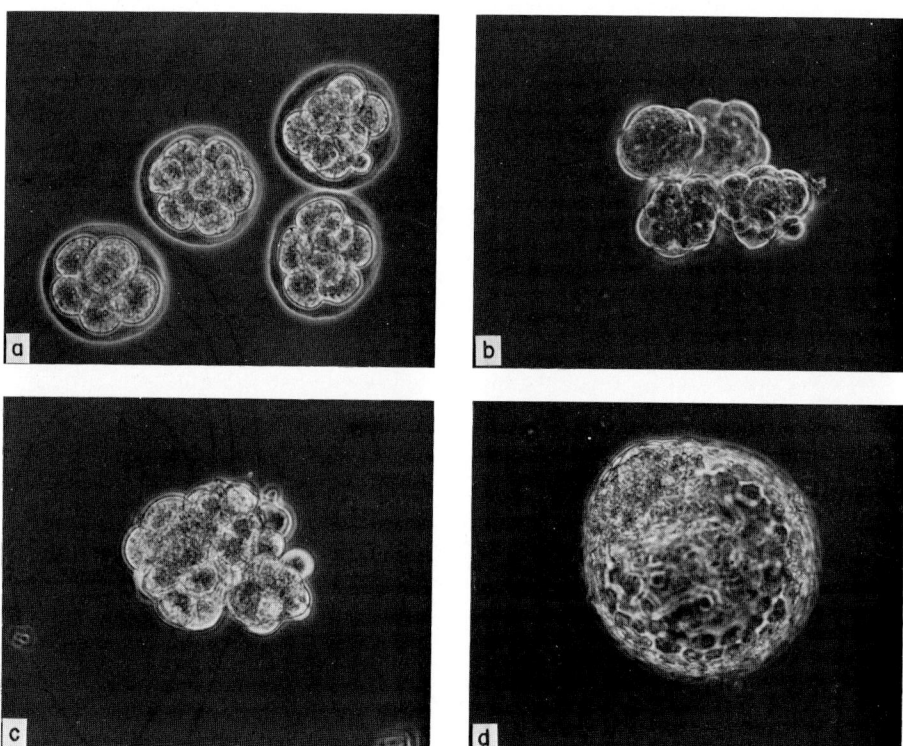

Fig. 9. Aggregation of four morulae to form a single, large blastocyst. (a) Isolated morulae before the removal of the zona pellucida. (b) The zona pellucida has been removed with acidic Tyrode's solution and the morulae cultured in the presence of 1% phytohemagglutinin for 2 minutes before being placed in contact. (c) After several hours, aggregation of the embryos is proceeding. (d) After 24 to 36 hours a single large blastocyst has formed from the four aggregated morulae.

after a period of culture, can be transplanted into the oviduct or uterus of a pseudopregnant foster mother for continued development.

During the manipulation to make aggregation chimeras, embryos must be maintained in a suitable culture medium that will allow their continued development. Biggers, et al., (1971) describe various culture systems and techniques, and McLaren (1976) compares some culture media used in different laboratories. Basically, a suitable culture medium consists of a Krebs–Ringer salt solution buffered with bicarbonate and supplemented with energy sources, bovine serum albumin, and antibiotics. Cultures are maintained at 37°C in an atmosphere of 5% CO_2 in air. The most convenient method for making chimeras is to keep the embryos in small drops of medium under mineral or paraffin oil (Brinster, 1963). A single culture dish can thus be used for a number of drops. For the initial explantation of the embryos, it may be convenient to use a phosphate-buffered medium [PB1 (Whittingham and Wales, 1969), modification described by Papaioannou and West (1981)] or phosphate-buffered saline, since the bicarbonate, buffered medium changes pH rapidly in a normal atmosphere. The embryos can then be transferred to the culture medium once they have been collected.

Cleaving embryos will reach the 2-cell stage approximately 1–1.5 days postcoitum (pc) and the 8- to 16-cell stage a day later, although the exact timing will depend on the particular strain of mouse used. Morulae will begin to compact and form tight junctions from about the 16-cell stage as a prelude to blastocyst formation (Ducibella et al., 1975), and once this begins to occur, aggregation of morulae becomes progressively more difficult. Thus, embryos should be removed from the oviduct at the 8-cell stage, approximately 2–2.5 days pc or even earlier. To do this, the oviducts are separated from the rest of the reproductive tract and placed in a small drop of medium in an embryological watchglass. The embryos are then flushed out of the oviduct by inserting a finely drawn Pasteur pipette into the ostium at the ovarian end and flushed out with medium. A mouth-controlled pipette is convenient for this procedure, since only a small volume of fluid needs to be flushed through the oviduct. If the flushing is not successful, the oviducts can be teased apart with fine forceps to free the embryos.

The zona pellucida can be removed either mechanically by pipetting in a pipette slightly smaller than the zona (Tarkowski, 1961), enzymatically with pronase (Mintz, 1962b), or, most simply of all, by a brief exposure to acidic Tyrode's solution (pH 2.5) (Nicolson et al., 1975). Less than 1 minute in the acidic Tyrode's solution is usually sufficient to dissolve the zona pellucida, and the embryos should be removed to a buffered medium as soon as it begins to disappear.

Successful aggregation of cleaving eggs is dependent on their developmental stage, presence of Ca^{2+} ions in the medium, and temperature. A heated stage or prewarmed medium (37°C) will facilitate aggregation. Zona-free embryos are

I. Making Chimeras

placed 2 (or more) per drop of medium and pushed together with a small glass rod or pipette. After a short incubation, pairs that have not stuck together may be pushed together once again. When working at room temperature, the addition of phytohaemagglutinin to the culture medium can be very helpful in getting the embryos to stick together initially (Mintz et al., 1973). The use of rabbit antimouse serum as a ligand to hold morulae together has also been used successfully (Palmer and Dewey, 1983). The paired embryos are then left in culture overnight where aggregation and blastocyst formation will occur in the composite embryo (Fig. 9c and d). The chimeric blastocysts can then be transferred to recipient foster mothers for development.

Foster mothers should be females made pseudopregnant by mating with vasectomized males. They can be used either at 0.5 days pc, when the chimeric embryos are transplanted to the oviducts through the ostium (Tarkowski, 1959), or at 2.5–3.5 days pc when the embryos can be transplanted to the uterine lumen through a small hole made with a needle near the uterotubal junction. The most favorable and convenient combination for uterine transfer is for the embryos to be 1 day more advanced than the foster mother [3.5 day pc embryos into 2.5 days pc recipient (McLaren and Michie, 1956)], although synchronous transfers to the uterus at 3.5 days pc and transfers to the oviduct as much 3 days asynchronous (3.5 day pc embryos into 0.5 day pc females) are also quite successful (V. E. Papaioannou, unpublished observation). Since they are zona-free, aggregated embryos should not be transferred to the oviducts until the blastocyst stage (Bronson and McLaren, 1970) unless they are embedded in agar (Willadsen, 1979). With asynchronous transfers, the embryos implant more nearly according to the foster mother's gestational age, although some developmental precocity can be observed throughout the early part of gestation (Aitken et al., 1977; Marsk, 1977). Various methods of embryo transfer have been described (McLaren and Michie, 1956; Mintz, 1967; Rafferty, 1970; Tarkowski, 1959; Whittingham, 1968; Marsk, 1977).

Using the technique of morula aggregation, chimeras may be made from the 2-cell stage through the 16-cell or even the late morula stage. The composites may be made with synchronous or asynchronous embryos (Rossant, 1975), with whole or partial embryos, and even with cells from embryonic tumors (Fujii and Martin, 1980; Stewart, 1980). More than two embryos can be aggregated to form large composite blastocysts (Markert and Petters, 1978; Mintz, 1971; Pedersen and Spindle, 1980). Since there is relatively little cell mingling during cleavage stages, blastomeres can also be positioned to some extent, thus influencing their later distribution in the chimera (Hillman et al., 1972). However, up to the morula stage, blastomeres are probably totipotent (Papaioannou et al., 1978) and cell mingling does occur later, so it is impossible to control or predict the exact cellular distribution of components in a chimera. The genotype of components can also greatly influence their extent and pattern of contribution (Mullen and

Whitten, 1971). Somewhat more control can be exercised with methods of chimera production using later embryos.

2. BLASTOCYST INJECTION

Once mammalian embryos have reached the blastocyst stage, two distinct tissues are present, the inner cell mass (ICM) and the trophectoderm, both of which have restricted potential and different fates in development (Gardner and Papaioannou, 1975). At this stage it is possible to introduce cells into the cavity of the blastocyst to obtain chimeras. Gardner (1968, 1971) first developed this method and has described a five-instrument, microsurgical method for introducing cells into the blastocoelic cavity (Gardner, 1978). Modified methods using two microinstruments have been described (Babinet, 1980; Moustafa and Brinster, 1972), and the various methods have been compared in detail by Papaioannou (1981a). Although a variety of different cell types can be introduced into the blastocyst using this method (Papaioannou, 1979, 1981a), the most commonly used tissue in making chimeras for developmental studies has been synchronous or asynchronous ICM. A whole ICM from an appropriately marked donor, or one or more dissociated cells from an ICM, will give a high yield of chimeras (Gardner and Lyon, 1971; Papaioannou and Gardner, 1979; Rossant and Lis, 1979). Blastocyst injection will be described here in general terms. For complete technical details of the methods, including the type of equipment used and the method for making the microinstruments, the reader is referred to Gardner (1978), Babinet (1980), and Papaioannou (1981a).

The blastocyst injection method has certain advantages over morula aggregation despite being technically more difficult. First of all, the zona pellucida does not need to be removed, allowing adaptation of this method to species such as the rabbit in which the zona is necessary for subsequent development. It also makes possible interspecific combinations that would otherwise fail to implant, since the trophoblast will be entirely composed of the host blastocyst type, unlike aggregation chimaeras where the trophoblast will be a mosaic of cells from both components. The beginning of cellular differentiation and the concomitant restriction of cell potential at the blastocyst stage also makes possible the analysis of the fate of somewhat later embryonic tissue with blastocyst injection than is possible with morula aggregation.

Embryos for injection can be obtained by flushing medium through each horn of the uterus of pregnant females approximately 3.5 days pc. Well-expanded blastocysts should be seen at this age, but there may be some variation in the developmental stage according to the mouse strain used. As an alternative, 2.5 day pc morulae can be cultured overnight to the blastocyst stage for use (Papaioannou et al., 1979). Blastocysts can be used for injection up to at least 4 days pc. They will tolerate long periods at room temperature in an appropriately buffered medium (e.g. PB1) and still retain viability (Gardner, 1972).

I. Making Chimeras

Cells for injection into blastocysts can be isolated from various embryonic sources (Papaioannou, 1981a; Snow, 1978). Inner cell mass cells can be isolated from donor blastocysts by dissection with a micromanipulator (Gardner, 1972) or more easily by immunosurgery (Solter and Knowles, 1975), a process that relies on complement-mediated killing of trophectoderm cells of blastocysts that have been previously exposed to an anti-mouse antiserum. The zonula occludens junctions between trophectoderm cells prevent antibody from reaching ICM cells so only trophectoderm is lysed by complement if the embryos have been thoroughly washed after the antiserum treatment (Fig. 10). A 30-minute treatment of the isolated ICMs with 0.25% trypsin (Gardner and Lyon, 1971) or 15 minutes in trypsin preceded by 10 minutes in 0.5% pronase (Gardner and Rossant, 1979) followed by gentle pipetting will dissociate ICMs into single cells that are capable of contributing to chimeras.

Blastocyst injections can be carried out under phase or brightfield optics with either an inverted or standard microscope. Since the procedure can take some time, the embryos should be kept in medium under paraffin oil to prevent evaporation, either in a dish or in hanging drops in a manipulation chamber. Depending on the number of microinstruments used, two or three micromanipulators are positioned around the microscope. Leitz micromanipulators with double instrument heads are very versatile, since the two instruments on the same head can be moved together in any plane and can also be moved relative to one another in three planes. Two such manipulators and a third more simple

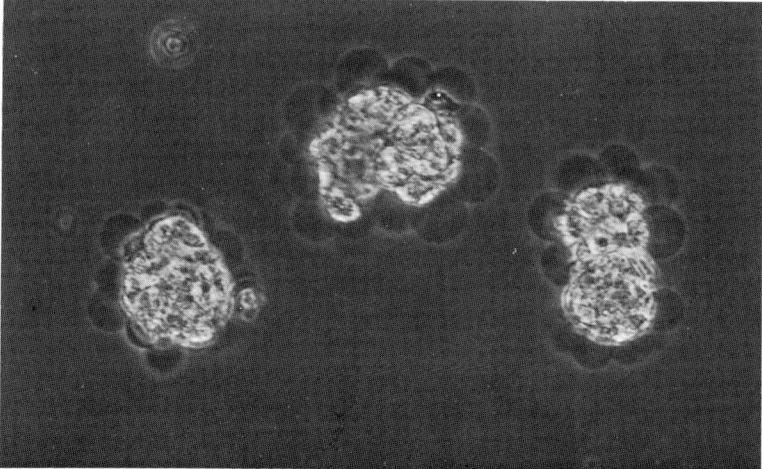

Fig. 10. Three blastocysts that have been treated with rabbit anti-mouse serum followed by complement for the immunosurgical removal of the trophectoderm. Only the outer, exposed layer of cells are lysing leaving the inner cell mass intact.

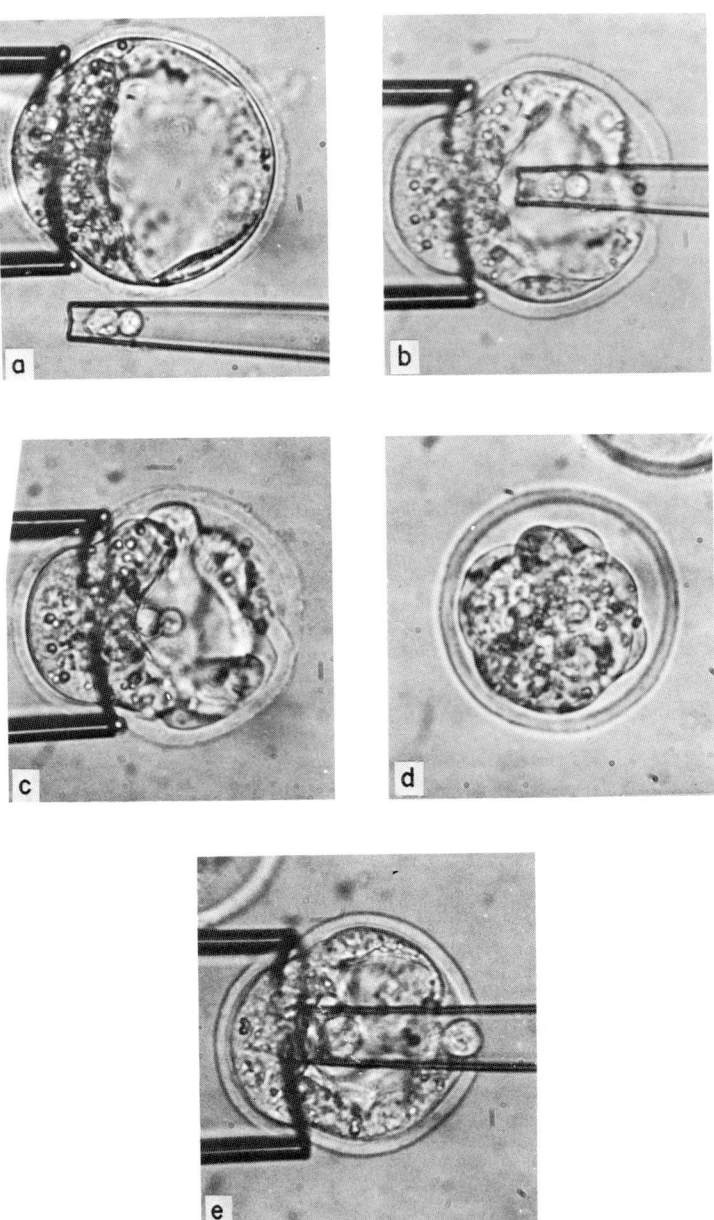

Fig. 11. Blastocyst injection using two microinstruments. (a) The blastocyst is held by the holding pipette with the ICM oriented toward the latter. The injection pipette is positioned at the same level as the ICM. (b) The injection pipette, containing two cells, is forced through the taut trophectoderm into the blastocoel. (c) The injected cells are released from the injection pipette before it is removed and are visible here at the surface of the ICM. (d) The same blastocyst a few minutes later. (e) Illustration of an injection using a larger pipette that could accommodate a small ICM. Reproduced from Babinet (1980).

1. Making Chimeras

manipulator holding a single instrument are used for the five-instrument injection method (Gardner, 1978; Papaioannou, 1981a).

Micropipettes are made from capillary tubing pulled by hand or with an electrode puller. Solid needles are made by first fusing a length of capillary tubing and then pulling it with an electrode puller. The instruments are then fashioned to their final shapes with the aid of a microforge (DeFonbrune). The critical dimensions of the microinstruments for various injection methods have been tabulated by Papaioannou (1981a). The micropipettes are connected to micrometer syringes with flexible, oil-filled tubing. Suction and expulsion are then controlled by means of these syringes to hold the embryos and to inject the cells. A DeFonbrune extraction and injection syringe or a Gilmont micrometer syringe are suitable for this purpose.

For the injection of cells with two microinstruments, only two simple micromanipulators are required. A blastocyst is held by suction on the end of one pipette while the cells are introduced from the opposite side with another pipette. For this procedure, the donor cells must be held inside the injection pipette and released into the blastocoel once the pipette has punctured the zona pellucida and trophectoderm. Some investigators have successfully used a beveled injection pipette (e.g., Mintz and Illmensee, 1975), but Babinet's (1980) alternative is to create tension in the trophectoderm by sucking the blastocyst partway into a holding pipette of diameter slightly smaller than the blastocyst (Fig. 11). The injection pipette, containing donor cells is then forced through the taut trophectoderm with a sharp, rapid movement. With this method, no bevel is required, and the injection pipette remains free of cytoplasmic debris so that instruments can be used repeatedly. Any number of isolated cells or even whole ICMs that can be completely sucked up into the injection pipette can be injected.

The five-instrument injection procedure, although requiring greater skill, is

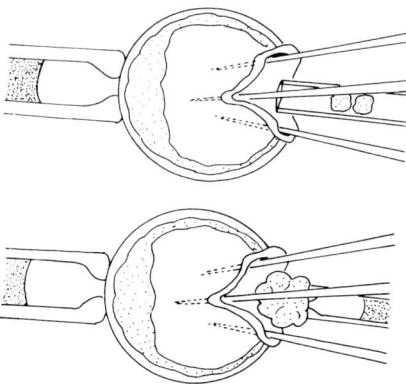

Fig. 12. Schematic representation of the injection into the blastocoel of isolated cells (top) or an ICM (bottom) through a triangular opening in the trophectoderm.

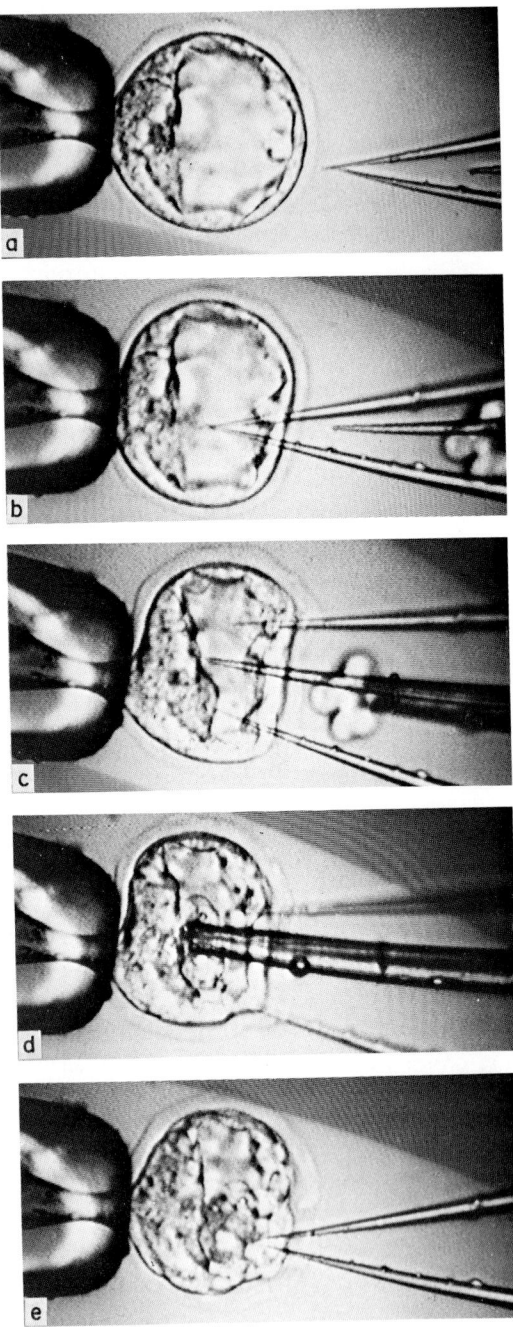

more versatile in terms of the amount of tissue that can be injected and the stage of blastocysts that can be used. For this procedure, the blastocyst is firmly held by a fire-polished holding pipette. Two solid needles are used to puncture the trophectoderm opposite the ICM and open a slit. A third needle is inserted into this slit, the needles are moved in opposite directions, and a triangular opening of a reasonable dimension is thus created in the trophectoderm (Fig. 12). An injection pipette, with cells either inside it or held by suction on its fire polished tip, are then inserted through this opening and released into the blastocoelic cavity. The hole is then closed and the instruments are withdrawn (Fig. 13).

Following injection of cells by either method, the trophectoderm will heal and the blastocyst will reexpand in culture, usually within a matter of 1–2 hours. These composite blastocysts can then be transferred to the oviducts or uteri of pseudopregnant foster mothers as described earlier.

3. RECONSTITUTION OF BLASTOCYSTS

For some developmental studies, it is advantageous to reconstruct a blastocyst making a chimera with trophectoderm of one cell type and ICM of another. Gardner, *et al.*, (1973) devised a means of reconstituting blastocysts using the five-instrument injection method to insert isolated ICMs into vesicles of pure trophectoderm that had been previously prepared by cutting blastocysts in half with razor blade fragments (Gardner, 1972). A simplified method has been recently described and used to make chimeras (Papaioannou, 1981a, b, 1982b) in which an isolated ICM is injected into a blastocyst before the host ICM, with overlying trophectoderm, is cut off (Figs. 14 and 15).

This reconstitution method (Fig. 14) depends on the tendency of fully expanded blastocysts to extrude or herniate through a slit made in the zona pellucida (Fig. 15) (Copp, 1979). While the blastocyst is firmly held in place with a holding pipette, a slit is made directly over the ICM using two needles. When the entire ICM with its overlying trophectoderm is fully extruded from the zona pellucida, the five-instrument injection method is used to insert an isolated ICM into that part of the trophectoderm remaining in the zona pellucida. This operation will cause the blastocyst to collapse and the torn edges of the zona pellucida will then serve as a clamp to hold the two ICMs apart. In the final step, the host ICM is cut off using crossed needles against a solid surface. Although this is most easily done with a micromanipulator, it is possible to use hand-held glass needles against the bottom of a watchglass (J. Rossant, personal communica-

---

Fig. 13. Injection of an ICM using the five-instrument method. (a) The blastocyst is held by suction on the holding pipette at the left. (b) The trophectoderm is punctured with two needles which are then spread apart and (c) a third needle is inserted. (d) The needles are spread to make a triangular opening through which an ICM is inserted. (e) The injection pipette and third needle are removed and the needles returned to their original positions and withdrawn. The injected ICM is visible in the blastocoelic cavity. Reproduced from Papaioannou (1981a).

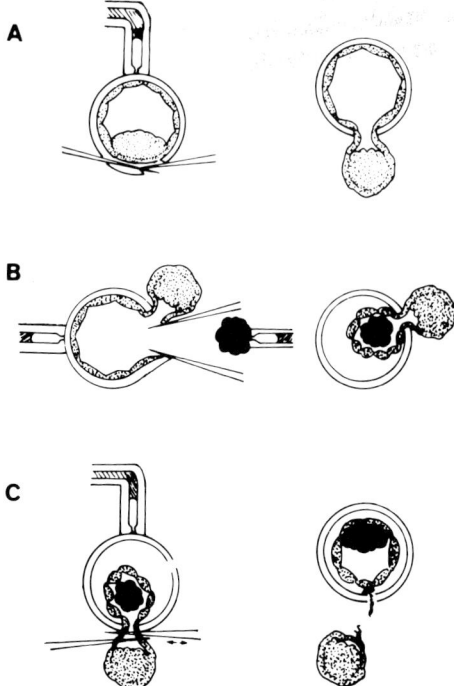

Fig. 14. Diagram of the reconstitution procedure. (a) A blastocyst is held in place by suction while a slit is made in the zona pellucida directly over the ICM with two needles. The embryo will herniate through this slit during culture at 37°C. (b) An ICM of a different genotype is injected into the herniated blastocyst through the same hole in the zona. (c) The extruded ICM end of the host blastocyst is cut off with straight needles leaving a reconstituted blastocyst in the zona pellucida with ICM of one type and trophectoderm of another. Reproduced from Papaioannou (1981a).

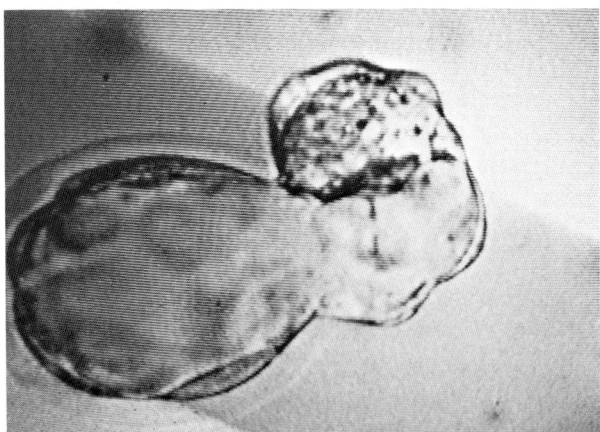

Fig. 15. Blastocyst with the ICM and overlying polar trophectoderm herniating out though a hole in the zona pellucida.

tion). Chimeras resulting from reconstituted blastocysts have been used to determine the fate of ICM and trophectoderm in later embryos (Gardner et al., 1973; Papaioannou, 1982b).

C. Making Chimeras after Implantation

1. IN UTERO

After implantation, the mammalian embryo is much less accessible for making chimeras since it cannot be replaced in the uterus once it has been removed. Also, the progressive restriction of cell potency with development as well as changes in the regulative capacity of the embryos may mean that donor cells will need to be precisely placed in a host and closely synchronized before they can successfully contribute to the formation of a chimera.

Some success in grafting cells to embryos *in utero* has been reported. Weissman, et al., (1978) used isolated donor cells from yolk sac blood islands of midgestation embryos. Between 10^4 and 10^5 of these cells were injected into synchronous, implanted embryos *in utero*. Microneedles with a solid glass point and a side hole were prepared with a microforge, filled with a cell suspension, and then inserted through the uterine wall and into a decidual swelling. Fiber optic lighting under the uterus provides a cool light source that helps in locating the embryo (V. E. Papaioannou, unpublished observations), but injection at a site three-quarters the distance between the mesometrium and the antimesometrial side of the uterus, in the center of the decidual swelling, will usually result in injection into the extraembryonic coelom or the yolk sac cavity of 8.5–10.5 day pc mouse embyros. In these experiments, the yolk sac blood island donor cells were shown to contribute to the host bone marrow hematopoietic cells and thymic lymphoid cells (Weissman et al., 1977, 1978). In similar studies using different markers, fetal blood island cells and primordial germ cells could not be shown to contribute to certain other adult tissues (V. E. Papaioannou and R. L. Gardner, unpublished observations).

Fleischman and Mintz (1979) have reported the production of hemopoietic chimeras following the injection of normal fetal liver cell suspensions into genetically anemic embryos. A beveled pipette, 20–30 μm in diameter, was used to inject cells into the placenta of midgestation embryos. Injection at a site one-quarter the distance from the mesometrial attachment to the antimesometrial side resulted in the rapid appearance of injected dye in the fetal circulation in control experiments and the successful production of chimeras when cells were injected. Other cell types have also been used with this method (Gootwine et al., 1982).

2. IN VITRO

Obviously the reliable placement of cells as well as their later detection, particularly if small, localized contributions are expected, are major problems in

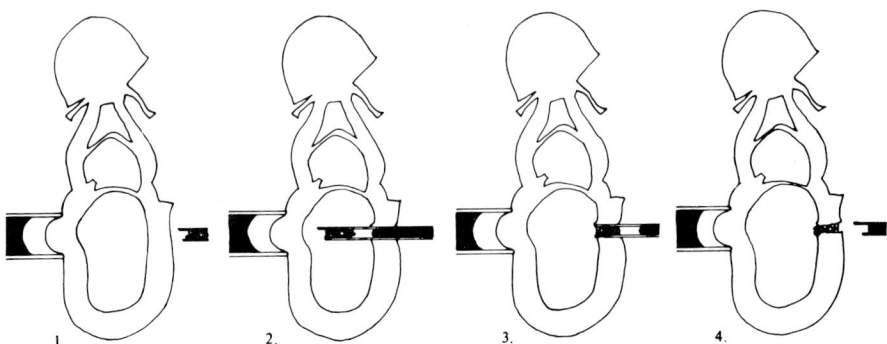

Fig. 16. Diagram of the injection technique for postimplantation embryos *in vitro*. (1) The egg cylinder is held by suction with a holding pipette opposite the intended site of injection. (2) The injection pipette, containing the donor tissue, is pushed through the egg cylinder wall into the amniotic cavity. (3) The injection pipette is withdrawn slowly and the donor tissue gently expelled into the ectoderm layer of the egg cylinder. (4) The injection pipette is removed and the embryo released from the holding pipette. Reproduced from Beddington (1981).

attempting to make chimeras with postimplantation embryos. An alternative to making chimeras *in utero* is to take advantage of the considerable advances that have been made in culturing intact postimplantation embryos (Buckley *et al.*, 1978; Sadler, 1979) in order to make chimeras for short-term developmental studies *in vitro* (Beddington, 1981).

Beddington (1981, 1982) used embryos dissected out of the uterus as egg cylinders at 7.5 days pc. Some were labeled with [^{3}H] thymidine to provide a cell-specific marker for the donor tissue; others served as the host embryos. The manipulations were done in drops of medium under paraffin oil using micromanipulators and a dissecting microscope. Pipettes of appropriate sizes were made with an electrode puller and a microforge. Cells from the donor embryo were sucked into an injection pipette. Then, while the host embryo was held firmly with a holding pipette, the injection pipette was pushed through the egg cylinder into the proamniotic cavity. As it was withdrawn, the donor tissue was expelled into the space created by the injection pipette (Fig. 16). Both orthotopic and heterotopic transplants can be done in this way. Analysis can be delayed for as long as normal development *in vitro* can be shown to occur, in this case, 36 hours after injection (Beddington, 1981, 1982).

IV. CONCLUSIONS

There has been a proliferation of techniques for the production of avian and mammalian chimeras in recent decades that has resulted in their increased use in studies of developmental biology. As this volume will document, chimera stud-

I. Making Chimeras

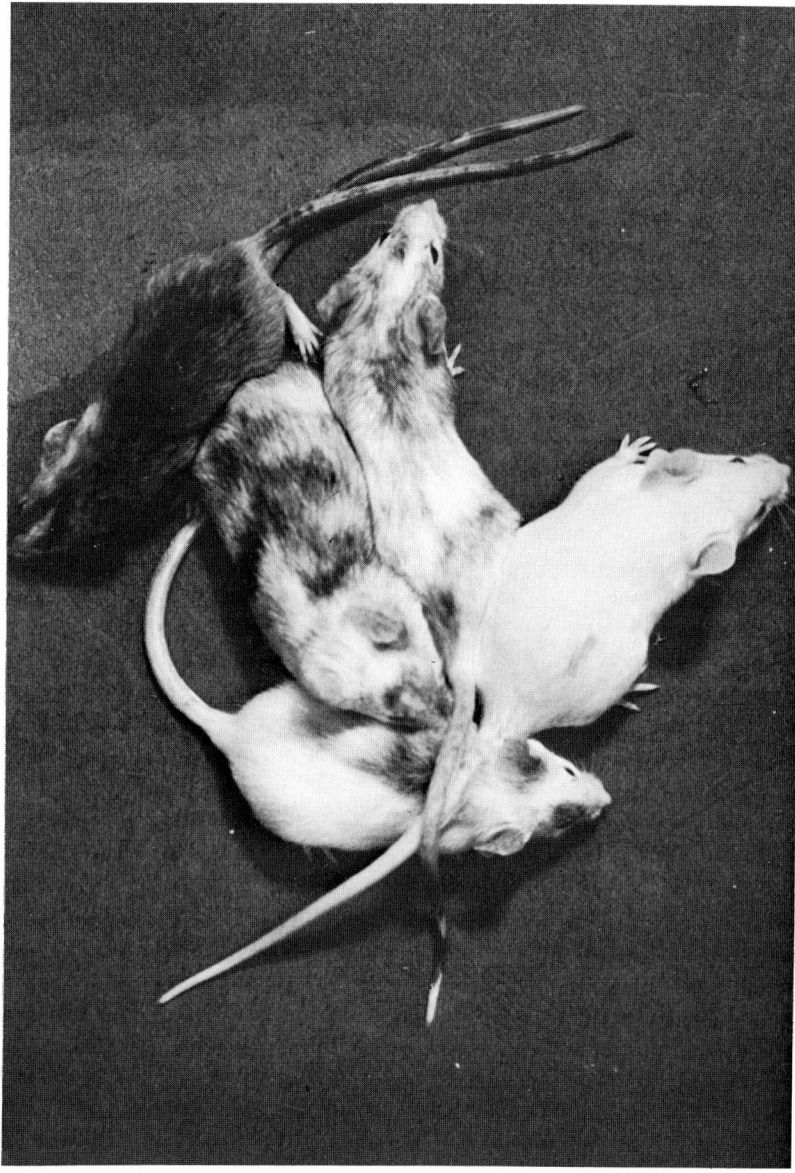

Fig. 17. Mouse chimeras made by combining embryonic cells from a pigmented and a nonpigmented strain.

ies have been applied to a range of problems that are difficult if not impossible to approach by other means. Crucial for this exploitation of mammalian chimeras have been improvements in culture techniques, based on a greater understanding of the requirements of embryos during development. They have allowed a wide range of experimental manipulations to be developed extending throughout a large part of embryonic and fetal development. Technical innovations, in particular the miniaturization of grafting and transplantation techniques made possible by micromanipulators, have opened for exploration the minute early mammalian embryo. Equally important for both avian and mammalian studies are new marker systems that allow the identification of individual cells in chimeras (Fig. 17), making possible a fine-grained analysis of the tissue make-up of the composite embryo. In birds, the combination of cells from two species has been particularly rewarding in that respect. In mammals a marker technique with equivalent qualities is actively sought. Further advances in these areas will surely lead to the full utilization of the potential of chimeras in the endeavor to understand the development of higher vertebrates.

ACKNOWLEDGMENTS

We wish to thank Cheryl Stafford and Lydie Perrin for help in preparation of the manuscript and Steve Halpern and Bernard Henri for photographic work. The work of V.E.P. was aided by Grant No. CD-137 from the American Cancer Society and a Basil O'Connor Starter Research Grant No. 5-379 from the March of Dimes. The work of F.D.L. is supported by the CNRS.

REFERENCES

Aitken, R. J., Bowman, P., and Gauld, I. (1977). *J. Embryol. Exp. Morphol.* **37**, 59–64.
Babinet, C. (1980). *Exp. Cell Res.* **130**, 15–19.
Babinet, C., and Bordenave, G. R. (1980). *J. Embryol. Exp. Morphol.* **60**, 429–440.
Beddington, R. S. P. (1981). *J. Embryol. Exp. Morphol.* **64**, 87–104.
Beddington, R. S. P. (1982). *J. Embryol. Exp. Morphol.* **69**, 265–285.
Biggers, J. D., Whitten, W. K., and Whittingham, D. G. (1971). *In* "Methods in Mammalian Embryology" (J. C. Daniel, Jr., ed.), pp. 86.–116. Freeman, San Francisco, California.
Billingham, R. E., Brent, L. and Medawar, P. B. (1956). *Philos. Trans. R. Soc. London, Ser. B* **239**, 357–414.
Brinster, R. L. (1963). *Exp. Cell Res.* **32**, 205–208.
Bronson, R. and McLaren, A. (1970). *J. Reprod. Fert.* **22**, 129.
Buckley, S. K. L., Steele, C. E., and New, D. A. T. (1978). *Dev. Biol.* **65**, 396–403.
Copp, A. J. (1979). *J. Embryol. Exp. Morphol.* **51**, 109–120.
Dieterlen-Lièvre, F., Martin, C., and Beaupain, D. (1979). *Folia Biol. (Prague)* **25**, 293–295.
Dossel, W. E. (1954). *Science* **120**, 262–263.
Dubois, R. (1968). *J. Embryol. Exp. Morphol.* **20**, 189–213.
Ducibella, T., Albertini, D. F., Anderson, E., and Biggers, J. D. (1975). *Dev. Biol.* **45**, 231–250.
Fleischman, R. A., and Mintz, B. (1979). *Proc. Natl. Acad. Sci. U.S.A.* **76**, 5736–5740.
Fontaine, J., and Le Douarin, N. (1977). *J. Embryol. Exp. Morphol.* **41**, 209–222.

1. Making Chimeras

Fujii, J. T., and Martin, G. R. (1980). *Dev. Biol.* **74,** 239–244.
Gardner, R. L. (1968). *Nature (London)* **220,** 596–597.
Gardner, R. L. (1971). *Adv. Biosci.* **6,** 279–296.
Gardner, R. L. (1972). *J. Embryol. Exp. Morphol.* **28,** 279–312.
Gardner, R. L. (1978). *In* "Methods in Mammalian Reproduction" (J. C. Daniel, Jr., ed.), pp. 137–165. Academic Press, New York.
Gardner, R. L., and Johnson, M. H. (1973). *Nature (London) New Biol.* **246,** 86–89.
Gardner, R. L., and Johnson M. H. (1975). *Ciba Found. Symp.* **29,** 183–200.
Gardner, R. L., and Lyon, M. F. (1971). *Nature (London)* **231,** 385–386.
Gardner, R. L., and Munro, A. J. (1974). *Nature (London)* **250** 146–147.
Gardner, R. L., and Papaioannou, V. E. (1975). *In* "The Early Development of Mammals" (M. Balls and A. E. Wild, eds.), pp. 107–132. Cambridge Univ. Press, London and New York.
Gardner, R. L., and Rossant, J. (1979). *J. Embryol. Exp. Morphol.* **52,** 141–152.
Gardner, R. L., Papaioannou, V. E., and Barton, S. C. (1973). *J. Embryol. Exp. Morphol.* **30,** 561–572.
Gootwine, E., Webb, C. G., and Sacks, L. (1982). *Nature (London)* **299,** 63–65.
Hajji, K., Martin, C., Perramon, A., and Dieterlen-Lièvre, F. (1983). In preparation.
Hamburger, V., and Hamilton, H. L. (1951). *J. Morphol.* **88,** 49–92.
Hasek, M. (1953). *Cesk. Biol.* **2,** 25.
Hillman, N., Sherman, M. I., and Graham, C. (1972). *J. Embryol. Exp. Morphol.* **28,** 263–278.
Lassila, O., Eskola, J., Toivanen, P., Martin, C., and Dieterlen-Lièvre, F. (1978). *Nature (London)* **272,** 353–354.
Lassila, O., Martin, C., Toivanen, P., and Dieterlen-Lièvre, F. (1982). *Blood* **59,** 377–381.
Le Douarin, N. (1969). *Bull. Biol. Fr. Belg.* **103,** 435–452.
Le Douarin, N., Coltey, M., and Guillemot, F. (1983). In preparation.
McLaren, A. (1976). "Mammalian Chimaeras." Cambridge Univ. Press, London and New York.
McLaren, A., and Michie, D. (1956). *J. Exp. Biol.* **33,** 394–416.
Markert, C. L., and Petters, R. M. (1978). *Science* **202,** 56–58.
Marsk, L. (1977). *J. Embryol. Exp. Morphol.* **39,** 129–137.
Martin, C. (1972). *C. R. Seances Soc. Biol. Ses Fil.* **166,** 283–285.
Martin, C. (1976). *J. Embryol. Exp. Morphol.* **35,** 485–498.
Martin, C., Lassila, O., Nurmi, T., Eskola, J., Dieterlen-Lièvre, F., and Toivanen, P. (1979). *Scand. J. Immunol.* **10,** 333–338.
Martin, C., Beaupain, D., and Dieterlen-Lièvre, F. (1980). *Dev. Biol.* **75,** 303–314.
Marzullo, G. (1970). *Nature (London)* **225,** 72–73.
Mayer, J. F., and Fritz, H. I. (1974). *J. Reprod. Fert:* 1. **39,** 1–9.
Micklem, H. S., and Loutit, J. F. (1966). "Tissue Grafting and Radiation". Academic Press, New York.
Mintz, B. (1962a). *Am. Zool.* **2,** 423 (abstr.).
Mintz, B. (1962b). *Science* **138,** 594–595.
Mintz, B. (1964). *J. Exp. Zool.* **157,** 273–292.
Mintz, B. (1967). *In* "Methods in Developmental Biology" (F. H. Wilt and N. K. Wessels, eds.), pp. 379–400. Crowell-Collier, New York.
Mintz, B. (1971). *In* "Methods in Mammalian Embryology" (J. C. Daniel, Jr., ed.), pp. 186–214. Freeman, San Francisco, California.
Mintz, B., and Illmensee, K. (1975). *Proc. Natl. Acad. Sci. U.S.A.* **72,** 3585–3589.
Mintz, B., Gearhart, J. D., and Guymont, A. O. (1973). *Dev. Biol.* **31,** 195–199.
Moore, M. A. S., and Owen, J. J. T. (1965). *Nature (London)* **208,** 956, 989–990.
Moore, M. A. S., and Owen, J. J. T. (1967). *J. Exp. Med.* **126,** 715–725.
Moustafa, L. A. (1974). *Proc. Soc. Exp. Biol. Med.* **147,** 485–488.

Moustafa, L. A., and Brinster, R. L. (1972). *J. Exp. Zool.* **181,** 193–202.
Mullen, R. J., and LaVail, M. M. (1976). *Science* **192,** 799–801.
Mullen, R. J., and Whitten W. K. (1971). *J. Exp. Zool.* **178,** 165.
Mulnard, M. J. (1973). *C. R. Hebd. Seances Acad. Sci.* **276,** 379–381.
Mystkowska, E. T. (1975). *J. Embryol. Exp. Morphol.* **33,** 731–744.
Nakamura, H., and Le Lièvre, C. (1982). *J. Embryol. Exp. Morphol.* **701,** 1–18.
New, D. A. T. (1955). *J. Embryol. Exp. Morphol.* **3,** 326–331.
Nicolson, G. L., Yanagimachi, R., and Yanagimachi, H. (1975). *J. Cell Biol.* **66,** 263–274.
Palmer, J., and Dewey, M. J. (1983). *Experientia* **39,** 196–198.
Papaioannou, V. E. (1979). *In* "Cell Lineage, Stem Cells and Cell Determination" (N. Le Douarin, ed.), pp. 141–155. Elsevier/North-Holland Biomedical Press, Amsterdam.
Papaioannou, V. E. (1981a). *Tech. Life Sci.: Physiol.* P1(1) 116/1-27.
Papaioannou, V. E. (1981b). *Prog. in Clin. Biol. Res.* **45,** 77–91.
Papaioannou, V. E. (1982a). *In* "The Mouse in Biomedical Research" (H. L. Foster, J. D. Small, and J. G. Fox, eds.), Vol., 4 pp. 147–209. Academic Press, New York.
Papaioannou, V. E. (1982b). *J. Embryol. Exp. Morphol.* **68,** 199–209.
Papaioannou, V. E., and Gardner, R. L. (1979). *J. Embryol. Exp. Morphol.* **52,** 153–163.
Papaioannou, V. E., and West, J. D. (1981). *Genet. Res.* **37,** 183–197.
Papaioannou, V. E., Rossant, J., and Gardner, R. L. (1978). *Symp. Br. Soc. Cell Biol.* **2,** 49–69.
Papaioannou, V. E., Evans, E. P., Gardner, R. L., and Graham, C. F. (1979). *J. Embryol. Exp. Morphol.* **54,** 277–295.
Pedersen, R. A., and Spindle, A. I. (1980). *Nature (London)* **284,** 550–552.
Rafferty, K. A., Jr. (1970). "Methods in Experimental Embryology of the Mouse." John Hopkins Press, Baltimore, Maryland.
Rossant, J. (1975). *J. Embryol. Exp. Morphol.* **33,** 979–990.
Rossant, J. (1976). *J. Embryol. Exp. Morphol.* **36,** 163–174.
Rossant, J., and Chapman, V. M. (1983). *J. Embryol. Exp. Morphol.* **73,** 193–205.
Rossant, J., and Frels, W. I. (1980). *Science* **208,** 419–421.
Rossant, J., and Lis, W. T. (1979). *Dev. Biol.* **70,** 249–254.
Rossant, J., Croy, B. A., Chapman, V. M., Siracusa, L., and Clarke, D. A. (1982). *J. Anim. Sci.* **55,** 1241–1248.
Rossant, J., Vigh, M., Siracusa, L. D., and Chapman, V. M. (1983). *J. Embryol. Exp. Morphol.* **73,** 179–191.
Sadler, T. W. (1979). *J. Embryol. Exp. Morphol.* **49,** 17–25.
Simon, D. (1960). *Arch. Anat. Microsc. Morphol. Exp.* **49,** 93–176.
Siracusa, L. D., Chapman, V. M., Bennett, K. L., Hastie, N. D., Pietras, D. F., and Rossant, J. (1983). *J. Embryol. Exp. Morphol.* **73,** 163–178.
Snow, M. H. L. (1978). *In* "Methods in Mammalian Reproduction" (J. C. Daniel, Jr., ed.), pp. 167–178. Academic Press, New York.
Solter, D., and Knowles, B. B. (1975). *Proc. Natl. Acad. Sci. U.S.A.* **72,** 5099–5102.
Spemann, H. (1962). "Embryonic Development and Induction." Hafner, New York.
Stern, M. S. (1973). *Nature (London)* **243,** 472–473.
Stewart, C. (1980). *J. Embryol. Exp. Morphol.* **58,** 289–302.
Tarkowski, A. K. (1959). *Acta Theriol.* **2,** 251–267.
Tarkowski, A. K. (1961). *Nature (London)* **190,** 857.
Tucker, E. M., Moor, R. M., and Rowson, L. E. A. (1974). *Immunology* **26,** 613–621.
Vakaet, L. (1974). *Ann. Biol. Anim., Biochim., Biophys.* **13,** 35–41.
Weissman, I. L., Barid, S., Gardner, R. L., Papaioannou, V. E., and Raschke, W. (1977). *Cold Spring Harbor Symp. Quant. Biol.* **41,** 9–21.
Weissman, I. L., Papaioannou, V. E. and Gardner, R. L. (1978). *Cold Spring Harbor Conf. Cell Proliferation* **5,** Book A, 33–47.

I. Making Chimeras

Whittingham, D. G. (1968). *Nature (London)* **220,** 592–593.
Whittingham, D. G., and Wales, R. G. (1969). *Aust. J. Biol. Sci.* **22,** 1065–1068.
Willadsen, S. (1979). *Nature (London)* **277,** 298.
Wolff, Et., and Haffen, K. (1952). *J. Exp. Zool.* **119,** 381–399.
Zeilmaker, G. H. (1973). *Nature (London)* **242,** 115–116.

CHAPTER II

Cell Markers

JOHN D. WEST

Radiobiology Unit
Medical Research Council
Harwell, Didcot
Oxon, England

I.	Introduction	39
II.	Applied Markers	40
III.	Genetic Markers	42
	A. Pigment	42
	B. Morphology and Histology	47
	C. Cytology	48
	D. Chromosomes	53
	E. Biochemistry and Immunology	54
	References	63

I. INTRODUCTION

Cell markers are used in almost all chimera experiments in order to verify that the experimental organism is in fact a chimera. Beyond this basic requirement, markers are used to investigate the deployment of the different (normally two) component cell populations of the chimera. This involves two questions: (1) What are the proportions of the component cell populations in various tissues? (2) How are they spatially arranged? Developmental biologists usually study experimental chimeras to learn something about the development of nonchimeric individuals. In order for the extrapolation from chimaera to nonchimera to be valid, it is important that genetic or other differences between component cell populations do not result in cell selection or nonrandom associations between the different cell populations.

The ideal cell marker would provide precise answers to the two questions, posed above, in all tissues of the body. Such a marker would fulfil the six criteria discussed by McLaren (1976) and Oster-Granite and Gearhart (1981). The ideal marker would be (1) *cell localised*, not secreted extracellularly; (2) *cell autonomous*, not transferred between cells or affecting other cells; (3) *stable* both within the first marked cells and all of their mitotic progeny; (4) *ubiquitous* throughout development among both the internal and external tissues of the body; (5) *easy to detect*, both grossly and in histological sections, without elaborate processing, and (6) *developmentally neutral*, not causing cell selection or influencing developmental processes such as cell mixing.

It is hardly surprising that no ideal marker exists that fulfils all six criteria. Inevitably the choice of cell marker is a compromise which depends on the range of tissues to be studied, the questions to be asked, and the precision required. For example, various genetically determined morphological markers could provide some degree of spatial information in certain tissues without being quantitative. On the other hand, many inherited biochemical markers allow the proportions of the two component cell populations to be estimated but few are useful for spatial analysis.

II. APPLIED MARKERS

Application of extrinsic markers, such as vital dyes, is a classic embryological technique (review by Weston, 1967), and this approach has been used to mark groups of cells before incorporating them into chimeras. Applied markers, however, are only suitable for relatively short-term studies because they are rapidly diluted by cell division. In addition care must be taken to ensure that the marker is developmentally neutral and that it stays within the descendants of the original marked cells.

Weston (1967) has reviewed the use of vital dyes, particulate markers, and radioisotopes (notably tritiated thymidine); other applied markers are listed in Table I. Weston considered vital dyes and particulate markers to be unreliable because they do not always remain localised, and vital dyes can be metabolised to colourless products. Toxicity may also be a problem with some applied markers. For example, tritiated thymidine can be toxic (Snow, 1973), and Hollyfield and Ward (1974) have shown that polystyrene microspheres are toxic to frog embryos.

The requirement for an applied cell marker to be cell-localised often conflicts with the need for a simple and efficient means of delivering the marker to the required cells. This is not a problem in the case of tritiated thymidine because it is a small molecule that is readily taken up by cells, but once inside it becomes incorporated into the DNA and hence is cell-localised. Other applied markers tend to be large molecules or particles that cannot pass through gap junctions. Such markers may be taken up by certain phagocytic cells or delivered to individual cells by microinjection (Stern and Wilson, 1972) or iontophoresis (Lo and Gilula, 1979). However, there is clearly a need for less cumbersome delivery

II. Cell Markers

TABLE I.

Applied Markers

Marker	Experimental system	Reference
1. Markers used in chimeric systems		
Vital dyes	Starfish chimeras	Mita and Satoh, 1982
	Axolotl, embryonic tissue transplantation	Detwiler, 1917
Tritiated thymidine	Mouse aggregation chimeras	Hillman et al., 1972; Garner and McLaren, 1974; Spindle, 1982
	Axolotl, embryonic tissue transplantation	Stocum, 1975
	Chick, embryonic tissue transplantation	Weston, 1963; Nicolet, 1970
	Mouse embryonic tissue transplantation (in vitro chimeras)	Beddington, 1981, 1982
	Parabiotic rats	Göthlin and Ericsson, 1973
Colloidal carbon	Hydra chimeras	Wanek and Campbell, 1982
Thorotrast (colloidal thorium)	Rat, macrophage transplantation	Göthlin and Ericsson, 1973
Fluorescein or rhodamine isothiocyanate	Mouse, lymphocyte transplantation	Butcher et al., 1980
	Mouse aggregation chimeras	Ziomek, 1982
Fluorescein–lysine–dextran	Xenopus, embryonic tissue transplantation	Gimlich and Cooke, 1983
2. Markers yet to be used in chimeric systems		
Oil droplets	Early mouse embryos	Stern and Wilson, 1972
Melanin granules	Early mouse embryos	Gardner, 1975
Polystyrene or latex microspheres	Frog embryos	Hollyfield and Ward, 1974
Horseradish peroxidase	Xenopus embryos	Jacobson and Hirose, 1978
	Early mouse embryos	Lo and Gilula, 1979; Balakier and Pedersen, 1982
Bolton-Hunter reagent	Xenopus and axolotl embryos	Katz et al., 1982; Smith and Malacinski, 1983
Fluorescent colloidal gold	—	Horisberger and Vonlanthen, 1979

systems, such as liposomes (Poste et al., 1976) or osmotic lysis of pinosomes (Okada and Rechsteiner, 1982), for applied markers. Regardless of the efficiency of delivery, the possibility that applied markers may be liberated from dying cells and then taken up by other cells remains a problem.

Despite the problems associated with applied markers, a number of them have been used for short-term studies involving chimeras as indicated in Table I.

III. GENETIC MARKERS

Genetically determined intrinsic markers are much more widely used in chimera experiments than the applied markers discussed above. One major advantage is that genetic differences are stably inherited. This does not mean, however, that all genetic markers are necessarily "stable" at the phenotypic level because gene expression may vary among cells in a tissue. In addition to the known marker genes, the component cell populations of a chimera may carry a number of unknown genetic differences, and these differences in genetic background may also pose problems. Genetic background differences can produce differences in cellular behaviour during development regardless of whether the different alleles of the marker gene itself are developmentally neutral. For example, Ivanyi (1978) lists 35 quantitative traits that are influenced by the mouse *H-2* histocompatibility system, and Bartlett and Edidin (1978) have shown that this system influences intercellular adhesion *in vitro*. Consistent differences in genetic background associated with a particular marker are most extreme in interspecific chimeras but also occur in intraspecific chimeras that are made from different inbred strains. Genetic background effects can be reduced in intraspecific chimeras by constructing them from congenic strains, which differ genetically at the marker locus and only a few closely linked loci. This approach is only really practicable for mice and *Drosophila* where congenic stocks are available or can be easily constructed. Alternatively, genetic background differences can be randomised by constructing chimeras from appropriately marked individuals taken from a single outbred stock.

A wide variety of potential genetic markers exists, particularly in *Drosophila* and mice. Genetic markers in *Drosophila* have been widely used to analyse mosaicism (Hall *et al.*, 1976), but much less work has been done on *Drosophila* chimeras because they are technically more difficult to produce. I have included examples from a variety of different species in the following discussion of genetically determined cell markers, but inevitably I have drawn most heavily on examples from the mouse.

A. Pigment

Various genetically determined pigment markers fulfil most of the requirements of an ideal marker, but their usefulness is restricted by the limited number of pigmented tissues. Probably because they are supremely easy to detect, pigment markers have been incorporated into a wide range of chimeric systems, as shown in Table II. Pigment markers in *Xenopus* and mouse chimeras are illustrated in Figs. 1 and 2.

II. Cell Markers

TABLE II.

Chimeric Systems Incorporating Pigment Markers

Chimeras	References
Hydra	Waneck and Campbell, 1982
Plants	Stewart, 1978
Drosophila	Zalokar, 1971, 1973
Xenopus	Volpe *et al.*, 1979
Axolotls	Maden, 1980; Pescitelli and Stocum, 1980
Chicks	Marzullo, 1970
Mice	Tarkowski, 1964
Deer mice	Klein and Markert, 1981
Rats	Mayer and Fritz, 1974
Rabbits	Gardner and Munro, 1974; Moustafa, 1974

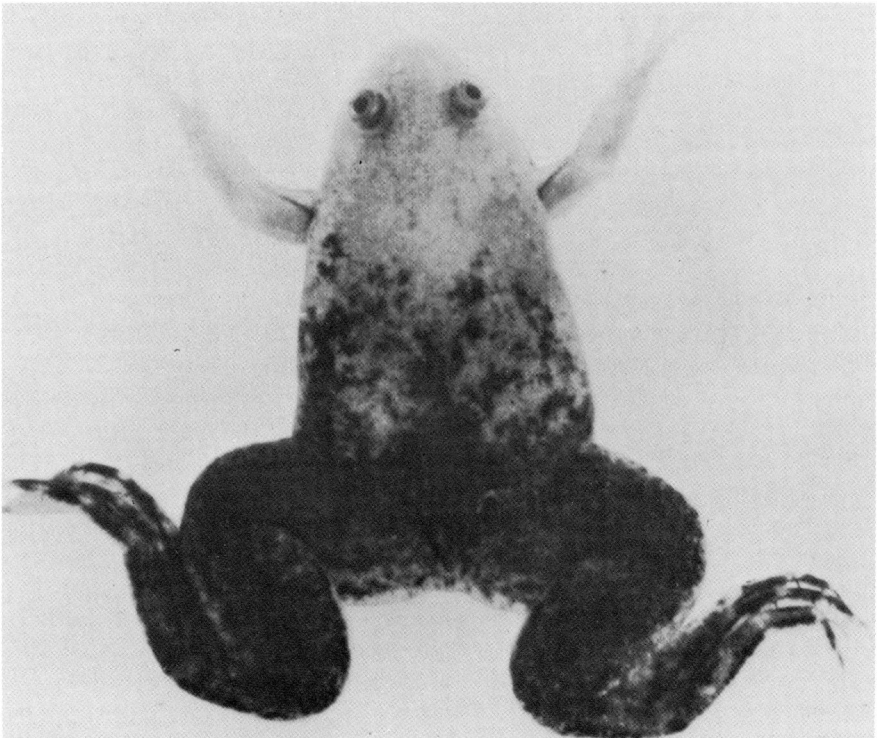

Fig. 1. An albino↔pigmented *Xenopus* chimera produced by joining anterior and posterior halves of two embryos. (Reprinted, by permission, from Volpe *et al.*, 1979.)

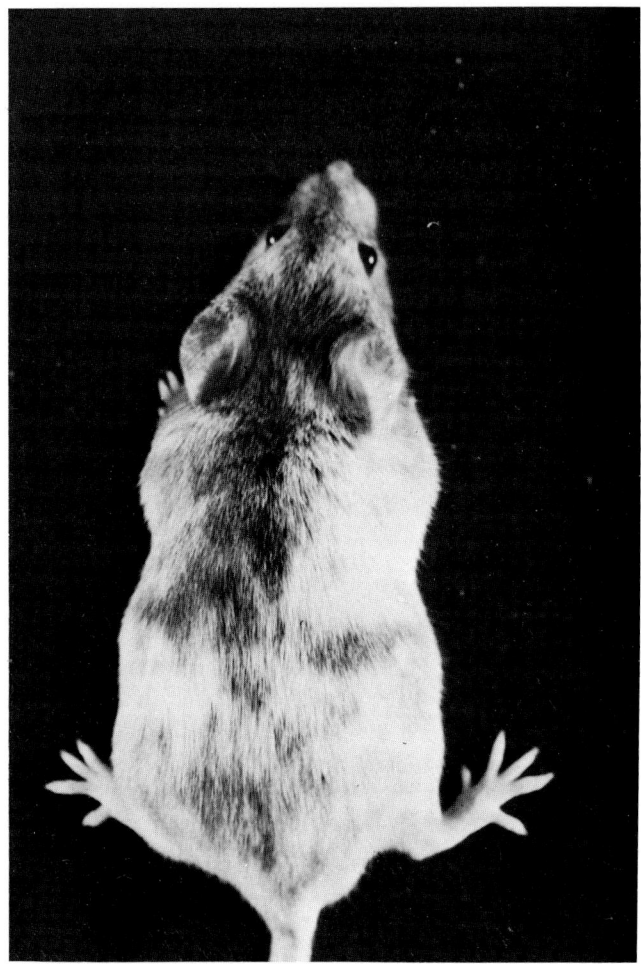

Fig. 2. An albino⇔pigmented mouse chimera. (Previously unpublished photograph kindly supplied by Professor R. L. Gardner.)

1. MOUSE ALBINO GENE

The albino gene (c) is the most commonly used pigment marker in mice. This gene causes a failure of melanin synthesis so that none of the melanocytes are pigmented. Albino has been widely used as a marker for chimeras in the coat (e.g., Mintz, 1967), as shown in Fig. 2, and the retinal pigment epithelium of the eye (e.g., Tarkowski, 1964). It has also been useful for other tissues of the eye, including the choroid, iris, and ciliary body (Tarkowski, 1964; Deol and Whitten, 1972), the Harderian gland, which lies in the orbit behind the eye (Tarkowski, 1964), and the membranous labyrinth of the inner ear (Deol and Whitten, 1972).

The retinal pigment epithelium arises from the embryonic optic cups, and the melanin granules remain cell-localised to provide an excellent spatial cell marker that can be used as early as $12\frac{1}{2}$ days postcoitum. All other melanocytes are migratory cells that arise from the neural crest. Epidermal melanocytes in the bulb of the hair follicle secrete melanin granules into the hair shaft and produce the coat pigmentation that can be seen soon after birth. McLaren and Bowman (1969) have shown that individual hairs from chimeric mice may be unevenly pigmented as a result of the two types of melanocytes colonising a single hair follicle. Thus, although coat pigmentation provides a useful spatial marker, it is not precise at the cellular level because the unit of pigmentation is multicellular. If other melanocytes derived from the neural crest also secrete their pigment extracellularly, this would reduce their usefulness as a precise spatial marker, but according to Searle (1968) secretion does not occur in the choroid or iris and normally seems to be restricted to the hair and skin melanocytes.

2. OTHER MOUSE PIGMENT GENES

The availability of other genes that affect melanocyte pigmentation enabled Gardner and Lyon (1971) to distinguish three melanocyte populations in chimeras using differences between normally pigmented melanocytes, those that were homozygous for pink-eyed dilution (p/p) and those that were either albino (c/c) or heterozygous for albino and the extreme dilution allele (c/c^e). Petters and Markert (1980) have gone one stage further and produced a chimaera with *four* genetically distinct populations of melanocytes, using differences between those with normal black pigmentation ($B/-$) and those that were homozygous for albino (c/c), recessive yellow (e/e), or both brown and dilute ($b/b\ d/d$).

All of the genes listed above are known to act directly on the melanocyte (Silvers, 1979) but other genes, such as agouti (A), act on the tissue environment to alter pigmentation. This gene produces a subterminal band of yellow phaeomelanin in certain hair types. The presence of this band is controlled by the genotype of the mesodermal component of the hair follicle rather than the genotype of the melanocytes themselves (reviewed by Silvers, 1979). Although agouti is a useful marker to indicate the presence of two cell populations in a chimera, it is not cell autonomous and may not be an accurate spatial marker. Mintz (1970, 1971) concludes that the narrow stripes seen in $A/A \leftrightarrow a/a$ chimeras reflect the distribution of patches of the mesodermal (dermal papilla) component of the hair follicles. However, McLaren (1976) argues that the regular stripes could equally be the product of some systemic wave pattern set up by the presence of two genetically distinct cell populations.

The dominant spotting gene (W) has also been incorporated into mouse chimeras. This gene appears to act directly on the migratory melanoblasts to cause their death but does not affect the retinal pigment epithelium (reviewed by Silvers, 1979). Although the action of this gene, which also affects erythropoiesis and gametogenesis, is too complex to warrant its recommendation as a marker, it can be employed to refine the use of albino (*c*) as a marker in the retinal pigment epithelium. Normally the overlying densely pigmented choroid masks the pigmentation in the retinal epithelium of adult eyes, but *W/W C/C* ↔ *w/w c/c* chimeras have unpigmented choroids that allow the variegation of the retinal pigment epithelium to be examined *in situ* (Gordon, 1977).

3. PIGMENT GENES IN OTHER VERTEBRATE SPECIES

Pigment markers have also been used for chimeras of other species. Albino has been used as a marker for rat chimeras (Mayer and Fritz, 1974; Yamamura and Markert, 1981) and rabbit chimeras (Gardner and Munro, 1974; Moustafa, 1974). Various other pigment genes, including agouti, pink eye dilution, and hooded, have been used in rats (Mullen and LaVail, 1976; Yamamura and Markert, 1981), and the blonde mutation has been used in deer mouse (*Peromyscus*) chimeras (Klein and Markert, 1981). Marzullo (1970) used feather pigmentation for analysing chick chimeras made by injecting cells from either Rhode Island Red or Barred Rock strains into White Leghorn embryos. Albino has also been used in *Xenopus* chimeras (Volpe *et al.*, 1979) as shown in Fig. 1, but in axolotls, although there is an albino (*a*) variant (reviewed by Frost and Malacinski, 1980), more use appears to have been made of the white (*d*) gene (e.g., Maden, 1980; Pescitelli and Stocum, 1980). The axolotl white gene appears to act in the tissue environment to inhibit the migration or survival of pigment cells (see Frost and Malacinski, 1980). Homozygous *d/d* axolotls have various amounts of pigment, and so this gene is a rather imprecise cell marker and suitable only for distinguishing the large regions of one cell population that are produced by grafting experiments.

4. *DROSOPHILA* PIGMENT GENES

Geneticists have accumulated a bewildering array of variants that alter the pigmentation of the *Drosophila* cuticle or eye. Examples of genes that are commonly used as markers in *Drosophila* mosaics include ebony (*e*) and yellow (*y*), which alter body and bristle colour, and white (*w*), brown (*bw*), maroon-like (*mal*), and vermilion (*v*), which affect eye colour. Zalokar (1973) produced chimeras by injecting donor cells marked with vermilion and brown into host embryos marked with yellow and white. On their own, neither the *y,w* host cells nor the *v,bw* donor cells are able to make brown eye pigment. (The *v* gene causes the absence of a pigment precursor necessary for brown eye pigmentation.) However, chimeric flies are able to make brown eye pigment because the product of the v^+ gene diffuses from the *y,w* host tissue into neighbouring regions of *v,bw* eye tissue where brown pigment is synthesised. This provides an elegant system for detecting chimeric flies but is likely to be less useful for spatial analysis unless it can be shown that all v,bw eye cells receive enough v^+ product to synthesise detectable amounts of brown pigment. Illmensee and Mahowald (1976) have also produced chimeric *Drosophila* and used a number of marker genes, including the pigment markers ebony, yellow, and maroon-like.

B. Morphology and Histology

On the whole, genes that affect morphology and histology do not provide good cell markers and have normally been incorporated into chimeras as the subject of the study rather than as independent cell markers. The outstanding exceptions to this are *Drosophila* genes, such as forked (*f*) that alters the shape of bristles and trichomes, and multiple wing hairs (*mwh*) that produces more than one trichome per cell. These genes are good cell markers because each structure arises from a single cell, and it may be more appropriate to consider them as cytological rather than morphological markers. One problem, however, is that not all the cells of the cuticle produce these structures so they are not perfect spatial markers.

1. HAIR MORPHOLOGY

Unlike *Drosophila* bristles, mouse hairs are multicellular in origin. Various hair structure variants such as fuzzy (*fz*), tabby (*Ta*), greasy (*Gs*), downless (*dl*), and waved-2 (*wa-2*) have been incorporated into mouse chimeras (McLaren and Bowman, 1969; Mintz, 1970, 1971; Cattanach *et al.*, 1972; Dunn, 1972; McLaren *et al.*, 1973; Green *et al.*, 1977). These chimeras show narrow bands of normal and variant hair similar to that described earlier for chimeras involving the agouti gene. McLaren (1976) has emphasised that it is not known whether the variegated phenotype directly reflects the underlying distribution of the two cell populations, and this issue has been discussed in the previous section with reference to agouti.

2. TISSUE MORPHOLOGY

Many genetic variants are known in the mouse that cause abnormal tissue morphology [see Green (1981) and the list of genes and their effects published in alternate issues of *Mouse News Letter*]. Some of these, such as short ear (*se*) and vestigial tail (*vt*) (Grüneberg and McLaren, 1972) and the skeletal differences between C57BL/6 and C3H inbred strains of mice (Moore and Mintz, 1972), have been incorporated into chimeras. Although they may indicate whether an individual is a chimera, they are inappropriate as independent cell markers.

Genes that cause cell death, tissue degeneration, or abnormal cell positioning have also been studied in chimeras and are listed in Table III. None of those listed are good cell markers, both because cell death is likely to be followed by movement of the surviving cells, leading to distortion of the spatial organisation

TABLE III.

Examples of Genes That Cause Cell Death, Tissue Degeneration, or Abnormal Cell Positioning in Mouse and Rat Chimeras

Gene	References
Mouse chimeras	
Various *t*-alleles	Mintz, 1964; Spiegelman, 1978
Retinal degeneration (*rd*)	Mintz and Sanyal, 1970; Wegmann *et al.*, 1971
Dystrophia-muscularis (*dy*)	Peterson, 1974, 1979
Purkinje cell degeneration (*pcd*)	Mullen, 1977a
Reeler (*rl*)	Mullen, 1977b
Staggerer (*sg*)	Mullen, 1977b
Weaver (*wv*)	Goldowitz and Mullen, 1980
Lurcher (*Lc*)	Wetts and Herrup, 1982
Rat chimeras	
Retinal dystrophy (*rdy*)	Mullen and LaVail, 1976

of the tissue, and because cell autonomy is hard to establish. Indeed, two of these genes are known not to be cell autonomous in their action. The mouse dystrophia-muscularis gene causes degeneration in the muscle but the primary site of gene action is extramuscular (Peterson, 1974, 1979), and the rat retinal dystrophy gene acts via the retinal pigment epithelium to cause degeneration in the underlying neural retina (Mullen and LaVail, 1976).

Gross morphological differences have also been used as regional markers for secondary chimeras involving grafts made between different species of *Hydra* (Wanek and Campbell, 1982) and axolotls (Pescitelli and Stocum, 1980), but this is of no use for primary chimaeras.

C. Cytology

Cytological markers include a wide range of genetic variants, including some pigment markers, such as albino in the retinal pigment epithelium, which have already been discussed. In this section we will consider cytological markers that are detectable in interphase cells using standard histological procedures. Chromosome markers, which can only be detected during cell division, and markers that require specific biochemical or immunological detection methods will be discussed separately below.

1. CHROMATIN

Perhaps the chimera marker that is best known to developmental biologists is the chick-quail chromatin marker (Le Douarin, 1969, 1973, 1976, 1980). Interphase nuclei of quail cells have one, or sometimes two, large, central hetero-

II. Cell Markers

chromatic masses that can be seen in sections stained with Feulgen and Rossenbeck's DNA stain or in sections prepared for electron microscopy. Chick cells (and most other animal cells) do not show the discrete masses, and this staining difference has been widely used as a cell marker where chick and quail cells have been associated to form chimeras (Fig. 3). This marker is relatively easy to detect in virtually all nucleated cell types and normally most, but probably not all, of the cells in a section of chimeric tissue can be identified as either chick or quail. Nevertheless, in a few tissues the quail heterochromatin shows some phenotypic variation. Jotereau and Le Douarin (1978) reported that the heterochromatin is often attached to the nuclear membrane of endothelial cells and more importantly, was undetectable in 35% of quail osteoclasts. The main disadvantage, however, is that the marker can only be used in interspecific chimeras, and chick and quail differ somewhat in their rate of development. (Quails hatch after 16 days, whereas chicks take 21 days to reach this stage.) Nevertheless this marker has proved to be extremely useful in studies of cell migration during embryogenesis (Le Douarin, 1976, 1980), where reciprocal grafting can help to control for species-specific effects such as differences in growth rate. The use of this marker is discussed in more detail in later chapters.

A system similar to the chick-quail marker was used by Triplett (1958) who grafted neural crest tissue between frogs and salamanders to produce interspecific chimeras. In this case the frog nuclei have large chromatin granules but the salamander nuclei do not. Nuclear chromatin is also clumped into discrete masses by the mouse ichthyosis gene (*ic*). This can be detected in various tissues (Green *et al.*, 1975; Meyers *et al.*, 1976; Goldowitz and Mullen, 1982) and has been incorporated in chimeras to study neural tissue (Goldowitz and Mullen, 1982). Unfortunately clumped chromatin is not seen in all *ic/ic* nuclei and also occurs in some $+/+$ cells, which severely limits its usefulness as a marker. [Goldowitz and Mullen (1982) recorded the clumped chromatin pattern in 24.2% of *ic/ic* cerebellar granule cells and 5.9% of the control $+/+$ cells.] Thiébaud (1983) has described a marker suitable for interspecific *Xenopus* chimeras that is based on differential fluorescent staining of interphase nuclei with quinacrine.

Mammalian sex chromatin is another possible nuclear marker that could be used in some XX $\leftrightarrow$ XY chimeras. Indeed Klinger and Schwarzacher (1962) used sex chromatin to map regions of a human XXY/XY mosaic conceptus that were predominantly XXY or predominantly XY. Unfortunately in the mouse, in particular, sex chromatin is a poor marker and, as with ichthyosis, there is considerable overlap in cellular phenotype between cells from XX and XY individuals. It is also possible that sex chromatin could confuse the scoring of the ichthyosis phenotype in mouse XX $\leftrightarrow$ XY chimeras that were marked with *ic*, but, unlike the ichthyosis chromatin, sex chromatin is located at the periphery of the nucleus.

Polymorphonuclear neutrophil granulocytes from females of some mammalian

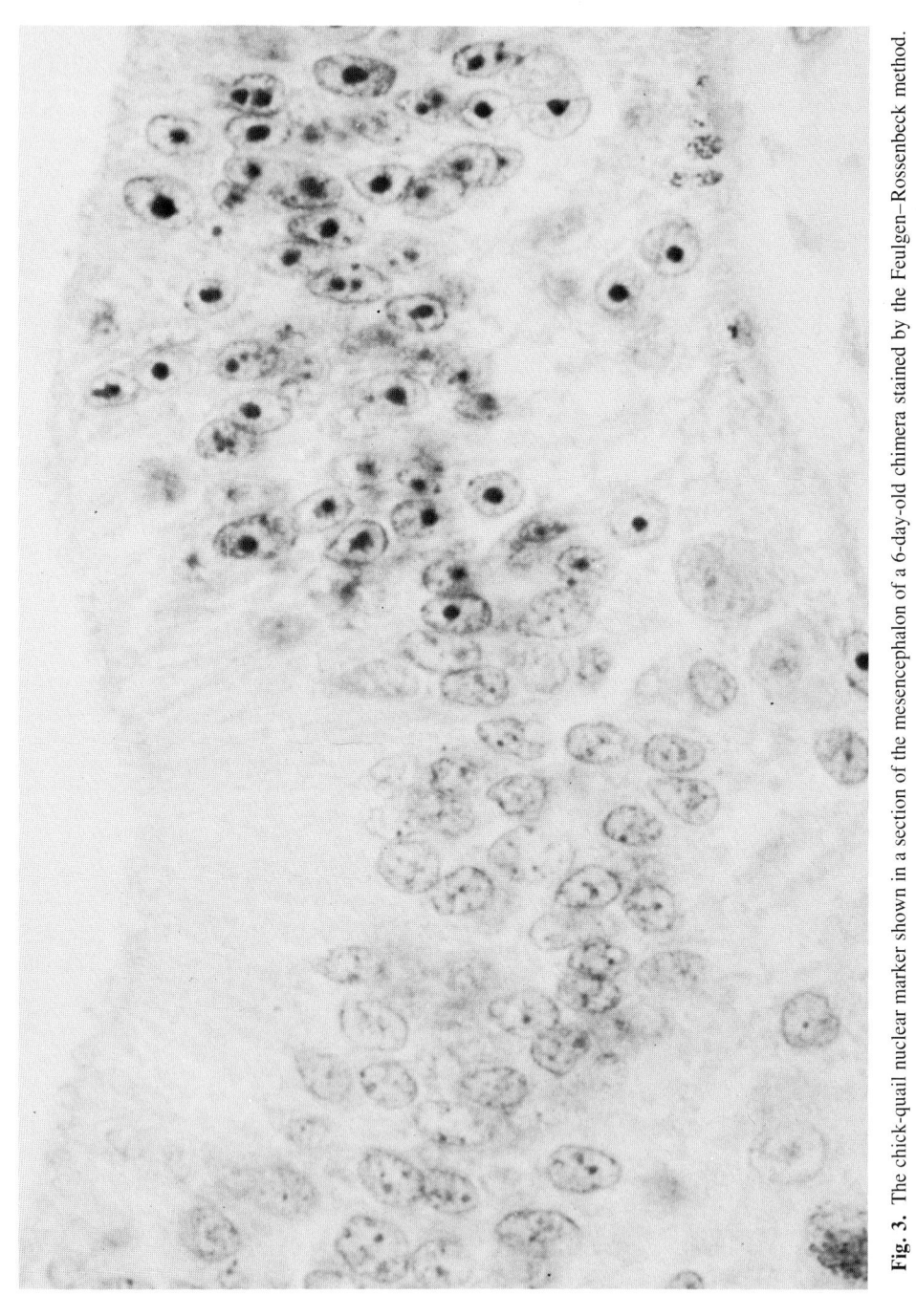

Fig. 3. The chick-quail nuclear marker shown in a section of the mesencephalon of a 6-day-old chimera stained by the Feulgen–Rossenbeck method. Large heterochromatic masses are present in the quail neuroblasts but not in the chick cells. (Previously unpublished photograph kindly provided by Professor N. Le Douarin.)

II. Cell Markers

species have a specific nuclear appendage known as the "drumstick" that is thought to contain the sex chromatin. Although drumsticks can only be recognised in about 3% of neutrophils from females and occasionally occur in males, this has provided a useful marker for female cells injected into irradiated male donors to produce radiation chimeras in rabbits, dogs, and monkeys (Porter, 1957; Porter and Couch, 1959; Magliulo et al., 1963). Rabbit radiation chimeras have also been produced where the donor cells carry the Pelger anomaly, which causes abnormal nuclear morphology of all nucleated blood and bone marrow cells and is not restricted to a small proportion of neutrophils (Czerski et al., 1960).

2. NUCLEOLI

The number of nucleoli has also been used as a cell marker. Blackler and Fischberg (1961) used the anucleolate *Xenopus* mutant in some elegant experiments where anucleolate germ cells were grafted into normal, binucleolate, *Xenopus* embryos. Slack (1980) has used nucleolar number as a cell marker for grafts made between triploid and diploid axolotls. Although the maximum number of nucleoli correlates with ploidy, nucleoli may fuse in some cells and so obscure the difference between diploid and triploid cells. According to Fankhauser and Humphrey (1943), most cells of triploid axolotls show three nucleoli, but in other species nucleolar fusion occurs more commonly. For this reason other methods, such as cell size and DNA content, have been used to recognise triploid cells.

3. CELL SIZE AND DNA CONTENT

Microdensitometry of nuclei in histological sections has been used to estimate DNA content and so identify triploid cells in *Xenopus* chimeras (e.g., Volpe et al., 1979). Cell size was used to recognise triploid cells in frog chimeras (Hollyfield, 1966) and tetraploid cells in *Xenopus* chimeras (Hunt et al., 1982), although there is some overlap in size between diploid and triploid cells. Nuclear size was also used as a marker by Raven (1937) who grafted cells from the ventral neural tube of the axolotl, *Amblystoma mexicanum*, to the newt, *Triton taeniatus*, which has smaller cells.

4. CELL SHAPE

The proportions of the two spermatozoa populations in XY ↔ XY mouse chimeras are normally estimated by simple breeding experiments, but they may also be estimated by direct examination of spermatozoa. Burgoyne (1975) used differences in the phenotype of spermatozoa from C3H and C57BL inbred strains to quantify the proportions of the two spermatozoa populations in C3H ↔ C57BL chimeras. However, this approach has yet to be applied to histological sections to map the location of different spermatozoa populations *in situ*.

5. TESTICULAR FEMINIZATION

The mouse X-linked gene testicular feminization, *Tfm*, causes a deficiency in androgen receptors. This gene is a potential cell marker in various androgen-responsive tissues (reviewed by Green, 1981) and has been used as such in epididymal cells. Drews and Alonso-Lozano (1974) manipulated the mouse phenotype using the gene sex reversed, *Sxr*, to produce phenotypically male mice that had two X chromosomes and were heterozygous for *Tfm*. These male X-inactivation mosaics had two distinct classes of epididymal cells. It is thought that *Tfm* affects cellular differentiation in a cell-autonomous way. Thus, it is argued that the smaller cells failed to differentiate in response to androgens because, after X chromosome inactivation, they expressed only the X chromosome carrying the *Tfm* allele.

As yet *Tfm* has not been used as a cell marker in chimeras, although Lyon *et al.* (1975) have made *Tfm*/Y ↔ +/Y chimeras and used them to produce *Tfm/Tfm* homozygotes (Lyon and Glenister, 1980). In principle, *Tfm* could provide a useful marker for both *Tfm*/Y ↔ +/Y and *Tfm/Tfm* ↔ +/Y male chimeras, but would probably be unsuitable for either +/+ ↔ *Tfm*/Y or *Tfm*/+ ↔ +/Y chimeras. (This is because +/+ ↔ *Tfm*/Y chimeras would be unlikely to develop into phenotypic males containing epididymides, and only a proportion of *Tfm*/+ cells would express *Tfm* after X inactivation had occurred.) These genetic restrictions make it unlikely that *Tfm* will be widely used as a cell marker, despite its pronounced effect on cellular differentiation in certain tissues.

6. CYTOPLASMIC INCLUSIONS

Few genetically determined cell markers involve cytoplasmic rather than nuclear characteristics. Mintz (1964) used the characteristic granular cytoplasm of cells of preimplantation C57BL/6 embryos to follow their fate immediately after aggregation with embryos from the ICR strain, and Mystkowska (1975) used a similar marker to identify bank vole cells in mouse ↔ bank vole chimeras.

Ultrastructural differences between mouse and rat cytoplasmic inclusions have been used to mark cells of these two species in chimeric rat ↔ mouse blastocysts (Tachi and Tachi, 1980), but the species-specific differences disappear shortly after implantation.

The beige gene, *bg*, has also been used as a cytoplasmic marker. At the gross level, beige lightens the pigment of the eye and coat, and at the cellular level it causes enlargement and clumping of the melanin granules. It has a similar effect on other cytoplasmic granules that have been stained with Sudan Black B in granulocytes, toluidine blue in mast cells (Fig. 4), and histochemical stains for specific lysosomal enzymes in other cell types. Oliver and Essner (1973) found enlarged lysosomes in 15 tissues of beige mice, using a histochemical stain for acid phosphatase, although the extent of the effect varied among tissues and among cells within a tissue. Beige has been used as a marker for neutrophil

II. Cell Markers

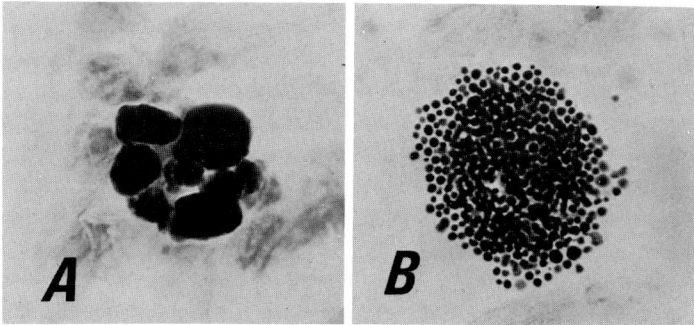

Fig. 4. The beige marker in mouse mast cells stained with acidified toluidine blue. (A) Mast cell of beige C57BL/6-$bg^J bg^J$ mouse; (B) Mast cell of normal C57BL/6 mouse. The cytoplasmic granules in the beige cell are considerably increased in size and decreased in number. (Reprinted by permission from Kitamura *et al.*, 1979. Copyright 1979 Macmillan Journals Limited.)

granulocytes, mast cells, and osteoclasts in radiation chimeras and chimeras produced by injecting *bg/bg* bone marrow cells into nonirradiated genetically anaemic hosts (Kitamura *et al.*, 1977, 1979; Ash *et al.*, 1981). Despite potential problems of variation in size of giant beige granules, this marker could be useful in a variety of tissues (Oliver and Essner, 1973). Dewey *et al.* (1976) have suggested that beige could be used in conjunction with genetic variants that cause differences in activity of lysosomal enzymes. It may also be possible to exaggerate the beige phenotype in some tissues with testosterone (Brandt *et al.*, 1975), but these possibilities remain largely unexplored.

D. Chromosomes

Chromosome markers can be used to estimate the proportions of the two component cell populations in any mitotically active tissue of a chimera. However, since this technique is restricted to dividing cells only a fraction of cells will be available for analysis, and these may not be representative of the whole tissue. This could lead to inaccurate estimates, particularly if mitotic activity was not uniform throughout the tissue or if the two component cell populations had different mitotic rates. A period of cell culture is often used before chromosome preparations are made, and this could allow cell selection to occur *in vitro*. Cells of a certain size or location may also be mechanically selected against during the preparation of tissue for chromosome analysis (Ford, 1966). Although different regions of a tissue can be analysed separately, chromosome markers cannot be used for spatial analysis *in situ*. The preparation totally disrupts the tissue and also destroys distinguishing features of different histological cell types. Despite these problems chromosomes provide cell markers for almost all cell types and have been quite widely used in chimeras.

Differences in chromosome number have been used in interspecific rat-→mouse and hamster→mouse radiation chimeras (Ford et al., 1956; van Bekkum, 1964), mouse↔bank vole aggregation chimeras (Mystkowska, 1975), and monosomy ↔ diploid and trisomy ↔ diploid mouse aggregation chimeras (Magnuson et al., 1982; Epstein et al., 1982). Chromosome number has also been used to identify diploid ↔ tetraploid mouse chimeras (Lu and Markert, 1980) and could be used in conjunction with the other triploid and tetraploid markers discussed earlier.

Sex chromosome differences have been employed in ZW ↔ ZZ chick chimeras (Moore and Owen, 1965; Lassila et al., 1978), XX ↔ XY sheep chimeras (Tucker et al., 1974), and XX ↔ XY mouse chimeras (McLaren, 1975), where the mouse Y chromosome was recognised by a combination of size, asynchronous replication (late-labelling), and absence of centromeric heterochromatin.

A large number of marker chromosomes are available in the mouse (reviewed by Searle, 1981), and these include the long X chromosome produced by Cattanach's insertion, Is(In7;X)1Ct, the distinctive small T6 chromosome, T(14;15)6Ca, and a variety of metacentric marker chromosomes produced by Robertsonian translocations. The T6 marker has been incorporated into radiation chimeras (Ford et al., 1956; Ford, 1966) and mouse aggregation and injection chimeras (Mystkowska and Tarkowski, 1968; Ford et al., 1975).

McLaren (1975, 1976) has also used a variant with reduced centromeric heterochromatin (C band) on the mouse chromosome 14 as a marker in aggregation chimeras. A number of similar variants have been reported for the mouse (Davisson, 1981), including a duplication of this region of chromosome 18, Dp(18Hc) (Evans et al., 1980), and these could be used as markers.

E. Biochemistry and Immunology

Geneticists continue to search for new variants that can be detected using biochemical or immunological methods. This is providing a wealth of variants that can be tested as suitable chimera markers and probably provides the best opportunity for discovering new marker systems, particularly for mice and *Drosophila*. Biochemical and immunological variants can be detected using "indirect" assay systems, which require destruction of the tissue organization, or "direct" *in situ* methods, which may provide spatial information.

1. ELECTROPHORESIS AND "TEST TUBE" ASSAYS

Various "indirect" markers have been used for chimeras. These include assays for haemoglobin solubility in mouse radiation chimeras (Popp and Cosgrove, 1959) and assays for β-glucuronidase activity in liver homogenates from mouse chimeras (Wegmann, 1970). Electrophoresis of haemoglobin variants has

II. Cell Markers

been used for mouse→mouse and rat→mouse radiation chimeras (reviewed by van Bekkum and de Vries, 1967), mouse aggregation chimeras (Wegmann and Gilman, 1970), and sheep injection chimeras (Tucker *et al.*, 1974). Electrophoretic variants of serum albumins and transferrins have also been used as markers for sheep aggregation and injection chimeras (Pighills *et al.*, 1968; Tucker *et al.*, 1974). Other electrophoretic variants used in mouse aggregation and injection chimeras include seminal vesicle protein (Mintz *et al.*, 1972), major urinary protein, which can be used to monitor liver chimerism in live mice (Mintz, 1974), and various enzymes, such as isocitrate dehydrogenase (Mintz and Baker, 1967), supernatant malic enzyme (Baker and Mintz, 1969), and glucose phosphate isomerase (Chapman *et al.*, 1972; Gearhart and Mintz, 1972), which are present in most body tissues. Immunological assays have been widely used to analyse various radiation chimeras (reviewed by van Bekkum and de Vries, 1967), sheep injection chimeras (Tucker *et al.*, 1974), and mouse chimeras (e.g., Mintz and Palm, 1969; Mintz and Silvers, 1970; Wegmann and Gilman, 1970; Warner *et al.*, 1977).

Electrophoretic variants offer the advantage that gene products from both cell populations can be separated and positively identified in the same assay, and their proportions may be quantified either subjectively (e.g., Wegmann and Gilman, 1970) or more objectively using densitometry (e.g., Peterson, 1974) or serial dilution (Klebe, 1975; Falconer *et al.*, 1981). Polymeric enzymes, such as isocitrate dehydrogenase and glucose phosphate isomerase, which are both dimers, form heteropolymer bands if different alleles are expressed in the same cell, and this has been used to demonstrate cell fusion in chimeric mouse skeletal muscle (Mintz and Baker, 1967). Electrophoresis of allozyme variants of glucose phosphate isomerase, GPI-1, is one of the most commonly used markers for mouse chimeras. This enzyme is abundant in all body tissues that have been examined, and modern electrophoretic techniques allow detection of a minor component representing only 1 or 2% of this enzyme in very small pieces of tissue (see, for example, Gearhart and Mintz, 1972; Peterson, 1979).

Although enzyme electrophoresis is a powerful method, it shares some of the disadvantages of other "indirect" markers and cannot be used to provide *in situ* spatial information. Also, the proportions of the two allozymes represent an average value for the whole sample, which will normally include blood contamination and cells that contain different amounts of enzyme, so the proportions of the two allozymes may not exactly reflect the proportion of the two component cell populations. Whether or not this is a serious objection depends on the degree of precision required. The problem of blood contamination is best overcome by perfusing each chimera with saline to flush out the blood before removing the tissues (Dewey *et al.*, 1976). If standardised dissection procedures result in predictable contamination levels, it might also be possible to apply a realistic correction factor based on the proportions of the two allozymes in the blood and an estimated standard contamination level for each organ, which could be derived from experiments with radiation chimeras. However, this approach has not yet been tried.

2. HISTOCHEMICAL MARKERS USED IN CHIMERAS AND MOSAICS

Some genetically determined enzyme activity variants, such as the very low activity of β-glucuronidase caused by the mouse $Gus\text{-}s^h$ gene, can provide excellent cell markers if they produce a large enough difference in activity. β-Glucuronidase assays of liver homogenates were used to infer the presence of patches of two populations of cells in chimeric liver (Wegmann, 1970), but the $Gus\text{-}s^h$ variant became a much more powerful marker when histochemical techniques were used to demonstrate these patches *in situ* (Condamine *et al.*, 1971; West, 1976b) (Fig. 5). Unfortunately, although β-glucuronidase histochemistry has been applied to several other tissues (Feder, 1976; Mullen, 1977a), this marker is not ubiquitous.

Several other mouse enzyme-activity variants have been tested as cell markers. The twofold difference in β-galactosidase activity determined by the *Bgl-s* locus has been used as a marker in the Purkinje cells of the cerebellum, kidney

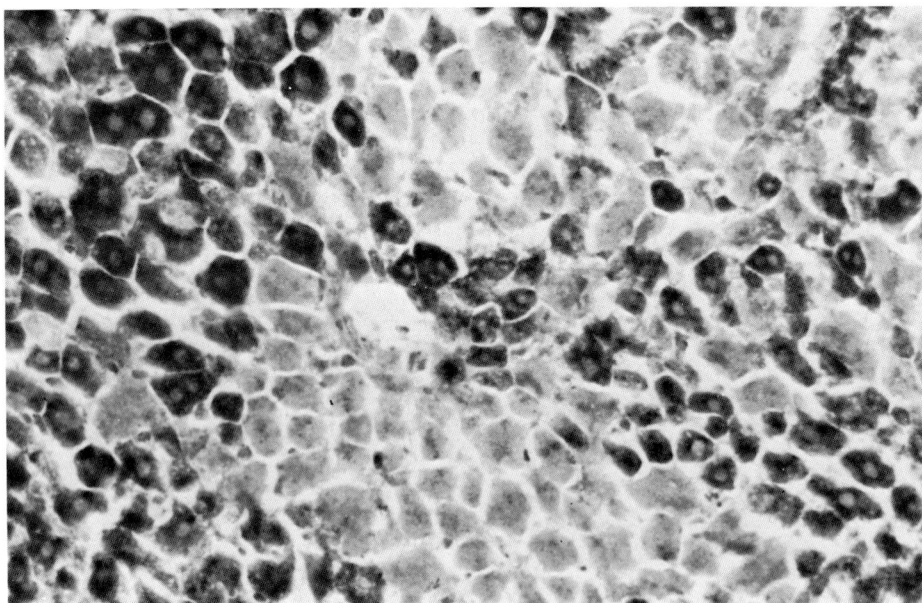

Fig. 5. The β-glucuronidase histochemical marker in the liver of a $Gus\text{-}s^h \leftrightarrow Gus\text{-}s^b$ mouse chimera. The darkly stained hepatocytes have normal levels of β-glucuronidase ($Gus\text{-}s^b$), whereas the paler cells have lower levels produced by the $Gus\text{-}s^h$ variant.

II. Cell Markers

TABLE IV.

Drosophila Enzyme Deficiency Markers That Have Been Used in Mosaics

Enzyme	Gene or chromosome deficiency symbol	Reference
Aldehyde oxidase	*mal*	Janning, 1972
Acid phosphatase	$Acph^{n11b}$	
6-Phosphogluconate dehydrogenase	Pgd^n	Kankel and Hall, 1976
Glucose-6-phosphate dehydrogenase	Zw^n	
Acetylcholinesterase	Ace^{m38} and $Df(3R)126d$	Ferrus and Kankel, 1981
α-Glycerophosphate dehydrogenase	α-*Gdh*	
Succinate dehydrogenase	*sdh*	Lawrence, 1981

tubules, pancreas and salivary gland (Dewey *et al.*, 1976; Dewey and Mintz, 1978). However, this is not applicable to all tissues, since β-galactosidase is not always sufficiently active, and the effect of *Bgl-s* on enzyme activity varies among tissues (Dewey and Mintz, 1978; Paigen, 1979). As already noted, Dewey *et al.* (1976) have suggested combining the effects of the beige marker with variants for lysosomal enzymes, such as β-glucuronidase and β-glactosidase.

Alkaline phosphatase has been used as a histochemical marker for granulocytes in various interspecific radiation chimeras (Shekarchi and Makinodon, 1959; van Bekkum and de Vries, 1967). Mouse granulocytes have little alkaline

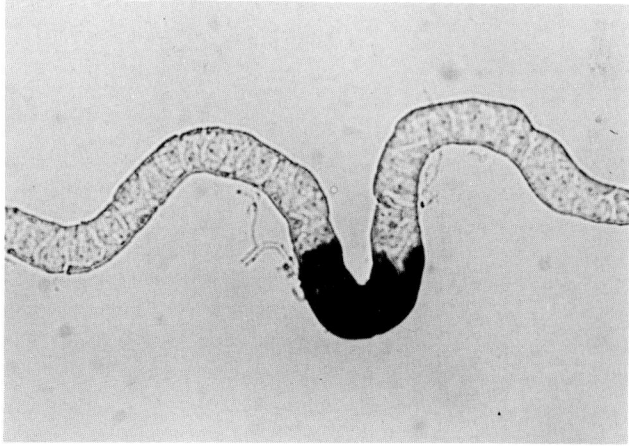

Fig. 6. The *Drosophila* aldehyde oxidase histochemical marker in the Malpighian tube of a larval *Drosophila* chimera. The stained cells are aldehyde oxidase positive, whereas the unstained cells are homozygous for the gene maroon-like, *mal,* and are aldehyde oxidase negative. (Reprinted by permission from Illmensee and Mahowald, 1976.)

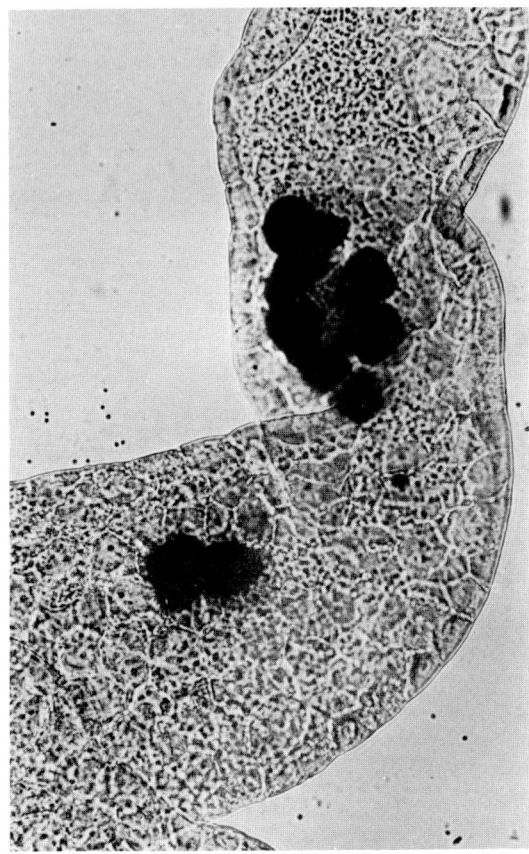

Fig. 7. The *Drosophila* aldehyde oxidase histochemical marker in the midgut region of a *Drosophila* chimera. The stained cells are positive for aldehyde oxidase, and the unstained cells are homozygous for the gene maroon-like, *mal*, and are aldehyde oxidase negative. (Reprinted by permission from Illmensee *et al.*, 1976.)

phosphatase so donor granulocytes from other species can be readily identified by staining spleen imprints or blood or bone marrow smears for this enzyme.

A number of *Drosophila* mutants or chromosome deficiencies, which either eliminate or greatly reduce the activity of various enzymes, have been incorporated into mosaics, and these are listed in Table IV. The aldehyde oxidase deficiency caused by the eye colour gene maroon-like, *mal*, has also been used as an elegant histochemical marker for *Drosophila* chimeras (Illmensee and Mahowald, 1976; Illmensee *et al.*, 1976), as shown in Figs. 6 and 7.

3. POTENTIAL HISTOCHEMICAL MARKERS

Although there are few histochemical markers that have been successfully used in chimeras, there are a number of genetic variants in the mouse that are

II. Cell Markers

potential histochemical markers. These include the X-linked sparse fur mutant, *spf*, which produces an unstable form of ornithine transcarbamylase (DeMars et al., 1976). Ornithine transcarbamylase has already been used as a histochemical marker in sections of liver from a woman who was heterozygous for a deficiency of this enzyme (Ricciuti et al., 1976) and, more recently, in livers of heterozygous *spf*/+ mice (Wareham et al., 1983). Although sparse fur may be a useful marker for the liver in mouse chimaeras and X-inactivation mosaics, ornithine transcarbamylase is insufficiently active in other tissues to permit a wider application as a marker.

There have been several attempts to use thermolabile enzyme variants as histochemical markers for mouse chimeras but so far without success. Dewey et al. (1976) were unable to demonstrate reduced activity of a thermolabile variant of α-glycerol-3-phosphate dehydrogenase in tissue sections, possibly because another emzyme in the cells used the same substrate and caused a positive reaction. Cell localisation can also be a problem: diffusion of one of the reaction products of glucose phosphate isomerase foiled attempts to use the thermolabile form of this enzyme, produced by the *Gpi-1^c* allele, as a histochemical marker in chimeras (J. D. West, unpublished). This study also showed that the conditions required to inactivate a thermolabile enzyme in tissue sections may differ between tissues and may not be the same as those used for tissue homogenates.

Several mouse "null alleles" have been reported that fail to produce a particular enzyme. These are listed in Table V and might be expected to provide excellent histochemical markers. In addition, mitochondrial malic emzyme is absent from adult mice that are heterozygous for two partially complementing deletions of the closely linked albino gene on chromosome 7 (Eicher et al., 1978).

Some null alleles, such as the *Gpi-1* null (Soares, 1979), are homozygous lethals and so need to be maintained as heterozygotes. This poses two problems. First, an independent marker would be needed to identify chimeras that had a homozygous, null/null population. Second, the homozygous genotype might be cell-lethal or at least selected against in the chimera. Other null alleles, such as *Mod-1^n*, are not homozygous lethals. It seems likely that another enzyme, such as the mitochondrial form of malic enzyme, assumes the role of the missing supernatant malic enzyme. This permits survival but could also produce a

TABLE V.

Mouse Null Alleles That Produce Enzyme Deficiencies

Enzyme	Gene symbol	Reference
Glucose phosphate isomerase	"*Gpi-1* null"	Soares, 1979
Phosphoglucomutase-1	*Pgm-1^n*	Soares, 1979; Johnson et al., 1981b
Supernatant malic enzyme	*Mod-1^n*	Johnson et al., 1981a
Xylose dehydrogenase	*Xld-1^c*	Newton et al., 1982

positive histochemical reaction for malic enzyme in a $Mod\text{-}1^n/Mod\text{-}1^n$ cell. Despite the presence of the mitochondrial enzyme, however, $Mod\text{-}1^n/Mod\text{-}1^n$ cells stain less intensely than normal cells, at least in some tissues, and preliminary experiments suggest that this could provide a useful histochemical marker (J. D. West, unpublished; R. L. Gardner, personal communication).

4. IMMUNOHISTOCHEMISTRY

Immunohistochemical techniques can also be used to detect antigenic differences *in situ*. In the past, technical difficulties, such as the restricted tissue distribution of many suitable antigens, the need for high titres of antibody and the fading of fluorochromes, have discouraged the use of these markers for chimeras. However, recent developments, such as improved methods of antigen detection and the production of monoclonal antibodies, some of which are commercially available, are likely to stimulate more interest in this area.

Immunofluorescent labelling has been used to detect both mouse and rat antigens in prenatal interspecific rat → mouse injection chimaeras (Gardner and Johnson, 1973). Immunofluorescence has also been used for mouse chimeras to detect *H-2* and other histocompatibility antigens in the kidney (Barnes *et al.*, 1974), cerebellar Purkinje cells (Dewey *et al.*, 1976), and other cell types including colon (Ponder, *et al.*, 1983).

An elegant immunohistochemical marker for mouse chimeras has been developed by Oster-Granite and Gearhart (1981). They prepared *allozyme-specific* antisera against the GPI-1B allozyme of glucose phosphate isomerase and used the peroxidase-antiperoxidase method to detect this allozyme in Purkinje cells of

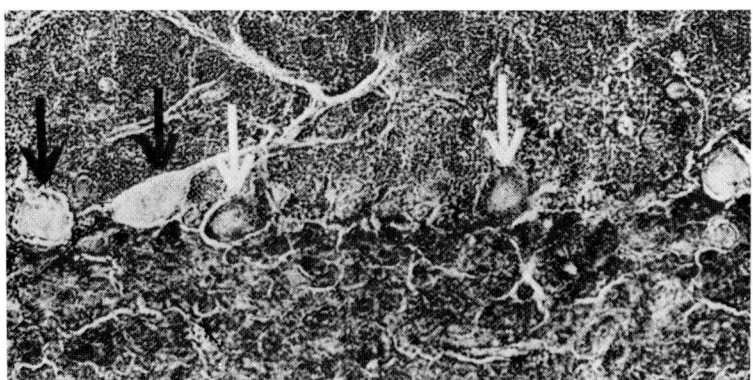

Fig. 8. The mouse glucose phosphate isomerase immunohistochemical marker in the Purkinje cells of a mouse chimera. Sections of cerebellar cortex were stained with purified anti-GPI-1B (allozyme-specific) antisera using a peroxidase-antiperoxidase (PAP) staining method. White arrows indicate stained (GPI-1B) Purkinje cells, and black arrows indicate unstained (GPI-1A) Purkinje cells. (Reprinted by permission from Oster-Granite and Gearhart, 1981.)

II. Cell Markers

mouse chimeras as shown in Fig. 8. Since glucose phosphate isomerase is a ubiquitous enzyme this antibody should prove to be useful in many more tissues.

A similar approach may also be possible using null alleles for various enzymes (described above). For example, the mouse *Mod-1^n* allele produces no protein that cross-reacts with antiserum to cytoplasmic malic enzyme (Lee *et al.*, 1980). Immunohistochemical techniques could, therefore, be used to detect malic enzyme *in situ* without the need to prepare allozyme-specific antisera.

One disadvantage of using enzymes as histochemical or immunohistochemical markers is the possibility that enzyme transfer may occur between cells. β-Glucuronidase, for example, is transferred between cells *in vitro* (Olsen *et al.*, 1981) and in mouse chimeras (Feder, 1976; Herrup and Mullen, 1979). In this case, however, the two cell phenotypes are still clearly distinguishable in chimeras. Another problem is that enzymes may not be uniformly distributed among the cells of a tissue so that even control sections may appear patchy.

5. LECTINS

Two new approaches have recently yielded extremely promising new biochemical markers for mouse chimeras. First peroxidase-conjugated *Dolichos biflorus* lectin (which binds specifically to *N*-acetylgalactosamine) has been used to demonstrate two populations of cells in fixed sections of vascular and intestinal epithelium from chimeric mice (B. A. J. Ponder and M. M. Wilkinson, personal communication; Ponder and Wilkinson, 1983). Although the polymorphisms discovered in this study have a limited tissue distribution, it seems likely that further research will reveal other polymorphisms for lectin staining, and this promises to be a very useful approach.

6. DNA HYBRIDIZATION *IN SITU*

The second new approach is based on differences in satellite DNA sequences between two mouse species. A radiolabelled cloned sequence of *Mus musculus* satellite DNA has been produced and used to label *M. musculus* nuclei specifically in interspecific, *Mus caroli* → *M. musculus* injection chimeras by DNA–DNA *in situ* hybridisation (Siracusa *et al.*, 1983; Rossant *et al.*, 1983). So far this marker has been used for embryonic chimeras and several tissues from adult chimeras, including brain, liver, kidney, and testis as shown in Fig. 9.

The main drawbacks of this technique are that it can only be applied to interspecific mouse chimeras and it requires a relatively sophisticated staining technique that relies on reagents that are not commercially available. The limitations of interspecific chimeras were mentioned at the beginning of this chapter and have been discussed with respect to *M. caroli* ↔ *M. musculus* chimeras by Rossant and Chapman (1983). They used pigmentation and the glucose phosphate isomerase electrophoretic marker to show that, although there may be some tissue-specific selection, the overall distribution of the two cell populations

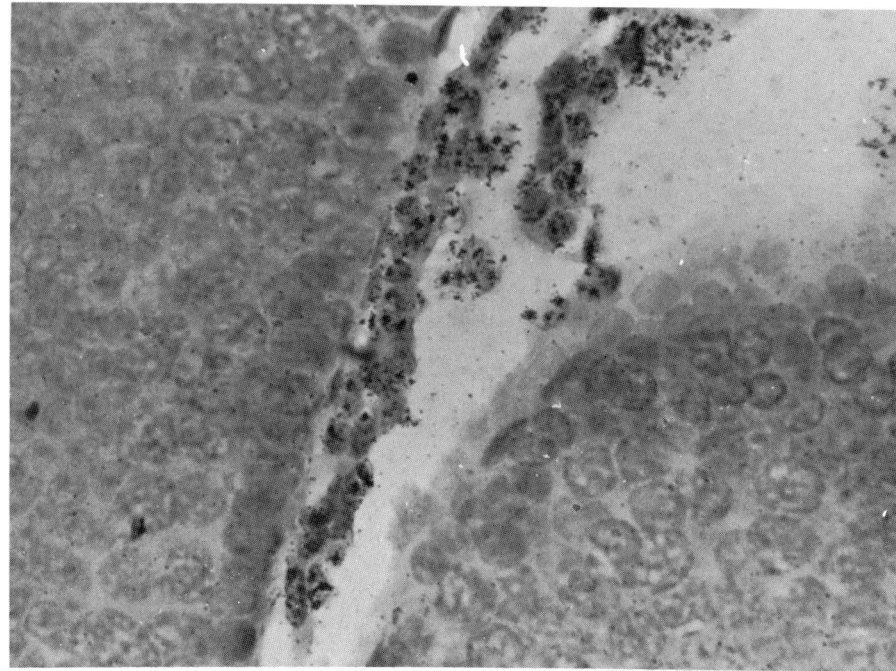

Fig. 9. The mouse DNA marker in the testis of an XX, *Mus musculus* ↔ XY, *Mus caroli* chimera. Testis sections were labelled by *in situ* hybridisation, with a radiolabelled cloned sequence of *M. musculus* satellite DNA, followed by autoradiography. In the region shown, the interstitial cells are almost all labelled (*M. musculus*), but the testis tubules contain only unlabelled (XY, *M. caroli*) cells. (Previously unpublished photograph kindly supplied by Dr. J. Rossant.)

in *M. caroli* ↔ *M. musculus* chimeras is broadly similar to that normally seen in intraspecific mouse chimeras. Cell mixing is sufficiently extensive to ensure that all tissues examined are normally chimeric, but it is not known whether the degree of cell mixing is comparable in interspecific and intraspecific mouse chimeras. This may become apparent from experiments with the DNA marker on tissues that have been analysed in intraspecific chimeras, and it could also be analysed by comparing pigment patterns in the retinal pigment epithelium of the two types of chimeras as described by West (1976a).

Despite some limitations, this DNA probe provides an exciting new cell-autonomous marker which, in principle, could be applied to any nucleated cell type and may prove to be as useful as the chick-quail nuclear marker. It is also likely to be useful for testing other mouse cell markers for cell autonomy. In the future the approach might be extended to intraspecific chimeras if appropriate DNA differences can be found or produced within a species. A number of recent experiments have shown that exogenous DNA can be stably incorporated into the

II. Cell Markers

mouse genome (e.g., Jähner and Jaenisch, 1980; Costantini and Lacy, 1981). Stable integration of multiple copies could provide a DNA marker than can be detected by *in situ* hybridisation.

Although not as straightforward as a good cytological marker, the mouse DNA marker, together with the possibility of improved histochemical or immunohistochemical markers and new lectin markers, promises a significant improvement in our ability to analyse cell deployment in mouse chimeras.

ACKNOWLEDGMENTS

I am particularly grateful to those who sent me illustrations or information for inclusion in this chapter. These include Drs. R. L. Gardner, K. Illmensee, Y. Kitamura, N. Le Douarin, M. L. Oster-Granite, B. A. J. Ponder, J. Rossant, J. M. W. Slack, P. V. Thorogood, and R. Tompkins. I also thank Dr. R. S. P. Beddington, Professor R. L. Gardner, Dr. M. F. Lyon, Dr. A. McLaren, Dr. B. A. J. Ponder, and Dr. D. Stephenson for constructive criticism of the manuscript and Mrs. P. Murphy for typing it.

REFERENCES

Ash, P., Loutit, J. F., and Townsend, K. M. S. (1981). *Clin. Orthop. Relat. Res.* **155,** 249–258.
Baker, W. W., and Mintz, B (1969). *Biochem. Genet.* **2,** 351–360.
Balakier, H., and Pedersen, R. A. (1982). *Dev. Biol.* **90,** 352–362.
Barnes, R. D., Holliday, J., and Tuffrey, M. (1974). *Immunology* **26,** 1195–1206.
Bartlett, P. F., and Edidin, M. (1978). *J. Cell Biol.* **77,** 377–388.
Beddington, R. S. P. (1981). *J. Embryol. Exp. Morphol.* **64,** 87–104.
Beddington, R. S. P. (1982). *J. Embryol. Exp. Morphol.* **69,** 265–285.
Blackler, A. W., and Fischberg, M. (1961). *J. Embryol. Exp. Morphol.* **10,** 641–651.
Brandt, E. J., Elliott, R. W., and Swank, R. T. (1975). *J. Cell Biol.* **67,** 774–788.
Burgoyne, P. S. (1975). *Dev. Biol.* **44,** 63–76.
Butcher, E. C., Scollay, R. G., and Weissman, I. L. (1980). *J. Immunol. Methods* **37,** 109–122.
Cattanach, B. M., Wolfe, H. G., and Lyon, M. F. (1972). *Genet. Res.* **19,** 213–228.
Chapman, V. M., Ansell, J. D., and McLaren, A. (1972). *Dev. Biol.* **29,** 48–54.
Condamine, H., Custer, R. P., and Mintz, B. (1971). *Proc. Natl. Acad. Sci. U.S.A.* **68,** 2032–2036.
Costantini, F., and Lacy, E. (1981). *Nature (London)* **294,** 92–94.
Czerski, P., Rosiek, O., and Sablinski, J. (1960). *Nature (London)* **187,** 955–956.
Davisson, M. T. (1981). *In* "Genetic Variants and Strains of the Laboratory Mouse" (M. C. Green, ed.), pp. 357–358. Fischer, Stuttgart.
DeMars, R., LeVan, S. L., Trend, B. L., and Russell, L. B. (1976). *Proc. Natl. Acad. Sci. U.S.A.* **73,** 1693–1697.
Deol, M. S., and Whitten, W. K. (1972). *Nature (London) New Biol.* **240,** 277–279.
Detwiler, S. R. (1917). *Anat. Rec.* **13,** 493–497.
Dewey, M. J., and Mintz, B. (1978). *Dev. Biol.* **66,** 550–559.
Dewey, M. J., Gervais, A. G., and Mintz, B. (1976). *Dev. Biol.* **50,** 68–81.
Drews, U., and Alonso-Lozano, V. (1974). *J. Embryol. Exp. Morphol.* **32,** 217–225.
Dunn, G. R. (1972). *J. Exp. Zool.* **181,** 1–16.

Eicher, E. M., Lewis, S. E., Turchin, H. A., and Gluecksohn-Waelsch, S. (1978). *Genet. Res.* **82**, 1–7.
Epstein, C. J., Smith, S. A., Zamora, T., Sawicki, J. A., Magnuson, T. R., and Cox, D. R. (1982). *Proc. Natl. Acad. Sci. U.S.A.* **79**, 4376–4380.
Evans, E. P., Burtenshaw, M. D., Brown, B. B., Papaioannou, V., and West, J. (1980). *Mouse Newsletter.* **62**, 70.
Falconer, D. S., Gauld, I. K., Roberts, R. C., and Williams, D. A. (1981). *Genet. Res.* **38**, 25–46.
Fankhauser, G., and Humphrey, R. R. (1943). *Proc. Natl. Acad. Sci. U.S.A.* **29**, 344–350.
Feder, N. (1976). *Nature (London)* **263**, 67–69.
Ferrus, A., and Kankel, D. R. (1981). *Dev. Biol.* **85**, 485–504.
Ford, C. E. (1966). *In* "Tissue Grafting and Radiation" (H. S. Micklem and J. F. Loutit), pp. 197–206. Academic Press, New York.
Ford, C. E., Hamerton, J. L., Barnes, D. W. H., and Loutit, J. F. (1956). *Nature (London)* **177**, 452–454.
Ford, C. E., Evans, E. P., and Gardner, R. L. (1975). *J. Embryol. Exp. Morphol.* **33**, 447–459.
Frost, S. K., and Malacinski, G. M. (1980). *Dev. Genet.* **1**, 271–294.
Gardner, R. L. (1975). *Symp. Soc. Dev. Biol.* **33**, 207–238.
Gardner, R. L., and Johnson, M. H. (1973). *Nature (London) New Biol.* **246**, 86–89.
Gardner, R. L., and Lyon, M. F. (1971). *Nature (London)* **231**, 385–386.
Gardner, R. L., and Munro, A. J. (1974). *Nature (London)* **250**, 146–147.
Garner, W., and McLaren, A. (1974). *J. Embryol. Exp. Morphol.* **32**, 495–503.
Gearhart, J. D., and Mintz, B. (1972). *Dev. Biol.* **29**, 55–64.
Gimlich, R. L., and Cooke, J. (1983). *Nature (London)* **306**, 471–473.
Goldowitz, D., and Mullen, R. J. (1980). *Neurosci. Abstr.* **6**, 743.
Goldowitz, D., and Mullen, R. J. (1982). *Dev. Biol.* **89**, 261–267.
Gordon, J. (1977). *Differentiation* **9**, 12–27.
Göthlin, G., and Ericsson, J. L. E. (1973). *Virchows Arch. A B* **12**, 318–329.
Green, M. C., (1981). "Genetic Variants and Strains of the Laboratory Mouse." Fischer, Stuttgart.
Green, M. C., Schultz, L. D., and Nedzi, L. A. (1975). *Transplantation* **20**, 172–175.
Green, M. C., Durham, D., Mayer, T. C., and Hoppe, P. C. (1977). *Genet. Res.* **29**, 279–284.
Grüneberg, H., and McLaren, A. (1972). *Proc. R. Soc. London, Ser. B* **182**, 9–23.
Hall, J. C., Gelbart, W. M., and Kankel, D. R. (1976). *In* "The Genetics and Biology of Drosophila" (M. Ashburner and E. Novitski, eds.), Vol. 1A, pp. 265–314. Academic Press, London.
Herrup, K., and Mullen, R. J. (1979). *J. Cell Sci.* **40**, 21–31.
Hillman, N., Sherman, M. I., and Graham, C. (1972). *J. Embryol. Exp. Morphol.* **28**, 263–278.
Hollyfield, J. G. (1966). *Dev. Biol.* **14**, 461–480.
Hollyfield, J. G., and Ward A. (1974). *J. Ultrastruct. Res.* **46**, 327–338.
Horisberger, M., and Vonlanthen, M. (1979). *Histochemistry* **64**, 115–118.
Hunt, R. K., Tompkins, R., Reinschmidt, D., Bodenstein, L., and Murphy, R. K. (1982). *Am. Zool.* **22**, 185–207.
Illmensee, K., and Mahowald, A. P. (1976). *Exp. Cell Res.* **97**, 127–140.
Illmensee, K., Mahowald, A. P., and Loomis, M. (1976). *Dev. Biol.* **49**, 40–65.
Ivanyi, P. (1978). *Proc. R. Soc. London, Ser. B* **202**, 117–158.
Jacobson, M., and Hirose, G. (1978). *Science* **202**, 637–639.
Jähner, D., and Jaenisch, R. (1980). *Nature (London)* **287**, 456–458.
Janning, W. (1972). *Naturwissenschaften* **59**, 516–517.
Johnson, F. M., Chasalow, F., Lewis, S. E., Barnett, L., and Lee, C.-Y. (1981a). *J. Hered.* **72**, 134–136.
Johnson, F. M., Hendren, R. W., Chasalow, F., Barnett, L. B., and Lewis, S. E. (1981b). *Biochem. Genet.* **19**, 599–615.

II. Cell Markers

Jotereau, F. V., and Le Douarin, N. M. (1978). *Dev. Biol.* **63,** 253–265.
Kankel, D. R., and Hall, J. C. (1976). *Dev. Biol.* **48,** 1–24.
Katz, M. J., Larek, R. J., Osdoby, P., Whittaker, J. R., and Caplan, A. I. (1982). *Dev. Biol.* **90,** 419–429.
Kitamura, Y., Shimada, M., Hatanaka, K., and Miyano, Y. (1977). *Nature (London)* **268,** 442–443.
Kitamura, Y., Matsuda, H., and Hatanaka, K. (1979). *Nature (London)* **281,** 154–155.
Klebe, R. J. (1975). *Biochem. Genet.* **13,** 805–812.
Klein, M. S., and Markert, C. L. (1981). *J. Exp. Zool.* **218,** 183–193.
Klinger, H. P., and Schwarzacher, H. G. (1962). *Cytogenetics* **1,** 266–290.
Lassila, O., Eskola, J., Toivanen, P., Martin, C., and Dieterlen-Lievre, F. (1978). *Nature (London)* **272,** 353–354.
Lawrence, P. A. (1981). *J. Embryol. Exp. Morphol.* **64,** 321–332.
Le Douarin, N. (1969). *Bull. Biol. Fr. Belg.* **103,** 435–452.
Le Douarin, N. (1973). *Dev. Biol.* **30,** 217–222.
Le Douarin, N. (1976). *CIBA Found. Symp.* **40** (new ser.), 71–101.
Le Douarin, N. M. (1980). *Nature (London)* **286,** 663–669.
Lee, C.-Y., Chasalow, F., Lee, S.-M., Lewis, S., and Johnson, F. M. (1980). *Mol. Cell. Biochem.* **30,** 143–149.
Lo, C. W., and Gilula, N. B. (1979). *Cell* **18,** 399–409.
Lu, T.-Y., and Markert, C. L. (1980). *Proc. Natl. Acad. Sci. U.S.A.* **77,** 6012–6016.
Lyon, M. F., and Glenister, P. H. (1980). *Proc. R. Soc. London, Ser. B* **208,** 1–12.
Lyon, M. F., Glenister, P. H., and Lamoreux, M. L. (1975). *Nature (London)* **258,** 620–622.
McLaren, A. (1975). *J. Embryol. Exp. Morphol.* **33,** 205–216.
McLaren, A. (1976). "Mammalian Chimaeras." Cambridge Univ. Press, London and New York.
McLaren, A., and Bowman, P. (1969). *Nature (London)* **224,** 238–240.
McLaren, A., Gauld, I. K., and Bowman, P. (1973). *Nature (London)* **241,** 180–183.
Maden, M. (1980). *J. Embryol. Exp. Morphol.* **56,** 201–209.
Magliulo, E., Crouch, B. G., and De Vries, M. J. (1963). *Blood* **21,** 620–625.
Magnuson, T., Smith, S., and Epstein, C. J. (1982). *J. Embryol. Exp. Morphol.* **69,** 223–236.
Marzullo, G. (1970). *Nature (London)* **225,** 72–73.
Mayer, J. F., Jr., and Fritz, H. I. (1974). *J. Reprod. Fertil.* **39,** 1–10.
Meyers, R. S., Klein, A. S., Eppig, J. J., and Eckhardt, R. A. (1976). *Genetics* **83,** s50 (abstr.).
Mintz, B. (1964). *J. Exp. Zool.* **157,** 273–292.
Mintz, B. (1967). *Proc. Natl. Acad. Sci. U.S.A.* **58,** 344–351.
Mintz, B. (1970). *Symp. Int. Soc. Cell Biol.* **9,** 15–42.
Mintz, B. (1971). *Symp. Soc. Exp. Biol.* **25,** 345–370.
Mintz, B. (1974). *Annu. Rev. Genet.* **8,** 411–470.
Mintz, B., and Baker, W. W. (1967). *Proc. Natl. Acad. Sci. U.S.A.* **58,** 592–598.
Mintz, B., and Palm, J. (1969). *J. Exp. Med.* **129,** 1013–1027.
Mintz, B., and Sanyal, S. (1970). *Genetics* **64,** Suppl., 43–44.
Mintz, B., and Silvers, W. K. (1970). *Transplantation* **9,** 497–505.
Mintz, B., Domon, M., Hungerford, D. A., and Morrow, J. (1972). *Science* **175,** 657–659.
Mita, I., and Satoh, N. (1982). *J. Exp. Zool.* **223,** 67–74.
Moore, M. A. S., and Owen, J. J. T. (1965). *Nature (London)* **208,** 956–990.
Moore, W. J., and Mintz, B. (1972). *Dev. Biol.* **27,** 55–70.
Moustafa, L. A. (1974). *Proc. Soc. Exp. Biol. Med.* **147,** 485–488.
Mullen, R. J. (1977a). *Nature (London)* **270,** 245–247.
Mullen, R. J. (1977b). *In* "Approaches to the Cell Biology of Neurons" (W. M. Cowan and J. A. Ferrendelli, eds.), pp. 47–65. Soc. Neurosci., Bethesda, Maryland.
Mullen, R. J., and LaVail, M. M. (1976). *Science* **192,** 799–801.
Mystkowska, E. T. (1975). *J. Embryol. Exp. Morphol.* **33,** 731–744.

Mystkowska, E. T., and Tarkowski, A. K. (1968). *J. Embryol. Exp. Morphol.* **20**, 33–52.
Newton, M. F., Nash, H. R., Peters, J., and Andrews, S. J. (1982). *Biochem. Genet.* **20**, 733–745.
Nicolet, G. (1970). *J. Embryol. Exp. Morphol.* **23**, 79–108.
Okada, C. Y., and Rechsteiner, M. (1982). *Cell* **29**, 33–41.
Oliver, C., and Essner, E. (1973). *J. Histochem. Cytochem.* **21**, 218–228.
Olsen, I., Dean, M. F., Harris, G., and Muir, H. (1981). *Nature (London)* **291**, 244–247.
Oster-Granite, M. L., and Gearhart, J. (1981). *Dev. Biol.* **85**, 199–208.
Paigen, K. (1979). *Annu. Rev. Genet.* **13**, 417–466.
Pescitelli, M. J., and Stocum, D. L. (1980). *Dev. Biol.* **79**, 255–275.
Peterson, A. C. (1974). *Nature (London)* **248**, 561–564.
Peterson, A. C. (1979). *Ann. N.Y. Acad. Sci.* **317**, 630–648.
Petters, R. M., and Markert, C. L. (1980). *J. Hered.* **71**, 71–74.
Pighills, E., Hancock, J. L., and Hall, J. G. (1968). *J. Reprod. Fertil.* **17**, 543–547.
Ponder, B. A. J., and Wilkinson, M. M. (1983). *Dev. Biol.* **96**, 535–541.
Ponder, B. A. J., Wilkinson, M. M., and Wood, M. (1983). *J. Embryol. Exp. Morphol.* **76**, 83–93.
Popp, R. A., and Cosgrove, G. E. (1959). *Proc. Soc. Exp. Biol. Med.* **101**, 754–758.
Porter, K. A. (1957). *Br. J. Exp. Pathol.* **38**, 401–412.
Porter, K. A., and Couch, N. P. (1959). *Br. J. Exp. Pathol.* **40**, 52–56.
Poste, G., Papahdjopoulos, D., and Vail, W. J. (1976). *Methods Cell Biol.* **14**, 33–71.
Raven, P. (1937). *J. Comp. Neurol.* **67**, 221–240.
Ricciuti, F. C., Gelehrter, T. D., and Rosenberg, L. E. (1976). *Ann. J. Hum. Genet.* **28**, 332–338.
Rossant, J., and Chapman, V. M. (1983). *J. Embryol. Exp. Morphol.* **73**, 193–205.
Rossant, J., Vijh, M., Siracusa, L. D., and Chapman, V. M. (1983). *J. Embryol. Exp. Morphol.* **73**, 187–191.
Searle, A. G. (1968). "Comparative Genetics of Coat Colour in Mammals," p. 52. Academic Press, New York.
Searle, A. G. (1981). *In* "Genetic Variants and Strains of the Laboratory Mouse" (M. C. Green, ed.), pp. 324–357. Fischer, Stuttgart.
Shekarchi, I. C., and Makinodon, T. (1959). *Proc. Soc. Exp. Biol. Med.* **100**, 414–417.
Silvers, W. K. (1979). "The Coat Colors of Mice. A Model for Mammalian Gene Action and Interaction," Springer-Verlag, Berlin and New York.
Siracusa, L. D., Chapman, V. M., Bennett, K. L., Hastie, N. D., Pietras, D. F., and Rossant, J. (1983). *J. Embryol. Exp. Morphol.* **73**, 163–178.
Slack, J. M. W. (1980). *J. Embryol. Exp. Morphol.* **58**, 265–288.
Smith, J. C., and Malacinski, G. M. (1983). *Dev. Biol.* **98**, 250–254.
Snow, M. H. L. (1973). *In* "The Cell Cycle in Development and Differentiation" (M. Balls and F. S. Billett, eds.), pp. 311–324. Cambridge Univ. Press, London and New York.
Soares, E. R. (1979). *Environ. Mutagen.* **1**, 19–25.
Spiegelman, M. (1978). *In* "Genetic Mosaics and Chimeras in Mammals" (L. B. Russell, ed), pp. 59–80. Plenum, New York.
Spindle, A. (1982). *J. Exp. Zool.* **219**, 361–367.
Stern, M. S., and Wilson, I. B. (1972). *J. Embryol. Exp. Morphol.* **28**, 247–254.
Stewart, R. N. (1978). *Symp. Soc. Dev. Biol.* **36**, 131–160.
Stocum, D. L. (1975). *Dev. Biol.* **45**, 112–136.
Tachi, S., and Tachi, C. (1980). *Dev. Biol.* **80**, 18–27.
Tarkowski, A. K. (1964). *J. Embryol. Exp. Morphol.* **12**, 575–585.
Thiébaud, C. H. (1983). *Dev. Biol.* **98**, 245–249.
Triplett, E. L. (1958). *J. Exp. Zool.* **138**, 283–311.
Tucker, E. M., Moor, R. M., and Rowson, L. E. A. (1974). *Immunology* **26**, 613–621.
van Bekkum, D. W. (1964). *Nature (London)* **202**, 1311–1312.

II. Cell Markers

van Bekkum, D. W., and de Vries, M. J. (1967). "Radiation Chimaeras." Academic Press, London.
Volpe, E. P., Tompkins, R., and Reinschmidt, D. (1979). *J. Exp. Zool.* **208,** 57–66.
Wanek, N., and Campbell, R. D. (1982). *J. Exp. Zool.* **221,** 37–47.
Wareham, K. A., Howell, S., Williams, D., and Williams, E. D. (1983). *Histochem. J.* **15,** 363–371.
Warner, C. M., McIvor, J. L., and Stephens, T. J. (1977). *Differentiation* **9,** 11–17.
Wegmann, T. G. (1970). *Nature (London)* **225,** 462–463.
Wegmann, T. G., and Gilman, J. (1970). *Dev. Biol.* **21,** 281–291.
Wegmann, T. G., LaVail, M. M., and Sidman, R. L. (1971). *Nature (London)* **230,** 333–334.
West, J. D. (1976a). *J. Embryol. Exp. Morphol.* **35,** 445–461.
West, J. D. (1976b). *J. Embryol. Exp. Morphol.* **36,** 151–161.
Weston, J. (1963). *Dev. Biol.* **6,** 279–310.
Weston, J. (1967). In "Methods in Developmental Biology" (F. H. Wilt and N. K. Wessells, eds.), pp. 723–736. Crowell-Collier, New York.
Wetts, R., and Herrup, K. (1982). *J. Embryol. Exp. Morphol.* **68,** 87–98.
Yamamura, K.-I. and Markert, C. L. (1981). *Dev. Genet.* **2,** 131–146.
Zalokar, M. (1971). *Proc. Natl. Acad. Sci. U.S.A.* **68,** 1539–1541.
Zalokar, M. (1973). *Dev. Biol.* **32,** 189–193.
Ziomek, C. A. (1982). *Wilhelm Roux's Arch. Dev. Biol.* **191,** 37–41.

2

Chimeras in Cell Lineage Studies

CHAPTER III

Early Development of Birds

L. VAKAET

Laboratorium voor Anatomie
Rijksuniversiteit Gent
Gent, Belgium

I.	General Introduction	71
II.	Somatic Cell Lineage Studies	72
	A. Introduction	72
	B. Origin of the Deep Layer in the Avian Blastoderm	73
	C. Disposition of the Anlage Fields before and during Ingression	76
III.	Chimeras in the Study of Embryonic Induction	81
	A. Introduction	81
	B. Neural Induction	82
	C. Induction of Secondary Axes	84
	References	87

I. GENERAL INTRODUCTION

Chimeras are made by grafts between members of the same species but of different genetic constitution (allografts) or between individuals of different species (xenografts). With Webster's Dictionary (1966) it is assumed that the distinctness of the cells of host and graft is one of the essential characteristics of a chimera. In avian allografts this distinctness has to be artificially produced, in the very early stages most often by isotopic labelling followed by autoradiography. Only autoradiographically marked grafts will be mentioned here. In some xenografts, such as those using the chick-quail marker system (Le Douarin, 1969), a natural marker is present (see Chapter I).

When chimeras are made with orthotopic grafts, they may permit the study of cell lineage. Cell lineage is a concept historically linked to the discovery that, during the embryonic development of molluscs and annelids, individual cells or groups of cells are invariably committed to the formation of certain structures in the larva and in the adult animal. In other classes, especially the vertebrates, embryonic development shows a greater plasticity. This distinction is, however, not essential, and probably Dalcq (1941) was right when he distinguished tachygenetic eggs (with early determined cells) and bradygenetic eggs (with later determined cells).

When chimeras are made with heterotopic grafts, they permit the study of determination, or of induction, often of both. No study of determination exclusively has yet been published in birds, but many studies have been undertaken of what is often erroneously called primary induction. This expression implies that this induction is the first to occur in development, an assumption that is at least improbable. The first studies of induction in birds concerned neural induction. Chimeras have been used to study not only neural induction, but also the induction of secondary axes: this will be dealt with in the Section III, C of this chapter.

The layers of the young blastoderm are indicated by the terms upper, middle, and deep layer, respectively. Cells of the upper layer make part of the dorsal surface of the blastoderm; cells of the deep layer make part of the ventral surface. Cells that do not touch either dorsal or ventral surface of the blastoderm are called middle layer cells. This nomenclature avoids the use of terms that imply determination. When the determination of some part of the layer is known, the appropriate term will be used.

The staging of intrauterine embryos follows the table of Eyal-Giladi and Kochav (1976), using roman numerals. For the blastoderm incubated during the first 24 hours the table of Vakaet (1970) will be followed, using arabic numerals. Blastoderms older than 24 hours are classified according to Hamburger and Hamilton (1951).

II. SOMATIC CELL LINEAGE STUDIES

A. Introduction

In birds, cell lineage of individual cells has not yet been achieved, but only cell lineage of cell groups. Cell groups in early development are Anlage fields of organs, whether they are determined or not. Cell lineage of Anlage fields permits study of their localization during development. The term "disposition" is used for the continuously changing localization of the Anlage fields during early development. Disposition is, therefore, a four-dimensional concept, combining the notions of time and space; it does not imply determination, but is a prerequisite for its study. Most studies of disposition have been aimed at setting up a map of presumptive organs, displaying for each organ the whole of the Anlage fields that will contribute to its formation. The terms are almost synonymous, a presumptive organ being the sum of all its Anlage fields. It should be clear that the maps are only still pictures from the film of development and may therefore vary markedly without being contradictory, provided the stage of development is taken into account.

The earliest studies of disposition used vital staining or particular markers such as charcoal or iron oxide. The introduction of autoradiography by Abercrombie and Causey (1950) extended the possibilities of studying allografts. The earlier studies were hampered by the impossibility of distinguishing host from graft in experimental combinations, and are therefore not considered to be genuine chimeras. The latest extension of possibilities was offered by xenografts, using the chick-quail marker system.

B. Origin of the Deep Layer in the Avian Blastoderm

The origin of the deep layer in the bird blastoderm is still incompletely elucidated, partly because its structure changes markedly during early development and partly because its development occurs in the uterus as well as during the first day of incubation. A brief description will be given of the intrauterine development of (1) the endophyll; (2) the hypoblast, which occurs during the first 10–12 hours of incubation; and finally (3) the definitive endoblast. Only the origin of the endoblast has been studied with chimeras, but a short description of the development of the endophyll and of the hypoblast is also necessary.

The purely descriptive studies of the origin of the endophyll have led to conflicting theories. The deep layer at oviposition is without doubt formed by the remnants of a vaster mass of cells derived from the deep blastomeres formed by cleavage. The most recent theory on the thinning out of the deep layer leading to the formation of the area pellucida is that of Eyal-Giladi and Kochav (1976), Kochav *et al.* (1980), and Fabian and Eyal-Giladi (1981), who think that most of the deep cells drop into the subgerminal cavity (Fig. 1) during stages IV and V of Eyal-Giladi and Kochav (1976). This theory, however, does not explain two facts observed with cinemicrography of unlaid blastoderms cultivated by the method of New (1955). First, the thinning out is accompanied by an expansion of the blastoderm, which may occur even though the blastoderm is situated with its ventral side upward, as it is under these culture conditions. Second, the number

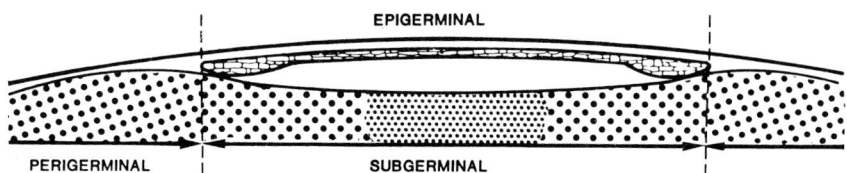

Fig. 1. The domains around the avian blastoderm and the terms used for the structures and spaces found in them.

of cells that would have fallen into the subgerminal cavity in normal development is not found in the cultured blastoderms when these are examined with scanning electron microscopy (L. Andries, personal communication).

Another possibility is that during area pellucida formation the originally deep blastomeres become inserted between those situated in the upper layer. This would easily explain both the expansion and the thinning out of the area pellucida and also the fact that, with hematoxylin and eosin staining, two cell populations are observed in the upper layer. The more eosinophilic cells show the same staining as the scarce cells of the deep layer. Recent scanning electron microscopic observations by Andries *et al.* (1983) are easily interpreted in terms of the insertion theory, but since no experimental evidence is at hand, it is useless to speculate further on this matter. If the chick↔quail marker system can be adapted to the study of unlaid blastoderms, a solution seems possible.

The hypoblast, which is formed during the first hours of incubation, offers better opportunities. Blastoderms of unlaid, manually retrieved eggs can be studied by New's culture technique. This technique, which is easier with blastoderms from laid eggs, allows a direct study of the deep layer because the ventral side of the blastoderm is turned upward. Based on cinemicrographic observations of cultured blastoderms marked by charcoal or iron oxide, a description of the formation of the hypoblast has been given (Vakaet, 1970). During the first 5 hours of incubation, the scattered endophyll cells that are found during the unincubated stage, and that are firmly attached to the ventral side of the upper layer, are inserted into a layer of cells that is loosely connected with the upper layer. Since no movements are visible with cinemicrography within the deep layer during this process, a contribution of cells from the upper layer is probable. Weinberger and Brick (1982a, b) arrive at the same hypothesis, from scanning electron microscopic observations (see also Fig. 2A and B).

The layer thus formed is not complete, as has been stated by Peter (1938). The completion of the hypoblast is achieved by a concentric ingrowth of cells of the deep layer, which forms the hypoblast of the area marginalis, at the periphery of the area pellucida. It penetrates into the area centralis only in the posterior quadrant of this area. Anteriorly, these hypoblast cells grow into the area centralis, where they complete the hypoblast. Meanwhile, they push the endophyll cells forward to form the endophyll wall or crescent, which is double-layered and composed of a mixture of endophyll and hypoblast cells.

The endoblast was long thought to form from the hypoblast. Hunt (1937a, b) was the first to suggest that it might be laid down by gastrulation movements. This was confirmed by total excision of the deep layer of stage 3–4 blastoderms (Vakaet, 1962a), after which the endoblast still forms normally and is centered around Hensen's node. This has been confirmed by Modak (1966), who excised the deep layer of the area pellucida at different stages. He combined this experi-

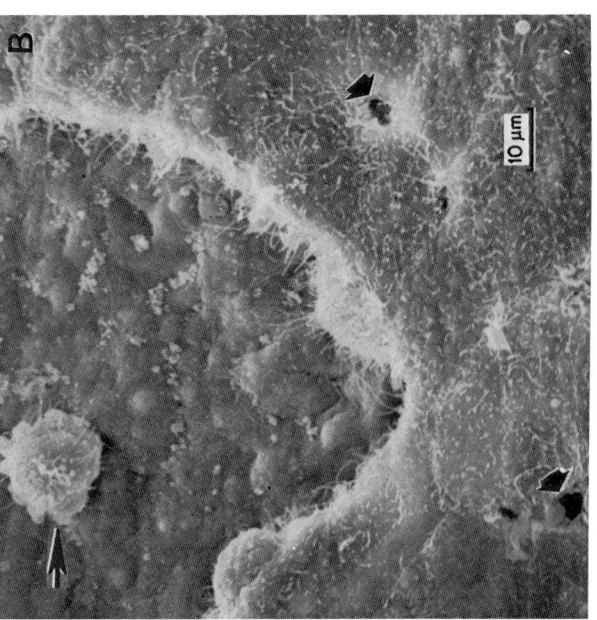
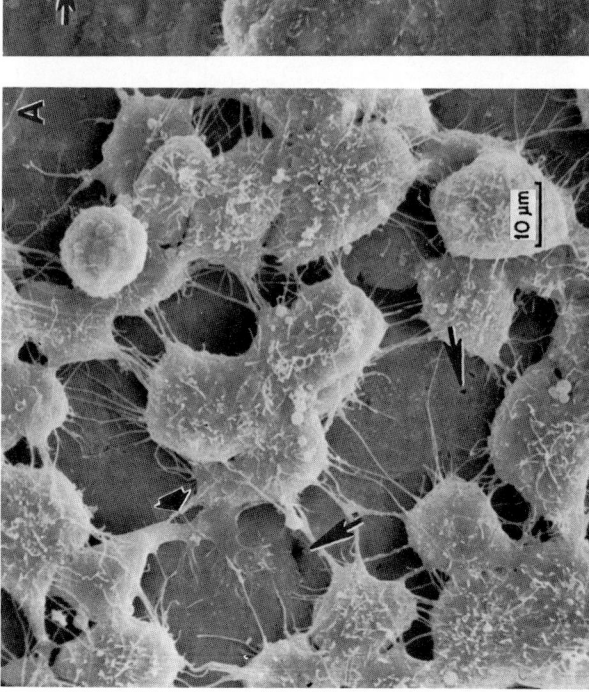

Fig. 2. Scanning electron micrographs of the ventral aspect of young blastoderms. (A) Unincubated with scattered deep layer cells, interconnected with thin filopodia and firmly attached to the ventral aspect of the upper layer with lamellipodia (broad arrow). Note the clefts (long arrows) between the ventral aspects of some upper layer cells, probably correlated with an incompletely formed basal lamina. (B) Incubated for 5 hours; the deep layer cells form a continuous but incomplete layer, interrupted with small (broad arrows) and larger holes. The long arrow in such a hole points to a cell that possibly leaves the upper layer and joins the deep layer, contributing to its completion. The ventral aspect of the upper layer shows no more clefts.

ment with carbon marking of the anterior part of the primitive streak, or with grafting tritiated thymidine labeled fragments of the same region. He concluded that the embryonic endoblast is formed by cells ingressing through the primitive streak, while the marginal hypoblast regenerates from the inner margin of the germ wall. Decisive evidence in favor of the formation of the definitive endoblast by ingression has been given by Fontaine and Le Douarin (1977) using xenografts. These authors excised the hypoblast of chick blastoderms at stage 2–6 and exchanged it orthotopically with the hypoblast of quail blastoderms of the same stage, using the culture system of New. The results showed that the endoblast was of host origin and situated like an island in graft hypoblast. Stern and Ireland (1981) studied endoderm formation in avian embryos during primitive streak stages by combining several techniques, including the use of chick↔quail chimeras. They were able to confirm that the deep layer of the early chick embryo receives contributions from both the marginal and the central regions of the area pellucida.

The disposition of the Anlage field of the endoblast before its final placement will be discussed in Section II, C.

C. Disposition of the Anlage Fields before and during Ingression

The results of the many attempts to set up maps of presumptive areas using vital stains or particles, summarized by Waddington (1952), illustrate the difficulties of deducing the disposition of the Anlage fields when there is no certainty that the movements of the markers and of the cells composing the fields are identical. It is remarkable that one tissue is absent from these maps: the definitive endoblast. Indeed, Hunt (1937a, b) had shown that the presumptive endoblast was originally located in the upper layer, but for many years his evidence was not accepted. From vital staining and explantation experiments on the chorioallantoic membrane, Hunt concluded that the endoblast takes part in gastrulation, but this question was only settled by the use of chimeras. Rosenquist (1966) confirmed the origin of the endoblast through ingression, and gave an estimate of its previous disposition within the upper layer.

Using orthotopic and isochronic tritiated thymidine labeled allografts, Rosenquist (1966) studied the morphogenesis of the upper layer of the chick blastoderm at stages 4–6. After reincubation of the chimeras up to stage 12 (Hamburger and Hamilton, 1951), he made systematic reconstructions of autoradiographed sections. From these, he was able to deduce the movements of the Anlage fields between the stage of intervention and that of fixation. The dispositions he found for the epiblast and the neurectoblast were in agreement with the description of Spratt (1952). The disposition of the endoblast, in the anterior half of the primitive streak at stage 4, confirmed previous findings. On the other

III. Early Development of Birds

hand, his description extended the results of Vakaet (1962a) and Modak (1966), in mapping more precisely the Anlage field of the endoblast at stage 4 and in describing the process of ingression through the primitive streak. Cells from this zone enter the streak, mix with cells from the opposite side while they descend together ventrally and migrate into the endoderm layer.

More generally, the work of Rosenquist (1966) confirmed in unequivocal fashion the invagination movements that had been ascertained with vital dye staining and carbon marking, and permitted the movements of large areas of upper layer to be traced into the primitive streak. These areas ingress ventrally and are distributed bilaterally into the deep layer and into the middle layer where they form the endoderm and the mesoderm, respectively. A disposition of the Anlage fields within the mesoblastic area was also given by Rosenquist (1966) (Fig. 3) and extended by Nicolet (1971).

Nicolet (1970), like Rosenquist, used tritiated thymidine labeled allografts, sometimes combined with the classical methods of marking. He clearly stated that only the upper layer takes an active part in building the embryo, giving rise to the embryonic endoblast, to the whole mesoblast and to all ectoblastic structures such as epiblast and nervous system. His most important contribution concerns the temporospatial disposition of the endoblastic and mesoblastic

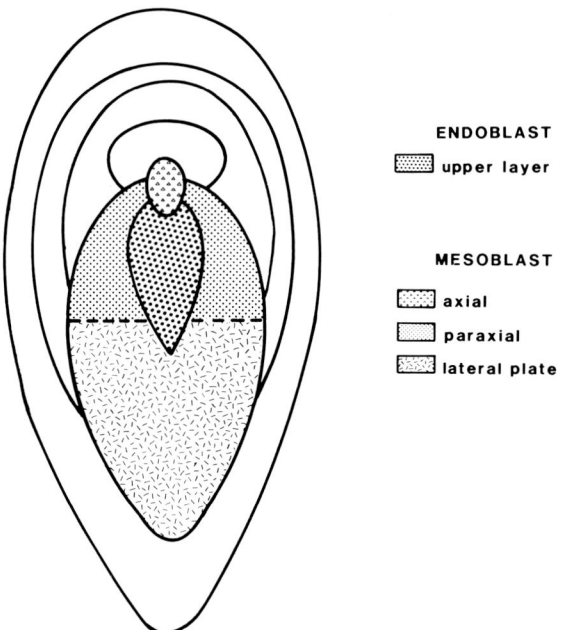

Fig. 3. Disposition of the Anlage fields of the endoblast and mesoblast in the upper layer, after Rosenquist (1966). This stage corresponds to stage 4 of Fig. 5.

Anlage fields during gastrular ingression. The endoblast ingresses through the anterior part of the primitive streak beginning at stage 4 and ending by stage 6. The anterior part of the primitive streak is composed of a mixture of endoblast and mesoblast cells, in which at stage 4 the endoblast and at stage 6 the mesoblast cells predominate. The mesoblastic Anlage fields in the upper layer that take part in the formation of the embryo are disposed one behind the other along the rostral half of the primitive streak: most anteriorly the axial mesoblast, followed by the paraxial mesoblast and finally the lateral mesoblast. The extraembryonic mesoblast occupies the flanks of the caudal part of the primitive streak. In the middle layer, these Anlage fields stretch out longitudinally, in such a way that the Anlage fields that ingress more anteriorly are disposed more medially within the embryo. This is shown in Fig. 5, which from stages 6 to 8 is a confirmation of the results of Nicolet (1970) and Vakaet (1962b). The stretching of the Anlage fields is brought about by the regression of Hensen's node and of the anterior half of the primitive streak (Spratt, 1947; Vakaet, 1960a).

The chick-quail marker system has been used for the study of the disposition of the Anlage fields in the young chick blastoderm (Vakaet, 1984). Isotopic and isochronic xenografts of quail upper layer fragments were made into chick blastoderms, the hosts were reincubated and the chimeras observed by cinemicrography. The blastoderms were at stages 1 to 5 (the period of initiation and elongation of the primitive streak) and at stages 6 and 7 (when shortening of the streak begins). Quail tissues are slightly more transparent than chick tissues, and this allowed the disposition of the Anlage fields to be observed during more than one stage, not only when they were moving in the upper layer but also in the primitive streak and, in favorable circumstances, for some time after their ingression. The conclusions are summarized in Figs. 4 and 5.

In Fig. 4 the extent and direction of the movements of xenografts in the upper layer are outlined from stages 2 to 5. As the primitive streak elongates, forming a rodlike structure, the upper layer converges on it. The elongation of the streak region (see also Vakaet, 1960b) is brought about by the anteroposterior stretching of the Anlage fields, after their integration into the primitive streak.

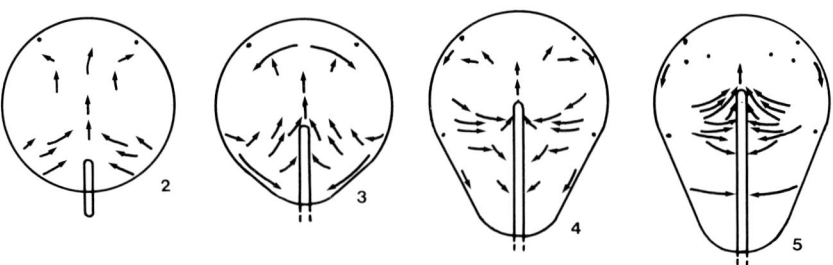

Fig. 4. Outline of movements (stages 2–5) observed in the upper layer of blastoderms carrying xenografts. See text for comments.

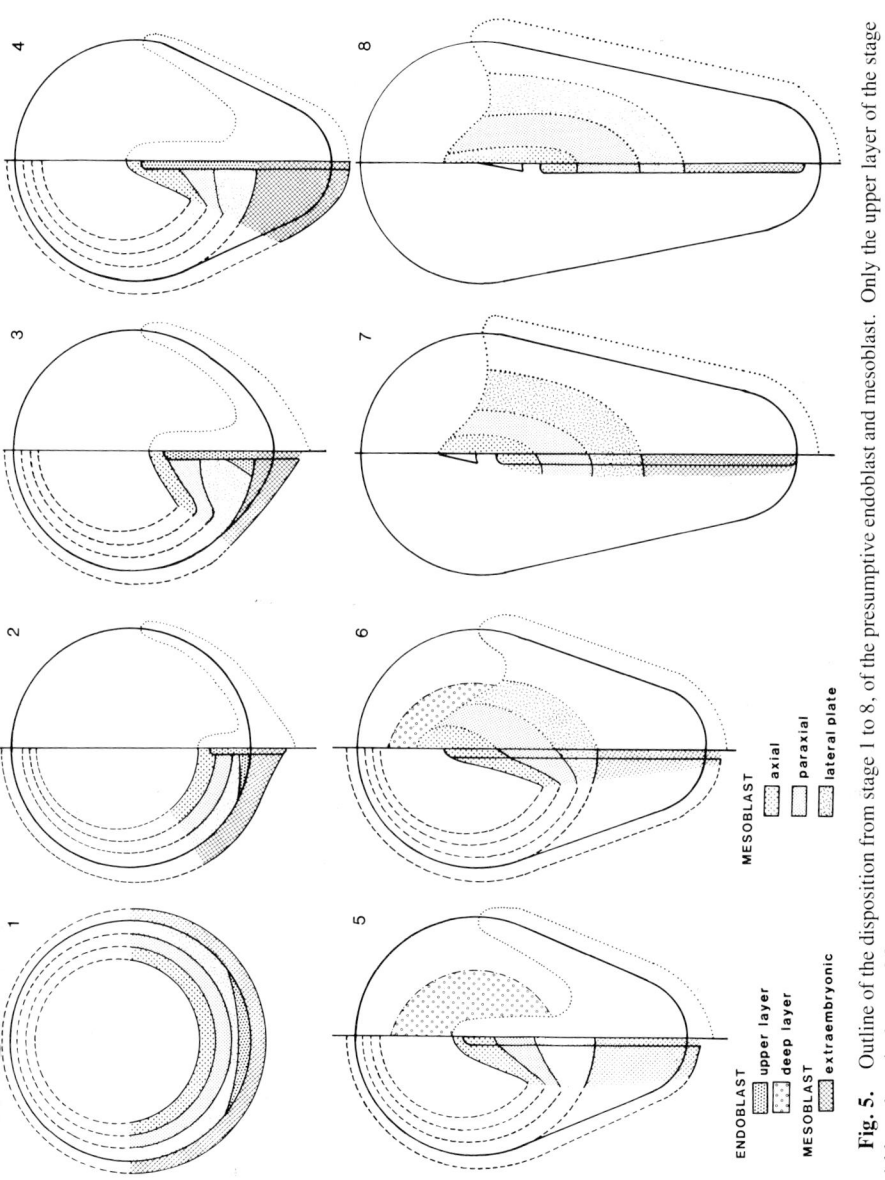

Fig. 5. Outline of the disposition from stage 1 to 8, of the presumptive endoblast and mesoblast. Only the upper layer of the stage 1 blastoderm is drawn. Of stages 2 to 8 the upper layer is drawn at the left-hand side of the midsagittal line across each blastoderm. The middle layer cells are represented at the right-hand side. In stages 5 and 6 the ingressed endoblast is also shown at the right. In stages 6 to 8 the disposition of the mesoblast fields after ingression is marked. See text for comments.

The divergent movement in the anterior periphery of the upper layer of the area centralis (the area marginalis being apparently immobile), combined with the other movements, duplicates the Polonaise movements described in the deep layer by Gräper (1929). Both whorl movements, in the upper layer and in the deep layer, occur at the same time during elongation of the streak, confirming the existence of the *mouvements simultanés* described by Pasteels (1937).

In Fig. 5, the disposition of the endoblast and of the mesoblastic tissues is outlined in blastoderms at stages 1 to 8. At stage 1 the disposition of the

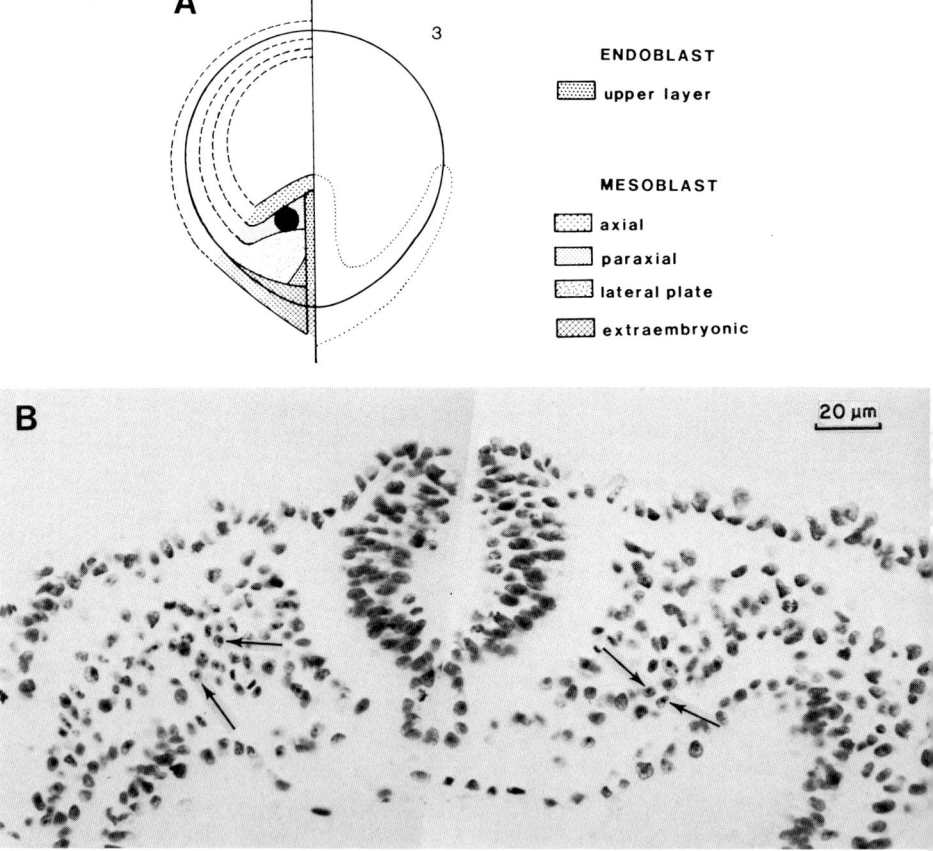

Fig. 6. (A) Chimera obtained by isotopically and isochronically grafting a fragment of the upper layer of a stage 3 quail blastoderm into the Anlage field of the paraxial mesoblast in a chick blastoderm (black spot). (B) Transverse section through the trunk region of the chimera after 22 hours of reincubation. Staining after Feulgen–Rossenbeck. Note the bilateral distribution of quail cells (arrows) in the chimera. More laterally grafted Anlage fields would be found more caudally after ingression.

presumptive organs is around the area centralis. As the anteroposterior axis is determined, however, only the posterior half of the area pellucida is disposed to form the embryonic axis. The tentative limits of the presumptive tissues are therefore only drawn in full for this posterior part that will undergo ingression. Characteristic of stage 1 is the disposition of the presumptive endoblast between lateral plate and extraembryonic Anlage fields. The movements of convergence push the presumptive organs toward the posterior midline, where only the two most peripheral fields, the extraembryonic mesoblast and the endoblast, will contribute to the primitive streak up to stage 4. The endoblast extends like the tip of a tongue between the Anlage fields of the embryonic mesoblast, separating the lateral and paraxial tissues completely into left and right halves, but only partly separating the axial tissues. The prechordial field is not divided, but ingresses at the tip of the primitive streak after stage 5. The presumptive notochordal, paraxial and lateral plate tissues elongate alongside the primitive streak. Further elongation of the primitive streak is brought about by the abrupt ingression of the endoblast at stage 5, leading to its disappearance from the upper layer as it is inserted into the deep layer to form the gut endoderm (see stages 5 and 6). The ingression of the endoblast is immediately followed by that of the prechordal mesoblast, situated rostrally to Hensen's node, leading to the final elongation of the primitive streak. The streak has a groove only from stage 5 on, as active ingression continues along its lips. The cleft mesoblast (axial, paraxial and lateral plate), disposed alongside the anterior half of the primitive streak ingresses as described by Nicolet (1971), so that from stage 7 on almost no mesoblastic Anlage fields are disposed in the upper layer. Those in the middle layer are disposed in their final position in the way already outlined by Nicolet, with those that were more centrally disposed in the upper layer lying more medially within the embryo.

This disposition is shown in Fig. 5, but another disposition is added by the use of chick↔quail chimeras. Within the different presumptive organ regions, the nearer the Anlage fields are initially disposed to the site of the primitive streak, the more rostral are the structures that they form, while those Anlage fields that are originally far from the streak become caudally situated in the embryo (Fig. 6).

III. CHIMERAS IN THE STUDY OF EMBRYONIC INDUCTION

A. Introduction

Morphological studies of induction imply the use of grafting techniques. Although xenografts have been used since the earliest study of induction in the bird blastoderm, when Waddington and Schmidt (1933) transplanted parts of the duck or chick primitive streak into the area pellucida of the other species, allografts have been used more often. In the earliest experiments, host and graft tissues were not distinguishable, and it appeared that even the use of tritiated thymidine labeled allografts did not allow the nature either of the evocating cells or of the evocating substance to be elucidated.

The question of the inducing cells in neural induction and induction of a secondary axis has been tackled, using the chick↔quail xenograft system. This allows the use of very small grafts, as every cell is identifiable even after the graft has been dispersed.

B. Neural Induction

The reviews of Waddington (1952) and Gallera (1971) both concluded that the question as to which cells evoke the normal neural reaction remained open. However, Gallera and Nicolet (1969) concluded from allograft experiments that the evocator substance for the normal reaction diffuses horizontally into the perinodal epiblast from the Anlage field of the endoblast, while this tissue is still in the upper layer. The experiments of McCallion and Shinde (1973) and of Cuevas and Orts Llorca (1974) confirmed that neural induction in birds is not species specific and may be brought about by chick-quail xenografts.

In an attempt to recognize the Anlage field of the tissue responsible for neural evocation, allografts were first used. Anterior parts of the primitive streak of blastoderms of different stages were grafted into the anterior endophyllic crescent (Vakaet, 1965). This led to the observation that Hensen's node of stage 4 blastoderms induced headlike structures, while Hensen's node of stage 6 and older blastoderms induced trunk-like formations. Since the grafts contained parts of all layers, which developed into different tissues after transplantation, and since the graft cells could not be distinguished from those of the host tissues,

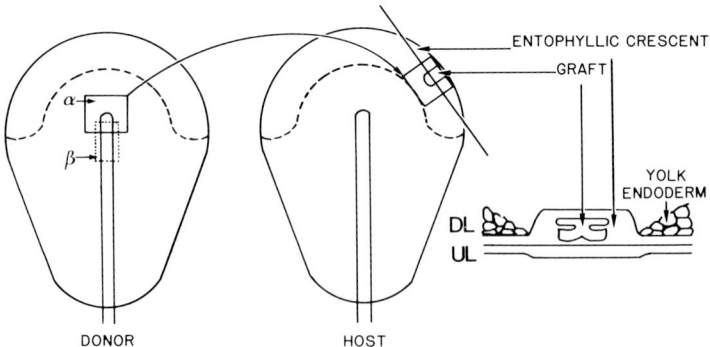

Fig. 7. Scheme of heterotopically xenografting fragments of primitive streak into the anterior entophyllic crescent. DL, deep layer; UL, upper layer. See text for comments.

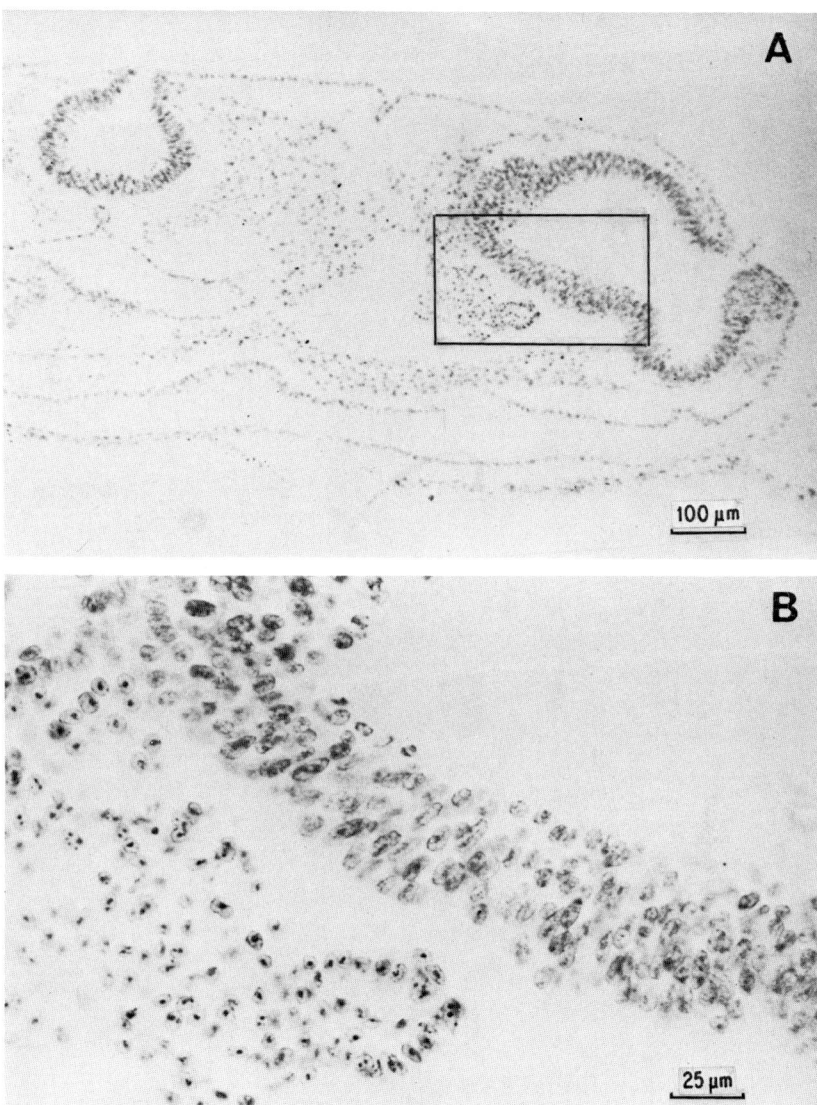

Fig. 8. Result of grafting a quail Hensen's node (a-type graft of Fig. 7) into a chick endophyll wall. Host and graft were at stage 5. Reincubation for 20 hours. Feulgen–Rossenbeck staining. (A) Two neural tubes are visible, at the left that of the host and at the right the induced neural tube. (B) Enlargement of box in (A). The evoking tissue is distinctly of quail nature and has formed, among other tissues, a notochord underlying the induced host neural tube.

these experiments were resumed when the chick-quail marker system became available. Blastoderms at stage 5 were then used for both host and graft (Vakaet, 1981).

In a first series of experiments, grafts of quail or chick Hensen's node, as represented in Fig. 7a (see α), were xenografted into the endophyll wall of the host. The resulting chimeras were incubated for 18 hours. After fixation, the host was examined for the presence of a secondary neural plate, while the graft was examined for the tissues that had differentiated from it, which should give information on the Anlage fields that were present in the graft at the moment of evocation of the reaction.

In some of these experiments, a secondary neural plate was formed although no Anlage fields of neural tissue were present in the graft. In these cases at least the evocation could apparently not be explained by homoiogenic induction of neural tissue by the same tissue. Only grafts containing Anlage fields of endoblast and the axial mesoblast remained as possible evocators of the neural reaction (Fig. 8).

To give more precision to the latter hypothesis, grafts as represented in Fig. 7 (see β) were tested. These grafts contain almost exclusively Anlage fields of the endoblast. As no neural induction occurred, it was concluded that neither the endoblast nor the lateral plate and area vasculosa that were sometimes found evokes neural reactions.

This series, based on elimination of the theoretical possibilities of evocation, only leaves the axial (prechordal and chordal) mesoblast as the evocator of the experimental neural plate reaction. Not all problems relating to neural induction have been solved, however. How far can the results of experimental neural induction be applied to normal neural induction? What is the nature of the evoking substance? How is the evoking stimulus transmitted? What is the significance of interspecies neural evocation?

C. Induction of Secondary Axes

Waddington (1933) was the first to describe the induction of secondary axes in the chick blastoderm by the translocation of the deep layer. He concluded from his experiments that the deep layer influences the morphogenetic movements in the upper layer in such a way that a secondary axis may be formed. These conclusions were backed by Spratt (1946), Vakaet (1967), and Eyal-Giladi and Wolk (1970). Yet the original paper is not convincing. The positive results that Waddington obtained were few (three double axes, with seven others showing a change in orientation of the axis, in a total of 39 experiments, all without histological control).

When it was found that the grafting of the midpiece of a primitive streak into the entophyllic crescent of a blastoderm at stage 5 sometimes induces the formation of a secondary primitive streak, it seemed likely that a second way of evoking secondary axes existed (Vakaet, 1964). Gallera and Nicolet (1969) drew attention to the fact that not every midpiece induces a primitive streak; some evoke instead the formation of a neural plate in the upper layer. Grafting a Nodus posterior into the entophyllic crescent, however, invariably induces a primitive streak (Vakaet, 1982). The graft disappears and a condensation that elongates and forms a small but typical primitive streak appears in the center of the empty area (Vakaet, 1982). Using chick↔quail xenografts, it has been demonstrated that only cells of the area vasculosa take part in the evoking process (Fig. 9).

The experiments of Waddington (1933) were repeated using xenografts. Care was taken not to graft middle layer cells together with the deep layer. In 27 such experiments no induction of primitive streaks was observed. After grafting the deep layer, a clearing up was observed in the graft area during subsequent incubation. The condensation, premonitory of the formation of a primitive streak, was not observed. Induction of a secondary primitive streak occurred in eight out of ten experiments in which, together with the deep layer, some middle layer cells were grafted.

From these observations it may be inferred that a secondary primitive streak in the avian blastoderm is evoked by middle layer cells. These middle layer cells were probably contaminants in the early experiments of Waddington (1933) and in those of Spratt (1946) and of Vakaet (1964). Waddington (1933) discussed this possibility but considered it improbable. In the xenograft experiments, evidence has been found that, without middle layer cells, an evocation of a secondary axis does not occur.

The disposition of the middle layer cells forming the graft is decisive. Only the middle layer cells ingressing through the posterior half of the primitive streak evoke a streak-forming reaction. The mesoblast cells that ingress through the anterior half of the primitive streak evoke a neural reaction in the upper layer, as discussed above. This could explain the results of Gallera and Nicolet (1969), who obtained induction either of a neural plate or of a primitive streak after grafting midpieces of primitive streaks, since the response could have depended on whether the graft was taken from the anterior or from the posterior half of the streak.

The experimental evocation of a secondary axis occurs, in all experiments carried out so far, by grafts of Anlage fields of the extraembryonic mesoblast that is disposed to form area vasculosa. These findings have depended on the technique of chick–quail xenografts, the power and limitations of which are thereby clearly demonstrated. On the one hand, it is a powerful help for the lineage of cell groups, but, on the other hand, it does not permit a direct study of the nature of the evoking substance. It may be of indirect help, when combined with other techniques, such as autoradiography, to characterize the biochemical nature of evocator substances.

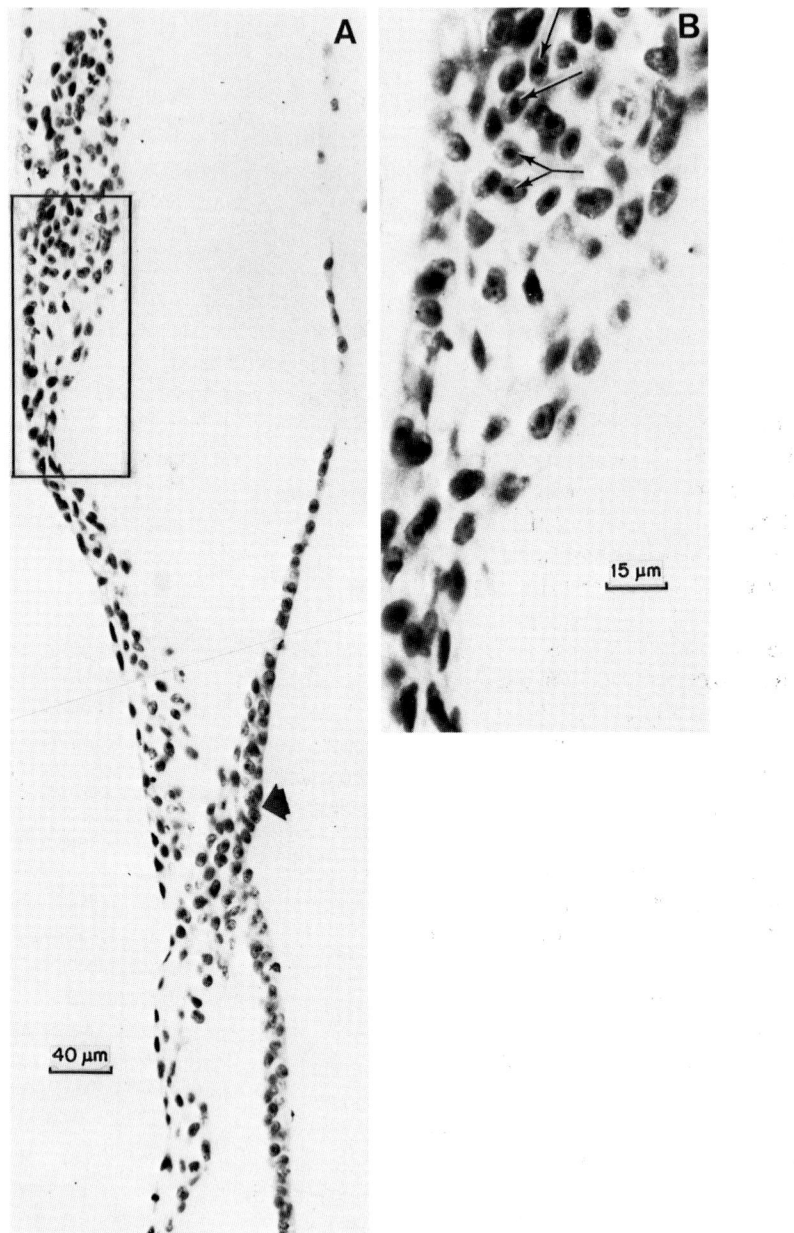

Fig. 9. Result of grafting a Nodus posterior of a quail blastoderm at stage 4 into the endophyll wall of a chick blastoderm at the same stage. Reincubation for 12 hours. Feulgen–Rossenbeck staining. (A) A typical primitive streak (broad arrow) has been induced, as is visible on this transverse section. The evoking cells are to be found at the periphery of the induced area. (B) Enlargement of a part of the area vasculosa (box in A), the only field in which quail cells (long arrows) are found in this chimera.

III. Early Development of Birds

ACKNOWLEDGMENTS

The original work referred to in this review was partly supported by Grant No. 3.9001.81 from the Belgian Fund for Medical Scientific Research to L. Vakaet and C. Vanroelen. The author is grateful to Mrs. E. Haest-Van Nueten for microsurgery, to Mrs. A. Seghers-Geldmeyer for scanning electron microscopy, to Mr. C. De Schepper and Mr. G. Van den Broeck for transmission electron microscopy, to Mr. F. De Bruyn for artwork, to Mr. J. Van Ermenghem for photographic processing, and to Mrs. N. Van den hende-Bol for typing the manuscript. I should also like to thank my staff members, Mr. L. Andries, Mr. F. Harrisson and Mr. C. Vanroelen for useful discussions.

REFERENCES

Abercrombie, M., and Causey, G. (1950). *Nature (London)* **166,** 229–230.
Andries, L., Vakaet, L., and Vanroelen, C. (1983). *Anat. Embryol.* **166,** 453–462.
Cuevas, P., and Orts Llorca, F. (1974). *Acta Anat.* **89,** 423–430.
Dalcq, A. (1941). "L'oeuf et son dynamisme organisateur." Albin Michel, Paris.
Eyal-Giladi, H., and Kochav, S. (1976). *Dev. Biol.* **49,** 321–337.
Eyal-Giladi, H., and Wolk, M. (1970). *Wilhelm Roux' Arch. Entwicklungsmech. Org.* **165,** 226–241.
Fabian, B., and Eyal-Giladi, H. (1981). *J. Embryol. Exp. Morphol.* **64,** 11–22.
Fontaine, J., and Le Douarin, N. M. (1977). *J. Embryol. Exp. Morphol.* **41,** 209–222.
Gallera, J. (1971). *Adv. Morphog.* **9,** 149–180.
Gallera, J., and Nicolet, G. (1969). *J. Embryol. Exp. Morphol.* **21,** 105–118.
Gräper, L. (1929). *Wilhelm Roux' Arch. Entwicklungsmech. Org.* **116,** 382–429.
Hamburger, V., and Hamilton, H. L. (1951). *J. Morphol.* **88,** 49–92.
Hunt, T. E. (1937a). *Anat. Rec.* **68,** 349–370.
Hunt, T. E. (1937b). *Anat. Rec.* **68,** 449–457.
Kochav, S., Ginsburg, M., and Eyal-Giladi, H. (1980). *Dev. Biol.* **79,** 296–308.
Le Douarin, N. (1969). *Bull. Biol. (Woods Hole, Mass.)* **103,** 435–452.
McCallion, D. J., and Shinde, V. A. (1973). *Experientia* **29,** 321–322.
Modak, S. P. (1966). *Rev. Suisse Zool.* **73,** 877–908.
New, D. A. T. (1955). *J. Embryol. Exp. Morphol.* **3,** 326–331.
Nicolet, G. (1970). *J. Embryol. Exp. Morphol.* **23,** 79–108.
Nicolet, G. (1971). *Adv. Morphog.* **9,** 231–262.
Pasteels, J. (1937). *Arch. Biol.* **48,** 381–488.
Peter, K. (1938). *Z. Mikrosk.-Anat. Forsch.* **43,** 362–415.
Rosenquist, G. C. (1966). *Contrib. Embryol. Carnegie Inst.* **38,** 71–110.
Spratt, N. T. (1946). *J. Exp. Zool.* **103,** 259–304.
Spratt, N. T. (1947). *J. Exp. Zool.* **104,** 69–100.
Spratt, N. T. (1952). *J. Exp. Zool.* **120,** 109–130.
Stern, C. D., and Ireland, G. W. (1981). *Anat. Embryol.* **163,** 245–263.
Vakaet, L. (1960a). *J. Embryol. Exp. Morphol.* **8,** 6–19.
Vakaet, L. (1960b). *J. Embryol. Exp. Morphol.* **8,** 321–326.
Vakaet, L. (1962a). *J. Embryol. Exp. Morphol.* **10,** 38–57.
Vakaet, L. (1962b). "Pregastrulatie en Gastrulatie der Vogelkiem." Arscia, Brussel.
Vakaet, L. (1964). *C. R. Seances Soc. Biol. Ses. Fil.* **158,** 1964.
Vakaet, L. (1965). *C. R. Seances Soc. Biol. Ses. Fil.* **159,** 232.
Vakaet, L. (1967). *Mem. Acad. R. Med. Belg.* **5,** 231–257.
Vakaet, L. (1970). *Arch. Biol.* **81,** 387–426.

Vakaet, L. (1981). *K. Acad. Gen. Belg.* **43,** 78–102.
Vakaet, L. (1982). *K. Acad. Gen. Belg.* **44,** 419–437.
Vakaet, L. (1984). To be published.
Waddington, C. H. (1933). *Wilhelm Roux' Arch. Entwicklungsmech. Org.* **128,** 502–521.
Waddington, C. H. (1952). "The Epigenetics of Birds." Cambridge Univ. Press, London and New York.
Waddington, C. H., and Schmidt, G. A. (1933). *Wilhelm Roux' Arch. Entwicklungsmech. Org.* **128,** 522–563.
Webster, N. (1966). "Webster's Third New International Dictionary of the English Language." G. & C. Merriam Co., Springfield, Illinois.
Weinberger, C., and Brick, I. (1982a). *Wilhelm Roux' Arch. Dev. Biol.* **191,** 119–126.
Weinberger, C., and Brick, I. (1982b). *Wilhelm Roux' Arch. Dev. Biol.* **191,** 127–133.

CHAPTER **IV**

Somatic Cell Lineages in Mammalian Chimeras

J. ROSSANT

Department of Biological Sciences
Brock University
St. Catharines, Ontario, Canada

I.	Introduction	89
II.	Allocation of Cells to Distinct Lineages	92
	A. Allocation of Cells to ICM and Trophectoderm: Nonchimera Studies	92
	B. Allocation of Cells to ICM and Trophectoderm: Chimera Studies	93
	C. Allocation of Cells to Primitive Endoderm and Primitive Ectoderm	95
	D. Allocation of Cells to Postimplantation Cell Lineages	95
	E. Allocation of Individual Cells within Broad Cell Lineages	96
	F. Retrospective Analysis of Cell Allocation	97
	G. Summary	101
III.	Restriction of Cells to Different Lineages	101
	A. Restriction of Cells to ICM or Trophectoderm Lineage	101
	B. Restriction of Cells to Primitive Endoderm or Ectoderm Lineages	103
	C. Restriction of Cells to Postimplantation Cell Lineages	104
IV.	Conclusions	106
	References	107

I. INTRODUCTION

Embryonic development involves both growth and differentiation. As cell division proceeds, cells become progressively more specialized and are allocated to distinct cell lineages, which give rise to the complex structures of the adult. The molecular mechanisms underlying the precise temporal and spatial pattern of cell lineage development in embryogenesis are still unclear, and, before detailed study of such mechanisms can be carried out, it is necessary to understand fully the cellular events involved in lineage development. Two complementary questions on cell lineage must be answered: when and where are cells allocated to different lineages in the embryo, and when and where are cells heritably restricted to a given lineage? The first question deals with cell fate in the intact embryo, and the second with cell potential outside the normal embryonic environment (Weiss, 1939).

Studies on cell allocation in various embryos have revealed different patterns of cell lineage development. In some invertebrates it has proved possible to follow complete embryonic cell lineages by direct observation of the intact embryos (Wilson, 1925; Reverberi, 1971; Davidson, 1976). The most striking recent example of such studies has been the complete elucidation of the embryonic and postembryonic cell lineages of the nematode worm, *Caenorhabditis elegans* (Sulston and Horwitz, 1977; Deppe *et al.*, 1978; Sulston *et al.*, 1983) by visual observation of cell division using Nomarski optics. In these studies and others, such as those on the leech in which extrinsic cell markers were used to follow cell lineages (Weisblat *et al.*, 1980; Zackson, 1982), cell lineage is almost completely invariant from one embryo to another and, indeed, varies little even between species (Sternberg and Horwitz, 1982). A given cell in any one embryo at any stage of development will give rise to the same structures in all embryos examined. Such invariance means that a highly stereotyped series of cell divisions occurs during cell lineage development. Limited detailed analysis of this sort is available for vertebrate embryos, although studies on the fate of horseradish peroxidase (HRP)-marked blastomeres in *Xenopus* larvae also reveal well-defined and relatively invariant domains derived from specific blastomeres, at least within the central nervous system (Hirose and Jacobson, 1979; Jacobson and Hirose, 1981). In mammals, there is little evidence of such invariant cell lineage, although the lack of any polarity in the early embryo makes it impossible to compare the lineage of the same blastomere in different embryos. Cells do become progressively allocated to broad cell lineages, but this seems to involve allocation of groups of cells rather than individual cells.

Invariant cell lineage in the intact embryo has often been interpreted as being the result of strict preprogramming of cell fate rather than the result of the effects of cell position and cell–cell interaction during development. Conversely, cell lineage development in embryos where lineages are less rigorously defined is often thought to be controlled more by position and cellular interactions than by a cell's past history. However, strictly speaking, no study on cell lineage in the intact embryo alone can determine the relative role of a cell's past lineage and its present position in establishing its fate, because these studies do not show when cells become heritably restricted to their fate. This can only be determined by

IV. Somatic Cell Lineages in Mammalian Chimeras

examining a cell's potential outside the normal embryonic environment. Two extremes of relationship between cell fate and cell potential can be imagined. In the first, the potential of any cell outside its normal environment is always equivalent to its fate in the intact embryo. Manipulation of nematode development, where cell lineages are almost invariant in the intact embryo and larva, has revealed that this is true of most cells at all stages of development beyond the 50-cell stage (Sulston and White, 1980; Sulston et al., 1983). There is some regulative capacity, but the system is generally very inflexible and shows limited ability to compensate for any tissue damage during development. The combination of invariance of cell lineage and inflexibility of cell potential is strong evidence that cell lineage rather than cell position and interaction is of overriding importance in determining the pattern of development in such a system. The opposite extreme would be an embryo in which cell potential was never restricted but cell fate in the intact embryo was determined solely by interrelationships between cells and by their response to external cues. In other words, cell lineage plays no role in cell fate. Such a system would allow considerable regulation for tissue damage, because cells would be equipotential at all stages. However, it would be very difficult to produce the complex progressive changes observed in development without any of the temporal cues provided by preprogramming of cells. Indeed, no embryo seems to use this approach. Cell lineage restriction in mammals seems to involve an intermediate type of system in which the disadvantages of both extremes are avoided. Successive restriction of cell potential does occur during development, removing the need for continued reinforcing cues to maintain a cell lineage. However, considerable lability may occur within broad cell lineages, allowing flexibility of response to changing conditions.

Different molecular mechanisms are likely to be involved in cell lineage restrictions that are based on preprogrammed decisions from those that are based on responses to a cell's present circumstances. Both kinds of event involve a temporal component (Johnson, 1981), but, in the former case, cells need not have any sense of their own spatial relationship to the rest of the embryo, while, in the latter case, such a sense of position is essential to orderly development. In this chapter, I shall review those experimental studies on the mammalian embryo, particularly those using genetic chimeras, which shed light on the relative importance of cell lineage versus cell position in this developing system. I shall not attempt a comprehensive overview of cell lineage development, since this has been presented elsewhere recently (Gardner, 1978, 1982a; Johnson et al., 1977; Johnson, 1981; McLaren, 1976; Rossant, 1977; Rossant and Papaioannou, 1977), nor shall I deal with any possible molecular mechanisms. Rather, the aim of this chapter is to assess critically our current state of knowledge on the cellular events underlying lineage development in order to indicate the directions in which the search for molecular mechanisms must proceed.

II. ALLOCATION OF CELLS TO DISTINCT LINEAGES

A. Allocation of Cells to ICM and Trophectoderm: Nonchimera Studies

Most studies on lineage allocation in the intact mammalian embryo have concentrated on the allocation of cells to the first two distinct lineages to become apparent in the embryo, the trophectoderm and the inner cell mass (ICM) of the blastocyst. A combination of visual observations (Graham and Deussen, 1978; Graham and Lehtonen, 1979; Lehtonen, 1980) and cell marker studies using oil droplets (Wilson *et al.,* 1972; Graham and Deussen, 1978) or HRP injecttions (Balakier and Pedersen, 1982) has indicated that allocation of cells to these lineages is not completely random. Cells that lie deep in the embryo during cleavage tend to end up in the ICM, and peripheral cells tend to form trophectoderm. Also, blastomeres divide asynchronously, and descendants of the first cell to divide from the 2-cell stage continue to divide ahead of other blastomeres and contribute preferentially to the ICM (Graham and Deussen, 1978; Kelly *et al.,* 1978). Cell lineage varies from one embryo to another, however, which suggests that allocation of cells to ICM or trophectoderm is a stochastic process, depending on the relative rates of division of different blastomeres and their continued interactions via cell contacts (Graham and Lehtonen, 1979; Lehtonen, 1980).

Most of these studies have concentrated on the fate of blastomeres up until the 16-cell stage of development, and have shown some regularity but no invariance in cell lineage. The exact time of final allocation of cells to either ICM or trophectoderm lineage in the intact embryo remains controversial. Johnson and his colleagues have proposed that ICM and trophectoderm lineages are established at the 16-cell stage by polarized cell division, generating a population of 6–8 inside cells, which will form the ICM, and 8–10 outside cells, which will form trophectoderm (Ziomek *et al.,* 1982a). Experiments in which fluorescein-labelled cells were included in the inside or outside population of morulae "reconstituted" from 6 apolar, inside cells and 10 polar, outside cells supported this hypothesis, since inside cells nearly always ended up in the ICM and outside cells in the trophectoderm (Ziomek and Johnson, 1982). However, this experiment does not strictly address the question of cell lineage in the undisturbed embryo, and the rigid establishment of two cell lineages at the 16-cell stage is not altogether compatible with some other observations. There is some disagreement on the numerology of the system. From sectioned material, which could suffer from fixation artefacts, Graham and Lehtonen (1979) estimate an average of two inside cells at the 16-cell stage, as opposed to the 6–8 estimated by Johnson and co-workers on the basis of various different properties of live cells (Handyside, 1980; Johnson and Ziomek, 1981; Reeve and Ziomek, 1981; Ziomek and Johnson, 1981). The former estimate suggests that continued recruitment of outside cell progeny to the ICM is necessary beyond the 16-cell stage in order to generate the observed number of ICM cells in the blastocyst and is also compatible with the observed differential contribution of early dividing blastomeres to the ICM (Graham and Deussen, 1978; Kelly *et al.*, 1978). The latter estimate does not require recruitment of outside cells but is not readily compatible with preferential contribution to the ICM from early dividing cells, since the ICM is thought to be generated by polarised cell division of nearly all blastomeres at the 8-cell stage (Ziomek *et al.*, 1982a). Resolution of these differences will require more detailed analysis of cell lineage in the intact embryo between the 16-cell and blastocyst stage using markers such as HRP. Initial studies using this marker have suggested a continued contribution of outside cells to the ICM beyond the 16-cell stage (Balakier and Pedersen, 1982), even if the results are adjusted for the possibility of labelling an outside–inside cell pair instead of a single outside cell (R. A. Pedersen, personal communication).

B. Allocation of Cells to ICM and Trophectoderm: Chimera Studies

Nearly all of these studies on allocation of cells to the early cell lineages have not made use of chimeras, but have used visual observation plus extrinsic markers to follow fate in the intact embryo. In theory, such studies are the only way truly to follow cell lineage, since they do not involve disturbing normal development (Weiss, 1939). The only proviso is that the markers used must not themselves influence cell fate. In practice, such studies are difficult beyond the blastocyst stage, because of rapid growth and dilution of any marker. Copp (1979) has used melanin granules successfully to trace the lineage of trophectoderm cells from the blastocyst, and R. A. Pedersen and co-workers (personal communication) are using HRP to mark postimplantation cells and follow their fate in egg cylinder embryos cultured *in vitro*. Nearly all other experimental analysis of cell lineage at the blastocyst stage and beyond has made use of chimeras in which genetically marked cells are introduced into the embryo and followed through later development (Chapter I). In such experiments, cell lineage development in the intact embryo is clearly not being followed, but provided that the condition of the undisturbed embryo is mimicked as closely as possible, these studies can provide extremely useful information not readily available by any other means. By definition, however, prospective studies of chimeras can only be used to determine the future fate of clearly defined cell types whose position in the intact embryo is known.

The first stage at which distinct cell types become morphologically apparent is the blastocyst stage, with the generation of ICM and trophectoderm, and chimera studies have shown that these two cell types have very distinct intrinsic fates from the earliest stages of the blastocyst. Entire early or late 3.5 day ICMs have been injected into blastocysts, where they incorporate into the host ICM and are, thus, in their usual environment (Gardner, 1975; Rossant and Lis, 1979a). Using glucosephosphate isomerase (GPI) isozymes as genetic markers (see West, Chapter II), it has been shown that the fate of the injected ICM cells is always to give rise to the fetus proper, plus extraembryonic endodermal and mesodermal structures. A situation, which more closely resembles the intact embryo and allows direct determination of the fate of both ICM and trophectoderm, is provided by blastocyst reconstitution experiments, in which the only difference from the intact embryo is that the original polar trophectoderm layer has been removed (Gardner *et al.*, 1973; Papaioannou, 1982). A vesicle of mural trophectoderm of one genotype is injected with an ICM of another genotype, and the fate of the two cell types is analysed later in development. Studies using GPI as the genetic marker indicated that the ICM was indeed restricted to the tissues determined by blastocyst injection, and that the trophectoderm layer gave rise to the ectoplacental cone, trophoblast giant cells, and extraembryonic ectoderm. However, there were one or two anomalous results (Papaioannou, 1982) which could be due to contamination in tissue dissection. Also, a minor contribution of one cell type to any tissue would not be detected by the marker used. We have used a new *in situ*

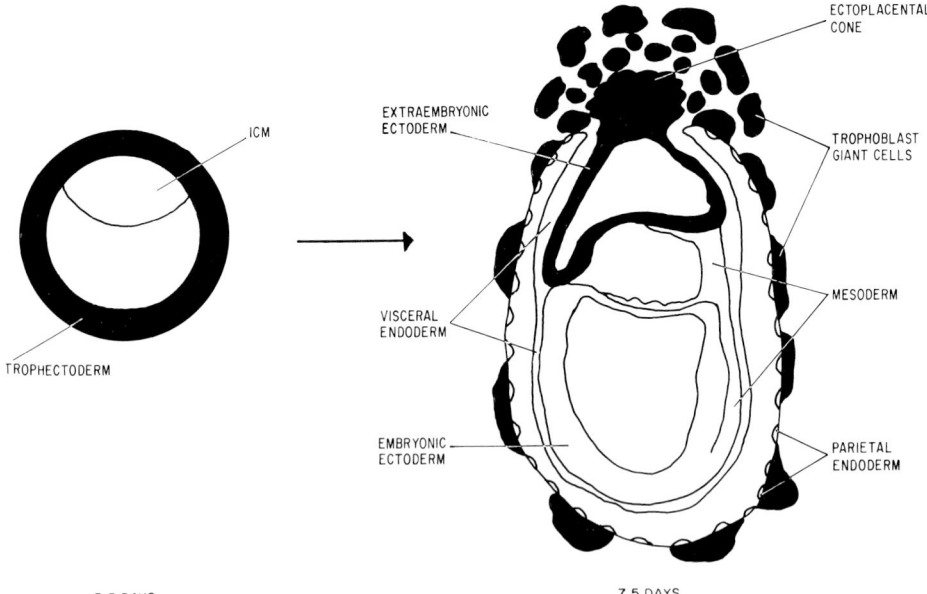

Fig. 1. Lineage derivatives of ICM and trophectoderm as revealed by *in situ* analysis of blastocysts reconstituted with *Mus musculus* trophectoderm and *M. caroli* ICM. Black tissues are trophectoderm derivatives and white tissues are ICM-derived.

marker system on sections of blastocysts reconstituted with *Mus caroli* ICM and *M. musculus* trophectoderm to confirm that the lineages assigned to ICM and trophectoderm by less direct means are correct and invariant (Siracusa et al., 1983; Rossant et al., 1983) (Fig. 1).

C. Allocation of Cells to Primitive Endoderm and Primitive Ectoderm

Blastocyst injection techniques have also been used to examine the fate of later embryonic cells. Primitive ectoderm and primitive endoderm cells from 4.5 day ICMs have been injected into 3.5 day blastocysts, and it has been shown that the primitive ectoderm gives rise to the fetus itself, including the definitive endoderm and the extraembryonic mesoderm, while the primitive endoderm gives rise to extraembryonic visceral (Gardner and Rossant, 1979) and parietal endoderm (Gardner, 1982b). It is not clear, however, how closely undisturbed cell fate is approximated by these studies in which the injected cells are asynchronous to the host embryo and are not strictly in their normal embryonic environment. The fate of primitive endoderm after blastocyst injection, for example, seems to be to contribute preferentially to parietal rather than visceral endoderm (Gardner, 1982b). Since this same "fate" occurs when known visceral endoderm cells from later embryos are injected into blastocysts (Gardner, 1982b), it seems likely that this distribution is not representative of the true fate of primitive endoderm cells, but is more indicative of their potential to form either visceral or parietal endoderm; their abnormal position in the injected blastocyst predisposes them toward the parietal pathway. Thus, these kinds of experiments give us a complicated mixture of approximate fate and actual potential. Injection of older cells into the blastocyst has been performed but must be considered to be so unlike normal development as to be solely related to cell potential (Rossant et al., 1979).

D. Allocation of Cells to Postimplantation Cell Lineages

Use of chimeras to study cell fate in later development has been hampered by the inaccessibility of the embryo to experimental manipulation once implanted in the uterus. However, improved culture techniques for postimplantation embryos have allowed some short-term study of marked cells in postimplantation chimeras. Beddington (1981, 1982) has shown that groups of [^{3}H]thymidine labelled cells from defined areas of one primitive-streak stage embryo can be injected into the same area in another embryo and their progeny detected through to the early somite stage embryo. She has used this technique to demonstrate that anterior embryonic ectoderm gives rise to head ectoderm and neurectoderm, distal ectoderm produces gut endoderm and mesodermal derivatives, and posterior ectoderm gives rise to mesodermal tissues exclusively, both embryonic and extraembryonic. The use of [³H]thymidine as a marker in these studies is not too satisfactory, since it has been shown that, although not lethal at the dose used, it may cause the marked cells to be at a selective disadvantage in chimeras (Kelly and Rossant, 1976). However, application of *in situ* marker techniques to such studies should make this a powerful tool for postimplantation lineage determination. The only limitations then become the difficulty of following the fate of single cells and the short culture period available. Both are technical limitations that may be overcome in future years.

E. Allocation of Individual Cells within Broad Cell Lineages

Chimeras have thus revealed that groups of cells become allocated to different broad cell lineages as development proceeds, but they have not revealed exactly when cell allocation occurs. In the preimplantation embryo, allocation has not been clearly demonstrated until after obvious morphological differentiation, when a difference in prospective fate is perhaps not surprising. In the postimplantation embryo different areas of the embryonic ectoderm show differences in fate before any obvious morphological differences (Beddington, 1981), but cell fate is not completely reproducible from one embryo to another. It is not clear whether this represents true lability of cell lineage or technical problems in identifying exactly the same cells in different embryos. The lack of reproducibility in cell fate prior to morphological differentiation in either the preimplantation or postimplantation embryo implies that the fate of groups of cells is not rigidly defined by cell lineage and is compatible with the notion that cell position may play an important role (Mintz, 1964; Tarkowski and Wroblewska, 1967). However, more information on cell lineage in the intact embryo and cell potential outside the intact embryo (Section III) is required before the relative importance of cell lineage and cell position can be determined.

There is also very little information available to date on allocation of individual cells within the broad cell lineages outlined. For example, when cells become allocated to the ICM lineage, are individual cells within the lineage allocated at random, or are they restricted to certain "compartments" (Crick and Lawrence, 1975) within the lineage? All the indirect evidence available suggests random allocation. When whole ICMs are injected into blastocysts, mosaicism is widespread in resulting adult mice (Gardner, 1971, 1975), even when the contributions of the injected ICM are small. This widespread mosaicism plus the high degree of covariance between mosaicism in many different tissues (Ford *et al.*, 1975) suggests that clones derived from individual ICM cells are randomly distributed and extensively mixed prior to final tissue allocation. There is little direct information available on the clonal development of single ICM cells in chimeras to confim this supposition. Gardner (Gardner and Lyon, 1971) has

IV. Somatic Cell Lineages in Mammalian Chimeras

made chimeras by injection of single ICM cells into blastocysts and shown extensive coat colour mosaicism similar to that observed after injection of whole ICMs. No analysis of internal tissues from such chimeras has been reported. Analysis of various tissues of late gestation embryos derived from single ICM cell injections has shown no obvious "compartmentalisation" of contributions by single ICM cells (Gardner, 1975; J. Rossant and R. L. Gardner, unpublished). However, these studies have only been performed using GPI analysis of complete organs, which may be of mixed tissue origin and/or contaminated with blood. The study of single cell clones marked by an *in situ* marker will be necessary to reveal any patterns in the clonal development of single ICM cells. Such studies should also be performed on later primitive ectoderm cells.

F. Retrospective Analysis of Cell Allocation

The difficulties of directly investigating allocation of cells to definitive tissue lineages in the postimplantation embryo have led to various attempts to use the patterns of mosaicism in adult chimeras to estimate progenitor cell numbers for a given tissue and, hence, perhaps give some indication of the time when the lineage was set aside in the fetus. A small progenitor cell number implies an early time of allocation, perhaps prior to any biochemical differentiation, while a larger progenitor cell number implies allocation closer to the time of tissue differentiation. Such studies are beset with problems, as has been discussed extensively elsewhere (McLaren, 1976; West 1978a, b). Two different, but related, approaches have been taken to estimate progenitor cell number from mosaic patterns in aggregation chimaeras, both of which have been used previously in *Drosophila* gynandromorph studies (Merriam, 1978). These approaches I will term mosaic frequency estimation and minimum descendant clone size estimation, using the terminology proposed by West (1978b).

1. MOSAIC FREQUENCY ESTIMATION

This estimate is based on determining the percentage of cases in which a given tissue is nonmosaic in a chimera. Clearly, the smaller the number of progenitor cells at the time the lineage has been set aside, the higher the likelihood of nonmosaicism in the tissue. Such as approach can indeed be useful for estimating relative sizes of precursor pools. For example, Gearhart and Mintz (1972) showed that single somites were derived from more than one cell since they were always mosaic in the chimeras, while Oster-Granite and Gearhart (1981) reported that the hippocampal formation in the brain tends to be of predominantly one genotype or another in otherwise balanced chimeras, which suggested origin from a very small group of progenitor cells. However, even this crude approach has its flaws. First, it is not always easy to establish whether a tissue is completely nonmosaic, unless *in situ* markers are used. Second, it is not immune to the problems of cell selection. Strain-specific, tissue-specific cell selection has been reported in many chimeric combinations (Mintz, 1970a; West, 1977; Peterson et al., 1979) and could result in a higher percentage of apparent non-mosaicism in certain tissues. The effects of cell selection appear to be progressive throughout adult life (Warner et al., 1977; Gearhart and Oster-Granite, 1981). Detailed analysis of mosaicism in *Mus musculus*↔*M. caroli* chimeras revealed no evidence of such tissue-specific effects in neonatal chimeras, but deviations did appear in older animals (Rossant and Chapman, 1983). It may, therefore, be more valid to look at embryonic or neonatal chimeras in order to assess the relative precursor pool sizes of different tissues. Recent analysis of chimeras derived from congenic C57B16 strains, where minimal cell selection is expected, has demonstrated the importance of cell selection in determining genotypic composition (Behringer et al., 1984). The phenomenon of chimeric drift in blood cell genotypic composition was not observed in these mice. Also, the mosaic distribution of a large group of these mice was not flat (Falconer and Avery, 1978) but close to a normal distribution, with most mice showing large contributions from both strains.

All these problems and more apply to attempts to extend the mosaic frequency method to give a quantitative estimate of precursor pool sizes. This method is based on the binomial expansion and states that a tissue will be nonmosaic in $1/2^{n-1}$ cases if it originates from n cells. The chief additional problem with this method is that it depends on the assumption that cells of the two genotypes are in equal proportions and randomly mixed at the time of tissue allocation (Lewis et al., 1972; McLaren, 1972). Random mixing of cells is not observed in the early stages of development of chimeras (Garner and McLaren, 1974). The effects of size regulation, which occurs shortly after the blastocyst stage (Buehr and McLaren, 1974; Lewis and Rossant, 1982), compound the problems of such estimates.

It is not yet clear whether this method can be applied to cell allocation at later stages of development, because we do not know how long cell growth remains clonal in the embryo. If extensive cell mixing and size regulation occur before allocation of cells to major organ systems, this method may have some validity, although it would still be subject to the difficulties of accurately determining mosaicism and the effects of cell selection and cell death. Adjustment would also have to be made for any deviation from 50% in the average contribution of each genotype to the chimera. Only a few workers have attempted to use mosaic frequency estimation as a quantitative measure of progenitor cell number (e.g., Mintz, 1970b; Wegmann and Gilman, 1970), and there have been no recent attempts to overcome all the problems involved.

2. MINIMUM DESCENDANT CLONE SIZE ESTIMATION

Another related but potentially more useful method of estimating progenitor cell numbers relies not on the overall frequency of mosaicism of a given structure

IV. Somatic Cell Lineages in Mammalian Chimeras

but on the smallest proportion of that structure that is genetically marked in a mosaic animal. This technique was first used by Stern (1940) to estimate the number of progenitor cells for the mesonotal portion of the wing disc in *Drosophila* gynandromorphs and has been extensively used since in gynandromorph studies (Garcia-Bellido and Merriam, 1969; Ripoll, 1972; Lawrence and Morata, 1977). The chief assumption of this technique is that once the progenitors for a given tissue are set aside, they divide and undergo cell death at equal rates, so that they contribute equally to the resulting adult structure. If this is so, then, for any given structure, there should be some minimum possible contribution of one genotype, which represents the contribution from one progenitor cell. If the minimum contribution observed is $1/n$ of the total structure, then there should also be mosaics with contributions of $2/n$, $3/n \to n/n$. If both of these conditions hold, then the number of progenitor cells at the time of tissue allocation is n. Estimates must not rely on the minimum contribution alone, since this is likely to be derived from an infrequent class of mosaics. The main advantage of this method over the mosaic frequency method is that it is not so strongly dependent on the degree of cell mixing or the relative contributions of the two genotypes at the time of tissue allocation. These factors should only affect the frequency of the different mosaic classes, not their existence. Even if clonal growth is coherent until the time of tissue allocation, a large enough group of chimeras should reveal animals with several of the different classes of possible contributions, provided that the coherent clones themselves are randomly distributed at the time of allocation.

In *Drosophila*, the coherent growth of cells throughout development makes estimation of the minimum descendant clone size relatively easy at least in cuticular structures, but in mammals, the technique has been less readily employed because of the difficulty in observing descendant clones in adult tissues. Even in two-dimensional tissues, such as coat pigmentation and retinal pigment epithelium, where *in situ* cell identification can be performed, cell mixing is apparent, so that simple measurements of areas of different genotypes are not possible. However, in both the tissues mentioned, overall patterns of distribution of pigment can be detected (Mintz, 1967; Mintz and Sanyal, 1970), which West (1978b) called "zones." These may represent the outlines of descendant clones that have later been obscured by cell mixing. Transverse bands in coat pigmentation are apparent in chimeras and can be made more obvious by introducing genes that inhibit cell migration, such as W^v (Whitten, 1978), and these probably represent the descendant clones derived from melanoblast precursors. Calculation of the progenitor cell marker for the melanoblasts has varied from 17 to 34 for each side (Mintz, 1967; Wolpert and Gingell, 1970), but clear determination of the minimal descendant clone has been hampered by the differential growth of descendant clones in different regions of the body (West and McLaren, 1976) and the degree of cell mixing. In the retinal pigment epithelium, Sanyal and Zeilmaker (1977) estimate 10 progenitor cells based on the minimum size of radiating pigmented segments and apparent integral multiples of this. In all these studies, the exact relationship between the bands observed and descendant clones is not clear.

In other tissues that have so far been analysed in detail with *in situ* markers, no remnants of "zones" of growth have been observed. The general observation has been extensive cell mixing resulting in small patch size in adult tissues (Condamine et al., 1971; Dewey et al., 1976; West, 1976; Mullen, 1977; Rossant et al., 1983; Ponder et al., 1983). The average patch size observed has no bearing on the number of descendant clones but is simply a function of the number of cell divisions since cell mingling ceased and the proportion of cells of the two genotype in the tissue (see West, 1975, for extensive discussion). The cerebellar Purkinje (PC) cells have been well studied in this regard, because individual cells can be scored for genotype using β-glucuronidase histochemistry (Mullen, 1977; Mullen and Herrup, 1979), glucosephosphate isomerase (GPI) isozyme immunohistochemistry (Oster-Granite and Gearhart, 1981) or intrinsic degeneration of PCs in lurcher$\leftrightarrow$normal chimeras (Wetts and Herrup, 1982). All these studies have revealed extensive intermingling of genotypes, with estimates of between 1 and 4 cells per average coherent clone in the adult cerebellum. However, Wetts and Herrup (1983) have recently performed a careful quantitative analysis of PC genotype in both lurcher$\leftrightarrow$normal and normal$\leftrightarrow$normal chimeras marked by β-glucuronidase. They counted all the cells of either genotype in several chimeras and found the smallest number of PCs of one genotype in one-half of the cerebellum to be 10,200. The other chimeras showed PC numbers of one genotype which were much closer to integral values of this number than expected by chance, indicating that 10,200 cells may represent the progeny of one progenitor cell. Total cell numbers and the integral multiples observed suggested a progenitor cell number of about eight. Knowledge of the cell division time in the mouse neural tube and the time of the last cell division in PCs enabled them also to estimate that the time of allocation of these eight cells was the neural plate stage of development, long before the first morphological sign of cerebellar formation. In this particular tissue, therefore, neither differential cell growth nor cell death nor migration seem to have obscured the underlying descendant clones.

This approach can be applied to other relatively simple tissues using ubiquitous *in situ* markers, but calculation of the exact proportion of cells of the two genotypes in a complex three-dimensional organ such as the liver may be next to impossible. Other indirect cell markers such as GPI could be used to estimate the proportion of one genotype in a given tissue, but this approach is hardly likely to be accurate enough to reveal any underlying integral multiples of mosaic contributions, given the sensitivity of the technique and the problems of getting pure tissues. Also, of course, the larger the actual progenitor cell number, the more difficult it will be to distinguish integral classes of contributions of one genotype.

G. Summary

Both prospective and retrospective studies of cell allocation in mammalian embryos have produced a woefully incomplete picture, when compared with some of the detailed cell lineages available for invertebrate embryos. However, useful information on the general patterns of lineage allocation have been obtained from chimeras and other experimental systems. Direct cell marking has revealed nonrandom but not invariant allocation of cells during cleavage development. Beyond this stage, all studies have relied on chimera formation, which only approximates undisturbed development. Prospective study of cell fate in chimeras has shown that cells become spatially allocated to different lineages as development proceeds, but it has revealed little on the timing of lineage allocation, since cells cannot usually be recognised and manipulated reproducibly until after some form of morphological differentiation has occurred. Retrospective studies of chimeras are intrinsically more difficult to interpret since they are based on a series of assumptions of somewhat dubious validity and, at best, they can only tell us indirectly when a cell lineage is established and nothing about the spatial allocation of cells to this lineage. Both prospective and retrospective studies are also complicated by the extensive cell mixing that occurs in mammalian development. There are still many areas of lineage allocation that are unclear, but the general picture seems to be that cells are progressively allocated to broad cell lineages but that such allocation is not completely invariant; individual cells within a lineage do not have their fate rigidly defined until fairly late in development. The relative lability of cell fate observed suggests that lineage alone cannot explain embryonic development. Studies on the potential of cells outside the embryonic environment confirm this and suggest an important role for cell position in lineage development.

III. RESTRICTION OF CELLS TO DIFFERENT LINEAGES

A. Restriction of Cells to ICM or Trophectoderm Lineage

Studies on cell lineage in the intact embryo have revealed no definitive allocation of cells to ICM or trophectoderm before at least the 16-cell stage. Chimera studies and other experimental manipulation have confirmed that no restriction in cell potential occurs up to this stage. Normal embryonic development has been achieved from half-embryos in mice (Tarkowski, 1959) and even smaller portions in rabbits (Moore et al., 1968) and sheep (Willadsen, 1979; and personal communication). Aggregation of two (Tarkowski, 1961; Mintz, 1962), three (Markert and Petters, 1978), four (Petters and Markert, 1980), and even larger numbers of eight-cell embryos (Hillman et al., 1972) is also compatible with normal development. These experiments all attest to the regulative properties of the early embryo, but do not strictly eliminate the possibility of cells sorting according to lineage specificity, followed by size regulation. The absence of cell sorting in aggregation chimeras has been reported in several studies (Mintz, 1965; Garner and McLaren, 1974), making this unlikely. The potential of randomly isolated single four-, eight- (Kelly, 1977), and 16-cell blastomeres (Rossant and Vijh, 1980) has been tested by aggregating them with other marked cells. In all cases, at least some of the cells gave rise to both ICM and trophectoderm, indicating that they were not restricted to one lineage or another. Other experiments indicate that a cell's position in the aggregate can influence its future fate (Hillman et al., 1972). However, none of the experiments preclude the possibility of lineage restriction of some cells at each stage.

Clearly, in the absence of experiments demonstrating the totipotentiality of every cell from an embryo, it becomes essential to study the potential of cells isolated, not randomly, but according to their expected fate in the intact embryo. There is general agreement that inside cells from the 16-cell stage tend to end up as ICM and outside cells as trophectoderm, although the invariance of this relationship is still in doubt (Section II, A). Inside cells isolated from 16-cell embryos (Ziomek et al., 1982b) and late morulae and early blastocysts (Handyside, 1978; Spindle, 1978; Rossant and Lis, 1979b) have all been shown to be capable of producing trophectoderm derivatives *in vitro* and *in vivo*. Indeed, complete postimplantation embryos can be formed by groups of apolar (presumed inside) cells at the 16-cell stage (Ziomek et al., 1982b) and immunosurgically isolated inside cells at the early blastocyst stage (Rossant and Lis, 1979b). Similar results were obtained from polar, presumed outside cells from 16-cell embryos (Ziomek et al., 1982b) and microsurgically isolated outside cells from late morulae (Rossant and Vijh, 1980). All these experiments indicate that cells are not restricted to a given lineage prior to the blastocyst stage. Cell lability may actually persist beyond the time of normal lineage allocation in the embryo, as illustrated most clearly by the ability of early ICMs to form trophectoderm derivatives, which they have not yet been observed to do in their normal blastocyst environment (Rossant and Lis, 1979a).

Although complete lineage restriction is not observed during cleavage, the behaviour of inside and outside cells does differ at the 16-cell stage and beyond. When fluorescein-labelled cells were added to various aggregates of polar and apolar 16-cell blastomeres, it was found that labelled polar cells always contributed at least one of their progeny to the trophectoderm layer, whatever their initial position in the embryo (Ziomek and Johnson, 1982). Labelled apolar cells could contribute to either ICM or trophectoderm, or both, depending on their initial position in the embryo. These and other experiments (Kimber et al., 1982; Randle, 1982) indicate that changes in cell behaviour anticipating cell fate do occur during cleavage, although final restriction of cell lineage is not apparent until the mature blastocyst stage. Inner cell mass cells taken from mature 3.5 day blastocysts cannot regenerate trophectoderm or contribute to trophectoderm derivatives when aggregated with morulae (Rossant, 1975a, b, 1976). Mural trophectoderm cells are also restricted by the blastocyst stage; they cannot regenerate ICM (Gardner and Johnson, 1972). The potential of polar trophectoderm has not been assessed because it is technically impossible to isolate. The blastocyst stage represents, therefore, the first stage at which the potential of groups of cells is restricted to their later lineages.

B. Restriction of Cells to Primitive Endoderm or Ectoderm Lineages

The next morphological differentiative event to occur is the formation of the layer of primitive endoderm on the blastocoelic surface of the ICM. Various pieces of evidence have suggested that, as for ICM and trophectoderm, cells are not strictly preprogrammed to become primitive endoderm or ectoderm, but are influenced by their position in the ICM (Gardner and Johnson, 1975; Rossant, 1975b), outside cells forming primitive endoderm and enclosed ones forming primitive ectoderm. By the time of morphological differentiation of the two cell types, they have apparently both become restricted in their potential to their future fates (Gardner and Rossant, 1979; Gardner, 1982b), although some limited ability of primitive ectoderm to regenerate endoderm has been reported (Pedersen *et al.*, 1977; Dziadek, 1979; Atienza-Samols and Sherman, 1979).

Preliminary experiments analysing the potential of single ICM cells from expanded 3.5 day blastocysts revealed that, in chimeras, single cells could contribute to either primitive ectoderm or primitive endoderm derivatives but not to both (J. Rossant and R. L. Gardner, unpublished). At this stage there is no morphological evidence of endoderm production (Nadijcka and Hillman, 1974), indicating that lineage restriction and allocation in the intact embryo may precede differentiation. Experiments in which isolated ICMs were surrounded by either embryonal carcinoma cells or endoderm cells derived from teratocarcinomas, and yet continued to generate endoderm on their outer surfaces, also support this contention (Rossant and Rosenstrauss, 1984). Thus, either uncommitted ICM cells have responded to external cues and become determined by the mature blastocyst stage, or the ectoderm–endoderm distinction is a preprogrammed step in the lineage of ICM cells, perhaps resulting from an asymmetric division of all early ICM cells.

A single early ICM cell can indeed contribute to both primitive ectoderm and endoderm after aggregation with an 8-cell embryo (J. Rossant and R. L. Gardner, unpublished). However, when the patterns of colonization of the two daughter cells of a single early ICM cell were examined, they revealed that both daughter cells could colonize the same tissues in some cases, rather than one giving rise to primitive ectoderm and the other to primitive endoderm (R. L.

Gardner, personal communication). This makes preprogrammed unequal cell division an unlikely explanation for the results observed. If the opposite explanation of early response to cell position were true, then one would predict that the two populations of cells from mature ICMs that contribute separately to primitive ectoderm and endoderm should correspond to inner and outer cells of the ICM, respectively. Also, all early ICM cells should be able to be reprogrammed to primitive ectoderm development if surrounded by other cells. These predictions are currently being tested.

It is interesting that the stage of ICM development at which cells are still not restricted to primitive ectoderm or endoderm is also the stage at which they can still regenerate trophectoderm under certain conditions (Rossant and Lis, 1979b). It is possible, therefore, that loss of the ability to form trophectoderm is concomitant with the lineage restriction to primitive endoderm and ectoderm. Clearly, this area requires further research.

C. Restriction of Cells to Postimplantation Cell Lineages

The two lineage restrictions described so far have both coincided fairly closely in time with morphological differentiation, and probably with lineage allocation. Examination of events in early postimplantation development reveals a more complex picture. Primitive endoderm cells differentiate into visceral and parietal cells, which have clearly distinct properties (Hogan and Tilly, 1981), but visceral cells, when injected into blastocysts, retain the capacity to differentiate into parietal cells (Gardner, 1982b). It is not yet known whether the reverse is also true, or for how long this ability persists. In the trophectoderm lineage, the diploid ectoplacental cone and extraembryonic ectoderm differ in their protein synthetic profiles from each other and from the trophoblast giant cells (Johnson and Rossant, 1981), and the extraembryonic ectoderm, at any rate, is unlikely to give rise to giant cells in normal, undisturbed development. However, extraembryonic ectoderm from embryos as old as 8.5 days postcoitum can form trophoblast giant cells in ectopic sites and *in vitro* (Rossant and Ofer, 1977), and 6.5 day extraembryonic ectoderm can contribute to extraembryonic ectoderm, ectoplacental cone, and trophoblast giant cells in chimeras after blastocyst injection (Rossant *et al.*, 1978; J. Rossant, unpublished). These results have been interpreted in terms of a stem-cell model for trophoblast development, in which extraembryonic ectoderm cells act as potential stem cells for all other trophoblast types (Copp, 1979; Rossant and Lis, 1981).

In both of these extraembryonic lineages, therefore, cells retain considerable lability even after clear morphological and molecular differentiation of cell populations has occurred. Such lability allows for regulation for tissue damage, which may occur more often to these outer cell layers than to the embryo itself. Since both primitive endoderm and trophectoderm are terminal cell types that play no further part in development of the fetus itself, strict lineage restriction within their derivative cell populations may not be necessary. Indeed, it may be selectively disadvantageous since inability to repair damage to these tissues, which provide nutrients and protect the fetus, would threaten the very survival of the fetus itself.

The time of onset of lineage restriction may, therefore, be determined by a balance between the advantage of not being restricted, in terms of regulative capacity, and the disadvantage, in terms of the difficulties of maintaining normal differentiative and morphogenetic patterns without inherited programming of cells. One might predict that the more complex differentiative events occurring in the postimplantation embryo itself would require earlier and more rigid determination of cells. There are very few data on this point. Beddington (1982) has shown, by heterotopic grafts of primitive ectoderm cells in chimeras, that the potential of different regions of the embryonic ectoderm at the primitive streak stage is not restricted to their normal fate. Ectopic graft studies (Skreb et al., 1976) have revealed similar findings. This shows that lineage restriction is certainly not a very early event. However, since there is no obvious morphological distinction between the different areas of ectoderm studied and no biochemical distinctions have yet been made, it is not yet clear that lability persists after differentiation, as is clearly shown for both primitive endoderm and trophec-

TABLE I.

Approximate Timing of Lineage Development in the Mouse Embryo

Cell lineage	Time of allocation	Time of differentiation[a]	Time of restriction
ICM	16 cell → EB[b]	EB	LB
Trophectoderm	16 cell → EB	EB	LB
Primitive ectoderm	LB → 4.5d	4.5d	LB → 4.5d
Primitive endoderm	LB → 4.5d	4.5d	LB → 4.5d
Parietal endoderm	5.5d?	5.5d	?
Visceral endoderm	5.5d?	5.5d	Not before 6.5d
Extraembryonic ectoderm	5.5d?	At least by 7.5d	Not before 9d
Ectoplacental cone	5.5d?	At least by 7.5d	7.5d?
Neurectoderm	7.5d	8.5d	After 7.5d
Gut endoderm plus notochord	7.5d	8.5d	After 7.5d
Mesoderm	7.5d	7.5d on	After 7.5d
Cerebellar Purkinje cells	8.5d?	11–12d?	?

[a] Differentiation assessed by morphological and/or biochemical criteria.
[b] Stages of development: EB, 3.5 day blastocyst, not fully expanded; LB, 3.5 day blastocyst, fully expanded; 4.5d, blastocyst just prior to implantation, 4.5 days *postcoitum*; 5.5d, early egg cylinder; 7.5d, primitive streak stage; 8.5d, early somite stage; 9d, late somite stage, chorioallantoic placenta established.

toderm tissues. Studies of the potential of later organ rudiments will be necessary to determine this. There are no direct data available from chimeras on the time of commitment of the major organs or tissue types in the embryo.

Bringing together the available data on the time of lineage allocation versus the time of lineage restriction reveals that there is no set temporal relationship between them (Table I). They may be more or less simultaneous, as is probably the case for the primitive ectoderm and endoderm lineages; restriction may follow some time after allocation, as is the case for the ICM lineage; allocation and differentiation may occur without restriction, as is possibly the case within some of the extraembryonic lineages.

IV. CONCLUSIONS

Parallel studies of allocation and restriction of cells to different lineages have shown that irreversible and heritable restriction of groups of cells to different lineages does occur as development proceeds (Fig. 2). However, the fate of individual cells does not appear to be rigidly determined by their lineage. The heritable restrictions observed seem to facilitate correct differentiation within a lineage by determining the range of possible responses to any positional cue. For example, cleavage stage cells respond to inside or outside position by differentiating into ICM and trophectoderm, respectively, while ICM cells, faced with apparently similar conditions, will differentiate into primitive ectoderm and endoderm. The exact balance between the roles of cell lineage and cell position in

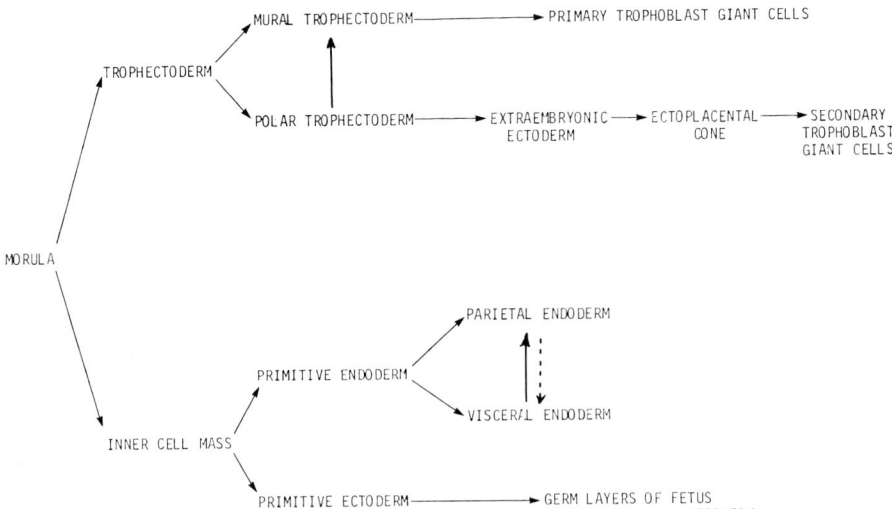

Fig. 2. Cell lineage restriction in the early mouse embryo.

mammalian embryonic development is still unclear, but further studies of cell lineage in the intact embryo and cell potential outside the normal embryonic environment should shed light on this problem. It is clear that a good understanding of the cellular events underlying cell lineage development is essential for investigations of the molecular mechanisms involved, and mammalian chimeras provide an invaluable tool for such studies.

ACKNOWLEDGMENTS

I should like to thank Dr. K. Herrup and Dr. R. A. Pedersen for useful discussion and Ms. M. Ferracuti for typing the manuscript. The author's own work described here was supported by the Canadian Natural Sciences and Engineering Research Council.

REFERENCES

Atienza-Somols, S. B., and Sherman, M. I. (1979). *J. Exp. Zool.* **208**, 67–96.
Balakier, H., and Pedersen, R. A. (1982). *Dev. Biol.* **90**, 352–362.
Beddington, R. S. P. (1981). *J. Embryol. Exp. Morphol.* **64**, 87–104.
Beddington, R. S. P. (1982). *J. Embryol. Exp. Morphol.* **69**, 265–285.
Behringer, R. R., Eldridge, P. W., and Dewey, M. J. (1984). *Dev. Biol.* **101**, 251–256.
Buehr, M., and McLaren, A. (1974). *J. Embryol. Exp. Morphol.* **31**, 229–234.
Condamine, H., Custer, R. P., and Mintz, B. (1971). *Proc. Natl. Acad. Sci. U.S.A.* **68**, 2032–2036.
Copp, A. J. (1979). *J. Embryol. Exp. Morphol.* **51**, 109–120.
Crick, F. H. C., and Lawrence, P. A. (1975). *Science* **189**, 340–347.
Davidson, E. H. (1976). "Gene Activity in Early Development," 2nd ed. Academic Press, New York.
Deppe, U., Schierenberg, E., Cole, T., Krieg, C., Schmitt, D., Yoder, B., and von Ehrenstein, G. (1978). *Proc. Natl. Acad. Sci. U.S.A.* **75**, 376–380.
Dewey, M. J., Gervais, A. G., and Mintz, B. (1976). *Dev. Biol.* **50**, 68–81.
Dziadek, M. (1979). *J. Embryol. Exp. Morphol.* **53**, 367–379.
Falconer, D. S., and Avery, P. J. (1978). *J. Embryol. Exp. Morphol.* **43**, 195–219.
Ford, C. E., Evans, E. P., and Gardner, R. L. (1975). *J. Embryol. Exp. Morphol.* **33**, 447–457.
Garcia-Bellido, A., and Merriam, J. R. (1969). *J. Exp. Zool.* **170**, 61–76.
Gardner, R. L. (1971). *Adv. Biosci.* **6**, 279–296.
Gardner, R. L. (1975). *Symp. Soc. Dev. Biol.* **33**, 207–238.
Gardner, R. L. (1978). *Results Probl. Cell Differ.* **9**, 205–242.
Gardner, R. L. (1982a). *In* "Cellular Controls in Differentiation" (C. W. Lloyd and D. A. Rees, eds.), pp. 257–278. Academic Press, London.
Gardner, R. L. (1982b). *J. Embryol. Exp. Morphol.* **68**, 175–198.
Gardner, R. L., and Johnson, M. H. (1972). *J. Embryol. Exp. Morphol.* **28**, 279–312.
Gardner, R. L., and Johnson, M. H. (1975). *Ciba Found. Symp.* **29**, 183–200.
Gardner, R. L., and Lyon, M. F. (1971). *Nature (London)* **231**, 385–386.
Gardner, R. L., and Rossant, J. (1979). *J. Embryol. Exp. Morphol.* **52**, 141–152.
Gardner, R. L., Papaioannou, V. E., and Barton, S. C. (1973). *J. Embryol. Exp. Morphol.* **30**, 561–572.
Garner, W., and McLaren, A. (1974). *J. Embryol. Exp. Morphol.* **32**, 495–503.
Gearhart, J. D., and Mintz, B. (1972). *Dev. Biol.* **29**, 27–37.

Gearhart, J. D., and Oster-Granite, M. L. (1981). *J. Hered.* **72,** 3–5.
Graham, C. F., and Deussen, Z. A. (1978). *J. Embryol. Exp. Morphol.* **48,** 53–72.
Graham, C. F., and Lehtonen, E. (1979). *J. Embryol. Exp. Morphol.* **49,** 277–294.
Handyside, A. H. (1978). *J. Embryol. Exp. Morphol.* **45,** 37–53.
Handyside, A. H. (1980). *J. Embryol. Exp. Morphol.* **60,** 99–116.
Hillman, N., Sherman, M. I., and Graham, C. F. (1972). *J. Embryol. Exp. Morphol.* **28,** 263–278.
Hirose, G., and Jacobson, M. (1979). *Dev. Biol.* **71,** 191–202.
Hogan, B. L. M., and Tilly, R. (1981). *J. Embryol. Exp. Morphol.* **62,** 379–394.
Jacobson, M., and Hirose, G. (1981). *J. Neurosci.* **1,** 271–284.
Johnson, M. H. (1981). *Biol. Rev. Cambridge Philos. Soc.* **56,** 463–498.
Johnson, M. H., and Rossant, J. (1981). *J. Embryol. Exp. Morphol.* **61,** 103–116.
Johnson, M. H., and Ziomek, C. A. (1981). *Cell* **24,** 71–80.
Johnson, M. H., Handyside, A. H., and Braude, P. R. (1977). *Dev. Mamm.* **2,** 67–98.
Kelly, S. J. (1977). *J. Exp. Zool.* **200,** 365–376.
Kelly, S. J., and Rossant, J. (1976). *J. Embryol. Exp. Morphol.* **35,** 95–106.
Kelly, S. J., Mulnard, J. G., and Graham, C. F. (1978). *J. Embryol. Exp. Morphol.* **48,** 37–51.
Kimber, S. J., Surani, M. A. H., and Barton S. C. (1982). *J. Embryol. Exp. Morphol.* **70,** 133–152.
Lawrence, P. A., and Morata, G. (1977). *Dev. Biol.* **56,** 40–51.
Lehtonen, E. (1980). *J. Embryol. Exp. Morphol.* **58,** 231–249.
Lewis, J. H., Summerbell, D., and Wolpert, L. (1972). *Nature (London)* **239,** 276–278.
Lewis, N. E., and Rossant, J. (1982). *J. Embryol. Exp. Morphol.* **72,** 169–181.
McLaren, A. (1972). *Nature (London)* **239,** 274–276.
McLaren, A. (1976). "Mammalian Chimaeras." Cambridge Univ. Press, London and New York.
Markert, C. L., and Petters, R. M. (1978). *Science* **202,** 56–58.
Merriam, J. R. (1978). *Results Probl. Cell Differ.* **9,** 71–96.
Mintz, B. (1962). *Am. Zool.* **2,** 432.
Mintz, B. (1964). *J. Exp. Zool.* **157,** 273–292.
Mintz, B. (1965). *Preimplantation Stages of Pregnancy, Ciba Found. Symp., 1965* pp. 194–207.
Mintz, B. (1967). *Proc. Natl. Acad. Sci. U.S.A.* **58,** 244–351.
Mintz, B. (1970a). *Annu. Symp. Fundam. Cancer Res. (Proc.)* **23,** 447–517.
Mintz, B. (1970b) *Symp. Int. Soc. Cell Biol.* **9,** 15–42.
Mintz, B., and Sanyal, S. (1970). *Genetics* **64,** Suppl., 43–44.
Moore, N. W., Adams, C. E., and Rowson, L. E. A. (1968). *J. Reprod. Fertil.* **17,** 527–531.
Mullen, R. J. (1977). *Nature (London)* **270,** 245–247.
Mullen, R. J., and Herrup, K. (1979). *In* "Neurogenetics: Genetic Approaches to the Nervous System" (X. O. Breakefield, ed.), pp. 173–196. Am. Elsevier, New York.
Nadijcka, M., and Hillman, N. (1974). *J. Embryol. Exp. Morphol.* **32,** 675–695.
Oster-Granite, M. L., and Gearhart, J. D. (1981). *Dev. Biol.* **85,** 199–208.
Papaioannou, V. E. (1982). *J. Embryol. Exp. Morphol.* **68,** 199–209.
Pedersen, R. A., Spindle, A. I., and Wiley, L. M. (1977). *Nature (London)* **270,** 435–437.
Peterson, A. C., Frair, P. M., Rayburn, H. R., and Gross, D. P. (1979). *Soc. Neurosci. Symp.* **4,** 258–273.
Petters, R. M., and Markert, C. L. (1980). *J. Hered.* **71,** 71–74.
Ponder, B. A. J., Wilkinson, M. M., and Wood, M. (1983). *J. Embryol. Exp. Morphol.* **76,** 83–93.
Randle, B. J. (1982). *J. Embryol. Exp. Morphol.* **70,** 261–278.
Reeve, W. J. D., and Ziomek, C. A. (1981). *J. Embryol. Exp. Morphol.* **62,** 339–350.
Reverberi, G. (1971). "Experimental Embryology of Marine and Freshwater Invertebrates." North-Holland Publ. Amsterdam.
Ripoll, P. (1972). *Wilhelm Roux' Arch. Entwicklungsmech. Org.* **169,** 200–215.
Rossant, J. (1975a). *J. Embryol. Exp. Morphol.* **33,** 979–990.
Rossant, J. (1975b). *J. Embryol. Exp. Morphol.* **33,** 991–1001.

IV. Somatic Cell Lineages in Mammalian Chimeras

Rossant, J. (1976). *J. Embryol. Exp. Morphol.* **36**, 163–174.
Rossant, J. (1977). *Dev. Mamm.* **2**, 119–150.
Rossant, J., and Chapman, V. E. (1983). *J. Embryol. Exp. Morphol.* **73**, 193–205.
Rossant, J., and Lis, W. T. (1979). *Dev. Biol.* **70**, 249–254.
Rossant, J., and Lis, W. T. (1979). *Dev. Biol.* **70**, 255–261.
Rossant, J., and Lis, W. T. (1981). *J. Embryol. Exp. Morphol.* **62**, 217–227.
Rossant, J., and Ofer, L. (1977). *J. Embryol. Exp. Morphol.* **39**, 183–194.
Rossant, J., and Papaioannou, V. E. (1977). *In* "Concepts in Mammalian Embryogenesis" (M. I. Sherman, ed.), pp. 1–36. MIT Press, Cambridge, Massachusetts.
Rossant, J., and Rosenstrauss, M. (1984). In preparation.
Rossant, J., and Vijh, K. M. (1980). *Dev. Biol.* **76**, 475–482.
Rossant, J., Alexandre, H. L., and Gardner, R. L. (1978). *J. Embryol. Exp. Morphol.* **48**, 239–247.
Rossant, J., Vijh, M., Siracusa, L. D., and Chapman, V. E. (1983). *J. Embryol. Exp. Morphol.* **73**, 179–191.
Sanyal, S., and Zeilmaker, G. H. (1977). *Nature (London)* **265**, 731–733.
Siracusa, L. D., Chapman, V. M., Bennett, K. L., Hastie, N. D., Pietras, D. F., and Rossant, J. (1983). *J. Embryol. Exp. Morphol.* **73**, 163–178.
Skreb,N., Svajger, A., and Levak-Svajger, B. (1976). *Ciba Found. Symp.* **40**, 27–46.
Spindle, A. I. (1978). *J. Exp. Zool.* **203**, 483–489.
Stern, C. (1940). *J. Morphol.* **67**, 107–122.
Sternberg, P. W., and Horwitz, H. R. (1982). *Dev. Biol.* **93**, 181–205.
Sulston, J., and Horwitz, R. (1977). *Dev. Biol.* **56**, 110–156.
Sulston, J. E., and White, J. (1980). *Dev. Biol.* **78**, 577–597.
Sulston, J. E., Schierenberg, E., White, J. G., and Thomson, J. N. (1983). *Dev. Biol.* **100**, 64–119.
Tarkowski, A. K. (1959). *Nature (London)* **184**, 1286–1287.
Tarkowski, A. K. (1961). *Nature (London)* **190**, 857–860.
Tarkowski, A. K., and Wroblewska, J. (1967). *J. Embryol. Exp. Morphol.* **18**, 155–180.
Warner, C. M., McIvor, J. L., and Stephens, T. J. (1977). *Transplantation* **24**, 183–193.
Wegmann, T. G., and Gilman, J. G. (1970). *Dev. Biol.* **21**, 281–291.
Weisblat, D. A., Zackson, S. C., Blair, S. S., and Young, J. D. (1980). *Science* **209**, 1538–1541.
Weiss, P. (1939). "Principles of Development." Holt, New York.
West, J. D. (1975). *J. Theor. Biol.* **50**, 153–160.
West, J. D. (1976). *J. Embryol. Exp. Morphol.* **36**, 151–161.
West, J. D. (1977). *Exp. Hematol.* **5**, 1–7.
West, J. D. (1978). *Dev. Mamm.* **3**, 413–460.
West, J. D. (1978). *In* "Genetic Mosaics and Chimeras in Mammals" (L. Russell, ed.), pp. 435–444. Plenum, New York.
West, J. D., and McLaren, A. (1976). *J. Embryol. Exp. Morphol.* **35**, 87–93.
Wetts, R., and Herrup, K. (1982). *J. Embryol. Exp. Morphol.* **68**, 87–98.
Wetts, R., and Herrup, K. (1983). *J. Neurosci.* **2**, 1494–1498.
Whitten, W. K. (1978). *In* "Genetic Mosaics and Chimeras in Mammals" (L. Russell, ed.), pp. 445–463. Plenum, New York.
Willadsen, S. M. (1979). *Nature (London)* **277**, 298–300.
Wilson, E. B. (1925). "The Cell in Development and Heredity." Macmillan, New York.
Wilson, I. B., Bolton, E., and Cuttler, R. H. (1972). *J. Embryol. Exp. Morphol.* **27**, 467–479.
Wolpert, L., and Gingell, D. (1970). *J. Theor. Biol.* **29**, 147–150.
Zackson, S. L. (1982). *Cell* **31**, 761–770.
Ziomek, C. A., and Johnson, M. H. (1981). *Wilhelm Roux' Arch. Dev. Biol.* **190**, 287–296.
Ziomek, C. A., and Johnson, M. H. (1982). *Dev. Biol.* **91**, 440–447.
Ziomek, C. A., Pratt, H. P. M., and Johnson, M. H. (1982a). *Symp. Br. Soc. Cell Biol.* **5**, 149–165.
Ziomek, C. A., Johnson, M. H., and Handyside, A. H. (1982b). *J. Exp. Zool.* **221**, 345–355.

CHAPTER V

Germ Cell Lineages

ANNE McLAREN

MCR Mammalian Development Unit
University College London
London, England

    I. Germ Cell Lineages 111
        A. Insects 111
        B. Amphibia 113
        C. Birds 114
        D. Mammals 115
   II. Conclusions 126
        References 127

I. GERM CELL LINEAGES

Experimental chimeras have been used to explore two aspects of the development of germ cells. Finding the origin of the germ cell lineage, including answering the question of whether a continuous germ line exists from the fertilized egg through to the adult, requires the use of germ cell markers for its investigation: genetically marked cell populations, brought together in a chimeric embryo, are particularly appropriate for this purpose. Second, some light has been thrown on the development of germ cells in the gonad, and the role that their own genetic constitution plays in determining their phenotypic sex, by studies of chimeras in which germ cells of one genetic type colonize a gonad of contrasting type. These two aspects will be treated together in this chapter, since the same experiment often turns out to be relevant to both questions.

A. Insects

Germ cells in *Drosophila* originate from the cytologically distinct pole cells that are located at the posterior pole of the blastoderm. That it was the cytoplasm in this region that was crucial for germ line development, rather than the nuclei that migrated into it, was suggested by the demonstration (Illmensee and Mahowald, 1974, 1976) that injection of posterior pole cytoplasm, containing characteristic polar granules, into the anterior pole or the midventral region of the cleavage embryo was followed by the formation of cells in the ectopic site that were indistinguishable in appearance from normal posterior pole cells.

However appearance was not enough. To show that the cytoplasm actually contained germ-line determinants, it was necessary to establish that the ectopically formed pole cells were capable of giving rise to normal fertile gametes. This was done by transplanting them to the posterior pole of another embryo, using genetic markers to distinguish donor from host gametes, and also to guard against the possibility that some nuclei had been transferred along with the cytoplasm from the original donor. When the chimeric flies bred, some of their gametes could be shown to originate from the transplanted cells, induced to develop as pole cells by the injected posterior pole cytoplasm (Illmensee and Mahowald, 1974, 1976). Control transfers of anterior or ventral cells that had not been exposed to posterior pole cytoplasm never resulted in germ line chimerism, but gave rise only to somatic structures.

Donor cells in the chimeric flies could be identified histochemically, as well as by breeding tests. After transplantation of ectopic "pole" cells, cells of donor origin were identified not only in the gonads, but also in larval and adult midgut tissue. If these represented functional gut cells, rather than merely presumptive germ cells that had become trapped in the gut during their migration to the gonads, it would suggest that pole cells normally have somatic as well as germinal descendants, i.e., that the germ cell and somatic lineages are not entirely separate in the *Drosophila* embryo.

The above experiments were done using posterior pole cytoplasm from fertilized eggs. To discover when the capacity to determine germ cells first appears during oogenesis, Illmensee *et al.* (1976) took posterior pole cytoplasm first from unfertilized eggs, and then from increasingly early stages of oogenesis, and repeated the experiment, injecting the cytoplasm into the anterior tip of cleaving embryos. It turned out that progeny of donor origin were only observed when the injected cytoplasm was derived from oocytes of stage 13 and later, i.e., after the breakdown of the nurse chamber and the release of its contents into the oocyte. When earlier stages were used, the donor cells differentiated into anterior cuticular structures, but no germ-line chimerism was found.

To discover whether the phenotypic sex of *Drosophila* germ cells was determined entirely by their own chromosome constitution, or whether sex reversal was possible, van Deusen (1976) transplanted genetically marked pole cells at random into host embryos. Of 15 germ-line chimeras detected, 7 were female and 8 were male, and in every case the chromosomal sex of the donor component was shown by progeny testing to be the same as that of the host. Female donor cells with an attached Y chromosome were used (maleness in *Drosophila* de-

V. Germ Cell Lineages

pends on the ratio of X chromosomes to autosomes, not on the presence of a Y chromosome), so that if sex reversal had occurred they would have been capable of forming motile sperm. The absence of any "heterosexual" chimeras suggests strongly that chromosomally male germ cells in an ovary, and female germ cells in a testis, cannot differentiate into functional gametes of the opposite sex.

B. Amphibia

In Anura, the germ cell lineage can be traced back to the early embryo. Cytologically distinguishable "germinal cytoplasm" acts as a marker, though whether or not it actually constitutes a germ cell determinant is still debated (see Smith et al., 1983). This characteristically staining cytoplasm is first seen in the fertilized egg and is segregated during cleavage, so that at the blastula stage the cells containing it are concentrated in the ventral region. The descendants of these cells are located in the endoderm at the neurula stage; when reciprocal grafts of this region were made between *Xenopus laevis* embryos distinguished by a nuclear marker, the donor cells gave rise to functional gametes, and no functional germ cells entered the gonads from any other source (Blackler and Fischberg, 1961; Blackler, 1962), proving the continuity of the germ cell lineage from the fertilized egg to the mature gametes.

The germ cell chimeras that resulted from these grafting experiments established that at least some aspects of the phenotype of the egg are determined by its own genotype and not by the environment of the ovary in which it develops. When grafts were made between subspecies of *Xenopus* in which the eggs differed in both size and colour, eggs of both host and graft phenotype were ovulated by the chimeric toads, with no intermediates (Blackler, 1962).

On the other hand, the direction of sexual differentiation of the germ cells is determined not by the genetic sex of the germ cells themselves, but by the sex of the gonad into which they migrate. This was shown for *Rana sylvatica* by Humphrey (1933), using a different embryo grafting technique to produce intersexual chimeras, and for *Xenopus laevis* by Blackler (1965) (erroneously cited as Blackler, 1962, in McLaren, 1976). In *Xenopus* the male is homozygous (ZZ) for the sex chromosomes, and the female is heterozygous (ZW). Blackler found that in 6 out of 11 female chimeras, the donor progeny (distinguished by a nuclear marker, sometimes also by egg colour) were all male, as would be expected if the ovary had been colonized by ZZ germ cells; while 2 out of 5 male chimeras produced a ratio of female to male donor progeny consistent with that expected from a testis populated by ZW germ cells (i.e., 3:1 since WW females are viable). The progeny of host origin included equal numbers of males and females, as also did the progeny of donor origin in the other 8 germ cell chimeras, in which presumably donor and host had been of the same genetic sex.

Blackler (1965) concluded from his studies on germ cell chimeras that functional gametes of both sexes can be formed from primordial germ cells of either genetic sex, according to the gonadal environment in which they develop, and hence that the sex chromosomes in *Xenopus* play no essential role in gametogenesis.

The existence of a continuous germ cell lineage in nonanuran amphibians is open to doubt. In urodeles, little or no evidence of germ plasm can be found, and the primordial germ cells appear to originate from the epiblast rather than from the endoderm. A recombination experiment in which embryonic chimeras were made by grafting animal pole tissue from an embryo of one species onto vegetal pole tissue of another species established that the primordial germ cell arose from the animal pole cells, though the vegetal pole tissue needed to be present as an inducer (see Nieuwkoop and Satasurya, 1979). As we shall see, with respect to the origin of the germ cell lineage, birds and mammals seem to resemble urodeles more than anurans.

C. Birds

The origin of primordial germ cells in birds has been explored by Eyal-Giladi *et al.* (1981), using chick-quail chimeras. At the blastoderm stage, the hypoblast of one species was combined with the epiblast of the other; the chimeric embryo was then cultured to the 3–6 somite stage, and the primordial germ cells examined in histological sections for the presence of the characteristic quail nuclear marker. It turned out that the overwhelming majority of germ cells were from whichever species had provided the epiblast. The authors conclude that the germ line in birds must be of epiblastic origin, rather than hypoblastic as had previously been believed. This conclusion has, however, been questioned (see Smith *et al.*, 1983), on the grounds that the germ line progenitors could have entered the epiblast from the hypoblast prior to chimera formation.

Germ cell chimerism can be produced in avian embryos either by grafting or by injecting primordial germ cells into the blood vessels of a host embryo, since avian germ cells normally travel in the bloodstream for part of their migration route to the genital ridges. These techniques have been used to show that the chemotactic attraction exerted by the early chick gonad on germ cells of its own species operates similarly on germ cells of both turkey (Reynaud, 1969) and quail (Tachinante, 1974a), irrespective of the sex of the germ cell. Even mouse primordial germ cells appear to migrate toward the chick gonad when grafts of mouse hind gut (Rogulska *et al.*, 1971) or gonad (Tachinante, 1974b) are placed in the coelomic cavity of early chick embryos. Quail gonads also exert a chemotactic attraction on chick germ cells in a chimeric embryo (Didier, 1977).

The further differentiation of germ cells in chimeric gonads was studied by Haffen (1968, 1975), who was able to rear reciprocally grafted chick embryos to the stage of sexual differentiation by transferring them to a succession of host embryos. In some of the reciprocal pairs, one chimeric embryo developed an ovary and the other a testis, indicating that the original grafts must have been between embryos of opposite sex. The direction of sexual differentiation of the gonads appeared to be determined by the chromosomal sex of the somatic component rather than by that of the germ cells (see Chapter XV). In the ovaries of such unlike-sexed pairs, the germ cells degenerated shortly after entering meiosis. These were presumably chromosomally male (ZZ) germ cells that had migrated into the gonad from the grafted germinal crescent region. Haffen concluded that, in the chick, male germ cells were incapable of undergoing oogenesis in an ovary. Chromosomally female (ZW) germ cells in a chimeric testis did not degenerate, but the embryos could not be maintained long enough to find out whether they would have successfully entered meiosis at the time appropriate for male germ cells. It is known, however, that ZW germ cells can give rise to functional spermatozoa when the left rudimentary gonad of a female chick is induced to develop as a testis after removal of the right ovary.

Female germ cells normally enter meiosis at 16 days of incubation in the chick, 11 days in the quail. Hajji *et al.* (1984) made chick-quail chimeras by grafting pieces of embryo *in ovo* at 2 days of incubation, using recessive sex-linked albino mutations to ascertain the genetic sex of the associated embryos. Quail germ cells that had colonized a chick ovary began to enter meiosis at 12–13 days of incubation, well before the normal time for the chick. The delay relative to quail controls could presumably be accounted for by the disturbance caused by the experimental procedure. Hajji *et al.* concluded that the ovarian stroma plays a permissive role, allowing germ cells to enter meiosis during embryonic life, in contrast to the inhibitory influence of the embryonic testis; but the actual timing of entry into meiosis must be programmed in the germ cells themselves.

D. Mammals

1. TRACING THE GERM LINE

Primordial germ cells can first be distinguished at about 8 days *postcoitum* in mice, and at equivalent stages in some other mammals, by their relatively high alkaline phosphatase activity. Before this time, the germ cell lineage cannot be identified, as no marker such as the polar granules of *Drosophila* or the germinal cytoplasm of *Anura* has yet been found to characterize the mammalian germ line. The only available experimental approach is to check whether individual cells from an early embryo have the potential to give rise to functional germ cells in a chimera.

Using this approach, Kelly (1977) disaggregated mouse embryos at the 4-cell stage, and reaggregated each of the four blastomeres with a separate, genetically marked carrier embryo. In at least two instances, 3 of the 4 blastomeres of a single embryo contributed to the germ line as well as to the somatic tissues of the resulting chimeric mice; none contributed to the germ line alone. Gardner (1977) took single donor cells from the inner cell mass of a mouse blastocyst $3\frac{1}{2}$ days *postcoitum*, or from the epiblast 24 hours later, and injected them into genetically marked $3\frac{1}{2}$-day blastocysts. The donor tissues contained a total of about 15 and 20 cells, respectively, at the stage when the cells for injection were taken. Of the resulting somatic chimeras that could validly be tested for a germ-line contribution, 4/6 from $3\frac{1}{2}$-day and 2/7 from $4\frac{1}{2}$-day donors proved also to be germ-line chimeras. Again, no germ-line chimeras without a somatic contribution were detected.

Thus if a germ-line determinant exists in mice, it cannot be segregated at either the first or second cleavage divisions; indeed, it must be present in a substantial proportion of inner cell mass and early epiblast cells, perhaps in all, and is not yet segregated to give a "germ cell only" lineage. It seems more likely (for discussion, see McLaren, 1981a) that no germ-line determinant in the "*Drosophila* pole plasm" sense exists in mammals, and that every epiblast cell retains up to the time of gastrulation the potential to develop either as a somatic or as a germ cell, depending on what position in the egg cylinder it occupies.

What about tissues other than epiblast? At one time, primordial germ cells were believed to originate in the yolk sac endoderm, a derivative of the primary endoderm; however Gardner and Rossant (1979) conclusively showed that injection chimeras had donor contributions to the germ cell lineage if the donor cells were taken from the epiblast at $4\frac{1}{2}$ days *postcoitum*, but not if they were taken from the primary endoderm at the same stage. This result confirmed the conclusion reached by Falconer and Avery (1978), who observed that in all published accounts of mouse aggregation chimeras, the proportions of the two chimera components in the germ cells were positively correlated with the proportions in somatic tissues known to be derived from the epiblast. Because of the geometry of the blastocyst and the lack of cell mingling between the two aggregated embryos, this result would be expected if the germ cells too were derived from the epiblast but not if they came from the primary or yolk sac endoderm.

2. "RESCUE" BY CHIMERISM

The use of chimeras to test the ability of cells to contribute to the germ line can throw light on the developmental potency not only of normal embryonic cells but also of those that would otherwise be prevented by genetic or other reasons from giving rise to functional gametes.

Unfertilized mouse eggs can be induced to develop parthenogenetically, without the intervention of a fertilizing sperm, by a variety of activating stimuli. The resulting parthenogenetic embryos never develop beyond mid-gestation. However, chimeras made between diploid parthenogenetic and normal embryos show

V. Germ Cell Lineages

evidence of participation of the parthenogenetic component not only in the somatic tissues of the adult (Stevens *et al.*, 1977; Surani *et al.*, 1977) but also in the germ line (Stevens, 1978). The demonstration that a female chimera of this type can produce fertile eggs from the parthenogenetic component proves that fertilization is not a necessary condition for the subsequent formation of germ cells, and that the oocyte itself is potentially totipotent in the sense that it can provide all that is required for the development of any cell type in the female body, including more oocytes.

Embryonic tumours (teratocarcinomas) can be tested for developmental potency in a similar way. Mintz and Illmensee (1975) took groups of five stem cells from a teratocarcinoma that had been derived some years earlier by transplanting a $6\frac{1}{2}$-day mouse embryo beneath the kidney capsule of a suitable host; they injected the stem cells into genetically marked blastocysts, and obtained adult chimeras in which the teratocarcinoma donor cells could be shown to have made a normal contribution to many somatic tissues. One chimeric male proved on mating to produce spermatozoa only of the teratocarcinoma donor genotype.

The first genetic mutant to be rescued by chimerism was *Jimpy* (*Jp*), an X-linked mutant that is lethal in hemizygous males. Aggregation chimeras were made between embryos from normal females and females carrying *Jimpy;* the viable X-linked marker *Tabby* was used to identify overt chimeras (Eicher and Hoppe, 1973). One male chimera proved on breeding to transmit the X chromosome bearing *Tabby* and *Jimpy;* thus *Jp*/Y cells must have been able not only to survive in the testis, but also to form functional spermatozoa. The existence of this male made possible for the first time a test of allelism between *Jimpy* and another X-linked lethal, *Myelin synthesis deficiency* (*Msd*). Abnormal offspring were produced whether the chimeric male was mated to *Jp*/+ or to *Msd*/+ females; the two mutants were therefore judged to be allelic.

XY mice carrying another X-linked mutant, *Testicular feminization* (*Tfm*), are viable but sterile. Because their cells are genetically incapable of responding to androgen, they develop testes but no other male characteristics. They are therefore phenotypically female, and spermatogenesis is arrested at the first meiotic division. Aggregation chimeras with a *Tfm*/Y as well as a normal XY component proved, however, to produce two populations of functional spermatozoa, *Tfm*-bearing as well as normal (Lyon *et al.*, 1975). Such a chimaeric male could be mated with a *Tfm*/+ female to produce mice homozygous for *Tfm*. Since the *Tfm*/Y germ cells were presumably as refractory to androgens in the chimeric testis as they were in the testis of a *Tfm*/Y male, the successful completion of spermatogenesis in the chimera suggests that the determining role of testosterone operates not on the male germ cells directly, but on some somatic component of the testis, probably the Sertoli cells. In the chimeric testis, a *Tfm*/Y spermatocyte could then only complete spermatogenesis if it was associated with a normal Sertoli cell.

Recessive mutations at the complex T/t locus that are lethal or semilethal in homozygous condition also affect sperm production and function. Male mice that carry two different alleles (t^x/t^y) may be viable but sterile; on the other hand, a single allele in heterozygous condition ($t^x/+$) is usually transmitted to more than 50% of the progeny. Papaioannou et al. (1979) have investigated the basis for this transmission ratio distortion, and for the sterility of the t^x/t^y genotype, using chimeras produced by aggregating embryos heterozygous for two different recessive mutations, t^{w2} and t^{w5}. Unlike t^{w2}/t^{w5} males that are always sterile, the two $t^{w2}/+ \leftrightarrow t^{w5}/+$ male chimeras identified were of normal fertility; moreover they transmitted each t mutation with the same degree of distortion that is characteristic of $t/+$ heterozygous males. Since the spermatogenic cells derived from the $t^{w2}/+$ and $t^{w5}/+$ components thus appear to function quite independently of one another in the chimeric testis, the authors conclude that sterility and distorted transmission ratios both depend on specific interactions between T/t alleles within diploid spermatogenic cells, or among postmeiotic derivatives of the same germ cell. The conclusion with respect to transmission ratios must, however, be regarded as a preliminary one, in view of the small scale of the study and the wide variations in the proportions of progeny derived from the two components in normal XY↔XY chimeras (see below, also McLaren, 1975a).

An unexpected finding turned up when interspecific chimeras made between *Mus musculus* and *Mus caroli* were bred (Rossant and Chapman, 1983). Three female germ-line chimeras were identified; these produced no litters when inseminated either naturally or artificially with *M. caroli* sperm, but mating with *M. musculus* males was successful. Pure *musculus* progeny numbered 122, compared with 15 hybrids; but whereas the *musculus* progeny showed a normal 1: 1 sex ratio, the 15 hybrid progeny were all male, both phenotypically and chromosomally. The reason for this striking departure from Haldane's rule, according to which the inviable or infertile sex in a distant cross is the heterogametic one, is not known. The chimeras gave a much higher rate of hybrid production than did earlier attempts to produce hybrids by artificially inseminating *M. musculus* females with *M. caroli* spermatozoa. This might reflect a difference in fertility or viability according to which species provides the egg and which the spermatozoon; alternatively, if the hybrid fetuses in *M. musculus* mothers were lost for immunological reasons, it might represent "rescue" of the hybrid genotype by an immunologically tolerant chimeric mother.

3. GERM CELL MIGRATION

In birds, as we have seen, primordial germ cells are carried round in the bloodstream in the course of their migration to the genital ridges, so that injection of foreign germ cells into the bloodstream at the appropriate time yields embryos that are chimeric with respect to the germ line but not to any somatic tissue other than blood. No comparable manoeuvre to obtain "pure" germinal chimeras has been devised in mammals, though a spontaneous human chimera is known (Mayr

V. Germ Cell Lineages

et al., 1979; see Chapter VII) in whom chimerism was present only in the germ line. She was identified by the lack of concordance between her own blood group and those of her progeny, and showed no evidence of chimerism in blood, or in skin fibroblasts.

Whether in mammals any primordial germ cells reach the genital ridges via the bloodstream rather than via the solid tissues of the hind gut is still an open question. Convincing pictures of what appear to be germ cells within foetal blood vessels have been published (e.g., Wartenberg, 1983). In those species where twinning is accompanied by vascular anastomosis (cattle, marmoset monkeys), karyotyping germ cells in the gonads of heterosexual twin pairs has yielded reports of XX primary spermatocytes at diakinesis–metaphase in the testes of XX/XY bull calves (e.g., Teplitz *et al.,* 1967), while in marmosets XY oocytes have been reported in females and XX spermatogonia and spermatocytes in males (e.g., Benirschke and Brownhill, 1963; Hampton, 1970). On the other hand, an extensive study of testicular preparations from seven XX/XY bulls twinned to freemartins, and one XX/XY male marmoset, failed to find any XX spermatogonia or primary spermatocytes (Ford and Evans, 1977), and raised the question as to whether the mitotic XX spreads previously identified were really from spermatogonia, and whether the meiotic cells were misidentified as XX rather than XY.

As we shall see later, studies on chimeric and mosaic mice suggest that XX germ cells in a testis degenerate shortly after birth, and certainly play no part in spermatogenesis. This finding throws into question the view that the strong female bias observed in some but not all progenies of bulls twinned with freemartins (for references, see Ford and Evans, 1977) is due to the persistence of XX germ cells migrating into the gonad via the bloodstream during foetal life. If germ cells of one twin really colonize the gonads of the other, progeny testing of bulls born twin to a genetically distinguishable bull calf should reveal progeny sired by the "wrong" twin, but studies so far have failed to find any such progeny (Stone *et al.,* 1960, 1964).

Persistence of XY germ cells in an ovary would not be in conflict with the findings in mice (see next section); however, XY cells detected in freemartin gonads (see Ford and Evans, 1977) are unlikely to represent germ cells derived from the male co-twin, since the chromosomes did not resemble those characteristic of spermatogonia, and oogonia do not normally survive in the masculinized freemartin ovary.

4. FATE OF GERM CELLS IN THE GONADS

Once the mammalian germ cells have colonized the genital ridges, they continue to proliferate mitotically for a period. In the female mouse, germ cells enter the prophase of meiosis after about 3 days of mitotic activity, and ovarian follicles form at about the time of birth. In the male, germ cell proliferation again ceases after about 3 days, but instead of entering meiosis the germ cells remain in arrest and do not resume mitosis until a couple of days after birth; meiotic figures are first seen about a week later.

So much for the normal course of events, where chromosomally XX germ cells develop in an ovary and XY germ cells in a testis. But what happens in the gonad of an XX↔XY chimera, which usually (see Chapter XV) develops as a testis but occasionally as an ovary or an ovotestis? Both XX and XY germ cells presumably colonize the gonad: do they then develop according to their own chromosomal sex or according to their somatic environment?

a. XX↔XY females. Female XX↔XY chimeras are uncommon, and breeding records have so far been published for only seven (Ford et al., 1975; Evans et al., 1977; Gearhart and Oster-Granite, 1981). Of 348 progeny examined, 347 were derived from the XX component, but one on genetical grounds could only have come from the XY component. The exceptional animal was a male, who proved to be XXY in chromosome constitution, with a Y chromosome characteristic of the XY chimera component (Ford et al., 1975). Thus an XY egg, derived either by meiotic nondisjunction from an XY germ cell or by earlier mitotic nondisjunction giving rise to an XXY cell line, must have been fertilized by an X-bearing spermatozoon. The ability of an XY germ cell to undergo functional sex reversal was confirmed when Evans et al. (1977) identified an XY oocyte at diakinesis in a cytological preparation from another fertile XX↔XY female.

Further confirmation that the presence in a germ cell of at least the testis-determining region of the Y chromosome does not prevent it from undergoing oogenesis comes from recent studies on *Sex-reversed (Sxr)* in mice. Female *Sxr* carriers (McLaren and Monk, 1982; Cattanach et al., 1982) that have this region translocated to the X chromosome and subsequently inactivated in a proportion of somatic cells, are fully fertile, with no evidence that their reproductive life is curtailed (A. McLaren, unpublished observation). Gordon (1976) made aggregation chimeras between embryos fathered by XY *Sxr* nonalbino carriers and embryos from normal albino matings. He failed to observe any progeny derived from an XX *Sxr* female germ cell, and concluded that the presence of *Sxr* must interfere with oogenesis. However, no independent evidence was available as to whether or not an XX *Sxr* component was present in any of the chimeric mice. Of the six female chimeras, the two that produced progeny only from the albino component were assumed to be XX↔XX *Sxr*, but this is an ill-grounded assumption, since single-component progenies are not uncommon among normal XX↔XX chimeras. The other 4 produced 1, 6, 11, and 72 young from the nonalbino component; failure to detect an XX *Sxr* son (expectation $\frac{1}{4}$) can only be regarded as compelling evidence for an XX rather than an XX *Sxr* nonalbino

V. Germ Cell Lineages

component in the last of these females. Bradbury (1983) has recently repeated Gordon's experiment and has succeeded in obtaining progeny from XX *Sxr* oocytes of an XX↔XX *Sxr* female chimera. Thus, the fate of XX *Sxr* germ cells in a chimera would be similar to that in an XX *Sxr* individual, namely, that in a testis they degenerate but in an ovary they give rise to fertile oocytes.

Even if the presence of an entire Y chromosome does not hinder a germ cell from undergoing oogenesis, XY relative to XX germ cells in an XX↔XY ovary may well be at a strong disadvantage because of the lack of a second X chromosome. Ford *et al.* (1975) and Evans *et al.* (1977) both argued that XY oocytes in an XX↔XY ovary would rarely be detected, even if they were at no selective disadvantage relative to XX germ cells, because only chimeras with a minority of XY cells develop as fertile females (see Chapter XV), also a Y-bearing oocyte fertilized by a Y-bearing spermatozoon would not survive. Since, however, both X chromosomes of an XX germ cell are expressed during oogenesis, presence of but a single X is likely to be disadvantageous, and indeed XO germ cells not only produce subnormal oocytes but also degenerate in abnormally large numbers soon after birth (see Burgoyne and Baker, 1981). The present tally of XY-derived progeny from XX↔XY females (1/348) may therefore be taken as real evidence for the selective disadvantage of XY relative to XX germ cells.

b. XX↔XY males. XX↔XY chimeras usually develop as males (see Chapter XV). Early studies established that the XX germ cell population did not contribute functional spermatozoa: circumstantial evidence was provided by the normal 1 : 1 sex ratio among chimera progeny (Mintz, 1968), and direct evidence by the absence of any young derived from the XX component of known XX↔XY male mice carrying genetic markers (Mystkowska and Tarkowski, 1968; McLaren, 1975a). Two XX↔XY chimeric rams only produced functional spermatozoa from one, presumably XY, component (Tucker *et al.*, 1978). XX germ cells not only fail to form spermatozoa, but are not even to be found among primary spermatocytes (Mystkowska and Tarkowski, 1968). By analogy with the situation in XX *Sxr* males, in which all the germ cells carry two X chromosomes, the great majority of XX germ cells in a testis start to develop in the male direction before birth, appearing as normal prospermatogonia in mitotic arrest within the testis cords, but degenerate shortly after birth at about the time that XY germ cells would be resuming mitotic proliferation (see McLaren, 1983). All the available evidence suggests that XX germ cells are incapable of undergoing spermatogenesis in mammals.

Mystkowska and Tarkowski (1970) were struck by the unexpected observation that the testes of male XX↔XY chimeras often contained some germ cells in meiotic prophase before birth, at a stage of foetal development when in the ovary the germ cells had entered meiosis, but in the normal testis only prospermatogonia in mitotic arrest were to be seen. Most of these meiotic germ cells degenerated before birth, but in one male chimera some growing oocytes were found in the testis as late as 5 days after birth. The authors speculated that the meiotic cells could consist either of XX germ cells that were entering meiosis autonomously, at the same time that they would have in an ovary, or alternatively of a mixed population of XX and XY germ cells induced to enter meiosis by some stimulus deriving from a local concentration of XX somatic cells in the chimeric gonad.

The observation of meiotic germ cells in the testes of foetal chimeras was confirmed by McLaren et al. (1972). At $16\frac{1}{2}$ and $17\frac{1}{2}$ days *postcoitum* zygotene and pachytene stages were seen; these failed to reach diakinesis and by $18\frac{1}{2}$ days *postcoitum* they were degenerating. Many oocytes in the normal ovary degenerate at a corresponding stage of foetal development. This study favoured the view that the meiotic germ cells were XX in chromosome constitution, since labelling of the cells with tritiated thymidine showed that DNA replication had taken place on the same day as in the normal ovary, but no sign was seen of the striking late-labelling pattern characteristic of meiotic prophase in male germ cells, nor of the prominent sex vesicle that normally contains the XY sex bivalent. On the other hand, the small proportion of germ cells entering meiosis (less than 5%), when 50% on average should have been XX, favoured some localized meiosis-inducing stimulus from XX somatic tissue.

An identical picture, of a few germ cells in meiotic prophase but the great majority developing in the male direction, as prospermatogonia, was seen by McLaren (1981b) in the foetal testes of XX *Sxr* males. Here all germ cells contain two X chromosomes, but the somatic tissue of the gonad, and the germ cells up to the time of reactivation of the silent X, are believed to be made up of a mosaic of female-determining and male-determining cells, according to whether or not the male-determining *Sxr* sequences on the X chromosome have been inactivated (McLaren and Monk, 1983; Cattanach et al., 1982). The situation is thus closely analogous to that of an XX↔XY chimera. The meiotic cells were not distributed randomly within the gonad, but were always located on the side of the testis adjacent to the body wall. This spatial regularity, which also seems to characterize the testes of XX↔XY chimeras, could reflect either a response to some meiosis-inducing substance diffusing to the gonad from the mesonephric region of the body wall, as suggested originally by Byskov (1974; Byskov and Saxén, 1976), or some characteristic feature of structure or composition in this region of the gonad.

The failure of germ cells to enter meiotic prophase in the foetal testes of XO *Sxr* individuals suggested to McLaren (1981b) that the presence of the second X chromosome in XX *Sxr* germ cells might increase their responsiveness to a meiosis-inducing stimulus. Recent findings on *Sxr*, however, have revealed that XO *Sxr* differ from XX *Sxr* individuals not only in lacking a second X chromosome, but also in having the testis-determining region attached to an active X chromosome in every cell rather than in only 50% or so of cells through random X inactivation. The evidence as to whether the second X chromosome increases responsiveness, or the presence of the testis-determining region on the active X *Sxr* or Y chromosome of a premeiotic germ cell decreases responsiveness, or whether the presence of genetically female-determining cells in the somatic tissue of the mosaic (XX *Sxr*) or chimeric (XX↔XY) testis facilitates entry into meiosis, is discussed by McLaren (1983). The last possibility would allow XY as well as XX germ cells to be included in the meiotic population. Further support for this view comes from a recent observation that meiotic germ cells can be found in the foetal testes of XX↔XY chimeras in which the XX component is homozygous for the *W* gene, so that few if any XX germ cells reach the gonads (A. McLaren and M. Buehr, unpublished observations). Direct tests to determine whether the meiotic germ cells in XX↔XY foetal testes are XY as well as XX should soon become feasible.

5. SINGLE-SEX CHIMERAS

As we have seen, XX↔XY chimeras very seldom produce progeny from more than one component, the XY component if they are males and the XX component if they are females. For XY↔XY and XX↔XX ("single-sex") chimeras, however, there is no *a priori* reason why the two populations of germ cells should not be represented equally in the gonad, giving rise to two types of progeny, one from each of the chimera components.

Indeed, the assumption is often made (e.g., Gordon, 1976) that any chimera breeding from one component only is of XX↔XY constitution. The available data on chromosomally sexed chimeras, sparse though they are, do not justify this assumption. In my own series (McLaren, 1975a), 3 out of 3 XX↔XX females and 4 out of 5 XY↔XY males gave progeny from one component only. On the other hand, both the XY↔XY chimeras of Mystkowska and Tarkowski (1968) bred from both components, as did one XX↔XX female and 2 out of 3 XY↔XY males reported by Ford *et al.* (1975). A chimeric XY↔XY ram also produced progeny from both components (Tucker *et al.*, 1978). If the 18 mice for which Gearhart and Oster-Granite (1981) examined more than 30 metaphase plates without finding one of opposite karyotype are all assumed to be single-sex chimeras, 5 out of 8 XX↔XX and 6 out of 10 XY↔XY bred from one component only. The proportion of single-sex chimeras that produce progeny from one component may of course vary from one series to another, depending perhaps on the strain combination involved.

The reason why any single-sex chimera produces progeny from one component only remains a mystery (for discussion, see McLaren, 1978). The most likely explanation is that, at least in some strain combinations, one or the other component is often by chance excluded from the germ cell lineage entirely. This could be because the number of cells initially set aside to become primordial germ cells is very small, so that the chance of all being from one component or the other is quite high. Recent evidence suggests, however, that the number of cells set aside to become primordial germ cells is not small (see Snow and Monk, 1983). Alternatively, it may be that the two components are not randomly mixed at the time of allocation, so that the whole region from which the germ cell progenitors are drawn may often contain cells of one component only. No published information yet exists on the degree of cell mingling in the epiblast of primitive-streak stage chimeric embryos.

Another type of explanation for the failure of both components to be represented among the functional gametes involves cell selection, i.e., the more rapid proliferation of germ cells of one component than of the other. Eight to ten successive cell divisions of the germ cell population separate the time when the primordial germ cells can first be recognized from the time when they enter meiosis in the female and mitotic arrest in the male, so even a small difference in mitotic rate would lead to one germ cell component outgrowing the other, in both sexes. Strain combinations where such a difference exists for tissues in general are known as unbalanced, and chimeras giving progeny from one component only do indeed tend to occur more frequently among unbalanced than among balanced strain combinations (see McLaren, 1978, Table 1). But this effect should always lead to the same component predominating, and in some of the studies in which a number of chimeras breeding from one component only have been reported (e.g., McLaren, 1975a) neither component predominated, but rather some of the progenies were all from one component and some all from the other.

Mintz (1968) published an interesting example, where selection may have been operating, of a C3H↔C57BL male chimera mated to C57BL females that produced 15% of progeny from his C57BL component in the first 10 litters that he fathered, 1% in the next 10 litters, and none thereafter. Eleven other breeding males of the same strain combination were said to show the same breeding pattern. Mintz suggested that although in the female, where the mitotic proliferative phase of gametogenesis is completed before birth, the adult germ cell population should be stable, germ cell selection might operate in the male to bring about a progressive shift in the proportion of the two types of progeny, since spermatogonial proliferation continues throughout adult life. This interpretation would not, however, be consistent with the model of spermatogonial kinetics according to which each stem cell divides to give one other stem cell and one cell that contributes to the spermatogenic population. A progressive change in breeding pattern with time could be due to factors that did not involve spermatogonial selection.

The number of progeny sired by each of the two components of a male chimera may in any case not provide an accurate reflection of the relative contribution of those components to the germ cell population. The ratio of progeny of

V. Germ Cell Lineages

the two types shows significant heterogeneity from one litter to another in some strain combinations, so that of two litters sired by the same male and born on successive days, hence conceived close together in time, one could be all of the one type and the other all or almost all of the other (McLaren, 1975a; Buehr and McLaren, 1984). Gearhart and Oster-Granite (1981) have commented on the same phenomenon; on the other hand, when the two chimera components differed at only a single gene locus, no significant heterogeneity was observed (Mystkowska, cited by McLaren, 1975a). I initially suggested that the heterogeneity might be due to nonrandom distribution of the two components between testes and/or among testis tubules, combined with nonrandom contribution of the various spermatogenic regions to an individual ejaculate. An explanation in terms of nonrandom distribution between testes was ruled out by removing one testis from an XY↔XY chimeric male and observing that the heterogeneity between litters remained (Buehr and McLaren, 1984). More recent data suggest that the heterogeneity may reflect variation in the time of mating, since the ratio of the two types of progeny in litters from a single chimeric male differs strikingly according to whether the females are paired with the male during the day or during the night (Buehr and McLaren, 1984).

6. GERM CELL MODIFICATION

Germ cells develop in very intimate association with somatic cells—Sertoli cells in the testis and follicle cells in the ovary. In a normal gonad, germ cells and somatic cells are of the same genotype, so any genetically determined influence exerted on the germ cells by the somatic environment would pass unnoticed. In a chimera, on the other hand, a germ cell of one genotype may be associated with a Sertoli cell or with follicle cells of contrasting genotype, so that it becomes possible to enquire whether any of the germ cell's genetic characteristics are imposed upon it from outside.

One aspect of spermatozoan phenotype that is known to be genetically determined is the shape of the head and midpiece. Burgoyne (1975) investigated whether this genetic determination acted in the germ cell itself or via the enveloping Sertoli cell, using chimeras made between C3H and C57BL mice. C57BL spermatozoa have shorter and broader heads but longer midpieces than do C3H, and these characteristics were shown to persist unchanged in the chimeric males. Thus a C57BL-XX↔C3H-XY male who was breeding entirely from his C3H component yielded spermatozoa that were all distributed as a single peak, corresponding to the dimensions characteristic of C3H males, with no indication of any spermatozoa of C57BL or intermediate shape. No evidence was found in any of the chimeras for any influence of the somatic environment on spermatozoan shape.

On the other hand, the influence of the somatic environment is clearly seen in those situations (e.g., testicular feminization) in which male germ cells are unable for genetic reasons to complete spermatogenesis, unless they are associated in a chimeric testis with somatic cells that lack the genetic defect. These instances of rescue by chimerism have already been described.

Genetically determined characteristics of the oocyte that have been analysed in a similar way include the granularity of the cytoplasm, the glucosephosphate isomerase 1 isozyme type, and the oocyte-specific expression of this isozyme. In chimeras made between strains differing in these three respects, A. McLaren and M. Buehr (unpublished observations, cited by McLaren, 1981c) found that oocytes developed autonomously, even when the follicle cells surrounding them could be directly shown by enzyme analysis to be largely or entirely of the contrasting type. Preliminary observations suggest, however, that the feature of DDK strain cytoplasm that renders it incompatible with spermatozoa of other strains may be influenced by the somatic environment of a chimeric ovary (see McLaren, 1981c).

Not only the phenotype but also the genotype of a germ cell could in principle be modified by exposure to genetically foreign cells in a chimeric gonad. For radiation chimeras, a claim has been made that genetic material from the donor strain can become incorporated into the sperm genome, so that the immunological properties of the progeny are modified (Kanazawa and Imai, 1974). T lymphocytes taken from the BALB/c progeny of BALB/c↔C57BL male chimeras proved in an *in vitro* cytotoxicity test to be more responsive to C57BL cells than were those of control BALB/c mice (McLaren *et al.*, 1981); however, in this situation an explanation in terms of some maternally mediated sensitization phenomenon seems more probable than an induced change in sperm genotype, since the BALB/c females mated to the chimeric males would have been exposed to C57BL antigens both in the semen and from the C57BL × BALB/c progeny that they carried. An examination of 1851 progeny from 20 chimeric males whose component strains differed with respect to nine defined genetic characters failed to reveal any indication that the genes for these characters were in any way modified in the course of spermatogenesis in the chimeric testes (McLaren, 1975b).

II. CONCLUSIONS

Transfer of cells between genetically distinguishable embryos, resulting in germ cell chimerism, has proved to be a powerful technique for tracing a continuous germ cell lineage in animals as distant as *Drosophila* and *Xenopus*. Similar techniques applied to urodeles, birds, and mammals suggest that in these groups the germ cell lineage is not continuous, and indicate from which tissue the primordial germ cells are derived. Germ cell chimerism in mice has been used to

V. Germ Cell Lineages

rescue cells that for genetic or other reasons are incapable of contributing to a functional germ line in any other way.

Embryonic grafting experiments in birds have demonstrated the importance of chemotactic influences in allowing germ cells to colonize the gonad, even across species barriers. When the germ cells differ genetically from the somatic component of the gonad, those phenotypic characteristics of the gametes that have been investigated in *Xenopus,* birds, and mice mostly develop autonomously, and the genotype of the germ cells also seems largely or entirely unaffected by their somatic environment. When the chromosomal sex of the germ cells differs from that of the gonad that they have colonized, a range of different outcomes is possible. In *Drosophila,* germ cells of either sex degenerate in a gonad of inappropriate sex; on the other hand, in anuran amphibia complete sex reversal of the germ cells in both directions is possible. In birds, chromosomally male germ cells degenerate in an ovary, but female germ cells in a testis can form functional spermatozoa, while in mice the reverse appears to be true.

In mammals, some problems have been illuminated by studies on chimeras but require further investigation if they are to be solved. The extent to which the initial direction of development of a germ cell, male or female, is determined by its own genotype rather than by the somatic environment of the gonad falls into this category. The reason why some single-sex chimeras produce progeny from one component only is still not known, and in general the relation between the presence of germ cells in the gonad and their functional capacity as judged by progeny testing requires further study. Such investigations on germ cell chimeras may throw light on possible relations among gametes of differing genotype in nonchimeric individuals.

REFERENCES

Benirschke, K., and Brownhill, L. E. (1963). *Cytogenetics* **2,** 331–341.
Blackler, A. W. (1962). *J. Embryol. Exp. Morphol.* **10,** 641–651.
Blackler, A. W. (1965). *J. Embryol. Exp. Morphol.* **13,** 51–61.
Blackler, A. W., and Fischberg, M. (1961). *J. Embryol. Exp. Morphol.* **9,** 634–641.
Burgoyne, P. S. (1975). *Dev. Biol.* **44,** 63–76.
Bradbury, M. W. (1983). *J. Exp. Zool.* **226,** 315–320.
Buehr, M., and McLaren, A. (1984). *J. Reprod. Fertil.* (in press).
Burgoyne, P. S., and Baker, T. G. (1981). *In* "The Development and Function of Reproductive Organs" (A. G. Byskov and H. Peters, eds.), pp. 122–128. Excerpta Medica, Amsterdam.
Byskov, A. G. (1974). *Nature (London)* **252,** 396–397.
Byskov, A. G., and Saxén, L. (1976). *Dev. Biol.* **52,** 193–200.
Cattanach, B. M., Evans, E. P., Burtenshaw, M., and Barlow, J. (1982). *Nature (London)* **300,** 445–446.
Didier, E. (1977). *C. R. Hebd. Seances Acad. Sci.* **284,** 671–678.

Eicher, E. M., and Hoppe, P. C. (1973). *J. Exp. Zool.* **183,** 181–184.
Evans, E. P., Ford, C. E., and Lyon, M. F. (1977). *Nature (London)* **267,** 430–431.
Eyal-Giladi, H., Ginsburg, M., and Farbarov, A. (1981). *J. Embryol. Exp. Morphol.* **65,** 139–147.
Falconer, D. S., and Avery, P. J. (1978). *J. Embryol. Exp. Morphol.* **43,** 195–219.
Ford, C. E., and Evans, E. P. (1977). *J. Reprod. Fertil.* **49,** 25–33.
Ford, C. E., Evans, E. P., Burtenshaw, M. D., Clegg, H. M., Tuffrey, M., and Barnes, R. D. (1975). *Proc. R. Soc. London, Ser. B* **190,** 187–197.
Gardner, R. L. (1977). *Int. Congr. Ser.—Excerpta Med.* **432,** 154–166.
Gardner, R. L., and Rossant, J. (1979). *J. Embryol. Exp. Morphol.* **52,** 141–152.
Gearhart, J., and Oster-Granite, M. L. (1981). *Biol. Reprod.* **24,** 713–722.
Gordon, J. (1976). *J. Exp. Zool.* **198,** 367–374.
Haffen, K. (1968). *C. R. Hebd. Seances Acad. Sci.* **267,** 511–513.
Haffen, K. (1975). *Am. Zool.* **15,** 257–272.
Hajji, K., Martin, C., Perramon, A., and Dieterlen-Lievre, F. (1984). Personal communication.
Hampton, S. M. H. (1970) Thesis, University of Texas, Houston.
Humphrey, R. R. (1933). *J. Exp. Zool.* **65,** 243–269.
Illmensee, K., and Mahowald, A. P. (1974). *Proc. Natl. Acad. Sci. U.S.A.* **71,** 1016–1020.
Illmensee, K., and Mahowald, A. P. (1976). *Exp. Cell Res.* **97,** 127–140.
Illmensee, K., Mahowald, A. P., and Loomis, M. R. (1976). *Dev. Biol.* **49,** 40–65.
Kanazawa, K., and Imai, A. (1974). *Jpn. J. Exp Med.* **44,** 227–234.
Kelly, S. J. (1977). *J. Exp. Zool.* **200,** 365–376.
Lyon, M. F., Glenister, P. H., and Lamoreux, M. L. (1975). *Nature (London)* **258,** 620–622.
McLaren, A. (1975a). *J. Embryol. Exp. Morphol.* **33,** 205–216.
McLaren, A. (1975b). *Genet. Res.* **25,** 83–87.
McLaren, A. (1976). "Mammalian Chimaeras." Cambridge Univ. Press, London and New York.
McLaren, A. (1978). In "Genetic Mosaics and Chimeras in Mammals" (L. B. Russell, ed.), pp. 125–134. Plenum, New York.
McLaren, A. (1981a). "Germ Cells and Soma: A New Look at an Old Problem." Yale Univ. Press, New Haven, Connecticut.
McLaren, A. (1981b). *J. Reprod. Fertil.* **61,** 461–467.
McLaren, A. (1981c). *J. Reprod. Fertil.* **62,** 591–596.
McLaren, A. (1983). In "Current Problems in Germ Cell Differentiation" (A. McLaren and C. C. Wylie, eds.), pp. 225–240. Cambridge Univ. Press, London and New York.
McLaren, A., and Monk, M. (1982). *Nature (London)* **300,** 446–448.
McLaren, A., Chandley, A. C., and Kofman-Alfaro, S. (1972) *J. Embryol. Exp. Morphol.* **27,** 515–524.
McLaren, A., Chandler, P., Buehr, M., Fierz, W., and Simpson, E. (1981). *Nature (London)* **290,** 513–514.
Mayr, W. R., Pausch, V., and Schnedl, W. (1979). *Nature (London)* **277,** 210–211.
Mintz, B. (1968). *J. Anim. Sci.* **27,** Suppl. 1, 51–60.
Mintz, B., and Illmensee, K. (1975). *Proc. Natl. Acad. Sci. U.S.A.* **72,** 3585–3589.
Mystkowka, E. T., and Tarkowski, A. K. (1968). *J. Embryol. Exp, Morphol.* **20,** 33–52.
Mystkowska, E. T., and Tarkowski, A. K. (1970). *J. Embryol. Exp. Morphol.* **23,** 395–405.
Nieuwkoop, P. D., and Satasurya, L. A. (1979). In "Primordial Germ Cells in the Chordates" Dev. Cell Biol. Ser., Vol. 7. (M. Abercrombie, D. R. Newth, and J. A. Torrey, eds.), Cambridge Univ. Press, London and New York.
Papaioannou, V. E., Artzt, K., Dooher, G. B., and Bennett, D. (1979). *Gamete Res.* **2,** 147–151.
Reynaud, G. (1969). *J. Embryol. Exp. Morphol.* **21,** 485–507.
Rogulska, T., Ozdzenski, W., and Komer, A. (1972). *J. Embryol. Exp. Morphol.* **25,** 155–164.
Rossant, J., and Chapman, V. M. (1983). *J. Embryol. Exp. Morphol.* **73,** 193–205.

V. Germ Cell Lineages

Smith, L. D., Michael, P., and Williams, M. A. (1983). *In* "Current Problems in Germ Cell Differentiation" (A. McLaren and C. C. Wylie, eds.), pp. 19–39. Cambridge Univ. Press, London and New York.
Snow, M. H. L., and Monk, M. (1983). *In* "Current Problems in Germ Cell Differentiation" (A. McLaren and C. C. Wylie, eds.), pp. 115–135. Cambridge Univ. Press, London and New York.
Stevens, L. C. (1978). *Nature (London)* **276,** 266–267.
Stevens, L. C., Varnum, D. S., and Eicher, E. M. (1977). *Nature (London)* **269,** 515–517.
Stone, W. H., Berman, D. T., Tyler, W. J., and Irwin, M. R. (1960). *J. Hered.* **51,** 136–140.
Stone, W. H., Friedman, J., and Fregin, A. (1964). *Proc. Natl. Acad. Sci. U.S.A.* **51,** 1036–1044.
Surani, M. A. H., Barton, S. C., and Kaufman, M. H. (1977). *Nature (London)* **270,** 601–603.
Tachinante, F. (1974a). *C. R. Hebd. Seances Acad. Sci.* **278,** 1895–1898.
Tachinante, F. (1974b). *C. R. Hebd. Seances Acad. Sci.* **278,** 3135–3138.
Teplitz, R. L., Moon, Y. S., and Basrur, P. K. (1967). *Chromosoma* **22,** 202–209.
Tucker, E. M., Dain, A. R., and Moor, R. M. (1978). *J. Reprod. Fertil.* **54,** 67–73.
van Deusen, E. B. (1976). *J. Embryol. Exp. Morphol.* **37,** 173–185.
Wartenberg, H. (1983). *Bibl. Anat.* **24,** 93–110.

3

Hemopoietic and Immune System

CHAPTER **VI**

Blood in Chimeras

FRANCOISE DIETERLEN-LIÈVRE

Institut d'Embryologie du CNRS
et du Collège de France
Nogent-sur-Marne, France

I.	Introduction	133
II.	Avian Chimeras	134
	A. Analysis of Red Cell Populations in Quail-Chick Yolk Sac Chimeras	135
	B. Tracing of Cells in Hemopoietic Organs of Chimeras	138
	C. Pattern of Colonization of Hemopoietic Organ Rudiments in "Complementary" Chimeras	144
	D. Evolution of Blood and Origin of Lymphocytes in Chick-Chick Chimeras	147
	E. The Emergence of Intraembryonic Stem Cells	148
III.	Amphibian Chimeras	152
	A. Extirpation Experiments	152
	B. Graft of Marked Rudiments	153
	C. Extrinsic Origin of Stem Cells	153
	D. Chimeric Complementary Embryos	154
IV.	Mammalian Chimeras	155
	A. Chimeric Organs	155
	B. Tetraparental Chimeras	156
	C. Blood Chimeras Obtained by Intraplacental Transplantation of Fetal Liver Cells	158
	D. Teratocarcinoma-Derived Blood Cells	160
V.	Conclusion	161
	References	161

I. INTRODUCTION

During ontogenesis, various hemopoietic sites function sequentially. They are responsible for the successive production of red cell lineages that differ in their morphology and in the assortment of their active globin genes. The blood-forming organs also turn out leukocytes, many of which, again, differentiate according to a precise chronological and spatial programme. Stem cells are central to these events. The fundamental concept that hemopoiesis involves interactions between two independent cell lineages, i.e., stromal and hemopoietic, derives, on the one hand, from embryological studies and particularly from observations in embryonic chimeras and, on the other hand, from the analysis of radiation chimeras (Till and McCulloch, 1961). The latter are adult animals, usually mice, that have been submitted to a strong irradiation dose and restored through injection of normal bone marrow cells. The irradiation knocks out all young, self-renewing hemopoietic cells, and if fresh bone marrow stem cells are not provided, replacement of blood cells is impaired and the animal dies. This is an important area of modern hematology, but it is outside my scope and will not be reviewed in the present chapter. In developmental biology, the main issue to which chimeras have contributed is the emergence of blood stem cells.

At the beginning of the century, Maximov (1909) put forward a monophyletic theory of hemopoiesis, according to which all blood cells derive from common multipotential ancestors. Experimental arguments supporting this idea were first provided by Moore and Owen (1965) in avian embryos. They showed that cells travelled between chick embryos parabiosed through their circulation. Relying on the chromosomal difference between male and female cell nuclei, they found cells from the partner in the thymus, spleen, bone marrow, and bursa of Fabricius, when the two embryos were of opposite sexes (see also Chapter VIII). The exchange of cells was extensive, especially if vascular anastomosis was established early during incubation, so they inferred that stem cells must arise outside the rudiments of the hemopoietic organs. The fully extrinsic origin of stem cells ensuring the hemopoietic function of the thymus, bursa of Fabricius, and bone marrow has been shown since in quail-chick chimeric organs (see Chapter VIII).

Since the only hemopoietic organ that appeared to have its own stem cells in the embryo was the yolk sac, Moore and Owen (1967a) extended Maximov's ideas by proposing that all stem cells arose in the yolk sac mesoderm and were carried to the embryo through the circulation. For about 10 years much of the thinking on the ontogeny of the hemopoietic system centered around this idea. However, this hypothesis was invalidated (Dieterlen-Lièvre, 1975) by observations in chimeras associating a quail embryo with a chick yolk sac and in amphibian chimeras grafted with marked ventral blood islands.

II. AVIAN CHIMERAS

The yolk sac is a double-walled vesicle that encloses the vitellus (Fig. 1). While the underlying endodermal layer takes up and transmits yolk nutrients to

VI. Blood in Chimeras

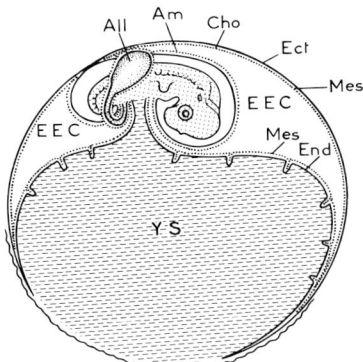

Fig. 1. Scheme of the 4-day chick embryo appendages. (Redrawn from Hamilton, 1952.) All, allantois; Am, amnios; Cho, chorion; EEC, extraembryonic coelom; Ect, ectoderm; End, endoderm; Mes, mesoderm; YS, yolk sac.

the embryo via the circulation, the overlying mesodermal layer gives rise to blood islands. This is the site where erythropoiesis first occurs in amniote embryos. A technique devised in 1972 by Claude Martin consists of extirpating the central area of the young chick blastoderm and replacing it with the corresponding area from a quail blastoderm (see Chapter I, Fig. 6 and 7).

The operations are performed during the second day of incubation, when blood cells are not yet circulating. The data from these chimeras established that yolk sac stem cells give rise to erythrocytes of the primitive lineage, but that when the secondary erythropoietic lineage differentiates, stem cells of intraembryonic origin become functional (Beaupain *et al.*, 1979). By using morphological cell markers (Dieterlen-Lièvre, 1975; Martin *et al.*, 1978), as well as immunocytological identification of quail or chick erythrocytes and electrophoretic studies of hemoglobins, it was possible to elucidate the complex movements of cells in the blood system as the embryo develops, and to determine the potentialities of these cells. These heterospecific chimeras, consisting of a quail embryo developing on a chick yolk sac, were studied between days 4 and 13 of incubation.

A. Analysis of Red Cell Populations in Quail-Chick Yolk Sac Chimeras

Antibodies specific for either quail or chick red cells were made in rabbits (Dieterlen-Lièvre *et al.*, 1976), and the proportions of red cells from the two species in the blood of chimeras were measured either by immunohemolysis or by immunofluorescence. It appeared that until 5 days of incubation all red blood cells were chick, i.e., they derived from stem cells formed in the yolk sac. On day 5, some chimeras possessed a few quail cells, making up at most 5% of their erythrocytes. Beginning on day 6, quail red cells began to flow into the circulation (Fig. 2), in some chimeras reaching 80% of erythrocytes by day 13. However, replacement of chick red cells by quail was always very irregular. Using the rabbit antisera in the presence of complement, it was possible to retrieve from the chimeras' bloods pure chick or quail populations, derived, respectively, from yolk sac stem cells and intraembryonic stem cells (Fig. 3). The hemoglobin patterns were studied in each of these two populations between days 5 and 13 of incubation by polyacrylamide gel electrophoresis (PAGE) (Beaupain *et al.*, 1979) or by isoelectric focusing (Beaupain *et al.*, 1980). More bands are resolved by the latter technique than by PAGE, but the results are comparable. The evolution of hemoglobin patterns in the chick embryo had been extensively studied in the chick (see Bruns and Ingram, 1973, for a review), but was not known in the quail. Our own study demonstrated an evolution parallel to that of the chick, with the appearance of definitive erythrocytes in the blood early on the fifth day of incubation. Chick erythrocytes from the chimeras, i.e., erythrocytes derived from yolk sac stem cells, belonged to the primitive lineage at 5 days of incubation and were a mixture of primitive and definitive cells at 7 days. Thus yolk sac stem cells gave rise to both lineages. On the other hand, at 5 days as discussed above, quail erythrocytes were present in minute quantities in some chimeras only; they belonged to the primitive series. At 7 days the quail population yielded different hemoglobin patterns in different individual chimeras. Commonly, only the definitive bands were found; the other pattern was similar to that of the controls, i.e., definitive plus primitive hemoglobins, but compared to the

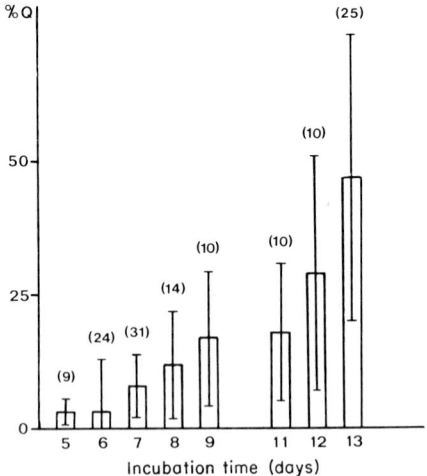

Fig. 2. Evolution of red blood cell composition in quail↔chick yolk sac chimeras. % Q, percentage of quail erythrocytes; numbers in parentheses are the numbers of chimeras examined.

VI. Blood in Chimeras

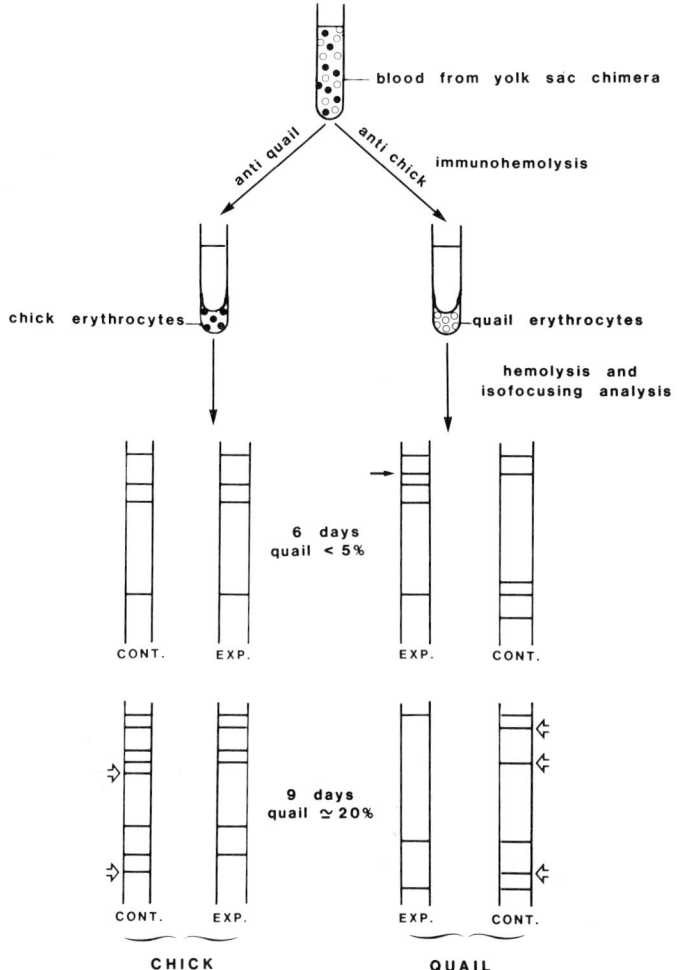

Fig. 3. Scheme of hemoglobin analysis in the yolk sac stem cell-derived population (chick) and the intraembryonic stem cell-derived population (quail). The microgels were prepared from individual chimeras (exp.) or control embryos (cont.). At 6 days, the pattern of the chick population of the chimera was identical to control chick; the quail population, being small, could not be completely purified and was contaminated by several chick bands; however, it definitely displayed a quail embryonic hemoglobin band (arrow). At 9 days both the chick and the quail populations from chimeras yielded less numerous bands than corresponding controls, because in the first the intraembryonic stem cell contribution was missing, while the latter was deprived of the yolk sac stem cell progeny. (Arrows point to control bands that are missing in experimental patterns.)

controls the primitive bands were attenuated. The quail erythrocytes from chimeras were "older," so to speak, than those in control quails. This was interpreted as follows: all primitive erythrocytes arise from yolk sac stem cells, whereas definitive erythrocytes derive from a minority of yolk sac stem cells and from increasing numbers of intraembryonic stem cells. Since the quail yolk sac is missing by virtue of the experimental design, quail primitive red cells will be lacking from the chimeras if all primitive red cells arise from yolk sac stem cells. Consequently at 7 days, quail erythrocytes all, or nearly all, belonged to the definitive lineage. Thus hemoglobin analysis corroborated closely the conclusions that could be drawn from the quantitative evolution of the quail-chick populations. Put together, the data demonstrated that the ontogenetic pattern of erythropoiesis is determined only by time; it is totally independent both of the stem cells' origins and of the organs where they mature. Erythroblasts that differentiate during the early period always synthesize primitive hemoglobins, whether they derive from yolk sac or intraembryonic stem cells. The hemoglobin switch then occurs, and all erythrocytes formed belong to the secondary lineage. Cytological tracing of cells in the hemopoietic organs or the yolk sac, analyzed in Section II, B, has further demonstrated that the stem cells may migrate extensively. These movements do not influence the pattern of their hemoglobin synthesis, provided they home to an erythropoietic environment.

B. Tracing of Cells in Hemopoietic Organs of Chimeras

Usually the lymphoid population in the thymus was completely quail. However in a total of 150 chimeras examined, two thymuses were composed of quail reticular cells and chick lymphoid cells. Thus the rule in yolk sac chimeras is that stem cells from an intraembryonic origin colonize the thymus rudiment. However, the two exceptional cases show that yolk sac cells have the potentiality to enter the thymus and undergo lymphopoiesis. Such a capacity was also demonstrated in another experimental set-up where a chick thymic endomesodermal rudiment at its attractive period was associated with a fragment of the extraembryonic area from a quail head-fold stage blastodisc (Jotereau and Houssaint, 1977). After a short organotypic culture period during which fusion of the tissues occurred, the association was allowed to develop in the somatopleure of a 3-day chick host (Fig. 4). Nine days after grafting, numerous quail thymocytes had developed in the graft. One observation in this experiment merits emphasis: precursor cells from the associated yolk sac entered the attractive thymus only after a 3-day culture delay. If precultured for 3 days, upon association with the thymus the extraembryonic area was immediately capable of yielding colonizing precursors. Thus, it seems that precursors with T lymphoid capacities are not immediately present in the very early yolk sac. Either they appear at a later stage than the stem cells that give rise to primitive erythrocytes, or these stem cells

VI. Blood in Chimeras

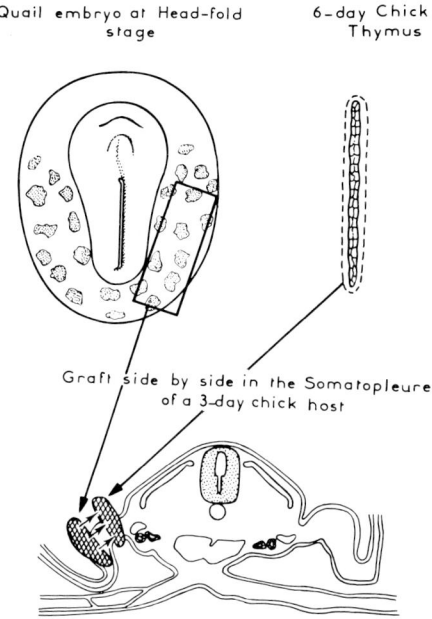

Fig. 4. Scheme of the association between early quail extraembryonic area and chick attractive thymus rudiment.

have to undergo a maturation process that renders them capable of colonizing the thymus.

It has not been possible to recognize clear-cut phases of colonization in the spleen by interspecies grafting of the rudiment. However early the rudiment is explanted, some hemopoietic stem cells seem already to be present. The spleen is a purely mesodermal rudiment with no epithelial component, and once hemopoietic stem cells enter it they become embedded in the mesenchyme and can no longer be discerned. Further, the rudiment differentiates from the dorsal mesentery, in a site where many intraembryonic stem cells arise (see below). Thus it is likely that stem cells are in or around the rudiment, or at any rate in close contact with it, from the earliest stage of development. In normal development, basophilic cells can be identified from 6 to 7 days onward, either in the sinusoids or among the mesenchymal cells. The hemopoietic processes carried out in the spleen during the incubation period are erythropoiesis in the sinusoids and granulopoiesis within the mesenchymal stroma. Shortly before hatching many granulocytes move out of the stroma, and great numbers of lymphocytes settle within its emptied meshes. By comparison with the thymus, there is a permanent turnover of hemopoietic cells. This movement intensifies around 11 days of development. In the yolk sac chimeras, transient chimerism occurred in the spleen at that time. Though the spleen of chimeras was usually populated by quail cells, in some embryos a few chick cells found their way to the spleen between 11 and 12 days, displaying the large nucleolus typical of young hemopoietic cells. At 13 days chick cells were rarely found.

At the 13-day stage when observation of these chimeras was terminated, differentiation of bursa (see Chapter VIII) and bone marrow was barely initiated. In the best developed bursas, lymphoid follicles had formed; these always contained quail cells. The origin of the various cell lineages in bone marrow had already been established (Jotereau and Le Douarin, 1978) by grafting either limb buds or femur rudiments of chick or quail onto the chorioallantois or in the somatopleura of the other species. The results demonstrated that the mesenchymal bone primordium gives rise to the chrondrocytes, osteocytes, bone marrow stromal cells, and the pericytes that encase the blood capillaries. On the one hand, the endothelial cells and hemopoietic cells are of hematogenous origin, as also are the osteoclasts, multinucleated cells responsible for bone resorption, whose relationship to osteocytes was a matter of debate. When limb buds were grafted, the osteoclasts were always from the host. On the other hand, when femurs from 9-day or older quail embryos were made to develop in the chick environment, osteoclasts could have nuclei from one species or the other or they might have nuclei of both species within the same cytoplasm. The difference in results between limb buds and femurs is of course one of timing. In the first instance, the rudiments were obtained prior to colonization and depended entirely on the host for both their vascular and hematogenous cell supply. In the second case, femurs taken after the start of colonization received hemopoietic cells from the host, some of which fused with donor cells yielding chimeric osteoclasts. It should be pointed out that quail osteoclast nuclei do not all display the typical heterochromatin mass and so are sometimes impossible to distinguish from chick nuclei. This is probably due to their functional or cell cycle activities, since all the nuclei in one multinucleated quail cell always have the same appearance. The experiments with quail limb buds in chick however prove that all osteoclasts are of extrinsic origin, since in that case all osteoclast nuclei have the chick appearance. This ambiguity probably explains why, in previous experiments relying on the quail-chick system, Kahn and Simmons (1975) concluded that although most osteoclasts were derived from hematogenous cells, some also arose from the *in situ* fusion of bone-derived cells. In the yolk sac chimeras, the bone marrow hemopoietic cell population, like that of thymus, spleen, and

Fig. 5. Yolk sac blood islands of control and chimeric 7-day embryos. Feulgen–Rossenbeck staining. × 1800. (a) Chick. (b) Chimera. (c) Quail. Arrows point to mesodermal or endothelial cells. In (a) and (b) these are chick and display a light, homogeneous nucleus. In (c) these nuclei have the heavy heterochromatic mass of the quail. The erythroid cells, encased in the blood islands, exhibit in (a) finely dotted chromatin and sometimes a large lightly staining nucleolus. In (b) and (c) the nuclei are typically quail, i.e., contain heavy, dark, chromatin clumps.

VI. Blood in Chimeras

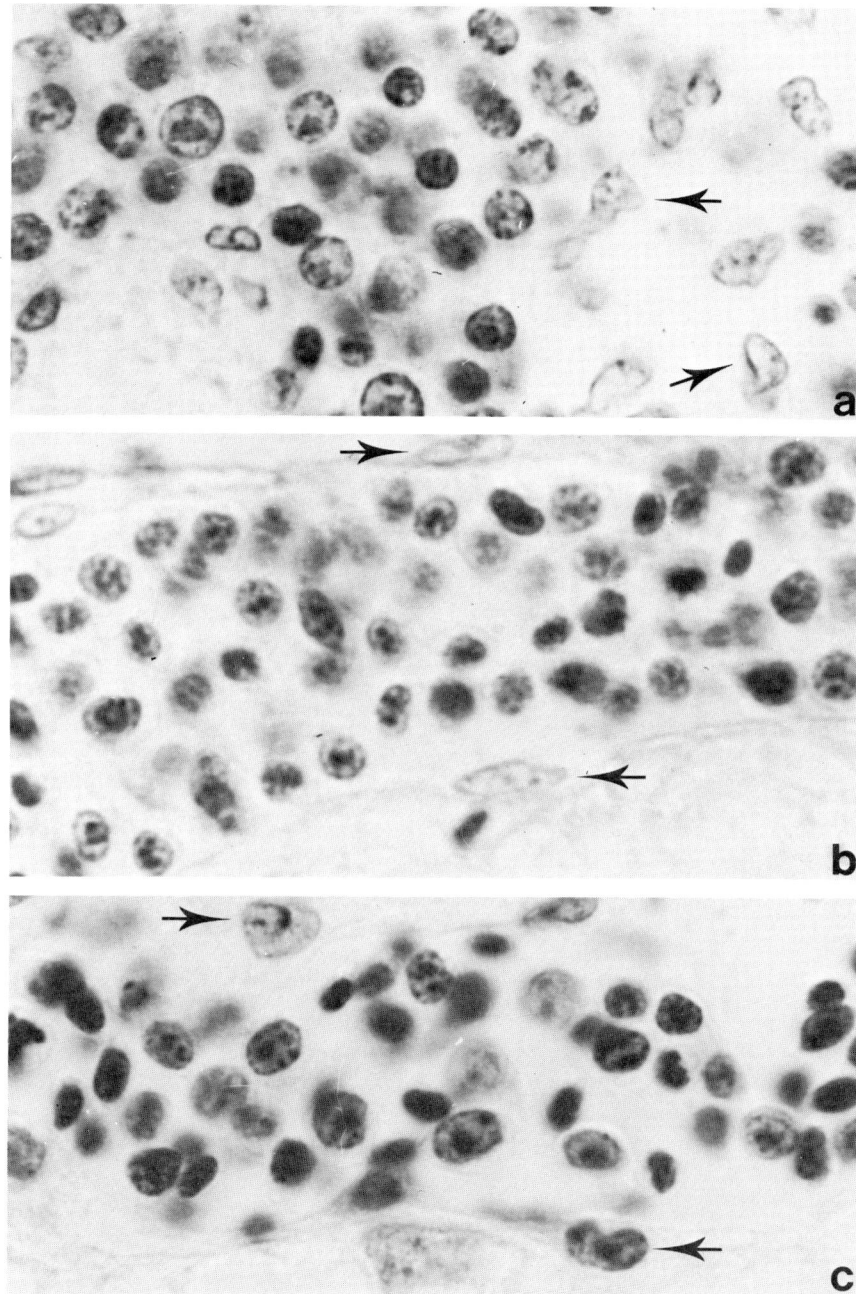

bursa, was always quail. In one exceptional case, a chick leg was found to have developed on the yolk sac, independently of the grafted chimera, and in this leg the bone marrow reticulum was chick, but the hemopoietic cells were quail. This particular case illustrates in a spectacular fashion that stem cells circulating in the vessels of the yolk sac at the time of bone marrow seeding arose from the embryo.

Until 7 days of age the blood islands in the yolk sac contained only chick erythroid cells, but from then on quail cells were also present. While the endothelium of the blood islands always remained chick, in some chimeras the whole erythropoietic population was quail (Fig. 5), indicating a movement of cells from the embryo to the yolk sac. Thus, carrying on its erythropoietic function, the yolk sac acts as a specific microenvironment where cells formed elsewhere pursue their differentiation.

In the blood of young avian embryos, many basophilic cells circulate among the erythrocytes. Their number declines sharply after 6–7 days of incubation, but they can still be found in organs where circulation is sluggish, such as the liver. In chimeras (Dieterlen-Lièvre and Martin, 1981), they were all chick until 4 days, but were 50% quail by 6 days of incubation (Fig. 6). The group of cells in Fig. 7 substantiates the traffic between the two compartments: chick cells from yolk sac to embryo, quail cells from embryo to yolk sac.

Thus a coherent picture emerges for the quail↔chick yolk sac chimeras. The first wave of hemopoiesis arises from yolk sac stem cells, which actually differentiate *in situ* in the blood islands. This wave develops synchronously between days 2 to 5 of incubation; it gives rise to primitive erythrocytes characterized by a specific set of hemoglobins. The important information yielded by chimeras is that the yolk sac is not the only site of stem cell formation. Stem cells also form within the central embryonic area of the blastodisc, and these become functional

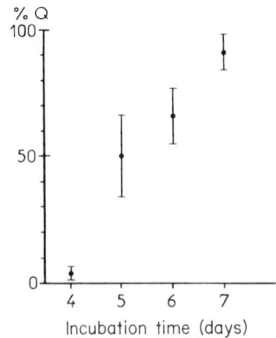

Fig. 6. Evolution of basophilic cells pertaining to quail versus chick in the blood of chimeras. These cells are precursors, probably committed to the erythroid lineage. Comparison with Fig. 2 shows that the percentage of quail among these cells increases earlier and more rapidly than among mature red cells. % Q, percentage of quail basophilic cells.

VI. Blood in Chimeras

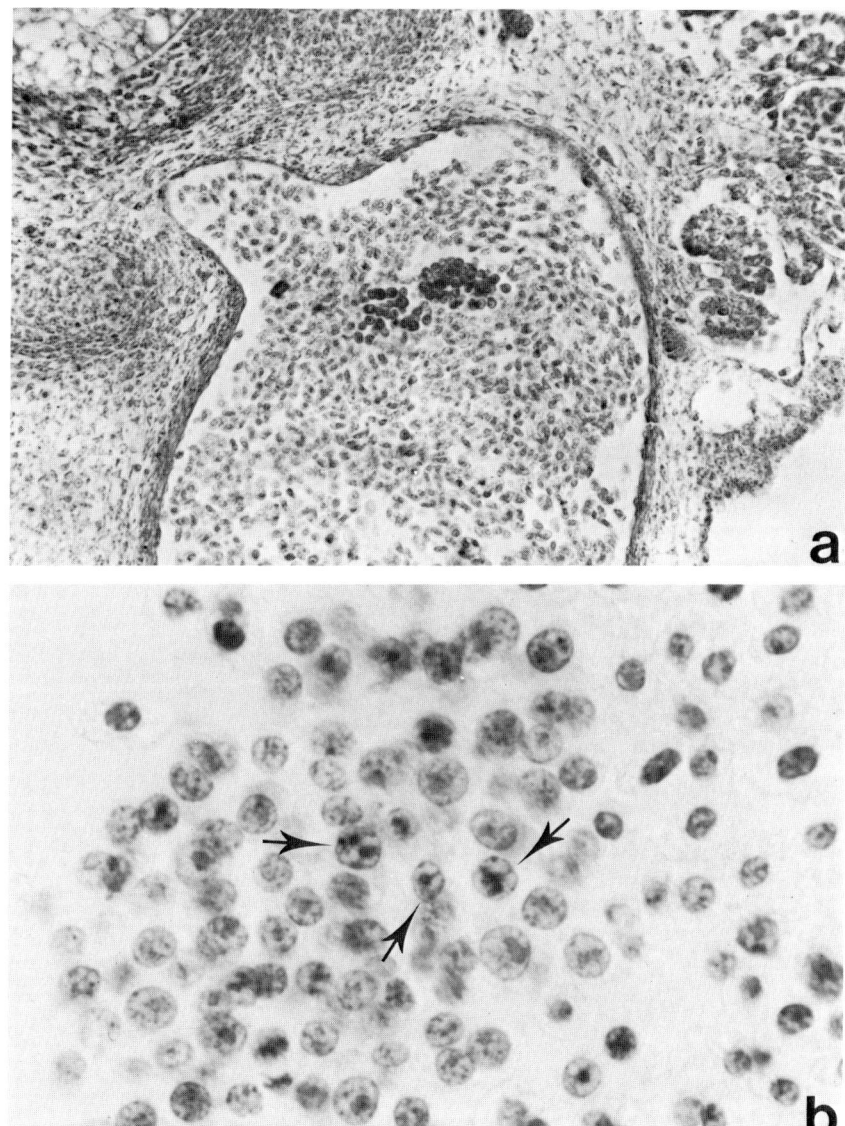

Fig. 7. A group of basophilic cells in the aorta of a 5-day chimera. (a) Methyl-green pyronine stain. × 160. (b) Feulgen-Rossenbeck poststaining of the same group of cells. A few are quail (arrows) among chick. × 1000.

from 5 days onward. In a few chimeras, a small intraembryonic contribution to primitive erythropoiesis exists, but massive differentiation of red cells from intraembryonic stem cells begins only at 5 days. Several witnesses of the replacement of one category of stem cells by the next are seen: first the appearance of quail basophilic cells in the blood, then the inflow of quail erythrocytes in the circulation and the appearance of quail erythroid cells in the yolk sac blood islands. When the first colonization wave occurs in the thymus, the erythrocytes are all chick, and the circulating basophilic cells are half quail, half chick. This suggests either that a direct circuit channels newly emerged stem cells toward the thymus, or that circulating basophilic cells may be early erythroblasts rather than totipotential stem cells.

C. Pattern of Colonization of Hemopoietic Organ Rudiments in "Complementary" Chimeras

That precise pathways channel stem cells toward the attractive organ rudiments became clear in quail↔chick chimeras (Martin et al., 1980) obtained by replacing only a segment of the central area in the chick blastodisc by its quail equivalent, instead of by grafting the whole area pellucida (see Chapter I for grafting patterns). For technical reasons, these chimeras were always constructed in the chicken egg, thus were always endowed with a chicken yolk sac. The level of the suture varied from the tenth to the eighteenth pair of somites; it was always made into the unsegmented mesoderm, immediately behind the last somite formed. "High" or "low" chimeras were obtained; in the first type head and neck belonged to one species (suture at level of somite 10); in the latter, head, neck, and wings were of one species on the abdomen of the other. Colonization of the thymus depended strikingly on the level of the suture and also on which species yielded the most anterior component. In quail "head–neck" chimeras (high chimeras), the thymus stroma, i.e., reticular epithelial cells derived from the endoderm and connective cells derived from the neural crest (see Chapter VIII), were quail and the lymphoid cells were chick. Spleen and bursa, on the other hand, were completely chick, both in stroma and hemopoietic cells. In quail "head–neck–wing" embryos (low chimeras), the thymus was entirely quail. This means that adding the wing level introduced a source of stem cells, sufficient to reverse completely the lymphoid population in the thymus. In these low chimeras, the bursa was still all chick, but the spleen sometimes had a chimeric hemopoietic population or was duplicated. Duplication could follow two patterns: either two regions, one chick and one quail, were enclosed within one another, making up an organ with a normal overall anatomy, or two spleens were present (Fig. 8). In both cases, it was striking to find that each region or each spleen had most of its hemopoietic cells of the same species as its stroma,

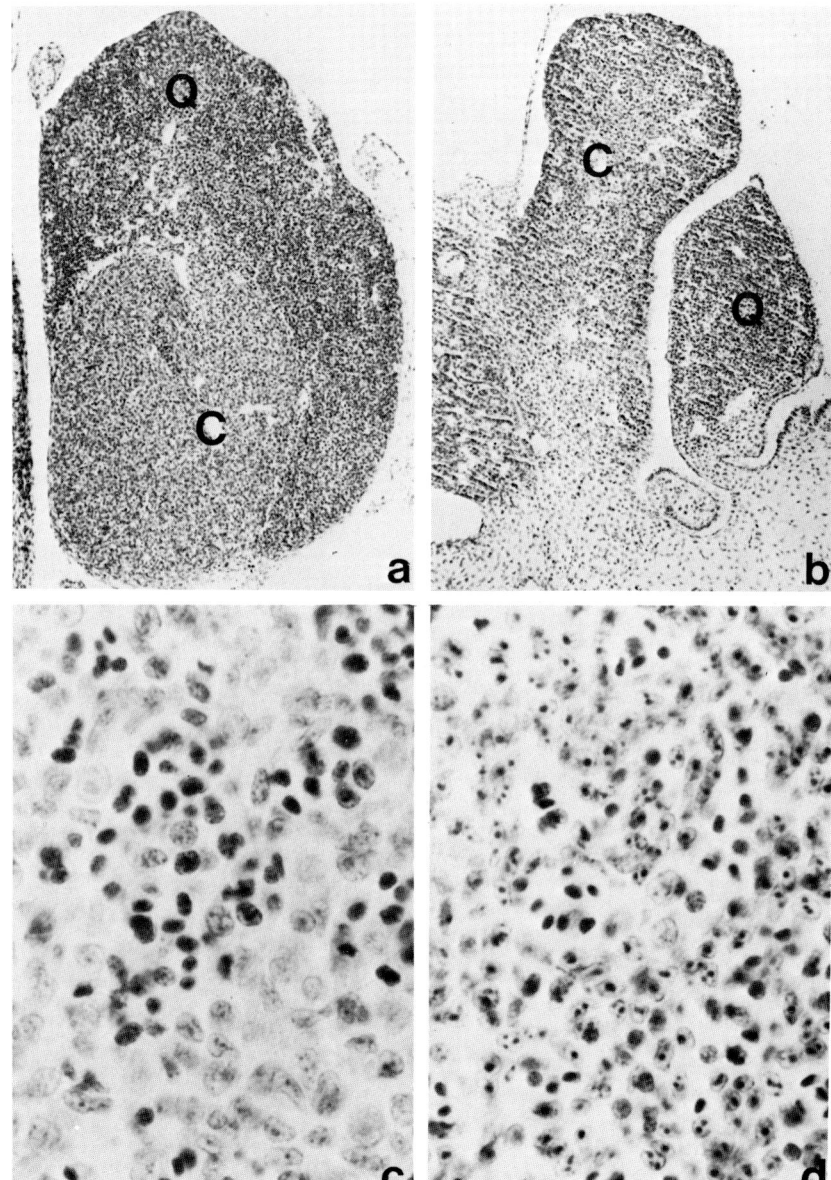

Fig. 8. Double spleen [(a) × 100] and duplicated speen [(b) × 90] in complementary chimeras. Q, quail part; C, chick part. The hemopoietic populations are mixed to a certain degree but tend to be mostly chick in the chick part [(c) × 760] and mostly quail in the quail part [(d) × 760].

indicating very little movement and mixing of cells in this particular region of the embryo.

The above description applies to anterior quail chimeras. In anterior chick chimeras paradoxical results were obtained. In head–neck chimeras of this combination, the thymus was always poorly developed; it was reduced to its stroma, the lymphoid population being absent. In the thymus of some of these chimeras lymphoid precursors were present but no thymocytes. Thus it looked as if the first wave of colonization (days 5–6 of incubation) had failed, while the second wave, which normally occurs at 11 to 12 days, had seeded the organ (see Chapter VIII for the pattern of colonization of the thymus).

However when "low" chimeras of this anterior chick combination were studied, expected results were obtained. The thymus was entirely chick; the spleen was quail with limited chimerism; the bursa was quail. Again introducing the wing level provided a source of stem cells. Why colonization did not occur in high anterior chick chimeras is still unknown. Perhaps vascular connections between the two components of the chimera were abnormal, or the stem cell source was not exactly in the same location in the two species. These two mechanisms are not very likely because diffuse foci of hemopoiesis (see below) occur in both quail and chick within the meshes of mesenchyme from 6 to 8 days of incubation. These foci are especially dense at the thorax level in chimeras as well as in normal embryos. Rather it is conceivable that the time difference in the developmental programmes of the two species is responsible for the faulty colonization. If quail totipotent stem cells emerged at a time when the chick stroma is not yet receptive, they would become committed to erythropoiesis or granulopoiesis, and hence when the chick thymic stroma becomes attractive, some 36 hours later than in the quail, stem cells with lymphoid potency would no longer be available. The presence of lymphoid precursors in otherwise empty thymic rudiments at 11–13 days suggests delayed colonization, maybe by newly emerged stem cells, but such later production of stem cells is entirely hypothetical at the present time.

These points may be resolved when the processes involved in the appearance of stem cells and in their migration are better understood. Whatever the case, these chimeras show that local pathways must be involved in the colonization of organ rudiments. If stem cells were released into the general circulatory pathways, some degree of chimerism should be achieved. Despite the demonstrated presence of yolk sac hemopoietic cells in the blood, their entry into intraembryonic organs remains limited.

Several aspects of the colonization patterns appear very similar to those in amphibian chimeras built on the same model (see Section III), suggesting that the mechanisms involved in the development of the hemopoietic system may be common.

VI. Blood in Chimeras

D. Evolution of Blood and Origin of Lymphocytes in Chick-Chick Chimeras

In quail-chick yolk sac chimeras, it is likely that interspecific differences are responsible for the irregular replacement of chick by quail red cells. The larger size of the chick yolk sac and the more rapid development of the quail embryo may play antagonistic roles; the interplay of these factors and perhaps others is difficult to assess. Further, since the quail-chick chimeras were not raised later than 13 days of incubation, one important point, i.e., whether yolk sac stem cells become extinct or participate in the definitive stem cell pool, could not be settled. This has been studied in chimeras constructed on the same experimental model but involving chick yolk sacs and embryos differing by sex chromosomes (Lassila et al., 1978), histocompatibility antigens (Lassila et al., 1982), or immunoglobin allotypes (Martin et al., 1979). In birds some MHC antigens (B-G and B-F) are expressed on the red cell surface and can be used as markers to follow cells of erythroid lineages. The homospecific chimeras were able to hatch and were raised until adulthood. Blood was taken at intervals. It was found (Table I) that the progeny of the yolk sac stem cells made up 80% of red cells at 7 days of incubation, but dropped precipitously thereafter. Indeed cells derived from the yolk sac disappeared completely after hatching. These studies also showed that thymic and bursal lymphocytes are entirely descended from intraembryonic stem cells.

TABLE I.

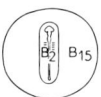

Origin of Erythroid Cells in Blood of Chick-Chick Yolk Sac Chimeras

Number of chimeras	Age	anti-B^2 (%)[a]	anti-B^{15} (%)	NCS[b] (%)
Embryo				
4	7 days	7.7	80.5	0.0
3	10 days	75.5	21.3	0.0
4	17–18 days	93.2	3.7	0.0
Posthatching				
1	2 weeks	98.5	0.0	0.0
2	4 weeks	99.2	0.0	0.0

[a] Percent of immunofluorescent cells after antiserum + fluorescein isothyocyanate conjugated anti-chicken IgG.
[b] NCS, normal chicken serum.

E. The Emergence of Intraembryonic Stem Cells

Thus definitive hemopoietic stem cells arise within the embryo. Understanding the ontogeny of the hemopoietic system revolves around several critical questions: (1) Where and when do these stem cells segregate from the mesoderm? (2) What is their relationship to yolk sac stem cells? Indeed as the mesodermal germ layer forms by ingression through the primitive streak during the process of gastrulation, it could be that cells of the hemopoietic lineage become committed in the whole blastodisc during a "unique event," such as was advocated by Moore and Owen (1967a). The peripheral cells would soon therafter become involved in erythropoiesis, while the central cells would be protected by their different cell surroundings from immediate differentiation, and through successive rounds of mitoses would acquire new properties, such as the capacity for self-renewal and the activation of new genes. Alternatively, while yolk sac stem cells arise early in development, intraembryonic ones may become committed at later stages, thus representing an entirely different population with different features from the start. (3) Another question concerns the relationship between hemopoietic and endothelial cells. At the beginning of the century, pioneers in the field of embryonic hemopoiesis postulated the existence of a common ancestral cell, the hemangioblast (Maximov, 1909; Dantschakoff, 1908).

These problems are currently under investigation using the chick dorsal mesentery as a grafting site for quail presumptive hemogenic rudiments (Dieterlen-Lièvre, 1980). As well as the quail-chick nuclear difference, the analysis relies on immunofluorescence staining by means of a monoclonal antibody, aMBI (Péault et al., 1983) that recognizes cells of the hemopoietic and endothelial lineages of the quail but not erythroid cells.

In normal development, diffuse hemopoiesis occurs in 6–9 day embryos of both species. Scattered basophilic cells are first present at 4 and 5 days within the mesenchyme. These cells become very numerous and conspicuous foci are present in the dorsal mesentery in the vicinity of the thoracic aorta (Dieterlen-Lièvre and Martin, 1981) (Fig. 9). Cells staining with aMBI appear in the lateral plate mesoderm in parallel with segmentation of the somite material, i.e., from head to tail. When grafted in the dorsal mesentery of the chick embryo, a quail rudiment containing stem cells, newborn bone marrow, for instance, gives rise to hemopoietic foci, the cells of which have quail nuclei and are aMBI-positive. The latter reaction shows that the cells belonged to the hemopoietic or endothelial lineage and arose from the graft. In this way it was shown that the 3- and 4-day quail aorta can give rise to erythroid and granuloid foci. Lateral plate mesoderm from earlier stages recovered with endoderm yielded aMBI-positive vessels and round cells, and also structures that resembled blood islands composed of aMBI negative erythroblasts encased in an aMBI-positive endothelium. The cells from lateral plate mesoderm, grafted alone, appeared extremely invasive and colo-

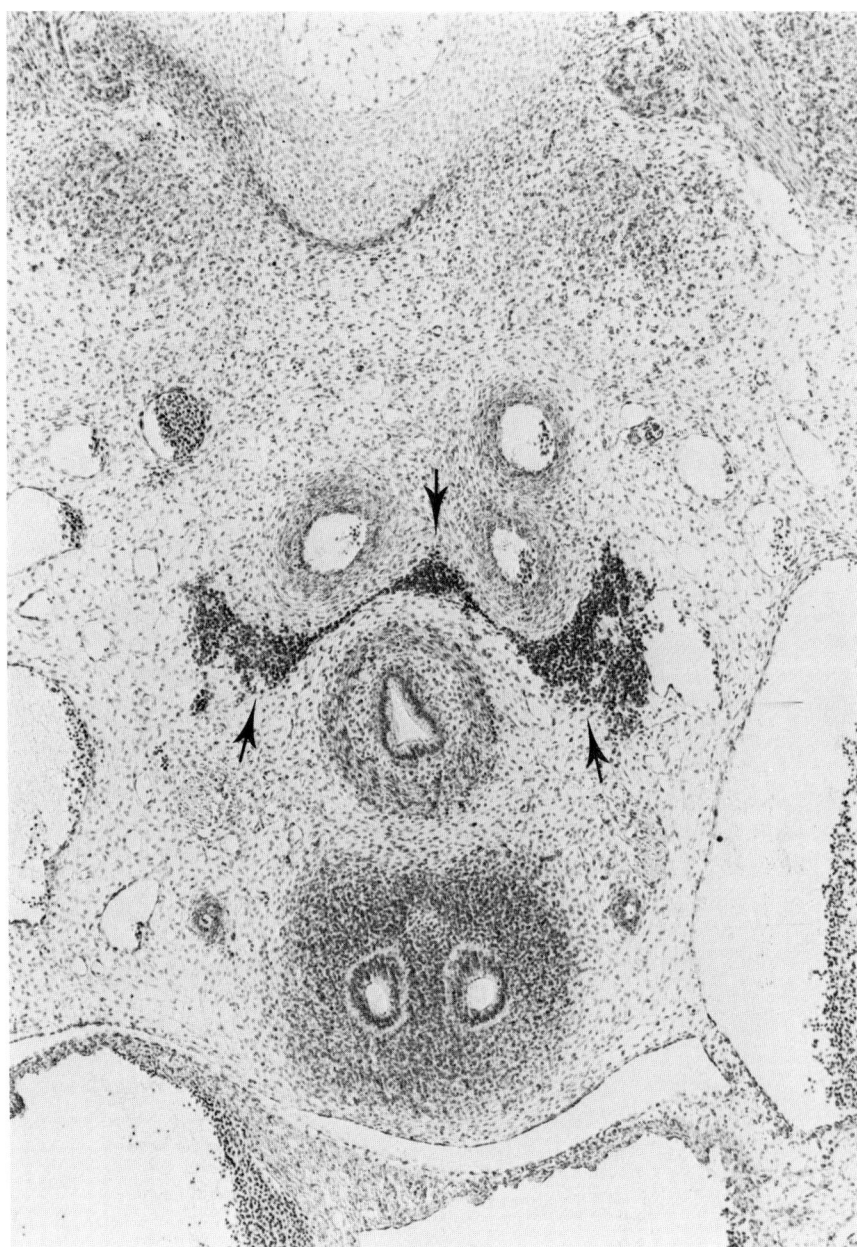

Fig. 9. Hemopoietic focus (arrows) in the dorsal mesentery of a 7-day chick embryo. It appears as a sinuous mass between oesophagus and pulmonary arteries and systemic aorta. Feulgen–Rossenbeck stain. × 110.

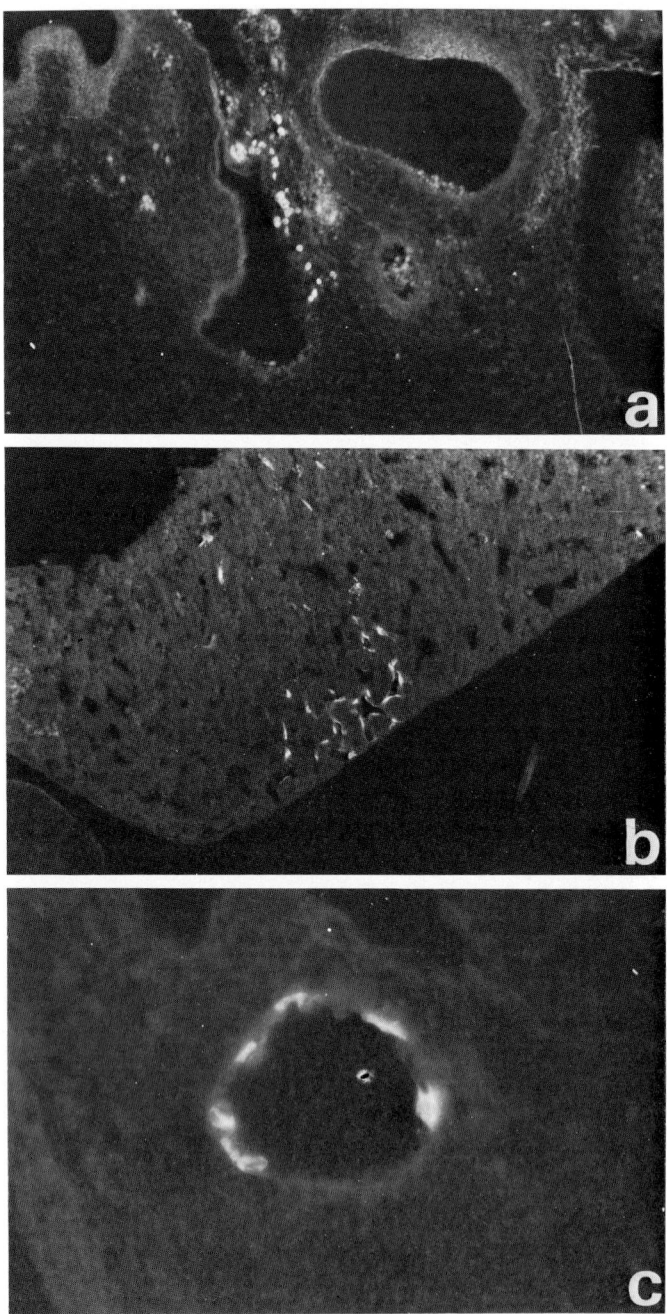

VI. Blood in Chimeras

nized even distant tissues of the host where they participated in vessel walls (Fig. 10c), sometimes coating the sinusoids of a large sector in the liver (Fig. 10b). They also homed to the dorsal mesentery, where they took on the round appearance and intense aMB1 surface staining typical of young hemopoietic cells (Fig. 10a).

Together with the appearance of aMB1-positive cells in young embryos, these experimental data suggest that cells committed to the hemangioblastic lineage arise from mesodermal precursors in the lateral plate during days 2–4 of incubation, according to a cephalocaudal gradient. This process is probably distinct from the emergence of yolk sac stem cells, which begins at least 24 hours earlier and proceeds in a caudocephalic direction.

The hemangioblasts appear to be endowed with invasive and recognition properties that enable them to collaborate with foreign endothelial cells in the construction of vessels. Even when incorporated into an endothelium, some of them may retain their hemogenic capacity, at least as long as this capacity is present in the embryo. Depending on the site where these hemangioblasts settle, they are able to multiply actively, giving rise for instance to the para-aortic foci in the dorsal mesentery.

In conclusion, ontogeny of the hemopoietic system in chick and quail involves a number of events that have been established and timed from chimeric organs or embryos. The chronological unfolding of these various processes, i.e., whether they are successive or simultaneous, appears identical in quail and chick, provided a different time scale is applied to the two species (Fig. 11). As described earlier (Dantschakoff, 1908; Maximov, 1909; Metcalf and Moore, 1971), basophilic cells are found in connection with the various incipient hemopoietic processes. In the blood of yolk sac chimeras, chick basophilic cells are progressively replaced by quail. The latter first make their appearance in the mesenchyme of the embryo, mostly in the vicinity of the thoracic and abdominal aorta, before the first colonization period of the thymus and the start of definitive hemopoiesis. It is thus tempting to ascribe to these cells a fundamental founding role for hemopoietic lineages. However, the following events still remain elusive, since basophilic cells in the blood and the para-aortic foci disappear before the permanent bone marrow stem cell pool becomes seeded. This latter process is about contemporary with the second thymic stem cell inflow. Thus the ontogeny of the hemopoietic system appears rhythmic, with periodic recurrence of similar

Fig. 10. aMB1 staining of quail cells of hemangioblastic lineage implanted in the chick embryo. Lateral plate mesoderm from 12–14 somite quail embryos was grafted into the coelom of 3-day chick embryos and allowed to grow for 3 days. (a) grafted cells have homed to the dorsal mesentery where they assume the typical round morphology of cells in the hemopoietic foci ($\times$ 86). (b) Some have invaded the liver where they coat some sinusoids ($\times$ 86). (c) Grafted cells participate in the lining of a venous capillary; the chick endothelial cells are nonfluorescent ($\times$ 408). aMB1 treatment was followed by fluoresceine-coupled rabbit anti-mouse serum.

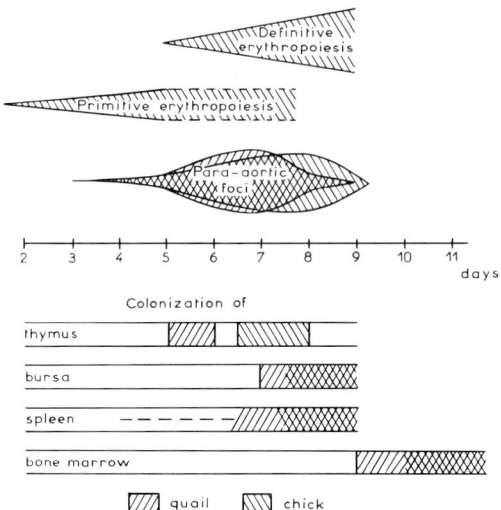

Fig. 11. Synopsis of characteristic events occurring during the development of quail and chick hemopoietic systems.

processes in different sites or organs. Studies on chimeras clearly show that stem cells arise during a first, early round in the yolk sac blood islands, and then form anew, essentially within the embryo. We do not know whether this secondary emergence event is the last, or whether another round occurs later.

III. AMPHIBIAN CHIMERAS

The analysis of blood formation in amphibian chimeras has revolved around the same problems, i.e., (1) whether organ rudiments have their own stem cells or become colonized at the outset of organogenesis; (2) where stem cells come from and what is the role of ventral blood islands. The method for obtaining chimeras is analogous to that used in birds: either rudiments are exchanged between marked embryos or complementary embryos are constructed surgically. The marking system relies on the ease with which polyploidy is induced by an osmotic shock applied to early embryos. The ontogeny of the hemopoietic system appears to be strikingly similar to that in birds.

A. Extirpation Experiments

The amphibian story began in the 1920s with extirpation experiments on urodele and anuran embryos (Federici, 1926; Goss, 1928; Slonimski, 1931).

VI. Blood in Chimeras

When the ventral blood island was removed prior to vascularization, the animals survived for some time, but red cells were drastically reduced or even absent when extirpation was complete. These experiments seemed to suggest that no erythroid lineage could be produced if the ventral blood island were missing and were taken as a strong argument for Moore and Owen's theory, i.e., that these islands were responsible for the production of stem cells destined to found the definitive hemopoietic system. However the bloodless larvae had a normal vascular system with normal endothelial linings.

B. Graft of Marked Rudiments

When Hollyfield in 1966 exchanged ventral blood islands between *Rana pipiens* embryos of different ploidies, no definitive erythrocytes from the marked grafts appeared in the tadpoles. This experiment was repeated a number of times to investigate the origin of thymic lymphoid precursors (Turpen *et al.*, 1973, 1975, 1979, 1981; Volpe *et al.*, 1977, 1979), but the ventral blood islands never contributed any cells to the thymus. Thus an experimental scheme comparable to that used in birds gave the same result, showing that definitive hemopoietic organs are not seeded by stem cells from the ventral blood islands. According to Hollyfield, if larvae deprived of blood islands remained bloodless, it is because they did not live till the time when definitive red cells normally replace the primitive ones.

In his 1966 experiments, Hollyfield also exchanged pronephric rudiments, initiating experiments that pointed to a possible alternative source of stem cells. Pronephric transplantation resulted in the appearance of red cells from the graft in two out of four experimental animals, suggesting that the kidney area was a region capable of providing red cells and maybe stem cells.

C. Extrinsic Origin of Stem Cells

For several years, the origin of stem cells in amphibians was a much disputed issue, centred around the ontogeny of the thymus. In the urodele *Pleurodeles,* host stem cells were found to colonize a grafted thymus (Deparis and Jaylet, 1975, 1976). The use of tetraploidy as a marker allowed the identification of interphase cells by their volume and of mitotic cells by their chromosome number. However, it could be argued that the large size of these cells made them less competitive than diploid ones. A marker chromosome was therefore used in later experiments, which confirmed the extrinsic origin of thymocyte precursors. Although mesenchyme surrounding the thymus endoderm was included, no donor phenotype lymphoid cells differentiated, indicating that the precursors do not originate in the immediate vicinity of the rudiment.

Working on anurans initially suggested the opposite interpretation (Turpen *et al.*, 1973, 1975; Volpe and Turpen, 1975). When thymic primordia were exchanged between diploid and triploid embryos of *Rana pipiens,* thymocyte precursors appeared to arise within the endomesodermal rudiment. The thymus was even attributed a special role in the ontogeny of the hemopoietic system, being considered at one time as supplying stem cells (Volpe and Turpen, 1975).

Recent studies on *Xenopus laevis* and *R. pipiens* have brought anurans into line (Volpe *et al.*, 1977, 1979; Tompkins *et al.*, 1979, 1980; Turpen *et al.*, 1981). It is now clear that a thymic rudiment obtained from a young enough donor is always colonized by lymphoid precursors from the host. Several technical factors may have been responsible for the early interpretations: (1) the marker, i.e. diploidy versus triploidy, is difficult to interpret; (2) the timing of colonization probably varies from species to species, and this event may occur very early in *R. pipiens;* (3) retrieval of rudiments without surrounding tissues is probably more difficult in amphibians than in birds.

D. Chimeric Complementary Embryos

The experimental scheme is identical to that described above for birds (Volpe and Turpen, 1975; Volpe *et al.*, 1979; Turpen *et al.*, 1979), though the level of recombination cannot be as accurately determined. In these chimeras, the anteriorly located thymus was colonized by stem cells originating from the posterior half. By varying the level of the suture, the authors were again led to indict the kidney region as a stem cell source. Transplantation of the mesonephros on one or both sides resulted in chimeric populations in the pronephric granulocytes, in the liver and spleen hemopoietic cells, and in circulating blood (Turpen *et al.*, 1981; Volpe *et al.*, 1981).

The explanted kidney area included not only nephrogenic tissue but some dorsal mesentery and dorsal aorta (Turpen *et al.*, 1983). Thus lateral plate is probably the mesodermal compartment that initially gives off stem cells for the hemopoietic system, as in birds. However, there still remains some dissent about the sites within lateral plate mesoderm where hemopoietic stem cells may arise. According to the anuran work these sites are dorsal. Deparis and Jaylet (1975; see also Turpen *et al.*, 1983), working with Pleurodeles, conclude that some stem cells originate from the mesoderm in the anterior region of the trunk, and that an inductive influence exerted by the hepatic endoderm mediates the formation of these cells.

A possibility that would reconcile both views is that lateral plate mesoderm has a widespread hemogenic capacity, perhaps under the influence of other cells. Depending on the experimental scheme, its timing and the species chosen, stem cell production could thus be traced to different sites.

IV. MAMMALIAN CHIMERAS

A. Chimeric Organs

Mammalian chimeric organs can be constructed by combining *in vitro* dissociated tissues or an attractive rudiment and a stem cell source. The first attempts at making chimeric organs were by Moore and Owen (1967b), who grafted mouse foetal thymus onto the chick chorioallantoic membrane (CAM) and found that the early rudiment was devoid of stem cells. Avian stem cells are incapable of colonizing mammalian rudiments, an observation first made by Auerbach (1961) and often confirmed since, which suggests that the recognition mechanisms involved in the development of the hemopoietic system do not operate between cells from these two classes. Chimeric mouse thymuses were obtained by Fontaine-Perus *et al.* (1981) by associating *in vitro* a thymus rudiment with yolk sac, fetal liver, spleen, or adult bone marrow. The organs, associated across a large-pored filter and cultured for several days, were eventually grafted on the CAM. Ten-day rudiments, cultivated or grafted alone, never became lymphoid. Twelve- to 14-day organs became heavily infiltrated with lymphocytes, which emigrated out 10 days after grafting. Lymphoid differentiation always occurred when the young rudiment was associated with bone marrow or a postcolonization thymus, but other organs were poor sources of precursors (Table II).

The extrinsic origin of precursors was borne out by using the two alleles of the Thy-1 antigen and glucosephosphate isomerase isozymes. This experiment dem-

TABLE II.

Efficiency of Various Organs as Lymphoid Precursor Donors[a]

Donor organ	Lymphoid differentiation
13-day limb bud	0/4[b]
8-day yolk sac	2/8
12-day fetal liver	4/12
17-day spleen	2/6
Adult bone marrow	5/6
13- to 19-day fetal thymus	77/78

[a] Modified from Fontaine-Perus *et al.* (1981).
[b] Number of cases with lymphoid differentiation/total number of associations.

onstrates that the early thymus rudiment is attractive to lymphoid precursors and is even capable of displacing cells which have previously homed to another thymus, but that hemopoietic organs differ very much in their content of competent stem cells.

A new avenue concerning the vascularization of the kidney rudiment has been opened recently by Saxén's group (Sariola *et al.,* 1983). Mouse metanephric anlagen were explanted onto the quail CAM (see also Chapter XV). The vessels that irrigated the developing kidney were always provided by the host, demonstrating the extrinsic origin of the vasculature in this organ, another instance where the independent origin of the hemangioblastic lineage is clear.

These experiments also showed that the angiogenic response evoked by the grafted rudiment is strictly dependent on its induction state. In whole 11-day kidneys, the host endothelial cells invaded the glomerulus, but when the rudiment was dissociated into metanephric mesenchyme and ureteric bud, the result was different. Isolated 11-day metanephric mesenchyme does not differentiate *in vitro:* from previous work it is known that transient contact with an inductor tissue, which can be removed before overt morphogenesis, permits the mesenchyme to differentiate. The vascular response elicited on the CAM by induced and uninduced mesenchyme was very different. Uninduced mesenchyme remained as a solid cell mass that did not become vascularized despite good survival. By contrast, induced mesenchyme was invaded by vessels. The vascular reponse was dependent on the length of the induction period prior to grafting. Sometimes a portion of the tissue remained undifferentiated and unvascularized. In most of these induced mesenchymes, however, glomeruli remained avascular, perhaps because of a disorderly development of the glomerulus structure. In any case, these experiments established that invasion by host capillaries was intimately linked with the formation of pretubular mesenchymal aggregates. Thus induced mesenchyme probably produces angiogenic factors responsible for endothelium ingrowth, whereas uninduced mesenchyme does not.

B. Tetraparental Chimeras

The most remarkable feature of these chimeras is immunological tolerance: two cell populations coexist through life. The accommodation of the immune system to this unusual circumstance is dealt with in Chapter IX. Chimerism of red blood cells will be considered here. The question that has attracted most attention is whether, in various strain combinations, one cell population tends to overgrow the other. In other words, is blood chimerism stable, or are there changes with time in the composition of the blood, i.e., are some genotypes more successful than others? And is the strain composition of hemopoietic organs and blood cell lineages identical? Studies on chimeras in mice and other

TABLE III.

"Chimeric Drift" in Tetraparental Animals[a]

References	Type of chimera	Markers	Observations and conclusion
Mouse			
Mintz and Palm (1965)	Aggregation chimera	H2 alleles	Two populations of erythrocytes
Mintz and Palm (1969)	Aggregation chimera	H2 alleles	C57BL/6 may have selective advantage over C3H
Wegmann and Gilman (1970)	Aggregation chimera	hb markers, Ig1 alleles, coat color	Concomitant chimerism in red cells, B lymphocytes and pigment cells, indicating that several cells are at the origin of each system. One mouse shows a shift in hb composition, without a shift in Ig1 composition
Gornish et al. (1972)	Aggregation chimera	T6 chromosome	Ratio of the two cell populations identical among hemopoietic organs but variable between individuals
Ford et al. (1974)	Aggregation chimera	T6 chromosome	Dominance of AKR over CBA mitoses
Ford et al. (1975)	Injection chimera	T6 chromosome	Two groups of lymphomyeloid tissues according to the proportions of host to donor cells, suggesting two distinct stem cell pools
Warner et al. (1977a)	Aggregation chimera	Surface alloantigens	Stable chimerism in some animals, chimeric drift in PBL of others. GVH-R responsible for the drift?
Warner et al. (1977b)	Aggregation chimera	Surface alloantigens and hb markers	Chimeric drift uncorrelated in WBC and RBC
Warner et al. (1977c)	Aggregation chimera	Surface alloantigens	WBC correlated to spleen composition but not to thymus in some chimeras
West (1977)	Aggregation chimera	hb markers, GPI and IDH isozymes	Chimeric drift favoring C57BL over C3H
Stephens and Warner (1980)	Aggregation chimera	Surface alloantigens, hb markers	Chimeric drift fluctuating back and forth between parental strains, independently in WBC and RBC
Sheep			
Tucker et al. (1974)	Injection chimera	Blood groups, albumin and transferrin markers	In 3 animals, the blastomere-derived population declines. In the fourth the blood is entirely and constantly derived from injected blastomeres.

[a] GVH-R, Graft-versus-host reaction; GPI, Glucosephosphate isomerase; hb, hemoglobin; H2, mouse major histocompatibility complex; Ig1, immunoglobulin allotypes; IDH, isocitrate dehydrogenase; PBL, peripheral blood leukocytes; RBC, red blood cells; WBC, white blood cells.

species, relying on a variety of cell markers (Table III) have suggested that red or white cell composition in the peripheral blood changed significantly with time, a phenomenon that has been termed "chimeric drift."

Stephens and Warner (1980) reconsidered this problem by studying nine tetraparental mice at weekly intervals over a period of 6 weeks. They showed that chimeric drift, though a reality, may occur in either parental direction and may fluctuate back and forth between the two parental types with no obvious periodicity. The changes in the white blood cell composition appeared not to be coordinated with those of red blood cells. The authors therefore suggest separate regulatory mechanisms for the two cell populations.

A convincing explanation for the chimeric drift phenomenon is provided in a recent study by Burton et al. (1982). These authors have studied the natural mosaicism of female mammals due to the coexistence in them of two cell populations differing with respect to their active X chromosome (Lyon, 1970). Female mice, heterozygous at the *Pgk-1* locus, were studied. This X chromosome locus encodes the enzyme phosphoglycerate kinase, which has two electrophoretic variants. The isozyme proportions were determined at 2–4-week intervals for 56 weeks. Large, and frequently sudden, variations in the amounts of A and B isozymes were evident in every individual female suggesting that, at each time point, blood cells are the progeny of only a few clones of cells and that the identity of these clones change rapidly. Using a mathematical model that takes into account the parameters of the mouse erythropoietic system, the authors made a computer simulation of the data. By assuming that three clones could provide the total steady-state erythrocyte requirement for 14 days, they obtained simulations that closely mimicked the actual data. For our present purpose, we shall mainly emphasize that the evolution of blood cell populations in normal females, as revealed by the isozyme marker, is closely similar to that in tetraparental mice.

C. Blood Chimeras Obtained by Intraplacental Transplantation of Fetal Liver Cells

Though tetraparental chimeras may possess hemopoietic cells with two different ancestries, they cannot be constructed according to a specific predetermined pattern. For instance it is not possible to restrict one of the genotypes to the hemopoietic lineage and the other to the microenvironmental stroma. In order to achieve this goal, Mintz and her co-workers introduced a new type of chimera, produced by injecting a suspension of fetal liver cells into the placenta of 11-day mouse fetuses (Fleishman and Mintz, 1979; Fleishman et al., 1982), the liver being the major hemopoietic organ of the mammalian fetus. Recipients with a genetically determined anemia were chosen, using different grades of the domi-

VI. Blood in Chimeras

nant spotting (*W*) mutation, the most severe of which is lethal when homozygous. All fetuses in each mother had to be injected in order to study the 25% of homozygotes resulting from the heterozygous matings. An array of markers were used to assess the contribution of the donor cells and their progeny to hemopoiesis and to probe the interplay of donor and recipient cells in the hemopoietic organs. These markers were strain-specific hemoglobin variants, isozymes of glucosephosphate isomerase and phosphoglucomutase, urinary proteins produced by hepatocytes, immunoglobulin allotypes, the "beige" giant granules of polymorphonuclear cells, and MHC antigens.

The anemic genotype was found to be a prerequisite for the establishment of donor stem cells. Normal littermates never yielded detectable donor red blood cells and the success of the transfer, as judged from the proportion of donor hemoglobin in the blood, tended to be related to the severity of the anemia, demonstrating either complete and permanent replacement of stem cells, or only limited and eventually transient contribution of donor cells. Thus the host stem cell defect was responsible for the existence of a niche permitting survival of the grafted cell population. The initial inoculum was small, because only $1-2 \times 10^5$ cells could be conveniently inoculated via the placental route; since only some of the animals were cured of their anemia, the cell dose was considered as limiting. It is thus probable that very few totipotent stem cells originally implanted, and that these then underwent long-term proliferation.

The replacement kinetics of white blood cells was compared to that of red. Though the extent of replacement of granulocytes was always much lower than that of erythrocytes, the percentage of the two populations in individual animals underwent nevertheless strikingly parallel evolution. Animals with complete and sustained erythroid replacement also produced substantial levels of donor-type immunoglobulin. The presence of donor-derived T cells could also be deduced from the proportion of white blood cells carrying donor MHC antigens in animals with a high level of erythroid replacement.

Several conclusions could be drawn from these studies: tolerance *to* and *of* the foreign cells was permanent; production of immunoglobulins occurred despite MHC discrepancy between Ig-secreting cells and T cells, on the one hand, and educating thymus, on the other. In view of the limited number of cells that probably implanted, the authors consider that common totipotent stem cells probably gave rise to both myeloid and lymphoid lineages. Histological study of the foetuses revealed that the injected cells first homed to and proliferated in the liver, which appeared strikingly different to the aplastic livers of nonrestored anemic foetuses. Thus the stem cells that thereafter colonized the bone marrow and spleen first passed through the liver.

It is interesting to compare these results with those obtained earlier in another study with similar aims. Weissman *et al.* (1978) injected yolk sac blood islands into the yolk sac cavity of 8- to 10-day mouse embryos. No donor red cells were found in the recipients. However *H2* or *Thy-1* markers did reveal a minute proportion of marked cells in spleen colonies deriving from bone marrow cells of some of these animals. Several factors may account for this limited success. The injection route was undoubtedly less efficient; the injected cells had to compete with the host's normal stem cells; finally, whereas Fleishman and Mintz used donor cells from the foetal liver, Weissman *et al.* used yolk sac cells. Despite accumulating evidence from culture experiments (Cudennec *et al.*, 1981; Wong *et al.*, 1982; Ripoche and Cudennec, 1983) that the yolk sac does produce stem cells capable of undergoing definitive erythropoiesis, there is still no critical evidence that (1) these are also totipotent; (2) they are capable of seeding the whole hemopoietic system in mammals; (3) in normal development they are the only stem cells to do so.

D. Teratocarcinoma-Derived Blood Cells

Mintz and Illmensee (1975) obtained chimeras by injecting cells from a teratocarcinoma line into a blastocyst. Some chimeras displayed the diluted coat color typical for the mutant steel (*Sl*). The gene, not known to be present in the original components, could be traced back to the animal that originally gave rise to the teratocarcinoma. This serendipitous observation led to the study of the hemopoietic system in these chimeras (Mintz and Cronmiller, 1978). Mutations at the steel locus alter the differentiation of three cell lineages, pigment, germ cells, and blood, by causing some yet uncharacterized microenvironmental deficiency. Steel thus appears as reciprocal to dominant spotting (*W*), which also affects these three lineages but produces a stem cell defect. The steel gene leads to macrocytic anemia and, when homozygous, to perinatal death, at least in the more severe alleles.

The blood of chimeras was phenotypically normal, even when a substantial number of red blood cells derived from the teratocarcinoma cells with the steel gene. Surprisingly this was so even in one mouse in which 80% of the cells of either the hemopoietic or stromal lineage seemed to be of the $Sl^J/+$ component. The red blood cell population was identified through the hemoglobin markers, while the constitution of organs, such as liver, spleen, and thymus, was inferred from GPI electrophoretic variants. The authors concluded that a minor fraction of normal cells in the microenvironment probably provided the necessary inductive effect, maybe by producing diffusible substances with long-range activity. However in view of this unexpected rescue, it could also be argued that the interpretation of the data was limited by the absence of individual cell markers and that blood cells, unavoidable in the hemopoietic organs, contributed some GPI that concealed or biased the true composition of stromal cells.

V. CONCLUSION

Chimeras, either as composite embryos or composite organs, have been powerful tools to elucidate the origin and movements of blood cells and the influences to which they are subjected during development.

The major lesson in the three classes of vertebrates that have been studied has been that blood and endothelium arise as an independent lineage that thereafter colonizes the mesodermal or endomesodermal frame of the hemopoietic organs. In birds and amphibians, the emergence of this hemangioblastic rudiment has been shown to occur independently in the primitive blood islands and in the embryo. Whether or not the yolk sac is the progenitor of definitive hemopoietic stem cells in mammals remains to be elucidated. This fundamental question has remained elusive up to now, partly due to the difficulty of making chimeras in the mammalian embryo during the organogenesis period. A future point of interest will be the developmental relationship between blood and endothelial cells. Delving into a deeper level of cell function, blood cells are also primary targets for studies on how relevant genes become activated during development.

ACKNOWLEDGMENTS

The original work reported in this chapter has been supported by the CNRS and by grants from DGRST and MRI.

REFERENCES

Auerbach, R. (1961). *Dev. Biol.* **3,** 336–354.
Beaupain, D., Martin, C., and Dieterlen-Lièvre, F. (1979) *Blood* **53,** 212–225.
Beaupain, D., Martin, C., and Dieterlen-Lièvre, F. (1980). *In* "*In Vivo* and *In Vitro* Erythropoiesis: The Friend System" (G. B. Rossi, ed.), pp. 21–32. Elsevier/North-Holland Biomedical Press, Amsterdam.
Bruns, G. A. P., and Ingram, V. M. (1973). *Philos. Trans. Soc. B* **266,** 225–305.
Burton, D. I., Ansell, J. D., Gray, R. A., and Micklem, H. S. (1982). *Nature (London)* **298,** 562–563.
Cudennec, C. A., Thiery, J. P., and Le Douarin, N. M. (1981) *Proc. Natl. Acad. Sci. U.S.A.* **78,** 2412–2416.
Dantschakoff, V. (1908). *Anat. Hefte* **37,** 471–589.
Deparis, P., and Jaylet, A. (1975). *J. Embryol. Exp. Morphol.* **33,** 665–683.
Deparis, P., and Jaylet, A. (1976). *Ann. Immunol. (Paris)* **127C,** 827–831.
Dieterlen-Lièvre, F. (1975). *J. Embryol. Exp. Morphol.* **33,** 607–619.
Dieterlen-Lièvre, F. (1980). *In* "*Adv. Physiol. Sci.*" **6,** (S. Hollau, G. Gardos and B. Sarkodi, eds.), pp. 139–145. Pergamon, New York.
Dieterlen-Lièvre, F., and Martin, C. (1981). *Dev. Biol.* **88,** 180–191.

Dieterlen-Lièvre, F., Beaupain, D., and Martin, C. (1976). *Ann. Immunol. (Paris)* **127c**, 857–863.
Federici, H. (1926). *Arch. Biol.* **36**, 466–487.
Fleishman, R. A., and Mintz, B. (1979). *Proc. Natl. Acad. Sci. U.S.A.* **11**, 5736–5740.
Fleishman, R. A., Custer, R. P., and Mintz, B. (1982). *Cell* **30**, 351–359.
Fontaine-Perus, J. C., Calman, F. M., Kaplan, C., and Le Douarin, N. M. (1981). *J. Immunol.* **126**, 2310–2316.
Ford, C. E., Evans, E. P., Burtenshaw, M. D., Clegg, H., Barnes, R. D., and Tuffrey, M. (1974). *Differentiation* **2**, 321–333.
Ford, C. E., Evans, E. P., and Gardner, R. L. (1975). *J. Embryol. Exp. Morphol.* **33**, 447–457.
Gornish, M., Webster, M. P., and Wegmann, T. G. (1972). *Nature (London) New Biol.* **237**, 249–251.
Goss, C. M. (1928). *J. Exp. Zool.* **52**, 45–63.
Hamilton, H. L. (1952). "Lillie's Development of the Chick." Holt, New York.
Hollyfield, J. G. (1966). *Dev. Biol.* **14**, 461–480.
Jotereau, F., and Houssaint, E. (1977). In "Developmental Immunobiology" (J. B. Solomon and J. D. Horton, eds.), pp. 123–130. Elsevier–North-Holland, Amsterdam.
Jotereau, F., and Le Douarin, N. M. (1978). *Dev. Biol.* **63**, 253–265.
Kahn, A. J., and Simmons, D. J. (1975). *Nature (London)* **258**, 325–327.
Lassila, O., Eskola, J., Toivanen, P., Martin, C., and Dieterlen-Lièvre, F. (1978). *Nature (London)* **272**, 353–354.
Lassila, O., Martin, C., Toivanen, P., and Dieterlen-Lièvre, F. (1982). *Blood* **59**, 377–381.
Lyon, M. (1970). *Philos. Trans. R. Soc. London, Ser. B* **259**, 41–52.
Martin, C. (1972). *C.R. Seances Soc. Biol.* **166**, 283–285.
Martin, C., Beuapain, D., and Dieterlen-Lièvre, F. (1978). *Cell Differ.* **7**, 115–130.
Martin, C., Lassila, O., Nurmi, T., Eskola, J., Dieterlen-Lièvre, and Toivanen, J. (1979). *Scand. J. Immunol.* **10**, 333–338.
Martin, C., Beaupain, D., and Dieterlen-Lièvre, F. (1980). *Dev. Biol.* **75**, 303–314.
Maximov, A. (1909). *Arch. Mikrosk. Anat. Entwicklungsgesch.* **73**, 444–561.
Metcalf, D., and Moore, M. A. S. (1971). "Haemopoietic Cells." North-Holland Publ., Amsterdam.
Mintz, B., and Cronmiller, C. (1978). *Proc. Natl. Acad. Sci. U.S.A.* **75**, 6247–6251.
Mintz, B., and Illmensee, K. (1975). *Proc. Natl. Acad. Sci. U.S.A.* **72**, 3585–3589.
Mintz, B., and Palm J. (1965). *J. Cell Biol.* **27**, 66A.
Mintz, B., and Palm, J. (1969). *J. Exp. Med.* **129**, 1013–1027.
Moore, M. A. S., and Owen, J. J. T. (1965). *Nature (London)* **208**, 956–990.
Moore, M. A. S., and Owen, J. J. T. (1967a). *Lancet* **2**, 658–659.
Moore, M. A. S., and Owen, J. J. T. (1967b). *J. Exp. Med.* **126**, 715–725.
Péault, B. M., Thiery, J. P., and Le Douarin, N. M. (1983). *Proc. Natl. Acad. Sci. U.S.A.* **80**, 2976–2980.
Ripoche, M. A., and Cudennec, C. A. (1983). *Cell Differ.* **13**, 125–131.
Sariola, H., Ekblom, P., Lehtonen, E., and Saxén, L. (1983). *Dev. Biol.* **96**, 427–435.
Slonimski, P. (1931). *Arch. Biol.* **42**, 415–477.
Stephens, T. J., and Warner, C. M. (1980). *Cell. Immunol.* **56**, 132–141.
Till, J. E., and McCulloch, E. A. (1961). *Radiat. Res.* **14**, 213–222.
Tompkins, R., Reinschmidt, D., and Volpe, E. P. (1979). *Dev. Comp. Immunol.* **3**, 635–642.
Tompkins, R., Volpe, E. P., and Reinschmidt, D. (1980). In "Development and Differentiation of Vertebrate Lymphocytes" (J. D. Horton, ed.), pp. 25–34. Elsevier/North-Holland, Amsterdam.
Tucker, E. M., Moor, R. M., and Rowson, L. E. A. (1974). *Immunology* **26**, 613–621.
Turpen, J. B., Volpe, E. P., and Cohen, N. (1975). *Am. Zool.* **15**, 51–61.

VI. Blood in Chimeras

Turpen, J. B., Volpe, E. P., and Cohen, N. (1973). *Science* **182,** 931–933.
Turpen, J. B., Turpen, C. J., and Flajnik, M. (1979). *Dev. Biol.* **69,** 466–479.
Turpen, J. B., Knudson, C. M., and Hoefen, P. S. (1981). *Dev. Biol.* **85,** 99–112.
Turpen, J. B., Cohen, N., Deparis, P., Jaylet, A., Tompkins, R., and Volpe, E. P. (1983). *In* "The Reticuloendothelial System" (N. Cohen and M. M. Sigel, eds.), Vol. 3, pp. 569–588. Plenum, New York.
Volpe, E. P., and Turpen, J. B. (1975). *Science* **190,** 1101–1103.
Volpe, E. P., Tompkins, R., and Reinschmidt, D. (1977). *In* "Developmental Immunolobiology" (J. B. Solomon and J. D. Horton, eds), pp. 109–114. Elsevier/North Holland, Amsterdam.
Volpe, E. P., Tompkins, R., and Reinschmidt, D. (1979). *J. Exp. Zool.* **208,** 57–66.
Volpe, E. P., Tompkins, R. and Reinschmidt, D. (1981). *In* "Aspects of Developmental and Comparative Immunuology" (J. B. Solomon, ed.), pp. 193–201. Pergamon, New York.
Warner, C. M., McIvor, J. L., and Stephens, T. J. (1977a). *Differentiation* **9,** 11–17.
Warner, C. M., McIvor, J. L., and Stephens, T. J. (1977b). *Transplantation* **24,** 183–193.
Warner, C. M., McIvor, J. L., and Stephens, T. J. (1977c). *Cell. Immunol.* **30,** 216–224.
Wegmann, T. B., and Gilman, J. G. (1970). *Dev. Biol.* **21,** 281–291.
Weissman, I., Papaioannou, V., and Gardner, R. (1978). *Cold Spring Harbor Conf. Cell Proliferation* **5,** 33–47.
West, J. D. (1977). *Exp. Hematol.* **5,** 1–7.
Wong, P. M. C., Clarke, B. J., Carr, H. O., and Chui, D. H. K. (1982). *Proc. Natl. Acad. Sci. U.S.A.* **79,** 2952–2956.

CHAPTER **VII**

Human Chimeras

PATRICIA TIPPETT

MRC Blood Group Unit
University College London
London, England

I. Introduction 165
II. Twin Chimeras 166
   A. Proportions of the Two Cell Lines 167
   B. Changes in Proportions of Two Populations of Cells 167
   C. Tolerance 169
   D. Fertility 169
III. Dispermic Chimeras 169
   A. Proportion of the Two Cell Lines 169
   B. Mechanisms 170
   C. Skin Colour 170
   D. Phenotypic Sex of Dispermic Chimeras 171
   E. Fertility of Dispermic Chimeras 171
IV. A Fertile XX/XY Dispermic Chimera 171
V. Twins with an Unusual XX/XY Chimerism 172
VI. Human Chimera Disclosed by Her Progeny 174
VII. Chimeras of Unestablished Type 175
   Proportions of the Two Cell Lines 175
VIII. XX/XY Mosaicism without Other Evidence of Chimerism .. 176
IX. Other Conditions Simulating Chimerism 176
X. Summary 177
   References 177

I. INTRODUCTION

Investigation of an unusual ABO blood group led Dunsford *et al.* (1953) to suggest chimerism to explain the donor's phenotype. Since then more than 70 natural chimeric propositi have been reported (Tippett, 1983).

Anderson *et al.* (1951) distinguished a chimera, an organism whose cells derive from two or more zygote lineages, from a mosaic, whose cells derive from a single zygote lineage. These definitions are used by all biologists. Human chimeras may be artificial (induced) or natural (spontaneous); natural chimeras include both twin, which have two populations in their blood only, and dispermic chimeras (Race and Sanger, 1975). McLaren (1976) considered these to be examples of secondary and primary chimerism, respectively.

At first human chimeras were identified by the presence of two cell lines of different blood (red cell) groups. More recently, red cell enzymes, HLA types, and chromosome banding have all proved useful in distinguishing chimeras from mosaics. Two twin propositi (Gilgenkrantz et al., 1981; Massimo et al., 1966) and one dispermic chimera (Fitzgerald et al., 1979) are recorded whose cell lines are distinguished by their chromosomes, but show no sign of two populations of red cells.

ABO is the most revealing red cell group because the phenotypes depend on antigens present on red cells and reciprocal agglutinins in the plasma. Many chimeras have been ascertained through an unusual ABO phenotype, sometimes a missing agglutinin and sometimes a frank mixture of A and O, or A and B, cells have brought them to notice. Not all examples of two populations of ABO groups will be spotted: a person with 99% O and 1% A will be caught because of the missing anti-A, but the converse mixture of 99% A and 1% O will escape detection. Race and Sanger (1975) calculated that because only one-third of European dizygotic twins differ in their ABO groups, two-thirds of the twin chimeras were missed. Dispermic chimeras who are fertile may also escape detection.

Study of chimeras, both twin and dispermic, has contributed to our understanding of the nature of the secretor process and the Lewis groups, the origin of ABO glycosyltransferases in serum, and the recognition that Xg, unlike the majority of X-borne loci, is not subject to inactivation.

II. TWIN CHIMERAS

Twin chimeras are due to blood vessel anastomoses between dissimilar twins in uterine life. The twins transfuse each other through these connections, and invading stem cells colonize the marrow of the host who for the rest of his life will make some cells of the genetic type of his twin together with cells of his own genetic pattern.

Of the twin chimera propositi known to the Medical Research Council Blood Group Unit by 1982 (Tippett, 1983), 22 had chimeric twins, 8 had twins who were not available or only inadequately tested, and 2 had a normal nonchimeric twin.

In four pairs (Massimo et al., 1966; Robinson et al., 1976; Gilgenkrantz et al., 1981; Iselius et al., 1979) the chimerism was observed through karyotyping, and in two of these pairs no blood group mixtures were observed. One pair was

VII. Human Chimeras

ascertained because anastomoses were seen at delivery (Nylander and Osunkoya, 1970). Of the 32 chimera propositi, 27 were ascertained through blood grouping, 26 through an apparently unusual ABO phenotype, and one because of a mixed field pattern of agglutination with anti-Rh (anti-D).

A. Proportions of the Two Cell Lines

The major cell line does not necessarily represent the true genetic cell line, nor is it always the same in both twins. In twins who differ in their ABO groups, the true genetic group can be determined from saliva, if they are secretors, and from serum glycosyltransferases even if they are nonsecretors. In male–female twins the percentage of the true genetic line of lymphocytes can be determined by karyotyping.

Table I shows six pairs of chimeric twins with different patterns of dominance of red cell or lymphocyte cell lines. They show that there is no regular pattern for the proportions of the two cell lines in twin chimera pairs since none of the patterns is more common than the others. Race and Sanger (1975) wondered if maternal antibodies, should they cross the placenta, might affect the reception of the graft. Although this hypothesis could explain the small A graft in Miss Wa. and in Mlle. C. Me., it does not explain the small B graft in Mr. Ba. or in John St. or the minority of AB cells in Mr. Y.F. (Table I) where the graft is compatible with the maternal ABO antibodies. Hosoi *et al.* (1977) suggest that if anastomosis is only brief each twin should have a majority of its own genetic cell lines. Twins like Mlle. C. Me. and M. C. Me (Muller *et al.*, 1974) suggest that the selective advantage for a red cell line is not necessarily the same as that for the lymphocyte line. The variety of patterns may reflect the cooperation of different effects, maternal antibodies, period and time of anastomosis, and establishment of tolerance, in the twins. The only pattern not recorded is both twins having majority cell lines of their grafts.

B. Changes in Proportions of Two Populations of Cells

Some adult twin chimeras show changes in the proportions of their red cell lines, suggesting that one cell line may have a slight advantage. For example the proportion of his own A_1 cells in Mr. Wa. has gradually decreased from 86% in 1957 to 63% in 1972 (Race and Sanger, 1975). The proportions of 99% O and 1% A have remained unaltered in his twin sister. In a Danish chimeric pair (Hansen *et al.*, 1978) the size of the graft of B cells has decreased from 27% to 10% in the male twin, while the proportion of 70% B to 30% A in the female twin has remained unaltered. Pausch *et al.* (1979) did not find any changes in proportions in an Austrian pair, both male, during 3 years.

It is hard to understand why the proportions should change in some adult

TABLE I.
Comparison of Proportions of Two Cell Populations in Chimeric Twins

Reference	Propositus and twin	Red cells[a]	Lymphocytes[a]	Remarks	Mother's ABO and secretor
Booth et al. (1957)	Miss Wa.	99% O, 1% A	98% XX, 2% XY	Own cells dominate in both cell lines in both twins	O secretor
	Mr. Wa.	86% A_1, 14% O	91% XY, 9% XX		
van der Hart and van Loghem (1967)	Mr. Ba.	99.8% O, 0.2% B	97% XY, 3% XX	Own cells dominate in red cells, XY dominate XX lymphocytes in both twins	B secretor
	Miss Ba.	80% B, 20% O	22% XX, 78% XY		
Ducos et al. (1970)	Mme. Ko.	90% O, 10% A_1B	87% XX, 13% XY	Own cells dominate in red cells, XX dominate XY lymphocytes in both twins	
	M. Heb.	65% A_1B, 35% O	34% XY, 66% XX		
Hosoi et al. (1977)	Mr. Y.F.	89% A_1, 11% A_1B	68% XY, 32% XX	One twin's red cells dominate in both twins but majority of own lymphocytes	AB nonsecretor
	Miss E.F.	12% A_1B, 88% A_1	60% XX, 40% XY		
Wrobel et al. (1974)	John St.	99% A_1, 1% B	80% XY, 20% XX	One twin's cell lines dominate in red cells and lymphocytes	AB secretor
	Judy St.	1% B, 99% A_1	46% XX, 54% XY		
Muller et al. (1974)	Mlle. C. Me.	79% B, 21% A_1	23% XX, 77% XY	One twin's red cell line dominates in both twins and the other twin's lymphocytes dominate in both	B secretor
	M. C. Me.	33% A_1, 67% B	72% XY, 28% XX		

[a] The true genetic line is given first.

chimeras and not in others. It is less surprising that the proportions should alter in chimeric twins who were babies when first tested. Such changes might reflect the loss of maternal environment and the development of their immune system.

C. Tolerance

All chimeras show tolerance for their twin's genetic line, that is they do not recognise the graft as foreign. This tolerance is demonstrated by lack of the agglutinin against the grafted cells (a group O twin with an A graft lacks anti-A in its plasma), by negative mixed lymphocyte culture (MLC) tests, and, in the only pair tested (Woodruff and Lennox, 1959), by mutual acceptance of skin grafts between pairs of chimeric twins.

D. Fertility

HumanXX↔XY twin chimeras are normally fertile unlike cattle XX↔XY twins: the female calf of an XX↔XY pair is always sterile, a freemartin. At least seven females of XX↔XY human chimeric pairs have children.

III. DISPERMIC CHIMERAS

Dispermic chimeras result from the fertilisation by two sperm of two maternal nuclei, the fusion of the two zygotes, and their growth into one body. McLaren (1976) calls the state "primary chimerism," and it has also been called generalised tissue mosaicism, whole-body chimerism, and tetragametic chimerism.

Tippett (1983) added 12 more dispermic chimeras to those reported by Race and Sanger (1975). These rare people are either hermaphrodites or intersexes, and/or have two cell lines or a mixture of skin colours showing that their chimerism is not limited to blood.

Of the 32 dispermic chimeras known, 19 (all XX/XY) were ascertained because of "sex upsets," such as hypospadias or hermaphroditism, 10 (3 XX/XX and 7 XX/XY) were ascertained in blood grouping laboratories, and the remaining 3 (an XX/XY girl with one hazel and one brown eye, a mongol boy and a boy (XX/XXY) with a triploid cell line) were found by cytogenetics laboratories. Ascertainment accounts for the excess of XX/XY dispermic chimeras.

A. Proportion of the Two Cell Lines

The proportion of the two cell lines may vary in different tissues. In the dispermic chimera of Delarue et al. (1969) the percentage of the triploid cell line varied from 5% in lymphocytes to 20% in muscle and 50% in liver, gonads, and skin. In the puzzling Swedish twins of Iselius *et al.* (1979), the proportions changed with time, see below.

B. Mechanisms

Ford (1969) and, later, McLaren (1976) classified dispermic chimeras. Several possible mechanisms were suggested involving accidents at different stages of oogenesis or an early fusion of two zygotes. They considered information given by pairs of allelic markers in determining the number and type of maternal nuclei involved. Many chimeras could be attributed to simultaneous fertilisation of an ovum by one sperm and the polar body produced at the second meiotic division by another sperm. Dewald *et al.* (1980) considered that this might be the most likely mechanism because fewer embryologic processes are involved. Chimeras formed by this mechanism would have a single maternal contribution for markers near the centromere of chromosomes, since crossing-over does not occur near the centromere, but could have different maternal contributions for markers such as blood groups that are further away from the centromere.

Dispermic chimeras with a single maternal contribution could represent the simultaneous fertilisation of two haploid mitotic nuclei, products of a precocious first cleavage division of an unfertilised ovum. The true hermaphrodite of Fitzgerald *et al.* (1979) and his parents were tested for eight red cell groups, HLA, four serum proteins, 23 red cell enzyme systems, and distinctive chromosome markers, however, only six markers were informative for the mother's contribution. The mother had given the same contribution to the XX and XY cell lines. The authors points out that with so few useful indicators of genetic difference, the mechanism cannot be definitely determined.

The suggestion of early fusion of two embryos to explain the chimera of de la Chapelle *et al.* (1974) with two maternal contributions is challenged by Dewald *et al.* (1980). Dewald and colleagues consider that there is no report of a whole-body chimera in which there are sufficient informative genetic markers conclusively to prove the specific mechanism by which the chimera arose.

C. Skin Colour

Patches of different skin colour show that two cell lines of a chimera are not confined to the blood. This was useful in showing two South African XX/XX females with two populations of red cells to be dispermic (Moores, 1973; Moores *et al.*, 1982).

Few chimeras have skin of different colour: only 5 of the 32 known dispermic chimeras are reported to have patchy skin colour, the two mentioned above, the dispermic chimera of mixed Negro, Caucasian, and Amerindian ancestry (Zuelzer *et al.*, 1964), the white "male" child of Corey *et al.* (1967), and the

VII. Human Chimeras

true hermaphrodite of Fitzgerald *et al.* (1979). Findlay and Moores (1980) summarize the observations in all five cases. They remark that all investigators of XX/XY chimeras have found the fibroblast karyotypes to be more mixed in the darker skin areas and more homogeneously XX or XY in the pale areas. They find that pigment patterns of chimeras can be divided into three main types: a flag-like rectangular pattern, a pattern of rounded units, and a striate pattern. To the common pigment demarcation lines, such as the dorsal and ventral midlines, observed in chimeras they add another which they call the "centaur" line. They conclude that it is impossible to predict the patterns of a dispermic chimera with two skin colours.

D. Phenotypic Sex of Dispermic Chimeras

Humans show the same tendency as animals for XX/XY chimeras to develop as males. Of the 27 XX/XY dispermic chimeras, 16 are phenotypically male, 7 are phenotypically female, and 4 are intersexes. The XX/XXY dispermic chimera was also phenotypically male.

E. Fertility of Dispermic Chimeras

Most of the known XX/XY dispermic chimeras are infertile, reflecting the method of ascertainment of these rare people. However, this is not invariable, three of the XX/XY dispermic chimeras found by blood grouping laboratories have children. This finding suggests that dispermic chimeras, even XX/XY ones, may as long as they are healthy and normally fertile remain undisclosed.

Dispermic chimeras of like sex are expected to be fertile; 3 XX/XX dispermic chimeras have proved themselves to be so by having children. Only one active germ cell line was observed in these three mothers. The only XY/XY dispermic chimera was a mongol boy.

The occurrence of two populations of fibroblasts, one XX and the other XY, proves that a chimera is a dispermic chimera. It is more difficult to demonstrate dispermism in fertile chimeras who have only XX or only XY cells. Skin colour has on two occasions shown that an XX/XX chimera was dispermic. No fertile XY/XY has, as yet, proved to be a dispermic chimera; two possible candidates are recorded as chimeras of unestablished type (Tippett, 1983).

IV. A FERTILE XX/XY DISPERMIC CHIMERA

The red cells of a blood donor, Mr. K. S., reported by Watkins *et al.* (1981), had very weak A and normal B antigens but his serum contained fairly strong A_1 and weak B transferases. This discrepancy between the ABO red cell group and the serum transferases led to further investigations of the donor and his family.

The B serum transferase of K.S. showed only 30% of normal level and his A transferase, which had all the characteristic of an A_1 transferase, about 50% of normal level. His wife and eldest son, both group A_2, had normal A_2 transferases; his second son, group B, had normal B transferase. The group B son had inherited his father's B gene, but the A_2 son had inherited neither A^1 nor B genes from his father. The level of glycosyltansferases in K.S.'s red cells differed from the serum levels; his red cell B transferase was of normal level and only a very low level of A transferase, again an A_1 transferase, was detected.

These observations raised the suspicion of chimerism which was confirmed by cytogenetic testing. K.S.'s fibroblasts showed two populations, 60% XX and 40% XY, although his lymphocytes were all XY. Some of his hair roots were positive for X chromatin and negative for Y bodies, while others were positive for Y bodies and negative for X chromatin.

Watkins et al. (1981) concluded that K.S. is a dispermic chimera whose bone marrow and reproductive organs are predominately XY and BO, whereas his skin and other tissues are partly XX and A^1O or A^1A^1.

Mr. K.S. was the second XX/XY dispermic chimera to demonstrate his sexual normality by having children. Mr. D.W. (Zuelzer et al., 1964) had a child after his dispermic condition was diagnosed (K.M. Beattie, personal communication). A third fertile XX/XY has one child (H. Seyfried, personal communication). Dispermic chimeras who are normal fertile individuals probably remain undisclosed unless they have mixtures of skin colours or are shown to have two red cell populations by blood grouping laboratories.

V. TWINS WITH AN UNUSUAL XX/XY CHIMERISM

An apparently monozygous pair of male twins were karyotyped because one twin had glandular hypospadias. Peripheral blood lymphocytes from both twins had an XX/XY chromosome constitution (Iselius et al., 1979). The twins and their parents were extensively tested using 11 red cell groups, secretions, ABH transferase assays, HLA groups, 7 biochemical markers, and Q-banding analysis of cultured lymphocytes.

The results of chromosome analysis of different tissues of the twins are summarized in Table II. Both twins presented an XX/XY chromosome constitution in the blood. Twin I showed only XY fibroblasts in his skin, but twin II had more XX than XY cells in tissues other than blood including testicular tissue. The proportion of XX lymphocytes decreased in both twins, but the frequency of XX cells dropped more rapidly in twin I than in twin II.

Two populations of cells were shown in the ABO groups only. The father was

VII. Human Chimeras

TABLE II.

The Two Cell Populations in an Unusual Pair of Chimeric Twins[a]

	Age at examination	Tissue	Karyotype		Red cells	
			XX	XY	A_1	B
Twin I	8 weeks	Skin arm	0	100		
	8 weeks	Blood	17	83	70	30
	20 weeks	Blood	6	94	70	30
	2.6 years	Blood	6	94	85	15
	7 years	Blood			90	10
Twin II	8 weeks	Skin arm	80	20		
	25 weeks	Skin scrotum	59	41		
	25 weeks	Testis left	65	35		
	8 weeks	Blood	13	87	60	40
	20 weeks	Blood	25	75	70	30
	2.6 years	Blood	6	94	85	15
	7 years	Blood			90	10

[a] After Iselius et al., 1979.

A_1B, mother group O, and both twins had 60–70% group A_1 and 40–30% group B cells (Table II). The proportion of B cells decreased until by 7 years both twins had only 10% B cells. The secretion studies and tranferase assays suggested that twin I was genetically A^1 and that his chimerism was limited to his blood, but that twin II was genetically B and chimerism was not limited to his blood.

The blood group results were in agreement with the chromosome analysis, which showed both twins to have XX/XY in their blood but only twin II to be XX/XY in other tissues.

The twins, considered monozygous at birth, were identical for chromosome markers, blood groups, HLA, isozymes and serum proteins. However, they do differ for markers in tissues other than blood, such as skin and secretory tissues. Considered separately twin I appears to be a twin chimera and twin II a dispermic chimera. Iselius et al. (1979) suggest two possible mechanisms to explain how these twins, so similar in appearance and genetic constitution, could have arisen. One possible explanation was that A_1, XY zygote became monozygotic twins and a B, XX dead triplet embryo fused with and was adsorbed by twin II's embryo. A second explanation was that the twins started as an XX/XY zygote and when the primitive streaks were laid down twin I was on a patch of A_1, XY cells and twin II, less luckily, on a mixture of B, XX and A_1, XY cells. Neither explanation satisfies all the facts, and the puzzle presented by these chimeric twin boys is not yet solved.

Two other dispermic chimeras had twin partners. Delarue et al. (1969) recorded that a macerated fetus was delivered at the same time as their dispermic chimera. The chimera, a male, had 5% of a triploid cell line XXY and 95% XX in his lymphocytes but in skin and gonads had 50% XXY. Unfortunately the macerated fetus was not karyotyped. The XX/XX baby girl, with a mixture of B and O cells, might have been called a twin chimera had not P.P. Moores (personal communication, 1969, cited by Race and Sanger, 1975, p. 533), found the boy twin to be neither B nor O, but straightforward A_2.

VI. HUMAN CHIMERA DISCLOSED BY HER PROGENY

Recognition of chimerism in a mother in an admirable study by Mayr et al. (1979) resolved an apparent mother–child exclusion in the ABO system. Mrs. Eiw., group AB, had delivered a group O child. To find the explanation of this apparent exclusion, Mayr et al. (1979) tested blood from Mr. Eiw. and three other children, as well as that of Mrs. Eiw. and the baby, for 21 genetic markers. The results were astonishing: none of the 4 children fitted genetically with their mother. The eldest child was excluded by Kidd and HLA; the second child and the baby were excluded by ABO, PGM_1, and HLA; and the third child by PGM_1 and HLA. As long as the father was indeed the father, all the children were also excluded by Bf and some children were also excluded by Rh or by ACP_1. The genes controlling these groups involve chromosomes 1, 2, 6, and 9, which ruled out explanations such as somatic mutation, inversion, or chromosomal accidents, since such mishaps are rare and unlikely to have occurred simultaneously to four chromosomes. This constellation of groups could not be explained by rare alleles. However, all the necessary halpotypes were found in the maternal grandparents.

An interchange of all four children, *a priori* improbable, was eliminated by questioning the relevant hospital staff. So Mayr et al. (1979) concluded that Mrs. Eiw. possessed two populations of different cells, one demonstrable in her blood and fibroblasts and the other in her gonads, and that Mrs. Eiw. was a chimera.

Mayr et al. (1979) searched for expression of the gonadal cell line in other tissues. Unidirectional mixed lymphocyte cultures showed that Mrs. Eiw.'s lymphocytes were stimulated by her mother's lymphocytes, even though Mrs. Eiw. carries both maternal HLA haplotypes in her body. This lack of immunological tolerance showed that Mrs. Eiw.'s gonadal HLA haplotype was not possessed by her cells involved in immune response. Similarly, cultured fibroblasts failed to demonstrate the ACP_1, PGM_1, or HLA type present in gonads but absent from the blood.

Mrs. Eiw. was XX, as shown by lymphocytes and by fibroblasts. Studies by $Q-$, $G-$, and $C-$ banding techniques of all family members did not reveal any

VII. Human Chimeras

numerical or structural abnormalities, nor any informative chromosome heteromorphisms. Without the investigation of cultured fibroblasts Mrs. Eiw., a dispermic chimera, would have been listed as a chimera of unestablished type (Mrs. Eiw. was erroneously included in this group; Tippett, 1983.)

This is the first human chimera not showing two populations of erythrocytes or lymphocytes or fibroblasts in spite of an extensive and very thorough investigation. Such chimeras, revealed only by their progeny, must indeed be rare.

VII. CHIMERAS OF UNESTABLISHED TYPE

Chimeras whose two recognisable cell lines originate from the bone marrow but who have no history of twinning may be twin chimeras whose twin resorbed early in fetal life or may be dispermic chimeras. Some XX/XX and XY/XY dispermic chimeras may be included in this group, since it is more difficult to prove such people to be dispermic. Ten chimeras of unestablished type are known (Tippett, 1983), all were found in blood grouping laboratories. Two of the four men and four of the six women have proved their sexual normality by having children.

Proportions of the Two Cell Lines

In all these chimeras one cell line dominates, although not necessarily the same cell line in fibroblasts as the one in lymphocytes. Table III summarizes the chimeras whose fibroblasts and lymphocytes were karyotyped. All except I.G., who was too young, had children. These chimeras show that one phenotype may dominate in one tissue and another phenotype in a different tissue. Lymphocyte karyotypes may be alien to their true sex; Mrs. P. M. with 96% XY, and Mrs. Co., with 100% XY, were both fertile females.

TABLE III.

Some Chimeras of Unestablished Type

Reference	Propositus	Red cells	Lymphocytes	Fibroblasts
Battey et al. (1974)	Mrs. P.M.	7% A_1 93% O	4% XX 96% XY	100% XX
Szymanski et al. (1977)	Mrs. Co.	16% A_1 84% O	100% XY	100% XX
Seyfried and Kusnieug (1975)	I.G. (girl)	95% A_1 5% B	100% XX	100% XX
Bird et al. (1982)	Mrs. C.C.	90% rr 10% R_1r	90% XX 10% XY	100% XX

Banding studies for chromosome markers and enzyme assays of cultured fibroblasts might identify some of these unestablished types as dispermic chimeras. Mayr et al. (1979) proved Mrs. Eiw. to be dispermic by enzyme tests of cultured fibroblasts.

VIII. XX/XY MOSAICISM WITHOUT OTHER EVIDENCE OF CHIMERISM

Race and Sanger (1975) recorded nine XX/XY mosaics who showed no other evidence of dispermy. Since 1975 the MRC Blood Group Unit has tested three more XX/XY propositi who have not shown two populations of red cells.

Perhaps the tests done were not extensive enough to show chimerism. However, accidents to the X chromosome are well known, and many different types of such mosaics are recorded: XX/XO, XX/XXX, XO/XY, XX/XXX, etc. Although many XX/XY have proved to be chimeras by showing two cell lines for other chromosomes, the same phenotype could arise from postzygotic misdivision of an XY or XXY zygote. Two true hermaphrodites in our records are XX/XY/XXY; this phenotype could be derived from an XXY zygote by two nondisjunctions. If the original cell line were lost, an XX/XY mosaic would result. Such mosaics would have genetically identical autosomes.

As Ford (1969) said "chimerism can be proved by genetic evidence" but "mosaicism (in individual cases) cannot be proved at all."

IX. OTHER CONDITIONS SIMULATING CHIMERISM

There are several explanations for two populations of red cells that must be eliminated before chimerism is established.

Clinical history will eliminate the mixed field pictures due to weakening of certain antigens in some patients with leukaemia or with myeloproliferative disease. Another disease change causes cells to become polyagglutinable, and these patients may appear to have two populations of red cells.

Family history may help to distinguish the rare people with an inherited type of ABO mosaicism. Rh mosaics, people with two populations of different Rh groups, are also excluded from chimeric status by family studies. In some propositi one of the red cell populations can be shown, from the Rh groups of the parents, not to be genetically possible. The mechanism by which these Rh mosaics arise is not yet understood.

Another donor with two populations of red cells, one CDe/cde, Fy(a+) and the other cde/cde, Fy(a−), would have been listed as a chimera by Jenkins and Marsh (1965) had not the parents' groups of CDe/CDe, Fy(a+b−) and

CDe/cde, Fy(a−b+) excluded the inheritance of an cde/cde, Fy(a−) cell line. This unusual phenotype cannot be explained by a small deletion or translocation of chromosome 1, since the *Rh* and *Fy* loci are far apart on the chromosome and the donor has two normal number 1 chromosomes.

A similar case of a healthy young female with a mixture of CDe/CDe, PGM_1 2-2 and cde/cde PGM_1 1-1 was considered by Muller (1978) to represent abnormal mitosis because her CDe/CDe, PGM_1 2-2 cell line was not genetically possible.

These two cases caution that propositi whose two populations are distinguished only by chromosome 1 markers should not be classed as chimeras until other differences are observed.

X. SUMMARY

Human chimeras show similarities to experimental animal chimeras, among them the variation of the proportions of the two cell lines in different tissues. However, there are also differences: spontaneous human twin chimeras differ from cattle chimeras since the human female is fertile. Human XX/XX dispermic chimeras are fertile but, so far, only one active cell line has been observed in the gonads. No fertile human XY/XY dispermic chimera is yet recorded. Most known XX/XY human dispermic chimeras are infertile and ascertained through cytogenetic testing because of developmental abnormalities, such as hermaphroditism; three XX/XY chimeric men ascertained in blood grouping laboratories were fertile.

There is still much more to be learnt from these interesting and rare people. It may soon be possible, in some cases, to determine the mechanism that has caused a dispermic chimera. In time the interaction of different cell lineages in the fetus may be understood, and cell migration and proliferation during development of an adult chimera may contribute to our reasoning about tolerance.

REFERENCES

Anderson, D., Billingham, R. E., Lampkin, G. H., and Medawar, P. B. (1951). *Heredity* **5**, 379–397.

Battey, D. A., Bird, G. W. G., McDermott, A., Mortimer, C. W., Mutchinick, O. M., and Wingham, J. (1974). *J. Med. Genet.* **11**, 283–287.

Bird, G. W. G., Wingham, J., Nicholson, G. S., Battey, D., Koster, H. G., and Webb, T. (1982). *J. Immunogenet.* **9**, 317–322.

Booth, P. B., Plaut, G., James, T. D., Ikin, E. W., Moores, P., Sanger, R., and Race, R. R. (1957). *Br. Med. J.* **1**, 1456–1458.

Corey, M. J., Miller, J. R., MacLean, J. R., and Chown, B. (1967). *Am. J. Hum. Genet.* **19**, 378–387.

de la Chapelle, A. Schröder, J., Rantanen, P., Thomasson, B., Niemi, M., Tiilikainen, A., Sanger, R., and Robson, B. (1974). *Ann. Hum. Genet.* **38**, 63–75.
Delarue, F., Liberge, G., Salmon, C., and Lejeune, J. (1969). *Rev. Fr. Transfus.* **13**, 129–134.
Dewald, G., Haymond, M. W., Spurbeck, J. L., and Breanndan Moore, S. (1980). *Science* **207**, 321–323.
Ducos, J., Colombies, P., Marty, Y., Blanc, M., Daver, J., and Edmond, J. (1970). *Rev. Fr. Transfus.* **13**, 261–266.
Dunsford, I., Bowley, C. C., Hutchison, A. M., Thompson, J. S., Sanger, R., and Race, R. R. (1953). *Br. Med. J.* **2**, 81.
Findlay, G. H., and Moores, P. P. (1980). *Br. J. Dermatol.* **103**, 489–498.
Fitzgerald, P. H., Donald, R. A., and Kirk, R. L. (1979). *Clin. Genet.* **15**, 89–96.
Ford, C. E. (1969). *Br. Med. Bull.* **25**, 104–109.
Gilgenkrantz, S., Marchal, C., Wendremaire, P., and Seger, M. (1981). In "Twin Research 3: Twin Biology and Multiple Pregnancy," pp. 141–153. Alan R. Liss, Inc., New York.
Hansen, H. E., Eriksen, B., Niebuhr, E., and Dabelsteen, E. (1978). *Hum. Hered.* **28**, 411–420.
Hosoi, T., Yahara, S., Kunitomo, K., Saji, H., and Ohtsuki, Y. (1977). *Vox Sang.* **32**, 339–341.
Iselius, L., Lambert, B., Lindsten, J., Tippett, P., Gavin, J., Daniels, G., Yates, A., Ritzén, M., and Sandstedt, B. (1979). *Ann. Hum. Genet.* **43**, 89–96.
Jenkins, W. J., and Marsh, W. L. (1965). *Transfusion* **5**, 6–10.
McLaren, A. (1976). "Mammalian Chimeras." Cambridge Univ. Press, London and New York.
Massimo, L., Gemme, G., Vianello, M. G., and Verri, B. (1966). *Acta. Genet. Med. (Roma)* **15**, 208–211.
Mayr, W. R., Pausch, V., and Schnedl, W. (1979). *Nature (London)* **277**, 210–211.
Moores, P. P. (1973). *Acta Haemat.* **50**, 299–304.
Moores, P. P., Findlay, G. H., Dunn, D., Watkins, W. M., Greenwell, P., and Bird, A. (1982). *Transfusion* **5**, 411.
Muller, A. (1978). Abstr. *Congr. Int. Soc. Blood Transf., 15th, 1978* p. 442.
Muller, A., de Grouchy, J., Garretta, M., André, J., Roubin, M., and Moullec, J. (1974). *Ann. Genet.* **17**, 23–28.
Nylander, P. P. S., and Osunkoya, B. O. (1970). *Obstet. Gynecol.* **36**, 621–625.
Pausch, V., Bleier, I., Dub, E., Kirnbauer, M., Weirather, M., Würger, G., and Mayr, W. R. (1979). *Vox Sang.* **36**, 85–92.
Race, R. R., and Sanger, R. (1975). "Blood Groups in Man", 6th ed. Blackwell, Oxford.
Robinson, E. A. E., North, D., Horsfield, G. I., and Kelly, F. (1976). *J. Med. Genet.* **13**, 528–530.
Seyfried, H., and Kusnieug, G. (1975). *Cong. Int. Soc. Blood Transfus., 14th, 1975* Abstracts, p. 95.
Szymanski, I. O., Tilley, C. A. Crookston, M. C., Greenwalt, T. J., and Moore, S. (1977). *J. Med. Genet.* **14**, 279–292.
Tippett, P. (1983). *Vox Sang.* **44**, 333–359.
van der Hart, M., and van Loghem, J. J. (1967). *Vox Sang.* **12**, 161–172.
Watkins, W. M., Yates, A. D., Greenwell, P., Bird, G. W. G., Gibson, M., Roy, T. C. F., Wingham, J., and Loeb, W. (1981). *J. Immunogenet.* **8**, 113–128.
Woodruff, M. F. A., and Lennox, H. (1959). *Lancet* **2**, 476–478.
Wrobel, D. M., McDonald, I., Race, C., and Watkins, W. M. (1974). *Vox Sang.* **27**, 395–402.
Zuelzer, W. W., Beattie, K. M., and Reisman, L. E. (1964). *Am. J. Hum. Genet.* **16**, 38–51.

CHAPTER VIII

Primary Lymphoid Organ Ontogeny in Birds

NICOLE M. LE DOUARIN,[1] FRANCINE V. JOTEREAU,[2] ELISABETH HOUSSAINT,[2] and JEAN-PAUL THIERY[1]

I.	Introduction	179
II.	Thymic Histogenesis	181
	A. Participation of the Three Germ Layers in Thymus Ontogeny	181
	B. Colonization of the Thymus Rudiment by Lymphocyte Precursors Proceeds by Waves	188
III.	Ontogeny of the Bursa of Fabricius	203
	A. Seeding of the Bursa of Fabricius by Extrinsic Hemopoietic Cells	204
	B. Cell–Cell Interactions between Endoderm and Mesoderm during Bursa Ontogeny and Their Developmental Significance	205
	C. Cell Migration and Morphogenesis in the Ontogeny of the Bursa of Fabricius: Correlations with Plasminogen Activator Activity	209
IV.	Concluding Remarks	212
	References	214

I. INTRODUCTION

The cells responsible for immune competence in the mature organism are mainly the lymphocytes and the macrophages. Active investigations carried out in birds and mammals during the two last decades have provided considerable insight into the cellular and molecular mechanisms underlying the immune function. Knowing how the immune system develops is also expected to provide explanations on the emergence of the immune recognition potential, including self and non-self discrimination, with all the consequences that alterations of this major function can represent in autoimmune diseases.

The avian embryo has yielded some landmark discoveries in the field of

[1]Institut d'Embryologie du CNRS et du Collège de France, Nogent-sur-Marne, France.
[2]Faculté des Sciences, INSERM, Nantes, France.

immune system development, although most modern immunology studies have been devoted to mouse and man. Choosing the mouse has obvious reasons. It is a small mammal, easy to breed, with a relatively short generation time, and mostly it is one of the best documented animal models from the genetic point of view. The interest in birds stems from their having a specialized organ for B lymphocyte differentiation, the *bursa of Fabricius;* the avian embryo is also accessible to experimentation in the egg during the whole period of development. Thus an array of experiments on primary lymphoid organ differentiation may be carried out *in vivo* that are impossible in mammals during intrauterine life.

Identification of primary lymphoid organ functions dates from the early 1960s when Miller (1961) found that removal of the mouse thymus on the day of birth impairs maturation of the immune function. It appeared that lymphocytes involved in delayed hypersensitivity reactions, allograft rejection and the initiation of graft-versus-host reactions differentiate in the thymus. Thymus-dependent lymphocytes were then called T lymphocytes. At the same period, a series of studies carried out in chicken showed that neonatally bursectomized birds failed to form humoral antibodies but maintained their capacity to reject skin allografts (Glick *et al.,* 1956; Warner and Szenberg, 1964; Cooper *et al.,* 1965).

Bursa-dependent lymphocytes (B lymphocytes) were thus considered responsible for humoral immunity. A large body of investigations has been devoted to the search of a bursa equivalent in mammals. Following the finding by Raff *et al.* (1976) that cytoplasmic Ig-positive cells (pre-B cells) appear in cultures of 13-day mouse foetal liver, this organ is now considered as the best candidate to fulfil this function prenatally, while in adult life the bone marrow is the site of B cell development.

The crucial period during which thymus and bursa influence immunological development is embryonic and early postnatal life, when lymphoid tissue is forming and immunological capacity is maturing. Undifferentiated hemopoietic cells (HC) home to these primary lymphoid organs where, under the influence of thymic and bursal environments, they become oriented, respectively, toward the T and B cell differentiation pathways. Lymphocytes then emigrate from the thymus and bursa and colonize the peripheral lymphoid system composed of spleen and of various gut-associated secondary lymphoid tissues. Sexual maturity marks the onset of primary lymphoid organ involution.

The developmental processes through which epitheliomesenchymal organs, such as the thymus and bursa, undergo lymphoid development have been a controversial issue for many years. Both the origin of lymphocytes and the respective contribution of endoderm, ectoderm, and mesenchyme to thymus and bursa histogenesis have been interpreted in various ways.

This chapter will describe how the use of chimeras providing cell markers have contributed to clarifying this problem. Second, new results concerning the developmental mechanisms ensuring stem cell seeding of the thymus and bursa primordia will be reported.

II. THYMIC HISTOGENESIS

A. Participation of the Three Germ Layers in Thymus Ontogeny

1. DOES SUPERFICIAL BRANCHIAL ECTODERM PLAY A SIGNIFICANT ROLE IN THYMUS HISTOGENESIS?

In birds the thymus rudiment originates from pharyngeal pouches III and IV. This view was first put forward by Verdun (1898) and confirmed by more recent observations (Venzke, 1942, 1952; Schrier and Hamilton, 1952; Hammond, 1954).

Like all the glandular organs derived from the embryonic pharynx, the thymus rudiment is originally formed by two tissue components: an endodermal epithelial bud that arises from the branchial pouches and is surrounded by sheets of mesenchymal cells. Soon after the first thickening of the endoderm is formed, the branchial pouch enters into contact with the superficial ectoderm of the branchial slit at the external border of the endodermal bud. This is why some authors have proposed a contribution of ectodermal cells to thymic histogenesis (Fraser and Hill, 1915; Hammond, 1954). However whether or not the ectoderm provides the thymic stroma with epithelial cells, and if so, which cells, has never been fully clarified. This is due to the inherent difficulty of recognizing endodermal from ectodermal epithelial cells at the precise junction of the two tissues in the branchial cleft and during thymic histogenetic processes. If any ectodermal contribution exists it cannot be important since the ventrolateral pharyngeal endoderm can be isolated as early as the 15-somite stage from chick or quail embryos. At this stage the branchial pouches are not yet formed, and the endoderm has not entered into contact with the ectoderm. If grafted ectopically into the somatopleure of a 3-day host embryo, the 15-somite pharyngeal endoderm develops into a variety of tissues, including thymus to which no ectodermal cells can have contributed (Le Douarin, 1967; Le Douarin *et al.*, 1968).

2. EXTRINSIC ORIGIN OF LYMPHOCYTES AND RESPECTIVE ROLES OF ENDODERM AND MESENCHYME IN THYMUS HISTOGENESIS

a. Historical Overview. It was originally postulated, from histological observations, that thymic lymphocytes were derived from an intrinsic cell component of the original thymus primordium. Two main variants of this idea were proposed according to which lymphopoietic capacity was ascribed either to the epithelial or to the mesenchymal parts of the rudiment.

The *transformation theory* initially proposed by Kölliker (1879) postulated that the epithelium gave rise to the lymphocytes as well as to the reticular cells.

This theory received support from many authors who claimed that transitional forms between endodermal cells and lymphocytes could be seen during development (Prenant, 1894; Beard, 1899, 1902; Bell, 1906; Stohr, 1906; Cheval, 1908; Dustin, 1920; Baillif, 1949; Ackerman and Knouff, 1964, 1965; King *et al.*, 1964; Weakley *et al.*, 1964; Ackerman, 1967; Sanel, 1967; Tachibana *et al.*, 1974).

According to the *substitution theory* formulated by Hammar (1905, 1908, 1910, 1911) and by Maximow (1909, 1912), thymic lymphocytes originated from mesenchymal cells that became large basophilic cells, then invaded the early epithelium anlage where they proliferated and differentiated into lymphocytes.

Experimental studies of thymus lymphopoiesis in bird chimeras using natural cell marking techniques later showed decisively that lymphocyte precursors were extrinsic to the primary lymphoid organ rudiments.

b. Experiments Using Sex Chromosomes as Cell Markers. In an elegant series of experiments, Moore and Owen (1967) joined chick embryos by vascular anastomoses at various times in development: either at 4–5 days through the yolk sacs or later (10–11 days) through the chorioallantoic membranes (CAM). When the parabionts were of different sexes, chromosome analysis revealed a high degree of chimerism in thymic cells in embryos joined early and only low levels or no chimerism when effective parabiosis started later. Moore and Owen concluded that an inflow of blood-borne HC was responsible for lymphoid differentiation in the chick thymus and that such an inflow took place early during thymus development and then stopped.

This view was supported later by culture and grafting experiments of the prelymphoid chicken thymic anlage (Owen and Ritter, 1969). When cultured in a cell-impermeable diffusion chamber on the chick CAM, the young thymus rudiment failed to become lymphoid, while older thymuses sustained lymphopoiesis. Lymphoid differentiation in the latter appeared to be correlated with the number of basophilic cells present in the thymus at the time of transplantation, cells presumed to be blood-borne HC that had already seeded the organ prior to its removal from the donor embryo.

Similar observations on bursa and other hemopoietic organs (see Metcalf and Moore, 1971, for a review) led these authors to formulate the *hematogenous theory* of blood-forming organ ontogeny. According to this view, none of the hemopoietic organs contains hemopoietic precursor cells. They have therefore to be seeded at a certain time in development by stem cells of extrinsic origin arising supposedly from yolk sac blood islands (Moore and Metcalf, 1970; Moore and Johnson, 1976).

Attractive as this hypothesis might be, it could not be fully accepted because of the characteristics of the labelling technique used. When the cell marker resides in the mitotic chromosomes, information can be gained only on cells dividing at the time of observation. Therefore, the above-mentioned experiments could not rule out that either the endoderm or the mesenchyme (or both) of the thymic and bursal rudiments had some lymphopoietic potency and could give rise to certain subpopulations of lymphocytes.

The quail-chick marker system finally proved the extrinsic origin of lymphocytes.

c. The Quail-Chick Marker in Lymphoid Tissues. Lymphocyte differentiation involves some structural modification of the chromatin that becomes more condensed than in most other somatic cells. In the reticular epithelial and connective cells of the chick and quail thymus and bursa, the interphase chromatin is dispersed, with only some small chromocenters in the chick cells, while in the quail cells one or two large patches of heterochromatin are associated with the nucleolus (Fig. 1). In lymphocytes of both species, the chromatin is more condensed than in other somatic cells, as can be seen in Fig.1. The quail and chick cells can, however, easily be told apart after either Feulgen–Rossenbeck staining of the DNA or electron microscopy.

d. The Use of the Quail-Chick Marker to Study Thymus Histogenesis. Three types of experiments have been performed in order to define the respective role of endodermal epithelium, mesenchymal cells, extrinsic HC, and other possible mesodermal elements in the thymic histogenetic process. They have consisted in labelling selectively either the endodermal and the mesenchymal components of the thymic rudiment or the precursor cells of the lymphocytes.

i. LABELLING OF THE ENDODERMAL THYMIC COMPONENT AND OF THE LYMPHOCYTES. These experiments are based primarily on the method first described by Le Douarin (1967), i.e., the graft of the rudiment under investigation into the somatopleure of a 3-day-old embryo (Fig. 2). The endoderm is taken from the presumptive or actual region of the third and fourth pharyngeal pouches in quail (or chick) embryos at 15- to 30-somite stages. It is then grafted into the somatopleure of 3-day chick (or quail) embryos. This results in the development of a normal, although ectopic, thymus in the body wall of the host. Screening of the thymic tissue in 5 μm serial sections stained for DNA revealed that the epithelial reticular cells (as controlled at the electron microscopic level through their desmosomal junctional complexes) are the only thymic cells with the nuclear marker of the donor epithelial endoderm. None of the lymphocytes, connective cells, or endothelial walls of the blood vessels possessed this marker; their nucleus was of host type. Therefore, thymic histogenesis involved the participation of the host mesenchyme at the graft site. This shows that the proper

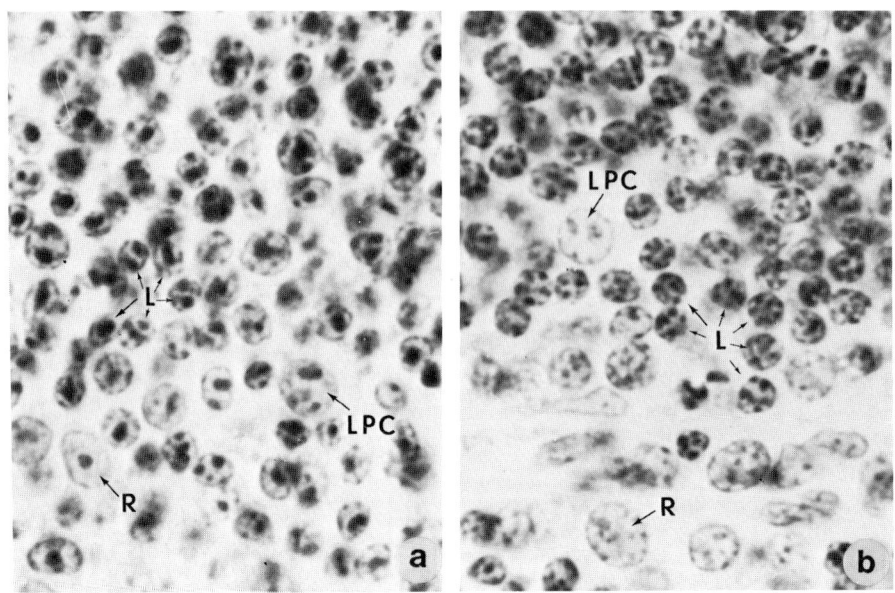
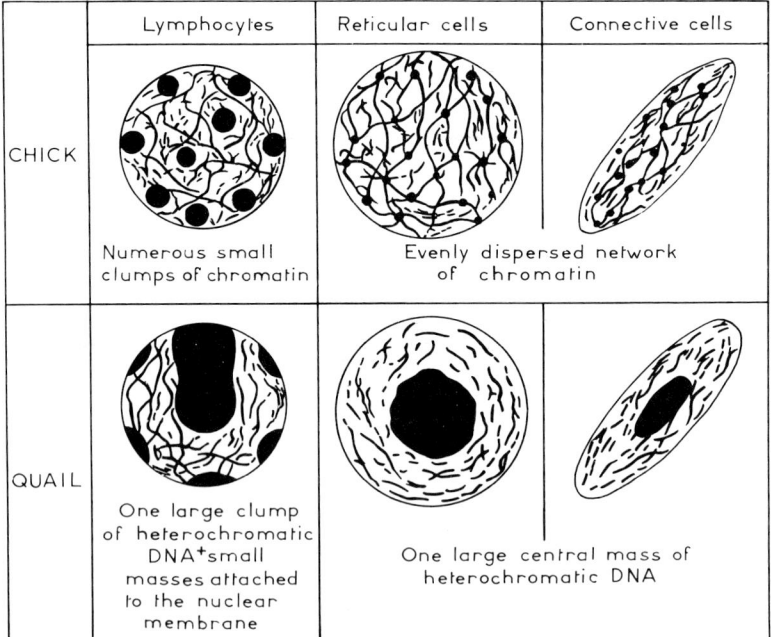

Fig. 1. Micrographs (a) and (b) and schematic drawings (c) showing the nuclear feature of quail and chick cell types in the thymus after Feulgen–Rossenbeck staining. In the quail thymus (a), lymphocyte (L) nuclei contain one large heterochromatic mass and several smaller ones attached to the nuclear membrane. In the chick thymus (b), several small chromocenters are dispersed in the nucleoplasm. In the quail, the lymphocyte precursor cells (LPC) show a large nucleus with a central Feulgen-positive condensation, whereas in the chick the nucleus is only lightly stained. Reticular (R)

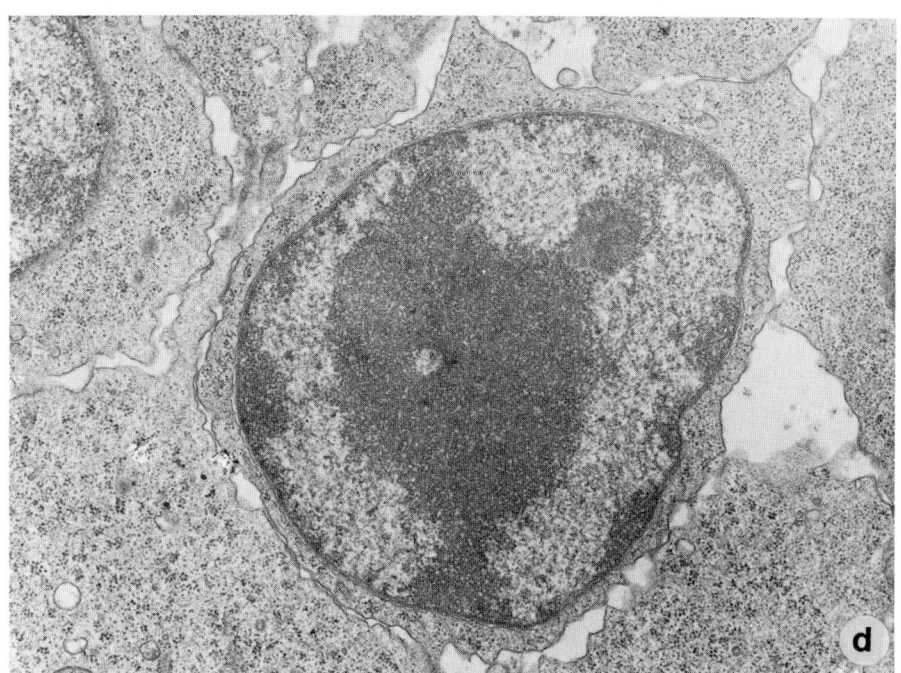
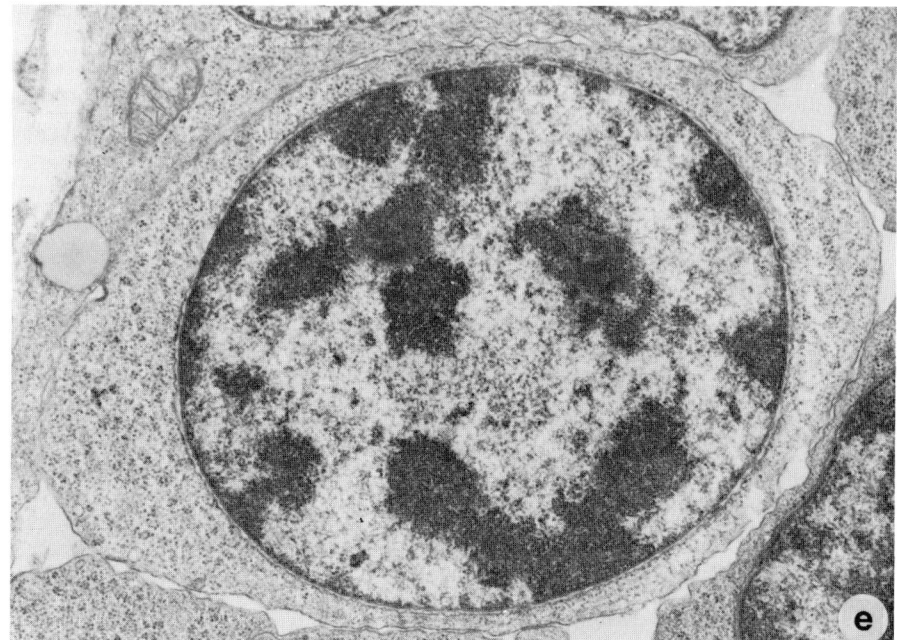

and connective cells of quail usually have one single heterochromatic mass. In the same cell types of the chick, the nucleus contains a fine network of evenly distributed chromatin. Feulgen–Rossenbeck staining. × 936. Electron microscopy of quail (d) and chick (e) lymphocytes. × 27600.

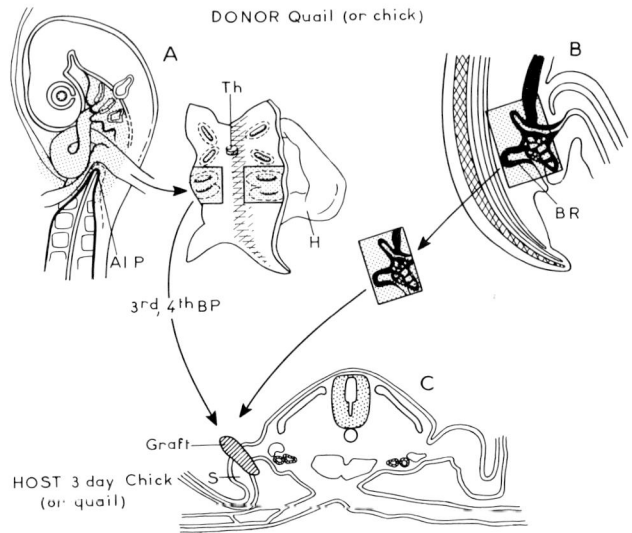

Fig. 2. Interspecific graft of the thymic and bursal rudiments into the somatopleure of a 3-day-old host. (A) Removal of the thymic rudiment from a 30-somite embryo from which the ventral side of the pharynx comprising the branchial arches and the cardiac primordium has been removed. Bilateral areas corresponding to the third and fourth pharyngeal pouches (endoderm + mesenchyme) have been selectively taken and grafted into the somatopleure of a 3-day-old host. (B) Removal of the bursal rudiment from a 5-day old embryo. (C) Graft of the thymic or bursal rudiments into the somatopleure of the host embryo at 3 days of incubation. The donor and the graft were always of different species, i.e., quail or chick. AIP, anterior intestinal portal; BP, branchial pouches; BR, bursal rudiment; H, heart; S, somatopleure; Th, thyroid rudiment.

mesectodermal component of the pharyngeal thymic rudiment can be replaced by mesodermal mesenchyme.

If the thymic primordium (i.e., the third and fourth branchial pouches including mesenchyme plus endoderm taken from the host around the 30-somite stage) is grafted interspecifically between quail and chick, lymphocytes are all host-type cells, indicating that their origin is extrinsic to the thymus primordium, and thus confirming fully the hematogenous hypothesis of lymphocyte precursor derivation.

Besides the lymphocytes, the thymus also includes macrophages and dendritic cells as described by Steinman and co-workers (see Steinman *et al.*, 1980, for a review). Preliminary results from our group indicate that precursors of these cells also have an extrinsic origin and colonize the thymic rudiment together with lymphoid precursors (Oliver and Le Douarin, 1984).

ii. SELECTIVE LABELLING OF THE MESENCHYMAL THYMIC COMPONENT *IN SITU*. In other studies in our laboratory dealing with the migration and differentiation of neural crest cells, it had been demonstrated that most mesenchymal cells of the

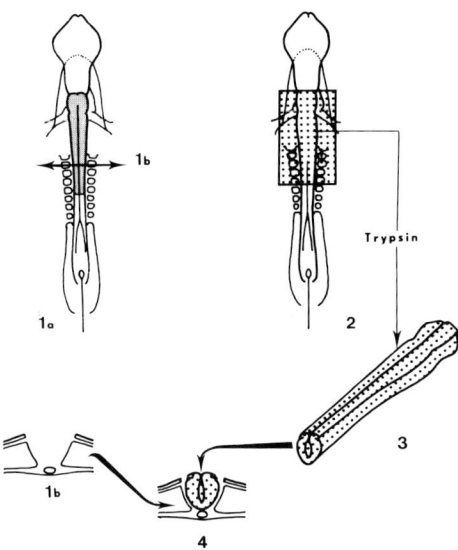

Fig. 3. Schematic drawing of the experimental procedure used to label selectively the mesenchyme of the thymic rudiment. The rhombencephalon of a 6- to 9-somite quail embryo is grafted isotopically and isochronically into a chick embryo. (1a) Chick embryo from which the rhombencephalon (dotted area) is excised. (1b) Transverse section at the level of excision showing that the somites, the notochord and the endoderm are left in situ. (2) The rhombencephalon is isolated from a quail embryo through enzymatic digestion. (3)–(4) Isotopic graft of the quail rhombencephalon into the chick host.

facial and hypobranchial regions were ectodermal in origin rather than mesodermal, as are all the other mesenchymes in the body (Le Lièvre and Le Douarin, 1975; see also Le Douarin, 1980, 1982, for reviews). Since the thymus is one of the glandular derivatives of the hyprobranchial region, it seemed of interest to investigate the role of the neural crest in thymic histogenesis.

Again through the use of the quail-chick marker system, it was demonstrated that the whole thymic mesenchyme arises from the rhombencephalic neural crest. Grafting interspecifically the rhombencephalon between quail and chick embryos in an isotopic and isochronic fashion (Fig. 3) resulted in the development of chimeric thymuses in which the only cells with donor-type nuclei were the connective cells lining the blood vessels. It was thus apparent that mesenchymal contribution to thymus histogenesis is restricted to connective cells of the interlobular spaces and to the wall (including the pericytes) of the blood vessels in cortex and medulla (Le Douarin and Jotereau, 1975).

iii. DI GEORGE'S SYNDROME INTERPRETED IN THE LIGHT OF THE CELL-MARKING EXPERIMENTS. Di George's syndrome (Di George, 1968), which affects derivatives of the third and fourth branchial arches in man, is characterized by a congenital absence (or strong reduction) of the thymus and parathyroids. This is associated with malformations of the large vascular trunks derived from the aortic arches and dysfunctions of the central nervous system at the bulbar level (Couly et al., 1982).

Failure of development of the thymus, parathyroids, and large vascular trunks of the heart must be related to alterations of neural crest development affecting either the migration process or the multiplication and differentiation of the rhombencephalic crest cells. Previous work from our laboratory had shown that the wall of the large arteries that were derived from the aortic arches, as well as the mesenchymal components of the pharyngeal glands, originated from the rhombencephalic neural crest. In the congenital human disease the neurocristopathy is linked with abnormalities in the development of the corresponding segment of the central nervous system. It is striking to see that the level of the rhombencephalon that has been found in birds to be the source of the neurectodermal mesenchyme of the third and fourth branchial arches is precisely the same that is affected in Di George's syndrome.

The reason why neural crest cell derivation affects histogenesis of thymus and parathyroid primordia, while their main cellular component is the pharyngeal endoderm, has to be looked for in the epitheliomesenchymal interactions that occur during the ontogeny of various organs (for references, see Le Douarin, 1964, Saxén et al., 1976).

The problem of thymic development has been documented in this respect by Auerbach (1960) and Le Douarin (1967), who demonstrated that thymus epithelium, isolated from the thymic mesenchymal component, stops growing and differentiating if cultured alone either *in vitro* or *in vivo*. When reassociated either with the proper thymic mesenchyme or with certain mesenchymes from other parts of the body, the isolated thymic endoderm can undergo normal histogenesis. In Di George's syndrome it is probable that the lack of mesenchymal components from the neural crest has not been compensated locally by mesenchyme of any other source and has resulted in the abortive development (more or less complete) of the thymus and parathyroid epithelial primordia.

B. Colonization of the Thymus Rudiment by Lymphocyte Precursors Proceeds by Waves

1. FIRST COLONIZATION OF THE THYMUS RUDIMENT

The time at which the thymus rudiment is first invaded by HC in ontogeny was determined by grafting the thymus anlage interspecifically between chick and quail embryos at increasingly late developmental stages. The total age of the thymus (i.e., age at grafting plus duration of the graft) at the time of observation was 14 days in all experimental series. It was demonstrated that the invasion of

VIII. Primary Lymphoid Organ Ontogeny in Birds 189

the thymus by HCs is a relatively rapid process that starts at 5 and 6.5 days of incubation, respectively, in quail and chick embryos and stops some 24 hours later in the quail and some 36 hours later in the chick. This first influx of cells into the thymus is followed by a refractory period characterized by a shut-off of the thymic receptivity for HCs. The duration of the nonreceptive period and whether receptivity is resumed and when have been further investigated by constructing various kinds of chimera.

2. EVIDENCE FOR A CYCLIC SUCCESSION OF RECEPTIVE AND NONRECEPTIVE PERIODS FOR STEM CELL ENTRY INTO THE EMBRYONIC AND EARLY POSTNATAL THYMUS

The quail thymus has been used so far in these investigations (Le Douarin and Jotereau, 1980; Jotereau and Le Douarin, 1982).

a. Interspecific Grafting Experiments

i. DEMONSTRATION OF A SECOND INFLUX OF HC INTO THE THYMUS. The principle of these experiments was to subject the thymus, taken from quail embryos after completion of the first colonization, to two successive grafts, the first host being a 3-day chick embryo and the second a 3-day quail embryo. The grafts had variable durations as indicated in Fig. 4, which shows the three different experimental series performed. In series I, the age of the thymus at the end of the first grafting period was 11 days, while in series II and III it was, respectively, 12 and 13 days. Chimerism analysis of the grafted thymuses was carried out when the total age of the organ was 20–21 days. It appeared that whatever the duration of the first graft (from 1 to 5 days, see Fig. 4) the decisive factor was the age of the thymus when it was transplanted into the second host.

None of the explants removed from the chick host at 11 days contained chick lymphoid cells when observed at 20 days. In contrast, when the first graft was prolonged until 12 or 13 days, 53 and 83%, respectively, of the explants had received chick HCs during the first engraftment period. It can be concluded that, at least under these experimental conditions, the quail embryonic thymus does not open to the second influx of HCs before the end of the eleventh day of incubation. Some of the thymuses of series II (i.e., when the residence in the first host ended at 12 days of thymus age) were found to contain nearly 100% of chick lymphocytes, indicating that within an approximate 24-hour period (between 11 and 12 days) enough lymphocyte precursors had seeded the thymus for complete (or nearly complete) renewal of the first generation of lymphocytes.

The fact that not all thymuses had been colonized in the first host, even when the graft was prolonged until 13 days, showed individual fluctuations of the seeding process timing. Whether they might, at least partly, be attributed to the graft conditions was further investigated. Quail thymuses taken from 12- and 13-

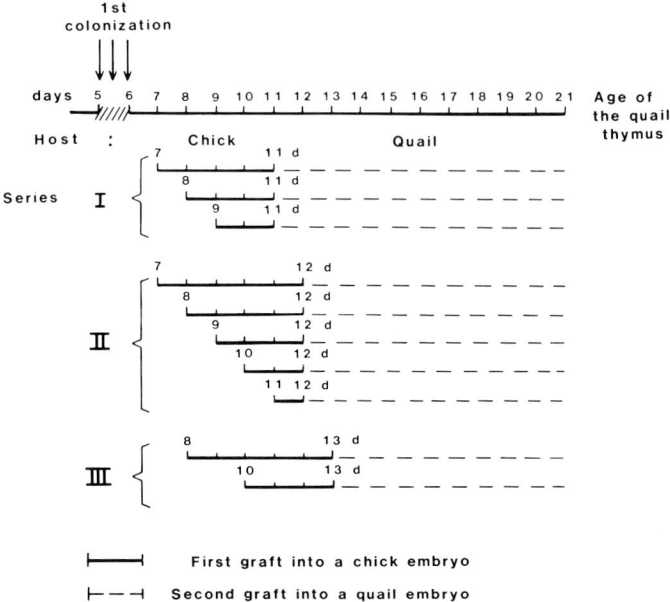

Fig. 4. Schematic drawing of the different experimental series: I, II, III. Embryonic quail thymuses were taken from 7- to 11-day old donors. They were transplanted into a 3-day chick embryo until they reached, respectively, 11, 12, or 13 days for the series I, II and III. Retransplantation was then realized in a 3-day quail embryo for various periods of time in order to leave the HC eventually immigrated from the first host to multiply and differentiate into chick lymphocytes.

day donors were grafted for periods of 2 to 3 days into chick embryos and then transferred into quail embryos for various periods of time. No chick lymphocytes were found in any of the 34 grafts examined, suggesting that the second influx of lymphocyte precursors had always been completed in the quail donor before grafting of the thymus into the chick. This discrepancy must be related to the fact that in the first experimental series, the quail thymuses were subjected to grafting during the period refractory to HC invasion. Such a manipulation must have disturbed the dynamics of cell division of the first wave of HC and, in a certain number of cases, delayed the second influx by 1 or 2 days. Therefore, it can be assumed that the second wave of HCs most often takes place during the twelfth day of incubation, and may be delayed for one more days under experimental conditions.

ii. EVIDENCE FOR A THIRD WAVE OF HCs INTO THE THYMUS. Signs of a third wave of HCs into the thymus could first be noted when 7- or 8-day-old quail thymuses had been transplanted into a chick embryo until the second colonization was completed (i.e., until 12 or 13 days) and then retransplanted into a quail

VIII. Primary Lymphoid Organ Ontogeny in Birds

host until they had reached 20 days of age. At that time, chick lymphocytes had replaced most thymocytes of the external cortex and of the medulla. Large cells with the morphology of quail HCs could then be seen as a fringe under the thymic capsule and scattered in the medulla around blood vessels. That these cells were lymphocyte precursors was confirmed when such thymuses were retransplanted for a third time into chick embryos for 8 days. A quail lymphoid population developed in the explants and completely replaced the previously established chick one, in both the cortex and medulla.

The timing of the third wave of lymphocyte precursors was subsequently studied by double transplantation experiments. Thymuses from 12- and 13-day-old quail embryos were grafted into a 3-day chick until they reached 14 to 18 days and then into a quail until 22 to 28 days. It appeared that thymus seeding did not take place unless the age of the grafted thymus was 17 days or more at the time of the second graft. Forty-seven percent of the explants were invaded by chick HC during the seventeenth day, and 80% had been seeded at the end of the eighteenth day in our experimental conditions.

After it has been colonized by the second wave of HCs, the quail thymus becomes again refractory to HC seeding for about 5 days. A third wave of HCs then takes place, which in our experiments started in most cases between 17 and 18 days and was completed in less than 2 days.

Thus, colonization of the quail thymus by lymphocyte precursors during embryonic and postnatal life appears as a cyclic process with a sequence of recep-

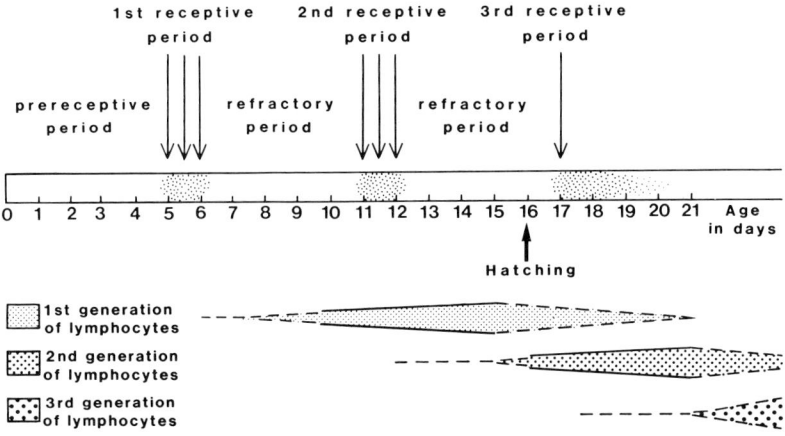

Fig. 5. Schematic representation of the three successive waves of HC entering the thymus in the embryo and in the early postnatal quail. The entry of HC is restricted to short periods separated by several days (five in our experimental conditions) during which no significant seeding takes place. The precise timing of the third wave of lymphocyte precursor cells into the thymus could not be determined precisely in our experimental conditions.

tive periods lasting about 24 hours, separated by refractory phases of about 5 days. As a corollary, the proliferative capabilities of the HC that colonize the young thymus appear to be limited to a definite and probably relatively small number of cell cycles. Figure 5 summarizes the data.

iii. CONFIRMATION OF THE CYCLIC RENEWAL PATTERN IN HOMOSPECIFIC COMBINATIONS. Since these experiments involved combinations between a thymic stroma and HC of different species and ages, it seemed necessary to control these results by using a technique that would not introduce these biases. This was made possible in the study of the second HC influx through the use of the sex chromosomes as cell markers in quail embryo parabioses (Le Douarin et al., 1984). Since these experiments are unpublished they will be described with some detail.

Birds are characterized by female heterogamety, in contrast to mammals where this is a characteristic of males. For this study, we have used a strain of quails (kindly provided for us by Dr. Perramon of the Institut National de la Recherche Agronomique) with a recessive albino mutation (a) affecting a pigmentation gene carried on the sex chromosome Z. The males homozygous for the albino gene Z^a/Z^a and the Z^a/W females are unpigmented both in the neural retina and in the feathers. The heterozygous Z^a/Z^+ males are normally pigmented, and melanin first appears on the neural retina at 3.5 days of incubation.

The embryos used for these experiments are the F_1 from the cross of female $Z^+/W \times$ male Z^a/Z^a, which yields 50% of pigmented males Z^+/Z^a and 50% of unpigmented females Z^a/W. Chromosome preparations were made according to the C-banding technique described by Sumner (1972) following colchicine treatment. Parabioses were established by transferring the content of two quail eggs at 2 days of incubation into the shell of an emptied chick egg (Fig. 6). At 2 days, the sex of the embryos cannot yet be identified, but 1.5 days later the associations of two embryos of the same sex can be discarded. Association of the embryos at 3.5 days results in death of most individuals through the rupture of the vitelline membrane surrounding the yolk, which is far more fragile at 3 than at 2 days of incubation. The vitelline membrane was very useful in that it prevented vascular anastomoses from being established between the yolk sacs of the two embryos. As a result, the first colonization of each parabiont thymus was by its own HCs. From 5–6 days, the CAMs started growing and met in the midline. Later on, vascular anastomoses became established through the CAM blood vessels (Fig. 6). Injection of a solution of India ink in PBS (1/1) in the bloodstream of one parabiont revealed that vascular connections arise between the two embryos during the eigth day of incubation. From this time onward, blood was chimeric, and the thymus could be seeded by cells of the opposite sex. In order to define the length of the refractory period separating the first from the second thymic seeding, the thymuses were removed from each embryo and grafted in the somatopleure of a 3.5 day host of the same sex (as determined by

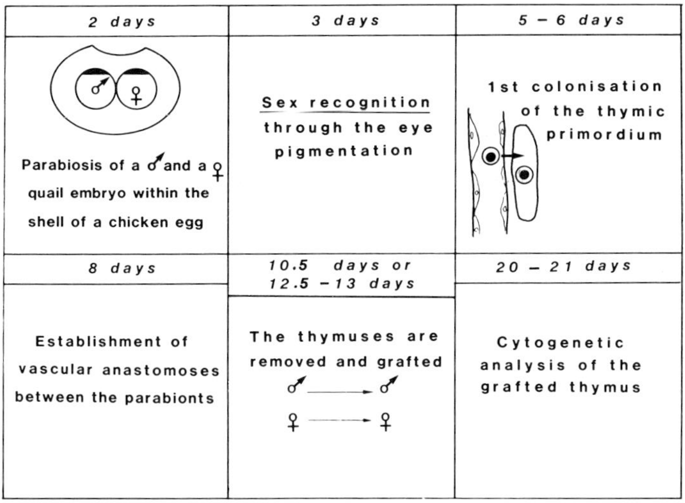

VIII. Primary Lymphoid Organ Ontogeny in Birds

2 days	3 days	5 – 6 days
Parabiosis of a ♂ and a ♀ quail embryo within the shell of a chicken egg	Sex recognition through the eye pigmentation	1st colonisation of the thymic primordium
8 days	10.5 days or 12.5 – 13 days	20 – 21 days
Establishment of vascular anastomoses between the parabionts	The thymuses are removed and grafted ♂ ——— ♂ ♀ ——— ♀	Cytogenetic analysis of the grafted thymus

Fig. 6. Outline of the experiment in which two quail embryos carrying an albino gene (*a*) on the heterochromosome Z, are associated by transfer into the empty shell of a chick egg. The males Z^a/Z^+ are normally pigmented, while the females Z^a/W are albino. The sex can be identified at 3 days through pigmentation of the eye. Since vascular anastomoses are not established before day 8 of gestation, the first thymic colonization concerns the HC of the embryo itself while the second one is due to a mixture of cells from the embryo and its parabiont. By transplanting the thymus of each embryo into a 3-day host of the same sex, either before or after the presumed time of the second colonization, we could confirm the results obtained in heterospecific grafting experiments (see table I for the results of the experiments described here).

pigmentation). Cytogenetic analysis of the cells in the grafts was carried out when the thymuses had reached 18–20 days. Two different periods were chosen to transfer the thymic rudiment: 10.5 and 12.5 days, that is before and after the presumed time of the second seeding. The results indicated in Table I show that practically no chimerism was found in the first series of thymuses which were

TABLE I.

Sex Chromosome Analysis of Thymuses of Parabiosed Quail Embryos

Age of the thymus at the end of parabiosis (days)	Mitoses of opposite sex (% chimerism)	
	Percentage	*n*
10.5	3.25	16
12.5	34	29

transplanted into an embryo of the same sex at 10.5 days, while a significant number of mitoses of the opposite sex were found in the 12.5 day series. This indicates that the second HC colonization of the thymus *in situ* in the quail takes place between about 11 and 12 days of incubation, thus confirming the results obtained in quail↔chick experiments.

3. DYNAMICS OF LYMPHOCYTE TURNOVER IN THE EMBRYONIC AND POSTNATAL QUAIL THYMUS: DEMONSTRATION OF A SEPARATE RENEWAL FOR CORTICAL AND MEDULLARY LYMPHOCYTES

The arrival of the second and third waves of HC and the expansion of their lymphoid progeny throughout the quail thymus was visualized through the use of the quail-chick marker. In experiments involving double transplantations of 7- to 8-day quail thymuses into a chick and then a quail host according to the pattern of series II and III indicated in Fig. 4, replacement of the first by the second generation of lymphocytes was followed. A few large chick HCs were observed both under the thymus capsule in the cortex and in the medulla close to blood vessels, at 13–14 days of thymic age (Fig. 7A). Injection of tritiated thymidine into the host made it possible to follow the multiplication of these cells. They started dividing only after about 3 days of residence in the graft, both in cortex and medulla, then their progeny simultaneously invaded the cortical and medullary areas, between 15 and 21 days of thymic age.

Invasion proceeded from the outer to the inner layers of the thymic cortex (Fig. 7B). In the medulla, lymphoid renewal started around the blood vessels, where groups of chick lymphocytes appeared. Thereafter, the replacement of quail by chick cells spread throughout the whole medullary area.

Replacement of the second by the third generation of thymic lymphocytes proceeded according to a similar pattern. The chronology of this renewal is indicated in Fig. 5.

a. The Chemotactism Hypothesis. The regularity of HC colonization of the thymic rudiment exemplified the problem of the mechanisms underlying such a complex series of events in which cell migration and cell–cell interactions are involved. This is why we have tried to analyze this phenomenon, concentrating mainly on the first influx of cells into the thymic primordium.

Two possibilities were envisaged to explain the periodicity of HC seeding. According to the first one, this timing depends on the availability of competent HCs at determined stages of the development. Another is a control by intrinsic factors related to the development of the thymus itself. In fact, we demonstrated that HC able to seed a thymus are available permanently in blood from at least 2 days before the onset of the first colonization. Therefore, it is the thymus itself that regulates its colonization by HCs.

VIII. Primary Lymphoid Organ Ontogeny in Birds

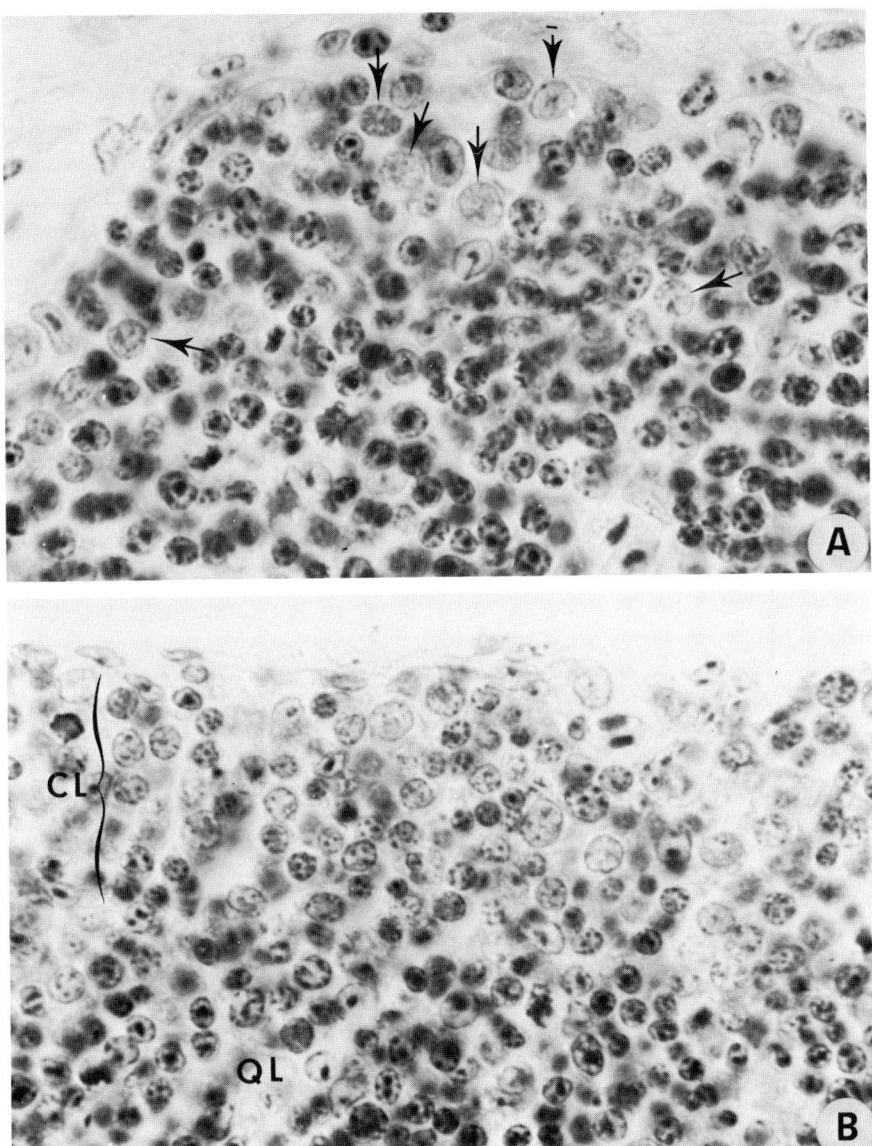

Fig. 7. Thymus of a 7-day quail embryo grafted into a chick embryo: chimerism analysis after Feulgen staining. (A) After 7 days in the chick, lymphoid precursors of the host (with chick nucleus morphology) are present in the external cortex (arrows). × 850. (B) After 10 days in the chick, chick lymphocytes (CL) have replaced quail lymphocytes (QL) in the external cortex. × 850.

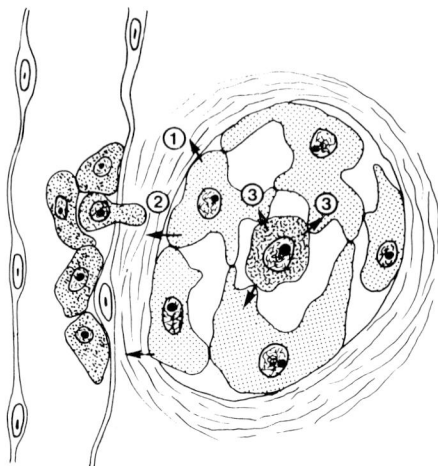

Fig. 8. Hypothesis for hemopoietic cell seeding in the thymic rudiment. The endodermal epithelial cells are supposed to produce a chemoattractant (1) that leads the blood borne HC to agglomerate in the vessels located close to the thymus anlage and thereafter to migrate and home to the thymus (2). A negative feedback (3) induced by the HC stops the production of the chemoattractant, and this explains the outset of the refractory periods that separate the first from the second HC influx to the thymus.

The chemotactism hypothesis of lymphoid organ seeding was formulated by one of us some years ago (Le Douarin, 1978, 1979). It was proposed that the mechanism responsible for thymic colonization might depend on the release of an attractive substance by the thymic stroma, at stages corresponding to the colonization periods. A feedback mechanism resulting from interactions between stromal and HC cells would ensure the arrest of such a production and explain the interruption of the HC inflow (Fig.8). This hypothesis is now supported by several experimental results of our group (Le Douarin et al., 1976; Jotereau et al., 1980; Ben Slimane et al., 1983) and also by results obtained in a mammalian model by Pyke and Bach (1979) and Fontaine-Pérus et al. (1981).

b. *In Vivo and Organ Culture Experiments.* Our first experiments supporting the chemotactic hypothesis consisted of the *in vivo* association of two explants: a hemopoietic donor organ of one species and a thymus or a bursa of the other, as a recipient organ, grafted side by side into the somatopleure of a 3-day chick or quail host according to the combination considered (see Fig. 9 and Table II). The rationale was that if the thymic colonization depends on an attractive mechanism, lymphopoietic cells present in the hemopoietic organ should be displaced toward the experimentally associated attractive thymus. Such was indeed the case.

For example, a 6.5-day chick thymus (taken just before the beginning of the first HC colonization) was grafted in association with quail yolk sac, bone marrow, spleen, or colonized thymuses and bursas of Fabricius. The results presented in table II, part A, show that HC present in these organs were able to migrate into the thymus rudiment where they differentiated into lymphocytes. Such migration did not occur toward a thymus taken at the refractory period that follows the first HC inflow (Table II, part B).

Having shown that a second wave of HC normally invades the quail thymus between 11 and 12 days, we have tested whether a quail thymus taken just before this period would also be able to displace HC from a chick hemopoietic organ. A chick bursa of Fabricius was associated with an 11-day quail thymus. In a certain number of cases HC migrated from the chick bursa to the quail thymus (Table II, part C). It is clear, therefore, that this experimentally induced migration relies upon properties exhibited by the thymus only during the periods of HC colonization.

However, it appears from the percentage of chimerism among the grafted thymuses that, even when the recipient explant was at a receptive period, seeding by cells coming from the associated quail explant did not occur in all cases

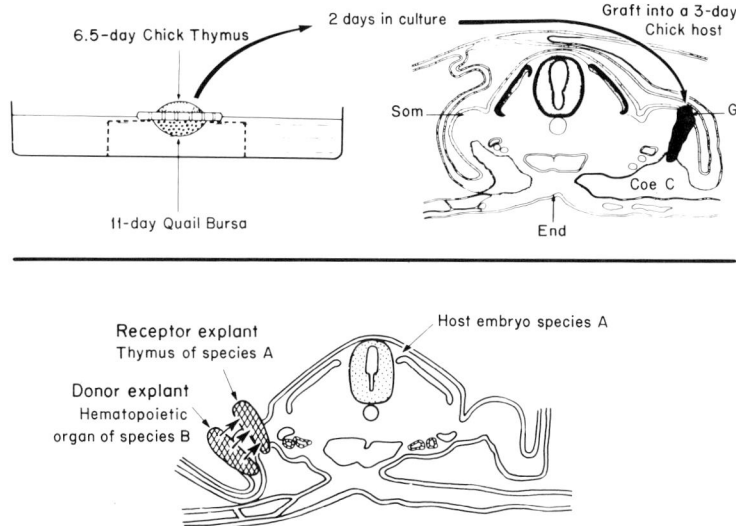

Fig. 9. Experimental procedures used to test the chemotactic hypothesis of HC seeding of the thymus (a) *in vitro* and (b) *in vivo*. In the *in vivo* experiments a thymus of species A and a donor hemopoietic organ of species B are grafted side by side into the somatopleure of 3-day-old A embryo. The two species considered are quail and chick designated as A or B according to the experimental series, the receptor explant and the host embryo being always of the same (quail or chick) species. Coe C, coelomic cavity; End, endoderm; G, graft; Som, somatopleure.

TABLE II.

Association in Graft Side by Side of Various Hemopoietic Donor Organs with Heterospecific Thymus Taken Either at a Receptive (A and C) or at a Refractory (B) Period

Recipient organ	Donor explant[a]						Number of chimeric explants	Percent of chimeric explants
	Chick	Quail						
	BF	YS	BM	Spl	Th	BF		
A. 6.5 day Chick thymus (first "receptive" period)		10[b]					6	60
			4				3	75
				6			5	83
					10		6	60
						18	10	56
B. 10 day chick thymus ("refractory" period)					8		0	0
						11	0	0
C. 11 day quail thymus (second "receptive" period)	13						6	46

[a] BF, lymphoid bursa of Fabricius from 12-day chick embryos and 11-day quail embryos; YS, yolk sac from embryos at headfold to 4-somite stages; BM, bone marrow from 14-day embryos; Sp, spleen fragments from 14-day embryos; Th, lymphoid thymuses from 8- or 14-day quail embryos.

[b] Number of explants observed.

(approximately in 50 to 80% of explants). This is probably related to the establishment of proper vascular connections between the two grafted tissues. Those occur at random, and in a certain number of cases may favour the influx of cells from the host embryo rather than from the associated explant.

Similar experiments performed in organ culture have also given rise to results supporting the chemotactism hypothesis (Fig. 9). Trans-filter culture experiments were realized in which a chick explant was placed on the upper side of a Nuclepore filter (with 5 μm pore diameter) while an 11-day quail bursa (donor organ) was cultured underneath the filter, according to a method previously described (Jotereau et al., 1980). The chick explant was a thymus taken either at its receptive period (6.5 days) or after the first colonization (10 days). For controls, liver and mesonephros tissues from 6.5-day chick embryos were used as possible receptive explants. The number of quail cells that crossed the filter was counted in the chick explant after various periods of times. It appeared that, as well as a small influx of quail cells found in control cultures, a large migration of bursal cells was induced by the receptive thymus (Fig. 10). Migrant quail cells were shown to be mostly lymphocyte precursors, since they differentiated into lymphocytes when the cultures were prolonged for 7 to 9 days or when the thymic explants were grafted into a chick embryo for a similar period of time. These experiments provide support for the view that the thymus actually produces at precise times of development a chemoattractant that induces an influx of

VIII. Primary Lymphoid Organ Ontogeny in Birds

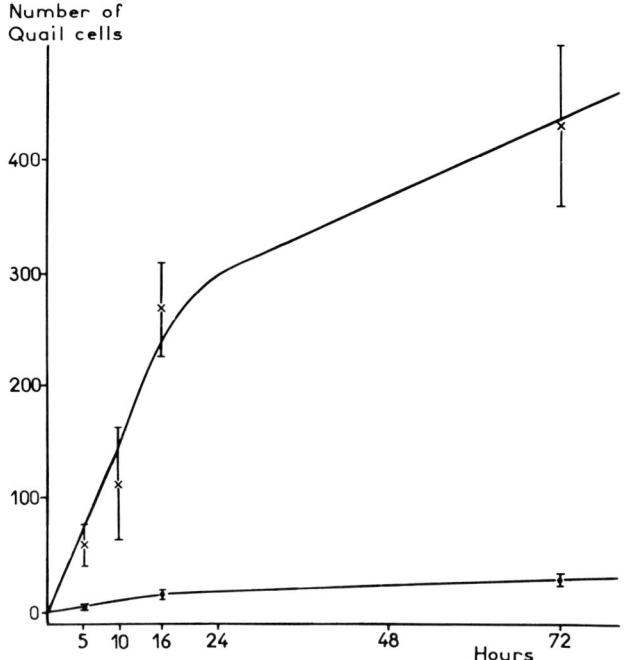

Fig. 10. Comparison of the number of quail cells that cross the filter and seed the chick organ in transfilter cultures (see Fig. 9) after 5, 10, 16, and 72 hours in culture. (×) Experiment in which the recipient organ is a 6.5-day chick thymus; (●) results obtained in three other experimental series (where chick recipient explant was either 6.5-day liver, 6.5-day mesonephros, or 10-day thymus of chick) are pooled (see Jotereau et al., 1980).

lymphocyte precursors. More information on this problem of HC immigration into the thymus has recently been provided by Thiéry and co-workers in our laboratory, through the use of a refined *in vitro* technique in which the actual cell movements are visualized through time-lapse cinematography (Ben Slimane *et al.*, 1983).

c. *The Use of Zigmond's (Modified) Chemotactism Chamber*

i. SOURCE OF LYMPHOID PRECURSORS. An abundant source of lymphoid precursors was found in embryonic bone marrow. By day 12 of incubation, almost 2×10^6 basophilic cells can be obtained free of erythroid cells from a single quail femur. When confronted *in vitro* with 6.5-day attractive chick thymuses (see Fig. 11), the basophilic cells penetrated the thymic epithelium, and after several days of grafting in the somatopleure they differentiated into T lymphocytes. These lymphoid precursors adhered to glass and migrated randomly at 12 μm / minute in RPMI 1640 supplemented with 10% fetal calf serum (standard medium).

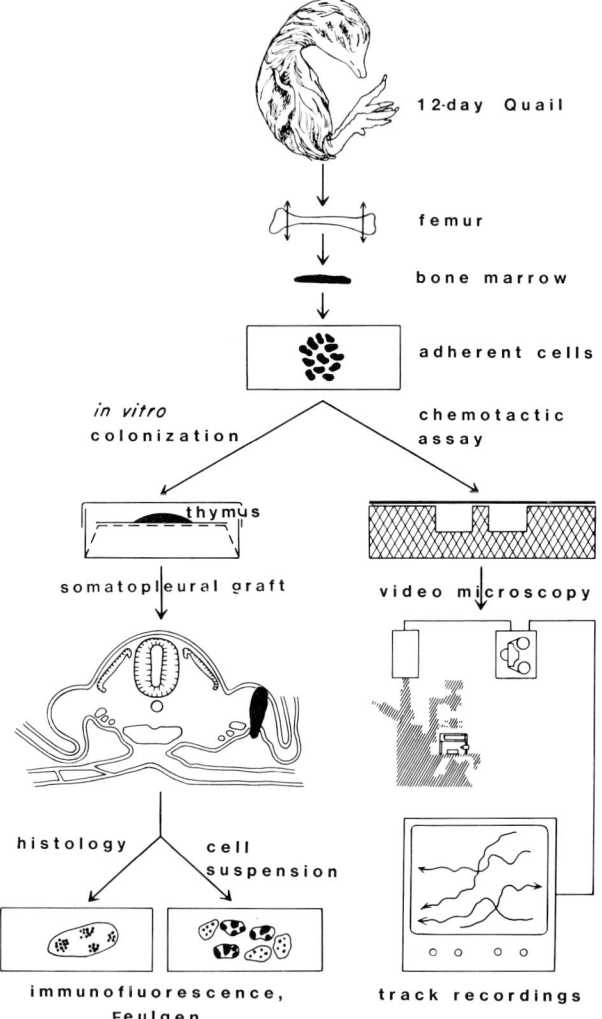

Fig. 11. Outline of the procedures used to define T lymphocyte potentialities and *in vitro* migration properties of the 12-day bone marrow adherent cells.

ii. THE CHEMOTACTISM CHAMBER. The adherent lymphoid precursors are placed in a thin film of medium (10 μm wide) separating the two compartments of the chemotactism chamber of Zigmond (1977). In this situation, a stable concentration gradient is formed rapidly when a diffusible substance is introduced into one of the compartments (Fig. 12).

Cells placed on the transparent bridge of the chamber (site of stable gradient)

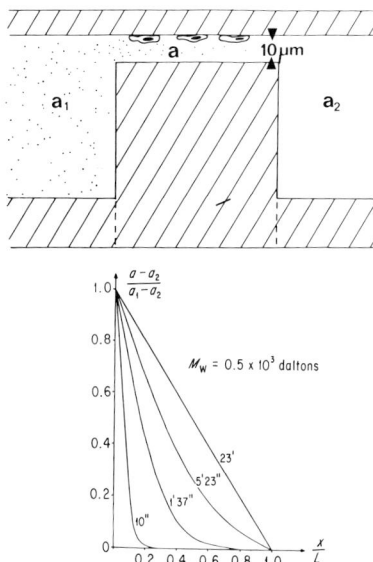

Fig. 12. Zigmond's chemotactism chamber. Transverse section. Cells attached to a glass coverslip are migrating in a narrow space between the two compartments under which relative concentration of a diffusible substance of 0.5×10^3 daltons computed according to diffusion equations formulated by Lauffenburger and Zigmond (1981).

can be readily filmed and their movement analyzed quantitatively. One of the major advantages of this chemotactism chamber resides in the fact that chemotaxis can be unambiguously distinguished from chemokinesis, i.e., increase in speed of locomotion of randomly migrating cells.

iii. ANALYSIS OF LYMPHOID PRECURSORS BEHAVIOR WITH THE ZIGMOND CHAMBER. Basophilic bone marrow precursors responded to 7.5-day chick thymuses introduced into one of the chamber compartments by an oriented movement toward these tissues. An average chemotactic index of 0.43 ± 0.22 was calculated according to McCutcheon (1946). In contrast, chick thymuses taken at 10 days during the first refractory period were unable to attract the lymphoid precursor cells ($I = 0.01 \pm 0.12$). Finally, 7.5-day mesonephros taken as a control tissue was also unable to orient the movement of lymphoid precursors.

It was noteworthy that both the attractive and the refractory thymuses improved the adherence of the cells to glass and increased their speed of locomotion from 12 to 25 μm / minute; such a chemokinetic effect was not observed with other embryonic tissues (Fig. 13).

When attractive thymuses were sealed in microdialysis membranes that retained molecules of more than 12,000 daltons, a similar average chemotactic

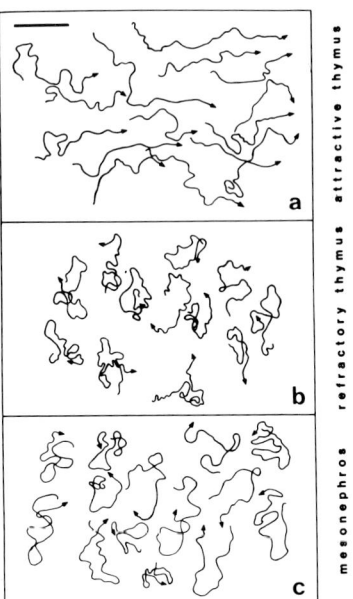

Fig. 13. Migration of 12-day adherent bone marrow cells in the presence of freshly explanted chick embryonic thymus and mesonephros. Cells located at the center of the bridge were followed for 15 minutes in play back of the recording. For clarity, only a few cell tracks are shown. (a) 7.5-Day attractive thymus. Most cells responded to the attraction but changed frequently their direction of movement. (b) 10-Day thymus. Cells were not attracted by the refractory thymus, but increased their speed of locomotion. Changes in their direction occurred more frequently than in control experiments or with the mesonephros. (c) Mesonephros. No preferred orientation was observed. Scale, 50 μm.

index was found; however precursor cells did not migrate faster and did not adhere better than in the presence of control medium. In a complementation experiment, free refractory thymuses and attractive thymuses, sealed in dialysis tubing and placed separately in the two compartments of the chamber, allowed the lymphoid precursors to migrate toward the attractive thymus with an adhesive chemokinetic and chemotactic behaviour identical to that obtained with free attractive thymuses (Fig. 14).

Ultrafiltration of conditioned media prepared from both attractive and refractory thymuses indicated that the chemotactic activity was associated with a factor of molecular weight between 500 and 1000, whereas adhesivity and chemokinesis was induced by one or several factors of molecular weights higher than 50,000. The molecular nature of both sets of factors remains unknown at the present time. Neither well-established chemotactic factors for human polymorphonuclear leukocytes nor adhesive glycoproteins of the extracellular matrix, such as collagens, laminin, and fibronectin, had an effect on lymphoid precursors.

VIII. Primary Lymphoid Organ Ontogeny in Birds

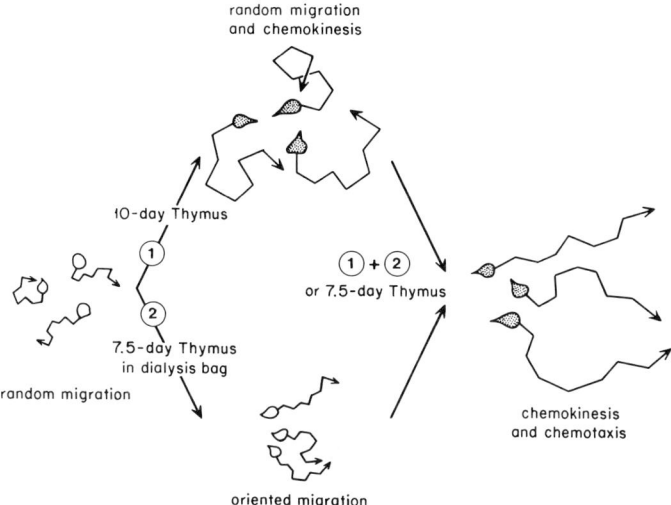

Fig. 14. Chemokinetic and chemotactic effects of thymus can be evidenced separately and can complement each other to provide the full response.

For the first time, it is established experimentally that chemotaxis is the mechanism responsible for the homing of a specific type of migratory embryonic cell. Our results substantiate the hypothesis formulated above; in addition they show that the possible feedback mechanism that shuts down the secretion of the chemotactic substance does not perturb the release of adhesive and chemokinetic factors from the thymus.

In vivo, a diffusible gradient of the small molecular weight chemotactic factor is likely to favour (i) the arrest and adherence of precursor cells to the luminal side of the endothelium, (ii) diapedesis, and (iii) proper directionality. The slope of the chemical gradient can easily be maintained from the source (thymic epithelium) to the sink (adjacent capillaries and jugular vein). (i) and (ii) are induced by chemotactic peptides and leukotrienes in the case of polymorphonuclear leukocytes (Russo *et al.*, 1981; Palmblad *et al.*, 1981)

Adhesive and chemokinetic factors may allow the cells to pass through the perithymic mesenchyme so as to accumulate around and finally penetrate within the thymic epithelium.

III. ONTOGENY OF THE BURSA OF FABRICIUS

The bursa of Fabricius arises as a dorsal endomesodermal diverticulum of the cloaca. From the twelfth day of incubation in the chick, the proliferation of endodermal cells leads to the formation of epithelial buds from which bursal follicles—the site of B cell differentiation—develop. Besides lymphopoiesis, the bursa is also involved in myelopoiesis, which develops in its mesenchymal component. Granulocytes and to a lesser extent erythrocytes actively differentiate in the bursa during embryonic life (mainly from day 8 to 18 in the chick and from day 7 to 14 in the quail).

A. Seeding of the Bursa of Fabricius by Extrinsic Hemopoietic Cells

By using the same interspecific grafting technique as described for the thymus (Fig. 2), the extrinsic origin of the hemopoietic cells that develop in the bursa was demonstrated (for details, see Le Douarin et al., 1975, Houssaint et al., 1976). In brief, when the endomesodermal bursal rudiment from quail was transplanted before the end of the seventh day of incubation to the somatopleure of a 3-day chick, lymphocytes and all other blood cells developing later in the organ originated from the chick host. In bursas taken from the quail donor 11 days onwards, lymphocytes and granulocytes were found to be entirely of quail type. Bursas taken between 7 and 11 days yielded a mixed hemopoietic population of quail and chick origin. The seeding process of the bursa in the quail embryo occurs therefore between 7 and 11 days of embryogenesis, that is just during the first refractory period of thymus colonization. Similar experiments carried out with chick bursas grafted into quail embryos demonstrated that HC seeding takes place between 7.5 and 14 days of incubation in the chick (Fig. 15).

Formation of follicles and seeding of the two bursal compartments (i.e., the mesenchyme and the endoderm) by HCs was studied in chick embryos. The seeding took place first in the mesenchyme, which is richly vascularized, and until 10 days the HCs were only found in the mesenchyme. Penetration of HCs in the epithelium started at the apical (more distal) region of the bursal endodermal bud and then proceeded toward its proximal end; later, follicle formation followed the same progression. Some of the HCs that colonized the organ did not reach the epithelium and became engaged in the myeloid differentiation pathway.

In all the experiments reported above, the grafted bursas were analysed for chimerism when they had reached the total age of 19–20 days. In other experiments, the grafting time was prolonged in order to investigate whether an eventual renewal of the lymphocyte precursors took place in the bursa as demonstrated for the thymus. Colonized quail bursas (taken at 11 days) were grafted in 3-day chicks either for 13 days in a single host or for 20 days in two successive hosts. No renewal was ever observed in any of the explants that still contained quail lymphocytes, even at 31 days of total age. These data support the hypothesis that the bursa of Fabricius is the site of a single hemopoietic cell seeding period that takes place during embryonic life after the first colonization of the thymus has been completed (E. Houssaint, unpublished results).

VIII. Primary Lymphoid Organ Ontogeny in Birds 205

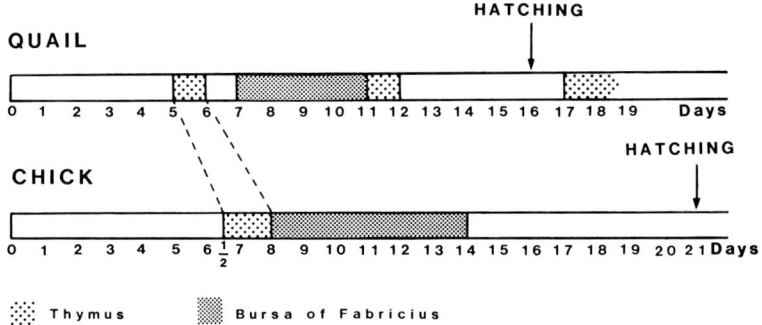

Fig. 15. Periods of HC seeding of the embryonic and neonatal primary lymphoid organs of quail and chick. In the quail thymus three periods of receptivity for HC have been identified. HC seeding of the bursa is unique and takes place between the first and second thymic invasions. Similarly in the chick HC colonization of the bursa is a single step process, and as in the quail it starts about 1 day after the end of the first thymic colonization. Other periods of HC colonization of chick thymus exist but their chronology has not yet been identified.

B. Cell–Cell Interactions between Endoderm and Mesoderm during Bursa Ontogeny and Their Developmental Significance

As mentioned above, thymus and bursa are derived from epitheliomesenchymal rudiments. Like other organs that share this characteristic, their development is highly dependent upon tissue interactions between epithelial and mesenchymal cells. These interactions usually exhibit a low degree of specificity, i.e., heterologous mesenchyme is able to trigger proliferation, differentiation, and even sometimes morphogenesis in an epithelium that usually reacts according to its own genetic programme (for reviews, see Le Douarin, 1964; Saxén et al., 1976).

The thymus fits perfectly with this model, since mesenchyme from various sources can participate in thymic histogenesis if experimentally associated with pharyngeal endodermal cells already committed toward the thymic developmental pathway (Le Douarin, 1967; Le Douarin et al., 1968; Auerbach, 1960). However, not all mesenchyme can fulfill this function; in particular, somite-derived tissues cannot, but all mesenchyme arising from the lateral plate is favourable for thymic endoderm differentiation (Le Douarin et al., 1968).

The developmental relationships between the epithelial and mesenchymal components of the bursa of Fabricius obey much stricter requirements than is the case for the thymus. Endoderm can be separated from mesoderm in the bursa by enzymatic digestion from 5 to 12 days of incubation in both quail and chick (Houssaint et al., 1976). Homo- or heterospecific recombinations of dissociated endoderm and mesenchyme can lead to normal development of the bursa when the recombined organs are grafted into a host embryo.

On the contrary, if the endoderm is associated with the somatopleural mesenchyme of a 3-day-old host, it survives but does not differentiate into follicles. The basophilic cells that have homed to the endoderm at the time of grafting leave the epithelium. Some can be found 2–3 days after grafting, in the surrounding host mesenchyme.

Several other sources of mesenchyme, such as the somites and intestine, were associated with the bursal endoderm, but none were able to promote the development of the epithelium; the latter neither retained lymphoid stem cells nor formed follicular buds (E. Houssaint and N. M. Le Douarin, unpublished).

The bursal mesenchyme from 5- to 7-day embryos was also associated with the thymic endoderm taken from the pharynx at 25- to 30-somite stages. No lymphoid organs resulted from these chimeras; the thymic epithelium, identified by its nuclei, developed poorly and did not give rise to reticular cells.

Since the bursal epithelium arises from the posterior gut, it seemed interesting to investigate the ability of bursal mesoderm to induce bursal differentiation in digestive endoderm. Endoderm isolated from the rectum of 5-day-old embryos associated with bursal mesenchyme at the same developmental stage did not result in bursal histogenesis. The gut epithelium did not develop follicles and was not colonized by lymphoid stem cells.

The developmental potential of the bursal mesenchyme, when separated from the endoderm, was also studied by isolating the mesenchyme of 10- to 12-day-old quail bursas and grafting it into the 3-day-old chick somatopleure. The hemopoietic stem cells present in the explant at the time of grafting began to disperse into the surrounding mesenchyme soon after grafting, and later disappeared. Basophilic cells were detectable in the quail mesenchyme for about 3 days after transplantation; whether they died or penetrated the host blood vessels could not be determined owing to their small number. When granulopoiesis was in progress in the bursa at the time of grafting, granulocytes could be seen for several days inside or in the vicinity of the explant.

Thus, tissue interactions between the mesenchymal and endodermal components of the bursal primordium are decisive for lymphoid differentiation of the organ. The differentiation signals exchanged between the epithelium and the mesenchyme are very specific, since no heterologous combinations resulted in bursal development. In addition, the cellular mechanisms that control stem cell homing depend upon reciprocal tissue interactions occurring between the two components of the rudiment.

These interactions were further investigated by using testosterone injection. The effect of testosterone on bursal development has been described frequently (Kirkpatrick and Andrews, 1944; Glick, 1957, 1964; Meyer *et al.*, 1959; Glick and Sadler, 1961; Warner and Burnet, 1961; Papermaster *et al.*, 1962). Treatment of embryos with testosterone at any developmental stage stops the development of the bursa and inhibits B cell differentiation. When large doses of testosterone propionate are injected into the egg (2.5 mg per egg), the lymphoid population of the bursa disappears completely within a few days; only granulocytes remain abundant in the mesenchyme when the injection is done from day 9 of incubation onward. The follicles either regress or do not grow, depending upon the age at which the hormone was administered.

In a series of grafting experiments, Moore and Owen (1966) showed that normal bursas transplanted into testosterone-treated host embryos underwent lymphoid differentiation, although the host bursa completely failed to develop. This indicates that bursal lymphoid precursor cells are not injured by testosterone treatment, but that stem cell homing is suppressed. Since this latter process depends on tissue interactions between the epithelium and the mesenchyme, it seemed interesting to learn whether testosterone prevents only one or both of the bursal tissue components from playing its role in bursal ontogeny. To answer this

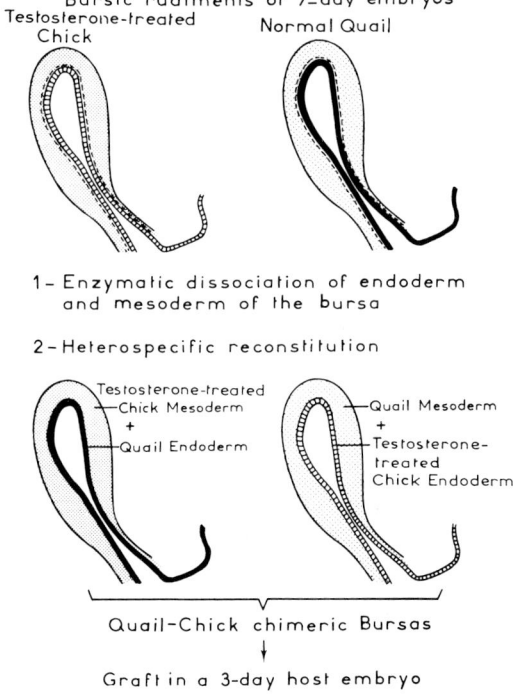

Fig. 16. Diagram showing the experimental procedure used to associate the two components of the bursa of Fabricius (mesoderm and endoderm), between a testosterone-treated chick bursa and a normal quail bursa at similar developmental stages.

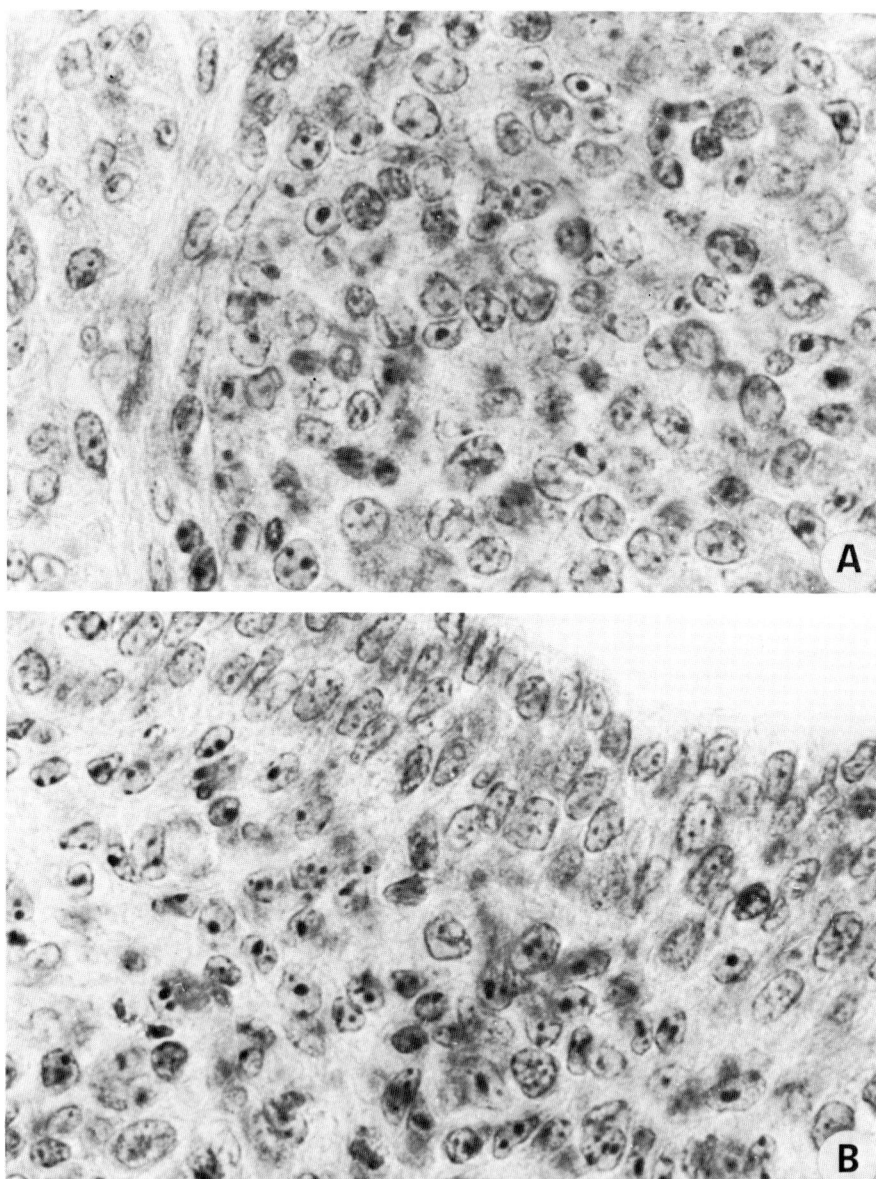

Fig. 17. Results of the experiment represented on Fig. 16. Feulgen–Rossenbeck staining. (A) Association of testosterone-treated chick mesoderm with normal quail endoderm followed by grafting in a 3-day chick host embryo: Development of lymphoid follicles with epithelial cells of the quail type and mesenchymal and lymphoid cells of the chick type. × 765. (B) Association of testosterone-treated chick endoderm with normal quail mesenchyme. In this case no follicle were formed and no lymphoid differentiation took place. × 765.

VIII. Primary Lymphoid Organ Ontogeny in Birds 209

question, the following experiment was devised (Le Douarin et al., 1980). After 7 days of incubation, quail and chick eggs were injected with 2.5 mg testosterone propionate. Two days later, the bursas were removed and dissociated into endoderm and mesenchyme. Normal bursas from 9-day-old quail and chick embryos were treated in the same way, and tissues were combined according to the scheme in Fig. 16.

The associations were grown for 12 hours in organ culture and were then grafted for 10 days into the somatopleure of 3-day chick embryos. Normal lymphoid development of the bursa occurred only in combinations involving normal endoderm plus testosterone-treated mesenchyme. In contrast, bursal histogenesis failed to occur in all the explants that were made up of treated endoderm plus normal mesenchyme. In these cases, the epithelium remained undifferentiated in a way similar to that of bursas in testosterone-treated whole embryos (Fig. 17).

In one case, the endoderm from a normal quail bursa was only partly replaced by testosterone-treated chick endoderm. Lymphoid follicle formation occurred normally in the untreated quail endoderm, whereas the neighbouring treated chick endoderm, although in contact with the same mesenchyme, did not differentiate. One can conclude that the endoderm loses its developmental capabilities when treated with testosterone and cannot respond to morphogenetic signals originating from the mesenchyme. In contrast, the mesenchyme was only transitorily affected by the androgen and remained capable of promoting the differentiation of an intact endoderm. It was also able to retain HC if it was associated with a normal bursal epithelium.

C. Cell Migration and Morphogenesis in the Ontogeny of the Bursa of Fabricius: Correlations with Plasminogen Activator Activity

The association of controlled extracellular proteolysis, mediated by plasminogen activator (PA), with tissue remodeling and cell migration has been demonstrated in various biological systems, such as mammary involution (Ossowski et al., 1979), trophoblast implantation (Strickland et al., 1976), and inflammation (Vassali et al., 1976). Bursal development, which involves hemopoietic cell immigration and ample morphogenetic movements for the formation of the epithelial plicae and lymphoid follicles, presents an interesting opportunity for further exploration of the relationships between such events and PA activity (Valinsky et al., 1981). Measurements of PA production in the whole bursa and in separated epithelium and mesenchyme revealed that it follows a developmental program and can be correlated with well-defined developmental events (Fig. 18).

Thus, in the chick, enzyme activity increases sharply from day 9 to day 13 of incubation, when immigration of hemopoietic stem cells and morphogenesis are

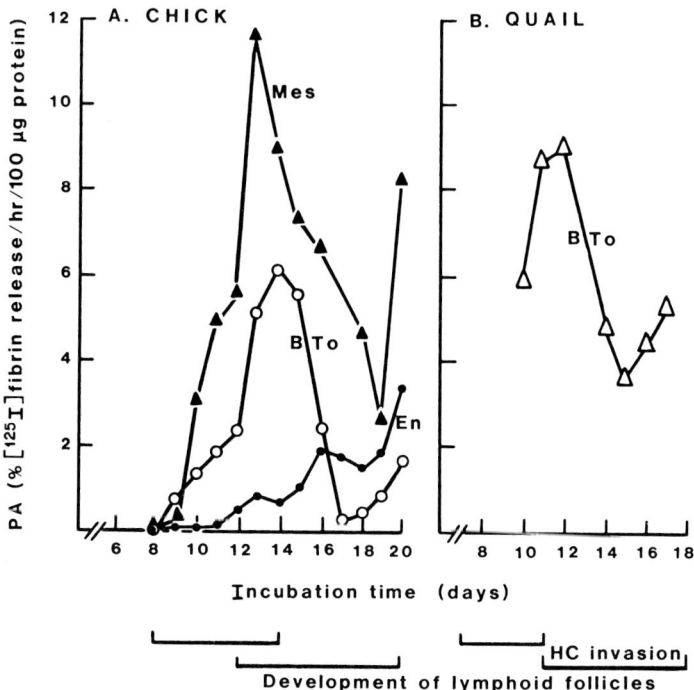

Fig. 18. Plasminogen activator (PA) production in the avian bursa of Fabricius. Whole bursas were removed from chick or quail embryos at the indicated times, homogenized in phosphate-buffered saline containing 0.5% Triton X-100 and assayed on [^{125}I] fibrin-coated wells as described in Valinsky et al. (1981). Dissociated bursal epithelium and mesenchyme were analyzed for PA content in a similar manner. PA activity is expressed as the percentage of [^{125}I] fibrin solubilized per hour per 100 μg protein. (A) PA content of whole chicken bursal homogenates (○—○) and homogenates of chicken bursal mesenchyme (▲—▲) and epithelium (●—●). (B) Activity in homogenates of whole quail bursas (△—△). Each data point represents an average of three separate determinations.

maximal and when myelopoiesis is particularly conspicuous in the mesodermal part of the organ. A second increase occurs at about 18 days, which may correspond to a phase of hemopoietic cell emigration. These findings represent another instance of the association of PA production with tissue remodeling and cell migration.

Plasminogen activator activity was initially localized mainly in the bursal mesenchyme, but increasing amounts of PA were observed in the epithelium during the second half of incubation. Of the many cell types in the mesenchyme potentially capable of PA synthesis, at least two cell populations can be identified as potential enzyme producers; these are the stromal cells, broadly defined, and HC and their derivatives. To identify the cells responsible for PA produc-

VIII. Primary Lymphoid Organ Ontogeny in Birds

tion, chimeric bursas were constructed in which the stroma (mesenchyme and endoderm derivatives) and the HCs were from different species (quail or chick). As mentioned above, quail-chick combination experiments have shown that all the hemopoietic stem cells present in the bursa, as well as their progeny, are of host origin when the organ is grafted before the time of its receptivity for hemopoietic cells. It has also been shown that these hemopoietic precursors are derived from the circulation, whereas all supporting elements, including the endothelial cells of the bursa, are of donor (that is graft) origin. Second, quail and chick PAs can be separated by SDS-polyacrylamide gel electrophoresis and identified in a zymographic gel procedure (Granelli-Piperno and Reich, 1978), thereby providing a marker for the cellular origin of the enzyme (Fig. 19).

Accordingly, quail bursas from 6.5-day donors, taken before they had been colonized by circulating stem cells, were grafted into the somatopleure of 3-day chick embryo hosts and incubated for an additional 4–6 days. The grafts were then removed, trimmed free of host tissue, homogenized and analyzed for PA. Reciprocal chimeras, that is, 8-day chick bursas grafted into quail hosts, were also prepared and incubated for a similar period before PA analysis.

As seen in Fig. 20, both kinds of chimeric bursa showed the presence of quail and chick PA activity in 12-day grafts. In other observations, both types of PAs were found in bursas removed and analyzed 5–12 days after grafting. Bursas removed 1–2 days after grafting (that is, prior to colonization) showed very little, if any, host PA activity.

At the stage corresponding to the experiments in Fig. 19, most of the bursal PA is produced by mesenchyme. Since the only host cells present in these grafts

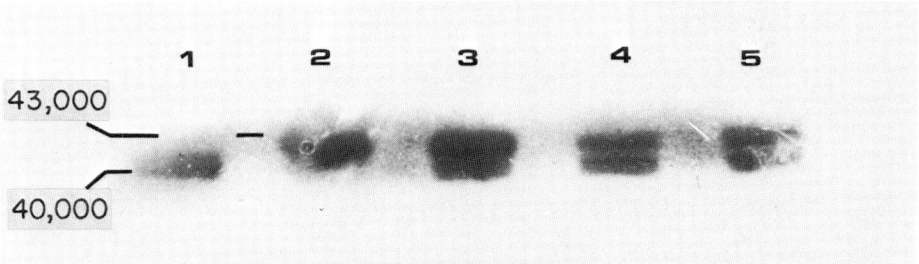

Fig. 19. Zymographic analysis of PA in chimeric bursas. Chimeric bursas, obtained from grafts of quail bursal rudiments into chick hosts and of chick rudiments into quail hosts, were homogenized and analyzed according to methods described in Valinsky *et al.* (1981). The figure shows zones of lysis produced in the casein–agar indicator layer of the zymograph. (Lane 1) PA activity in standard 12 day chick bursa; (lane 2) PA activity in standard 11 day quail bursa; (lane 3) PA activities in a mixture of standards; (lane 4) PA activities in a 6.5-day quail bursa grafted for 6 days in a 3-day chick host (total age of graft 12.5 days); (lane 5) PA activities in an 8-day chick bursa grafted for 4 days in a 3-day quail host (total age of the graft 12 days). Apparent molecular weights were determined by comparison with the electrophoretic mobilities of standard proteins.

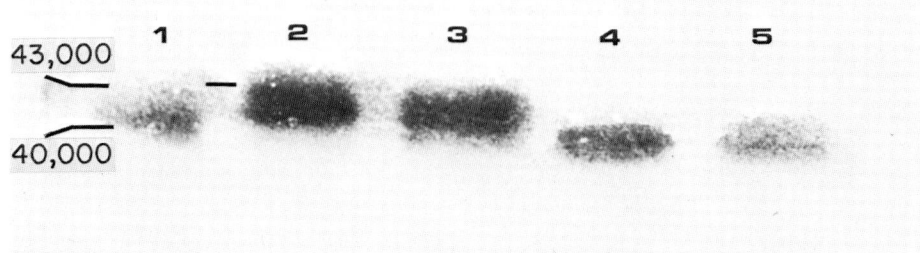

Fig. 20. Plasminogen activator in the bursa of Fabricius early in the receptive period for colonization by hemopoietic cells. Interspecific bursal grafts were made between quail and chick embryos. Grafts were removed, homogenized in buffer containing protease inhibitors and subjected to electrophoresis. Caseinolytic zones were allowed to develop as described in Valinsky et al. (1981). (Lane 1) standard chicken PA; (lane 2) standard quail PA; (lane 3) quail bursa (6 days) grafted for 2 days in a chick embryo (total age of graft 8 days); (lanes 4 and 5) chick bursa (8 days) grafted for 1 and 2 days, respectively, in quail embryos (total age of grafts 9 and 10 days).

are the hemopoietic stem cells, it follows that both the migrating stem cells (host), mesodermal cells, and, to a minor extent, endodermal stromal elements of the donor synthesize PA during this period of bursal morphogenesis.

Since chick PA can be detected in chimeric bursas within 3 days of grafting, before significant cell proliferation and myeloid differentiation has occurred, it appears that the HCs synthesize the enzyme from the time of their arrival, that is, during the migratory phase of bursal colonization. This finding is analogous to previous observations on inflammatory and other cells, where PA synthesis and action of plasmin (which results from limited proteolysis of plasminogen by PA) have also been correlated with migration (Vassalli et al., 1976).

Interspecific grafts, particularly those between quail and chick, have been valuable in the morphological analysis of many developmental phenomena (Le Douarin, 1978, 1982). The electrophoretic differences between the PAs add a biochemical marker that may extend the useful range of such grafts to a variety of other problems in embryology. Specifically, experimental designs resembling those used in this work could be applied to test for PA production accompanying migration of primordial germ cells, hemopoietic cell colonization of other organs, or neural crest cell dispersion throughout the embryo.

IV. CONCLUDING REMARKS

The use of cell markers, provided either by mitotic chromosomes or by the quail-chick chimera system, has been in many respects a fruitful approach to study the ontogeny of primary lymphoid organs. However, concerning the question of whether the thymus and the bursa are seeded by pluripotent hemopoietic cells or by cells already engaged toward the T or B lymphocyte differentiation pathway, no definite answer can be provided at the present time. If the two lines were segregated from the earliest steps of embryogenesis, then two categories of lymphocyte precursors would circulate in blood, and each of them would be selectively sensitive to signals emanating from the thymus and the bursa, respectively. The experiments described above, where HC already localized in the bursa of Fabricius remigrate into a "receptive" thymus, do not fit with such an hypothesis. On the contrary, they argue in favour of a single seeding mechanism for both bursa and thymus, triggered by a chemoattractant to which cells that seed either organ indifferently are able to respond.

Furthermore, according to N. M. Le Douarin and M. Coltey (unpublished observations), at least some of the cells migrating from the bursa to the thymus acquire in this novel microenvironment specific T cell markers, thus supporting the idea that, at least at the early stages of development, the cells which home to the bursa and to the thymus are multipotential rather than engaged into a specific lymphocyte differentiation pathway.

However, other experiments indicate that multipotential cells able to fulfill both these functions disappear from the bloodstream and even from the bone marrow toward the end of the incubation period. As shown by Weber and his co-worker (Weber and Mausner, 1977; Weber and Foglia, 1980), cells from these two hematogenous compartments can provide cells seeding the thymus but not the bursa a few days before hatching. Around the time of hatching, Toivanen and colleagues (1972) could find stem cells capable of populating the bursa only in the bursa itself. This observation was confirmed by Houssaint *et al.* (1983), who showed that precolonized bursas grafted syngeneically into newly hatched birds failed to be colonized, while thymuses were normally seeded in similar conditions.

Further investigations are needed to clarify this question, namely, whether the cells that seed thymuses during the third influx of HC (i.e., just after birth) have different developmental potencies from those of the first and second embryonic influx.

Selective manipulation of the cells present in the primary lymphoid organ rudiments probably still offers interesting prospects for research on immune system ontogeny. For instance, lymphocyte functional development could be analyzed in chimeric organs or embryos. A better insight into the mechanism of self and non-self repertoire acquisition during the intrathymic T cell differentiation process, could be expected from these approaches.

ACKNOWLEDGMENTS

This work was supported by the Centre National de la Recherche Scientifique, the Délégation Générale à la Recherche Scientifique et Technique, the Association pour le Développement de la Recherche sur le Cancer, and by the NIH (Grant 2 RO1 DEO 4257-06).

REFERENCES

Ackerman, G. A. (1967). *Anat. Rec.* **158,** 387–400.
Ackerman, G. A., and Knouff, R. A. (1964). *Anat. Rec.* **149,** 191–216.
Ackerman, G. A., and Knouff, R. A. (1965). *Anat. Rec.* **152,** 35–54.
Auerbach, R. (1960). *Dev. Biol.* **2,** 271–284.
Baillif, R. N. (1949). *Am. J. Anat.* **84,** 457–510.
Beard, T. (1899). *Lancet* **1,** 144–193.
Beard, T. (1902). *Zool. Jahrb.* **17,** 403–480.
Bell, E. T. (1906). *Am. J. Anat.* **5,** 29–62.
Ben Slimane, S., Houllier, F., Tucker, G., and Thiery, J. P. (1983). *Cell Differ.* **13,** 1–24.
Cheval, M. (1908). *Bibliogr. Anat.* **17,** 189–201.
Cooper, M. D., Peterson, D. A., and Good, R. A. (1965). *Nature (London)* **205,** 143–146.
Couly, G., Lagrue, A., and Griscelli, C. (1982). *Rev. Stomat. Chir. Max. Fac.* **84,** 103–108.
Di George, A. M. (1968). *Birth Defects, Orig. Artic. Ser.* **4,** 116.
Dustin, A. P. (1920). *Arch. Biol.* **30,** 601–693.
Fontaine-Pérus, J. C., Calman, F. M., Kaplan, C., and Le Douarin, N. M. (1981). *J. Immunol.* **126,** 2310–2316.
Fraser, E. A., and Hill, J. P. (1915). *Philos. Trans. Soc. London, Ser. B* **207,** 1–85.
Glick, B. (1957). *Poult. Sci.* **36,** 18.
Glick, B. (1964). In "The Thymus In Immunobiology" (R. A. Good and A. E. Gabrielsen, eds), pp. 345–358. Harper (Hoeber), New York.
Glick, B., and Sadler, C. R. (1961). *Poult. Sci.* **40,** 185.
Glick, B., Chang, T. S., and Japp, R. G. (1956). *Poult. Sci.* **35,** 224–245.
Granelli-Piperno, A., and Reich, E. (1978). *J. Exp. Med.* **148,** 223–234.
Hammar, J. A. (1905). *Anat. Anz.* **27,** 23–30.
Hammar, J. A. (1908). *Arch. Mikrosk. Anat.* **73,** 1–68.
Hammar, J. A. (1910). *Ergebn. Anat. Entwicklungsgesch.* **19,** 1–274.
Hammar, J. A. (1911). *Anat. Hefte* **43,** 201–242.
Hammond, S. W. (1954). *J. Morphol.* **95,** 501–522.
Houssaint, E., Belo, M., and Le Douarin, N. M. (1976). *Dev. Biol.* **53,** 250–264.
Houssaint, E., Torano, A., and Ivanyi, J. (1983). *Eur. J. Immunol.* **13,** 590–595.
Jotereau, F. V., and Le Douarin, N. M. (1982). *J. Immunol.* **129,** 1869–1877.
Jotereau, F. V., Houssaint, E., and Le Douarin, N. M. (1980). *Eur. J. Immunol.* **10,** 620–627.
King, J., Ackerman, G. A., and Knouff, R. A. (1964). *Anat. Rec.* **148,** 300–301.
Kirkpatrick, C. M., and Andrews, F. N. (1944). *Endocrinology* **34,** 340–345.
Kölliker, A. (1879). In "Entwicklungsgeschichte des Menschen und der höheren Tiere," 2nd ed., pp. 815–880. Engelman, Leipzig.
Lauffenburger, D. A., and Zigmond, S. H. (1981). *J. Immunol. Methods* **40,** 45–60.
Le Douarin, N. M. (1964). *Bull. Biol. Fr. Belg.* **98,** 544–676.
Le Douarin, N. M. (1967). *C. R. Hebd. Seances Acad. Sci.* **264,** 940–942.
Le Douarin, N. M. (1978). *Cold Spring Harbor Conf. Cell. Proliferation* **5,** 5–31.
Le Douarin, N. M. (1979). In "Mechanisms of Cell Change" (J. E. Ebert and T. S. Okada, eds.), pp. 293–326. Wiley, New York.
Le Douarin, N. M. (1980). *Curr. Top. Dev. Biol.* **16,** 31–85.
Le Douarin, N. M. (1982). "The Neural Crest." Cambridge Univ. Press, London and New York.
Le Douarin, N. M. and Jotereau, F. V. (1975). *J. Exp. Med.* **142,** 17–40.
Le Douarin, N. M. and Jotereau, F. V. (1980). In "Immunology 80': Progress in Immunology IV" (M. Fougereau and J. Dausset, eds.), Vol. 1, pp. 285–302. Academic Press, London.
Le Douarin, N. M., Bussonnet, C., and Chaumont, F. (1968). *Ann. Embryol. Morphog.* **1,** 29–40.

Le Douarin, N. M., Houssaint, E., Jotereau, F. V., and Belo, M. (1975). *Proc. Natl. Acad. Sci. U.S.A.* **72,** 2701–2705.
Le Douarin, N. M., Jotereau, F. V., and Houssaint, E. (1976). *In* "Phylogeny of Thymus and Bone Marrow-Bursa Cells" (R. K. Wright and E. L. Cooper, eds.), pp. 217–226. Elsevier/North-Holland Biomedical Press, Amsterdam.
Le Douarin, N. M., Michel, G., and Baulieu, E. E. (1980). *Dev. Biol.* **75,** 288–302.
Le Douarin, N. M., Coltey, M., and Guillemot, F. P. (1984). In preparation.
Le Lièvre, C. S., and Le Douarin, N. M. (1975). *J. Embryol. Exp. Morphol.* **34,** 125–154.
McCutcheon, M. (1946). *Physiol. Rev.* **26,** 319–336.
Maximow, A. (1909). *Arch. Mikrosk. Anat.* **74,** 525–621.
Maximow, A. (1912). *Arch. Mikrosk. Anat.* **79,** 560–611.
Metcalf, D., and Moore, M. A. S. (1971). "Haemopoietic Cells". North-Holland Publ., Amsterdam.
Meyer, R. K., Rao, M. A., and Aspinall, R. L. (1959). *Endocrinology* **64,** 890–897.
Miller, J. F. (1961). *Lancet* **2,** 748–749.
Moore, M. A. S., and Johnson, G. R. (1976). *In* "Stem Cells Renewing Cell Populations" (A. B. Cairnie, P. K. Lala, and D. G. Osmon, eds.), pp. 323–330. Academic Press, New York.
Moore, M. A. S., and Metcalf, D. (1970). *Br. J. Haematol.* **18,** 279–296.
Moore, M. A. S., and Owen, J. J. T. (1966). *Dev. Biol.* **14,** 40–51.
Moore, M. A. S., and Owen, J. J. T. (1967). *J. Exp. Med.* **126,** 715–723.
Oliver, P. D., and Le Douarin, N. M. (1984). *J. Immunol.*, **132,** 1748–1755.
Ossowski, L., Biegel, D., and Reich, E. (1979). *Cell* **16,** 929–940.
Owen, J. J. T., and Ritter, M. A. (1969). *J. Exp. Med.* **129,** 431–437.
Palmblad, J., Malmsten, C. L., Uden, A. M., Ridmark, O., Engstedt, L., and Samuelson, B. (1981). *Blood* **58,** 658–661.
Papermaster, B. W., Friedman, D. I., and Good, R. A. (1962). *Proc. Soc. Exp. Biol. Med.* **110,** 62–64.
Prenant, A. (1894). *Cellule* **10,** 85–184.
Pyke, K. W., and Bach, J. F. (1979). *Eur. J. Immunol.* **9,** 317–323.
Raff, M. C., Megson, M., Owen, J. J. T., and Cooper, M. D. (1976). *Nature (London)* **259,** 224–226.
Russo, R. G., Liotta, L. A., Thorgeirsson, U., Brundage, R., and Schiffman, E. (1981). *J. Cell Biol.* **91,** 459–467.
Sanel, F. T. (1967). *Z. Zellforsch. Mikrosk. Anat.* **83,** 8–29.
Saxén, L., Karkinen-Jaaskelainen, M., Lehtonen, E., Nordling, S., and Wartiovaara, J. (1976). *In* "The Cell Surface in Animal Embryogenesis and Development" (G. Poste and G. L. Nicolson, eds.), pp. 331–407. North-Holland Publ., Amsterdam.
Schrier, J. E., and Hamilton, H. L. (1952). *J. Exp. Zool.* **119,** 165–188.
Steinman, R. M., Chen, L. L., Witmer, M. D., Kaplan, G., Nussenzweig, M. C., Adams, J. C., and Cohn, Z. A. (1980). *In* "Mononuclear Phagocytes: Functional Aspects" (R. Van Furth, ed.), Part 2, pp. 1781–1807. Nijhoff, The Hague.
Stohr, P. (1906). *Anat. Hefte* **31,** 409–458.
Strickland, S., Reich, E., and Sherman, M. I. (1976). *Cell* **9,** 213–240.
Sumner, A. T. (1972). *Exp. Cell Res.* **75,** 304–306.
Tachibana, F., Imai, Y., and Kojima, M. (1974). *J. Reticuloendothel. Soc.* **15,** 475–496.
Toivanen, P., Toivanen, A., and Good, R. A. (1972). *J. Immunol.* **109,** 1058–1070.
Valinsky, J. E., Reich, E., and Le Douarin, N. M. (1981). *Cell* **25,** 471–476.
Vassalli, J. D., Hamilton, J., and Reich, E. (1976). *Cell* **8,** 271–291.
Venzke, W. G. (1942). *Iowa State Coll. J. Sci.* **17,** 145–148.
Venzke, W. G. (1952). *Am J. Vet. Res.* **13,** 395–404.

Verdun, P. (1898). *C. R. Seances Soc. Biol. Ses Fil.* **5,** 243–244.
Warner, N. L., and Burnet, F. M. (1961). *Aust. J. Biol. Sci.* **14,** 580–587.
Warner, N. L., and Szenberg, A. (1964). *Annu. Rev. Microbiol.* **18,** 253–268.
Weakley, B. S., Patt, D. I., and Shepro, D. (1964). *J. Morphol.* **115,** 319–354.
Weber, W. T., and Foglia, L. (1980). *Cell. Immunol.* **52,** 84–95.
Weber, W. T., and Mausner, R. (1977). *In* "Avian Immunology" (A. A. Benedict, ed.), pp. 47–59. Plenum, New York.
Zigmond, S. H. (1977). *J. Cell Biol.* **75,** 606–616.

CHAPTER IX

The Use of Chimeric Mice in Immunology: How Does the Immune System Know Self and Non-Self?

TAKESHI MATSUNAGA

*Department of General and Oncologic Surgery,
and Division of Biology
City of Hope National Medical Center
Duarte, California*

I.	Introduction—A Problem of Recognition of Self and Non-Self by the Immune System	217
II.	Various Chimeric Mice Used for Experiments	219
	A. Long-Term Chimeras	219
	B. Short-Term Chimeras	221
III.	Allogeneic MHC Tolerance in Chimeras and Some Other Problems	221
	A. Study of Allogeneic H-2 Tolerance in EA Chimeras	221
	B. Clonal Elimination versus Suppression	223
	C. Special Case of MHC Antigens and Lamarckian View of Tolerance	224
IV.	MHC Restriction, Immune Response (*Ir*) Gene, and Cell Interaction	226
	A. Early Debates—Is MHC a Barrier for Cell Interaction?	226
	B. Impact of *H-2* Restriction of Cytotoxic T Cells	228
	C. Immune Response (*Ir*) Genes in the Light of MHC Restriction	229
	D. A New Dispute: Are T Cells Selected by Thymus MHC?	231
	E. Chimeras and Idiotype Network	233
	References	235

I. INTRODUCTION—A PROBLEM OF RECOGNITION OF SELF AND NON-SELF BY THE IMMUNE SYSTEM

The vast potential of the immune system can be illustrated by the fact that a single inbred mouse is potentially capable of producing thousands of *different* antibody molecules against a simple haptenic determinant (Kreth and Williamson, 1973). This is because the immune system is equipped with genetic mechanisms that allow somatic generation of antibody diversity (Cohn, 1968, Tonegawa, 1976) in a random fashion, mechanisms adopted in evolution to cope with rapidly mutating antigens of viruses and bacteria. It is this genetic ability that has endowed the immune system with foresight, a Promethean characteristic (Ohno et al., 1981).

However, this powerful sword is double edged. As the generation of diversity is random with regard to binding specificity, anti-self clones are bound to arise. This means auxiliary mechanisms are required to eliminate or suppress such antiself clones. This usually works. For example, a person of blood group A may contain serum antibodies against group B antigens, but not against group A antigens.

Although this problem of self–non-self discrimination, known as self tolerance, was realized at the turn of the last century, and was expressed as "horror autotoxicus" by Erhlich (1900), it was Owen's finding in 1945 of blood group chimerism in cattle dizygotic twins that led Burnet to say "recognition of self is something that needs to be learned and is not an inherent genetic quality of organism" (Burnet and Fenner, 1949). His prediction was corroborated by Medawar's experiments showing that transplantation tolerance can occur in cattle dizygotic twins (Anderson et al., 1951) or can be induced by introducing allogeneic cells into immunologically immature newborn mice (Billingham et al., 1953).

Self tolerance is generally understood, thanks to the theory of clonal selection (Burnet, 1959), as resulting from physical or functional elimination of anti-self clones arising in ontogeny. This concept of clonal elimination has survived many years of experimental tests. More recently, suppressor T cells (Gershon and Kondo, 1970), a functional subset of T lymphocytes, have appeared on the scene and are implicated in some aspects of self tolerance. First, I shall discuss the problem of tolerance in chimeric mice in relation to self tolerance in general.

Recent experience has taught us another puzzling aspect of immunological recognition. This is known as "MHC (major histocompatibility complex) restriction." The MHC restriction has been established with cytotoxic T cells (T_c) and helper T cells (T_h). The original finding of Zinkernagel and Doherty (1974a) was that virus-specific T_c cells must share *H-2 K/D* [MHC class I antigens (Klein, 1979) in mice] determinants with virus-infected target cells for cell lysis reaction to occur. In other words, T_c cells must recognize in some way both viral and H-2 antigens on the target cells (Zinkernagel and Doherty, 1974b) (Fig. 1).

The manner in which helper T cells recognize foreign epitopes (antigenic determinants) is the same as for T_c cells, except that the H-2 Ia antigen [MHC class II molecules (Klein, 1979)] is used as a restriction entity. Thus, we face a little paradox in that T cells must recognize self MHC antigens in order to

IX. The Use of Chimeric Mice in Immunology

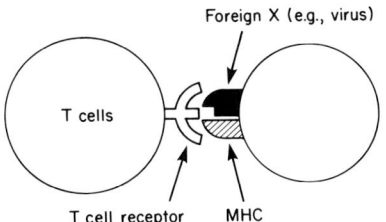

Fig. 1. When cytotoxic T cells or helper T cells recognize some foreign antigen X, such as a virus, they recognize MHC antigens of their own type at the same time. The foreign antigen X and the MHC molecules probably interact with each other to create new determinants.

recognize the foreign epitope X. MHC antigens are a set of cell surface glycoproteins (class I = MW 45,000 + 12,000 dimer, class II = MW 28,000 + 33,000 dimer). In humans, the class I antigens are known as HLA-A,B,C antigens. Because of the presence of a high degree of genetic polymorphism of these loci, two persons rarely have identical HLA types. Indeed, these antigens have long been known as classical transplantation antigens (see Klein, 1979).

It might be said that the MHC restriction is a device to (a) cope with intracellular pathogens such as viruses and (b) keep the immune response under vigorous control using regulatory T cells (helper and suppressor). I shall argue that regulatory T cells were adopted as secondary fail-safe mechanisms when the immune system faced two mutually dependent requirements, namely, (a) to expand ability for the generation of recognition diversity and (b) to improve ability for self–non-self discrimination.

Chimeric mice have been very useful in studies of self-tolerance, T cell recognition, MHC restriction, and cell-to-cell interaction. This chapter is not meant as a full survey of chimeric studies, but rather to point out and discuss some aspects of these immunolgical problems, and to show how experiments with chimeras have helped in their analysis.

II. VARIOUS CHIMERIC MICE USED FOR EXPERIMENTS

A. Long-Term Chimeras

Embryo-aggregation chimeras [EA chimeras (Tarkowski, 1961; McLaren and Bowman, 1969), allophenic mice (Mintz, 1967), tetraparental mice] are derived from two genetically different mouse embryos aggregated at the 8-cell stage by manipulation *in vitro*. Chimeric embryos are surgically transferred to the uteri of surrogate mother mice for further development (Fig. 2a). As the cell mixing procedure is performed at so early a stage of ontogeny, all tissues become chimeric, including lymphoid cells. The procedure for creating EA chimeras

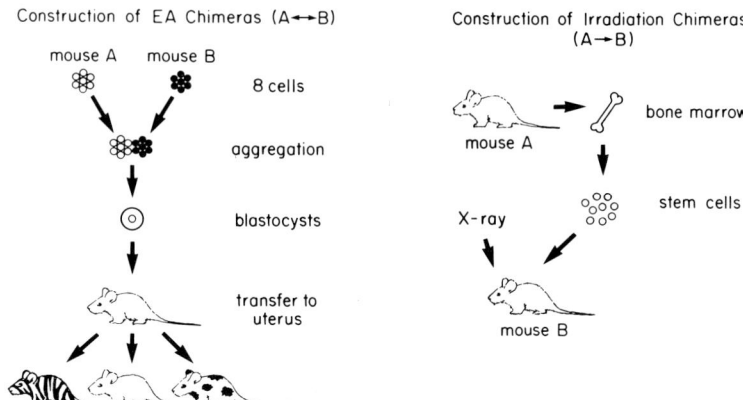

Fig. 2. EA chimeras can be made by aggregating two eight-cell embryos that differ at the MHC region (a). Irradiation chimeras can be constructed by injecting lymphocyte stem cells into host mice whose immune system has been destroyed by X irradiation (b).

does not really distort the normal paths of embryogenesis, as opposed to artificial manipulation used for making chimeras of other kinds (e.g., X-ray irradiation). Indeed, nature may practise embryo aggregation in humans, albeit rarely (de la Chaplle et al., 1974). EA chimeras constructed between H-2 incompatible strains can be designated as $H\text{-}2^a \leftrightarrow H\text{-}2^b$, for example.

Irradiation bone marrow chimeras are made by introducing hematopoietic stem cells residing in bone marrow of adult mice or in liver of fetal mice (14 days old or so) into host mice whose immune tissues (i.e., lymphocytes, macrophages, and stem cells) are entirely destroyed by an appropriate dose of X-ray irradiation (von Boehmer et al., 1975a) (Fig. 2b). A few months are needed before the host's immune system is repopulated by donor cells and starts to function. The great versatility and usefulness of irradiation chimeras lies in the varied combinations of host–donor strains that can be made. For example, (A × B) F_1 host mice can be reconstituted with parental A or B, or a mixture of A and B stem cells. These are designated as A (or B)→(A × B)F_1, A + B→(A × B)F_1, respectively. Entirely H-2 incompatible strain combinations A→B can also be made (Matzinger and Mirkwood, 1978). It should be mentioned that these irradiation chimeras have radioresistant thymic tissues (e.g., thymus epithelium) of host origin, which may influence T cell differentiation and selection of donor origin (see Section IV, D).

In both EA chimeras and irradiation bone marrow chimeras, lymphocyte and macrophage (and other blood cells) chimerism is stably maintained (Matsunaga et al., 1980), owing to the presence of stem cells of different genotypes (sometimes more than two genotypes) in a single animal. They can be conveniently called "long-term chimeras." Neonatally induced tolerant mice can be made by injecting allogeneic lymphoid cells into newborn mice and can be classified as long-term chimeras.

B. Short-Term Chimeras

Immunologists are familiar with various lymphoid chimeras that can be made by transfer of cell populations containing *mature and functional* lymphocytes and/or macrophages into other mice. The immune system of recipient animals can be made inactive either by X-ray irradiation or genetic defects. Donor cells can be selected, purified, or stimulated with antigens in various ways before transfer to hosts. Thus, host mice can be considered as *in vivo* test tubes. In these host mice, immune response can be measure in relation to various parameters of input cell types, *H-2* types, or antigen stimulation.

III. ALLOGENEIC MHC TOLERANCE IN CHIMERAS AND SOME OTHER PROBLEMS

A. Study of Allogeneic H-2 Tolerance in EA Chimeras

In using MHC antigens to study tolerance, some properties of the MHC may be mentioned (Klein, 1979). MHC class I antigens are ubiquitously expressed in many tissues, appearing early in ontogeny, whereas MHC class II antigens are expressed mainly in cells in the immune tissues (B cells, macrophages, some T cells).

Class I and, to a lesser extent, class II antigens are extremely polymorphic. If the *H-2* type of the grafted skin is different from the *H-2* type of the host mouse (allogeneic *H-2*), the strongest allograft rejection, which is a cell-mediated immunity, is observed (Gorer *et al.*, 1948). Our starting point is to accept that mice are tolerant of self (syngeneic) H-2 antigens, as they do not reject autografts or grafts from the same strain.

The allograft response has an *in vitro* correlate. In mixed lymphocyte reaction (MLR), proliferation of T cells, mainly T_h lineage, can be observed *in vitro* when responder cells of host mice react against mainly allogeneic class II MHC antigens on the stimulator cells, the antigen of foreign grafts (Bach, 1970; Meo *et al.*, 1975). These antigens (Ia in mice) on macrophages [antigen-presenting cell (APC)] are presented to T cells and trigger proliferative responses. There is some evidence that MLR responder cells may be involved in allograft responses (Liew and Simpson, 1980; Loveland *et al.*, 1981). Cytotoxic T cells are also generated during MLR (Nabholz *et al.*, 1974). They recognize *H-2 K/D* determinants on allogeneic cells, and lyse them upon contact (Nabholz *et al.*, 1974). Both MLR and cytotoxic responses can be elicited with unprimed responder cells (i.e., primary response). A primary reaction of this magnitude cannot be detected with syngeneic H-2 stimulator cells, because animals are tolerant of syngeneic H-2.

When strains of two different H-2 types are combined in EA chimeras, allogeneic H-2 tolerance is generated for both types. During the past 20 years or so, thousands of EA chimeras of H-2 incompatible combinations have been produced (Mintz, 1974; McLaren, 1976). They exhibit chimerism in many tissues, yet they are, in general, immunologically and physiologically normal. Early experiments showed the presence of H-2 chimerism in red blood cells (Mintz and Palm, 1969), and the specific nature of tolerance in terms of allograft response (Mintz and Silvers, 1967, 1970). These results are in good agreement with those of Medawar, Billingham, and Brent with the neonatal tolerance system (Billingham et al., 1953) and those of Hasek with parabiosed chickens (Hasek, 1953). However, EA chimeras are unique and different from other chimeras in that H-2 chimerism is present in all tissues starting in early ontogeny, i.e., allogeneic H-2 is present already in the 8-cell embryo, and, consequently, tolerance is almost always "complete" as opposed to occasional "incomplete" tolerance in other chimeras. The difference between EA chimeras and F_1 hybrid animals lies in the fact that two different H-2 antigens are generally expressed in separate cell populations in the former, whereas in F_1 mice, both H-2 antigens are found on the same cells, including lymphocytes.

In studying mechanisms of tolerance in EA chimeras, MLR and T_c *in vitro* responses have been employed. Initially, the concept of clonal elimination was challenged, as it was claimed that anti-allogeneic H-2 T_c cells (by microcytotoxicity test) (Wegmann et al., 1971) or MLR responding cells (Phillips et al., 1971) are present in EA chimeras, and that the effector activity of these "antiself" cells is checked by serum blocking factors (anti-H-2 antibodies) (Wegmann et al., 1971; Phillips et al., 1971), agents known often to enhance tumor growth, or by the intervention of suppressor T cells (Phillips and Wegmann, 1973). A problem of this proposal, while stimulating, is that it has proved difficult to reproduce similar results (Matsunaga, 1973; Festenstein et al., 1975). Thus, in our better controlled experiments, where the determination of H-2 chimerism and separation of one or the other H-2 population of lymphocytes were also included in the protocol, individual chimeras were separately tested, and neither alloreactive cells nor suppression was found (Matsunaga et al., 1980), i.e., chimeric lymphocytes were specifically unreactive against allogeneic parental cells in MLR or T_c response. These results are in agreement with those of other laboratories (Festenstein et al., 1975; Barnes and Graham, 1976; A. Harris, personal communication) but at variance with the early claims and interpretations (Wegmann et al., 1971; Phillips et al., 1971). In one of our chimeras, where the ratio of H-2 lymphoid cell population was extremely unbalanced, some evidence of T_c cell reactivity against allogeneic H-2 cells, which made up a minor population, was obtained (Matsunaga et al., 1980). We were led to conclude that a major mechanism of *H-2* tolerance in EA chimeras is clonal elimination, but suppression can operate under many circumstances (see below) as a secondary fail-safe mechanism (Matsunaga *et al.*, 1980).

B. Clonal Elimination versus Suppression

Absence of evidence for suppression in EA chimeras is not evidence of absence of suppression. Clonal elimination is not operating in a vacuum; like all other biological functions, it works only under certain conditions; it is subject to variables and prone to errors. Thus, clonal elimination depends on the differential susceptibility of clones, affinity of antigen-binding receptors, concentration and qualities of antigens (e.g., the special case of MHC, see Section III, C), availability of other interacting cells, etc., all of which change continually in time or in location. Clonal elimination is a process continued throughout the whole life span of animals. In the case of EA chimeras, some of these variables (antigen concentration, clonal size) are different from one individual to another because of the consequence of varied ratio in *H-2* chimerism. This fact certainly makes escape from clonal elimination easier in chimeras than in normal animals. It is conceivable that observation of occasional self-reactive cells in EA chimeras may be attributable to this circumstance. Adding to these variables, somatic mutation continues to generate anti-self clones at any stage of lymphocyte differentiation. All these anti-self clones might, under certain conditions, be quickly expanded and endanger the animal through the pathological effects of autoimmunity.

It is indeed this situation that nature has come to rescue by bringing a secondary, failsafe mechanism, a suppressor T cell system (Gershon and Kondo, 1970). Suppressor T cells almost always accompany the normal response against foreign epitopes, and contribute to the immunoregulatory circuit (Gershon, 1974). We may easily envisage that autoimmunity can be kept to a low and harmless level by suppressor mechanisms. In certain cases, autoimmunity has been shown to be caused by the breakdown of suppressor mechanisms (Cantor *et al.*, 1978). Thus, self–non-self discrimination is generated and maintained by dual mechanisms, clonal elimination and suppression (Matsunaga *et al.*, 1980).

In planning and interpreting many experiments on tolerance, the question of "clonal elimination versus suppression" has always been asked (e.g., Weigle, 1980), but the emphasis may now be shifted to ask, "which kind of condition favors one or the other of the two mechanisms?" My guess is that, in dealing with thousands of self antigens, the immune system has no clear-cut decision. In other words, we are likely to continue to find either or both mechanisms operating at the same time, even in the same animal, depending on the experimental conditions we choose. For instance, *H-2* tolerance in irradiation bone marrow chimeras (von Boehmer *et al.*, 1975a), and in EA chimeras can be fully explained by clonal elimination but in other MHC tolerance systems, suppressor mechanisms are often claimed (Kilshaw et al., 1975; Dorsch and Rosen, 1977). It is obvious that these experimental conditions are so different that a meaningful comparison is made difficult.

Do helper T cells have something to do with self–non-self discrimination? As B cells or cytotoxic T cells usually need signals from helper cells to be activated, clonal elimination of anti-self helper T cells functions as another fail safe mechanism. Consequently, anti-self B cells or cytotoxic T cell precursors, even if they are present, will remain harmless. Since helper T cells are restricted by MHC class II antigens, self antigen X plus self class II MHC are the key antigens maintaining this kind of regulation.

We still do not know whether and how MHC antigens are used by suppressor T cells. In the context of the present discussion, clonal elimination of anti-self suppressor T cells would be disadvantageous. It is possible that for this reason, suppressor T cells might recognize idiotypic determinants on helper T cells as interacting markers (Eichmann, 1975), since such determinants usually do not induce clonal elimination due to their low concentration (see Section IV, E).

What happens to animals that are deprived of these regulatory T cells? Nude mice, defective in thymus and T cell function (Kindred, 1979), have no suppressor T cells, and self tolerance at B cell level has to rely on clonal elimination only. As helper T cells are also lacking or very small in number in nude mice, amplification of anti-self antibody responses, even if it occurred, would also be minimal. This "nude mice" situation might have lasted for a while in evolution, since it is hard to imagine that helper or suppressor cells appeared in phylogeny before the emergence of antibody-forming B cells or cytotoxic T cells, i.e., those cells that are helped or suppressed. Consequently, self tolerance was achieved without them, i.e., by clonal elimination only. This is another argument that both mechanisms are operating in normal animals, with suppression being the secondary failsafe device.

C. Special Case of MHC Antigens and Lamarckian View of Tolerance

The concept of altered self, which implies that a single T cell receptor recognizes self-MHC + X (see Fig. 1), is now considered more favorably than five years ago (Matzinger, 1981), as experimental evidence weighs more in favor of it (see Section IV, D). This brings up one basic question: What is it that we talk about as self and allogeneic MHC antigens? In the altered self recognition, it is implicit that self MHC somehow interact with foreign epitopes, or both antigens come into close proximity, to create new determinants. This process must also occur between self MHC and self X antigens, unless MHC itself has a certain degree of self–non-self discriminatory ability.

IX. The Use of Chimeric Mice in Immunology

Male specific H-Y antigens are recognized by female cytotoxic T cells in an *H-2* restricted manner. It follows that tolerance of H-Y antigens in male mice is also *H-2* restricted. We may have similar cases with other self X antigens, cellular and soluble, and such self MHC plus self X antigens could potentially generate a large heterogeneity, of the order of hundreds or more. If those heterogeneous antigens are introduced into another animal of different MHC type, do they not provoke a strong response because of the multitude of foreignness? This might be the basis for the "alloaggression" in which T cell responses are two or more magnitudes higher than the usual anti-foreign B cell responses, in terms of precursor frequency (Matzinger and Bevan, 1977). This point is illustrated in Fig. 3. The fact that anti-minor histocompatibility T_c usually needs antigen priming to be demonstrated would mean that polymorphic differences at all minor loci do not significantly contribute to anti-allo MHC responses. For this reason, the experiments (e.g., Langhorne and Lindahl, 1982) that aim to test this hypothesis (i.e., alloaggression = self-X plus allo-MHC) have not yet been done properly.

The problem of self–non-self discrimination, as can be seen from the previous discussion, is not yet resolved and therefore remains an ample source of new theories, both intelligent and excessively ambitious. A recent example was the claim by Gorczynski and Steele (1980) that allograft tolerance is inherited by offsprings *paternally*. Immunologists can give reasonable explanations for *maternal* inheritance of immunological status in mammals, since the immune systems of mother and developing fetus are known to interact. So this proposal was taken seriously in some laboratories and subjected to vigorous experimental scrutiny. Unfortunately, the results threw doubt on the status of the original claim (Brent *et al.*, 1981). Furthermore, in other sets of experiments, homo-

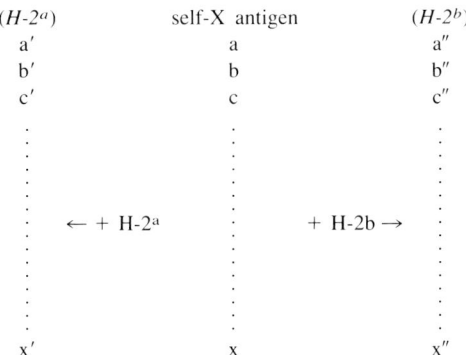

Fig. 3. In H-2^a mice, a set of altered self antigens are generated by self MHC plus self X antigens, which are designated as a', b', c', . . . , x'. Likewise, in H-2^b mice, a'', b'', c'', . . . , x'' are present. The immune system of H-2^a mice is tolerant of a', b', c', . . . , x' but not of a'', b'', c'', . . . , x'' and so reacts with them.

zygous progeny of *male* EA chimeras, an ideal model of *H-2* tolerant mice, were shown to make anti-H-2 cytotoxic T cell responses no less vigorously than normal homozygous progeny (McLaren *et al.*, 1981). In other words, the original claim was not confirmed. Consequently, the grandiose theory of Steele with a Lamarckian flavor should for the time being be considered only with a grain of salt.

IV. MHC RESTRICTION, IMMUNE RESPONSE (*Ir*) GENE, AND CELL INTERACTION

A. Early Debates—Is MHC a Barrier for Cell Interaction?

The discovery that immune response (*Ir*) genes were linked to mouse *H-2* regions (McDevitt *et al.*, 1972) provided another route through which the MHC was brought to the mainstream of immunology. Thus, *H-2 I* region alleles determine whether a mouse makes a good antibody response (high responder) or not (non- or low responder) against a given antigen. Since *Ir* gene control is antigen specific to some extent, it was once proposed that *Ir* gene products function as T cell antigen receptors (Benacerraf and McDevitt, 1972). However, since the immune response is a network of specific cell interactions, *Ir* gene specificity could be expressed at any level of these reactions.

As early as 1972, Kindred and Schreffler observed that nude mice that genetically lack a thymus can make anti-sheep red blood cell (SRBC) antibody responses if they are reconstituted with thymocytes from normal mice. However, they found that the *H-2* region must be shared by both the host nude mice and the thymoctyes transferred from other mice (Kindred and Schreffler, 1972). Since it was known that helper cells are T cells (thymus-derived) (Miller and Mitchell, 1968), it was suggested that T cell–B cell collaboration may not work it they are *H-2* incompatible.

This finding was further elaborated by Katz and his colleagues using hapten-carrier adoptive transfer to measure T cell–B cell interaction (Katz *et al.*, 1973). Spleens from mice primed with DNP-KLH (DNP dinitrophenyl; KLH, keyhole limpet hemocyanin) were taken as an anti-DNP B cell source, and transferred to X-ray irradiated host mice together with spleens taken from mice primed with another carrier (BGG, bovine γ globulin), as a helper T-cell source. When the animals were challenged with DNP-BGG, a maximal response was generated as a result of collaboration of B cells and BGG-specific helper T cells. To study whether the *H-2* region is a barrier for such reactions, various strain combinations were chosen for B cell and helper T cell sources, and F_1 animals were used as recipients. Good T cell–B cell collaboration was always found when the *IA* region of *H-2* (Katz *et al.*, 1974), which is one of the *Ir* gene regions, but not *K* or *D* regions, was shared by B cells and helper T cells.

IX. The Use of Chimeric Mice in Immunology

In these experiments, however, allogeneic interaction (such as MLR) may take place between *H-2* or *IA* incompatible cells. This possibility, hard to exclude, might complicate the interpretation.

Sprent *et al.* subsequently used manipulated T cell populations as helper cells: in these experiments, anti-allogeneic *H-2* reactivity is deleted by manipulation of "*in vivo* filtration" (Sprent and von Boehmer, 1976). When X-irradiated F_1 mice are injected with lymphocytes of one parent, cells reactive with alloantigen of the host mice are temporarily trapped in the spleen, leaving the rest of the cells in circulation. Thoracic duct lymphocytes taken from these mice are almost all T cells and show no anti-host *H-2* reactivity. When these "filtered" T cells were used as a helper cell source in an anti-SRBC antibody response in adoptive transfer, it was again found that T cell–B cell collaboration did not work when they differ with respect to the *H-2* region (within the *IA* region) (Sprent and von Boehmer, 1976).

In contrast, however, clearly opposite results have been obtained from experiments with long-term chimeras, i.e., EA chimeras and irradiation bone marrow chimeras. Bechtol *et al.* (1974) analysed the *Ir* gene- controlled anti-(T,G)-A-L synthetic polypeptide antibody response in EA chimeras. Examination of *Ig* constant-region allotypes of anti-(T,G)-A-L serum produced in EA chimeras derived from high and low responder strains revealed, in many chimeras, the presence of both allotypes, indicating that high as well as low responder B cells contributed to the response (Bechtol *et al.*, 1974). Nonspecific help was considered unlikely as an explanation, since EA chimeras that share *H-2* [(high × low)$F_1 \leftrightarrow$low], or that combine two low responder strains differing for the *H-2* region, gave results that support the interpretation that specific help was delivered by responder helper T cells to nonresponder B cells, despite the fact they do not share the *H-2* region (Bechtol and McDevitt, 1976).

Although their results were not in accordance with similar experiments done later with irradiation chimeras (Press and McDevitt, 1977) and with EA chimeras using other antigens (Warner *et al.*, 1976, 1977), Bechtol's experiments represent the first extensive use of long-term chimeras for immunological studies and played an important role in the development of ideas on *Ir* gene action.

More direct evidence to demonstrate T–B interaction of *H-2* incompatible cells came from irradiation bone marrow chimeras (von Boehmer *et al.*, 1975b). In this experiment, helper activity was directly measured using adoptive transfer. A + B→(A × B)F_1 chimeras were primed with SRBC, T cells of the chimeras were removed, and one population (e.g., strain A) was separated from other parental (i.e., strain B) populations by anti-H-2 plus complement treatment. When such helper cells were mixed with B cells of various *H-2* type, anti-SRBC response was seen with B cells of either A or B strain (i.e., parental strains), but not with *H-2* unrelated strains (von Boehmer *et al.*, 1975b). The same conclusion was reached in similar experiments where hapten-carrier adoptive transfer was employed to measure helper cells of bone marrow chimeras (Waldman et al., 1976).

It is clear that T cells from these bone marrow chimeras are able to interact with B cells across an *H-2* barrier. The fact that they cannot collaborate with *H-2* unrelated B cells indicates that those helper T cells are still *H-2* restricted and suggests that some kind of active process of *H-2* recognition has taken place, How, then, can two sets of contradictory results, one from short-term chimeras and the other from long-term chimeras, be reconciled?

B. Impact of *H-2* Restriction of Cytotoxic T Cells

Similar findings for the need of *H-2* identity between effector and target cell have been made in virus-specific cytotoxic T cell responses. Mice infected with lymphocytic choriomeningitis virus generate cytotoxic T cells that can interact and lyse virus-infected target cells *in vitro*. However, when the *H-2* type of the target cells is changed, there is no lysis, despite the fact that the cells were infected with the same virus (Zinkernagel and Doherty, 1974a). The *H-2* determinants responsible for this restriction were found to be K or D but not *I* region products (Zinkernagel and Doherty, 1975). The finding was widely confirmed with many other viruses (Zinkernagel and Doherty, 1979), minor histocompatibility antigen (Bevan, 1975), *H-Y* antigen (Gordon et al., 1975) and hapten-conjugated cell antigen (Shearer et al., 1975).

A notion proposed by Zinkernagel and Doherty (1974b) was that a single T cell can recognize in some way both virus and H-2 antigens. This can be done in two different ways. Virus (or foreign epitope X in general) and H-2 antigen may interact with each other on the plasma membrane and create new determinants. T cells may recognize such determinants as "altered self antigens", which implies that they possess a single receptor. This is the "altered self" model. Alternatively, one can assume that a T cell is endowed with two different receptor molecules, and virus and H-2 are recognized by them separately. (This is the "two receptor" or "dual recognition" model.) In any case, their proposal was different from other models regarding the function of MHC molecules [e.g., H-2 is the T cell receptor (Benacerraf and McDevitt, 1972), like–like interaction by H-2 (Katz, 1977)]. Later experiments showed that their interpretation was essentially correct.

Shevach and Rosenthal had earlier suggested that *Ir* genes in guinea pigs may operate at macrophage (APC) level in antigen presentation for T cells (Shevach and Rosenthal, 1973). Macrophages from responder, but not from nonresponder, strains could specifically stimulate (responder × nonresponder)F_1 T cells in the presence of antigen *in vitro*. This finding was extended by Erb and Feldmann (1975) in mice, who showed that it is helper T cells that respond to antigens presented with *Ir* gene products. These results can be explained by the concept of

H-2 restriction, except that Ia antigens (i.e., *Ir* gene products) rather than H-2 K/D antigens function as restrictive entities for responding T cells. How can *H-2* restriction operating at one level (e.g., macrophage–T_h cells) be functionally related to the other level (T-B cells)?

In Sprent's experiments, (A × B)F_1 T cells, transferred into and primed with SRBC in X-irradiated parental A strain hosts, were first removed from thoracic ducts and then transferred to the second X-irradiated F_1 hosts with antigen and collaborating B cells. In this circumstance, (A × B)F_1 helper T cells primed in A strain *H-2* hosts could collaborate with B lymphocytes only of A strain *H-2* type, but not of B strain *H-2* type (Sprent, 1978). This result strongly indicates that *H-2* restriction is determined by the antigen-presenting macrophages in strain A (i.e. the first hosts), and as a consequence, such cell populations could collaborate only with B cells of A strain *H-2* type.

The questions raised in the debates described in the previous section (MHC barrier for T–B collaboration in chimeras) can now be answered. In EA chimeras and A + B→(A × B)F_1 irradiation chimeras, antigen priming was done in the presence of antigen-presenting macrophages of both parental types. Helper T cells restricted to both parental *H-2* types were therefore induced. T–B collaborations can be seen, then, either with syngeneic *H-2* or with allogeneic *H-2* cells, but still in an *H-2* restricted manner. Incidentally, all these findings have made the idea of like–like (MHC versus MHC) (Katz, 1977) interaction less attractive.

C. Immune Response (*Ir*) Genes in the Light of MHC Restriction

How can *Ir* gene-controlled nonresponsiveness be explained by the concept of MHC restriction so far discussed? There are two possible explanations. The first is that the defect in nonresponders lies in the T cell repertoire, i.e., specific T cell clones are already eliminated. The phenotypic linkage between clonal elimination and MHC region could be a result of diversity generation superimposed by MHC antigens (Jerne, 1970), or the clones in question may cross-react with other self antigens, so that self tolerance is induced (Schwartz, 1978). For example, foreign antigen X plus self MHC-A might cross-react with self antigens in A strain, but X plus self MHC-B has no cross-reactive antigen in B strains. Thus, strain A would be a nonresponder and strain B a responder. (A × B)F_1 is a responder because antigen X plus MHC-B is still a non-self antigen. Thus, dominance of the *Ir* gene can be explained. The second possibility is that nonresponsiveness is due to failure of antigen presentation (Benacerraf, 1978; Matsunaga et al., 1979; Simpson and Matsunaga, 1979). For example, nonresponder MHC molecules fail to associate effectively with foreign antigen X and cannot be recognized by T cells as foreign.

Recently, Ishii *et al.* (1982) showed that nonresponder macrophages could present antigens effectively to allogeneic T cells of responder strains. This supports the first possibility (i.e., clonal elimination) as the mechanism of *Ir* gene nonresponsiveness. However, it should be noted that the two hypotheses are not mutually exclusive. With sophisticated mechanisms, such as the immune system, where the final outcome of the response is dependent on many complicated cellular and molecular interactions, nonresponsiveness can arise from a failure of any part of the whole system.

On the other hand, the concept of MHC restriction would predict that class I antigens also behave like *Ir* genes, i.e., nonresponsiveness determined by the *H-2 K/D* region should be found. This turns out to be the case (Simpson and Gordon, 1977).

Cytotoxic T cells recognizing male-specific H-Y antigens are *H-2 K/D* restricted (Gordon *et al.*, 1975). Yet it was soon realized that anti-H-Y responses could never be generated in association with certain sets of *K* or *D* determinants (Simpson and Gordon, 1977). Furthermore, *I* region *Ir* genes were also involved in regulating anti-H-Y responses. Certain mouse strains are anti-H-Y T_c nonresponders, because they have nonresponder *I* region *Ir* genes (Hurme *et al.*, 1978). To understand the mechanisms of *Ir* gene control in anti-H-Y T_c response, we studied EA chimeras (H-2^d↔H-2^k) and irradiation chimeras [H-2^d + H-2^k→(H-2^d × H-2^k)F_1]. Both parental strains are anti-H-Y T_c nonresponders. H-2^d mice (BALB/c) are nonresponders because they have the wrong K/D antigens (K^d, D^d), although I^d is a responder I region Ir gene. On the contrary, in H-2^k mice (C3H or CBA), K^k, D^k is the right restriction antigen for H-Y but I^k is a nonresponder *Ir* gene. When these chimeras were primed with H-Y on H-2^k cells, anti-H-Y T_c were generated, but they were K^d (i.e., BALB/c) in origin (Matsunaga and Simpson, 1978). It was postulated that in those chimeras, only H-2 I^d restricted T_h cells were induced by H-2 I^d macrophages (responder *Ir* type) which present H-Y antigen (see Fig. 4). Such T_h cells would interact only with H-2^d type T_c cells. Our results are in agreement with those of von Boehmer and Haas (1979), who showed H-2 I^b T_h restriction in mice of the H-2^b haplotype. However, whether T_h–T_c interaction is *H-2* restricted or not is still an open question, as other workers have failed to demonstrate such restriction (Raulet and Bevan, 1981). An inherent problem here is that *H-2* restriction is demonstrable only when *H-2* polymorphic determinants are predominantly involved in the restrictive recognition, i.e., a failure of *H-2* restriction does not rule out that T_h–T_c interaction is restricted to shared determinants of *H-2* molecules (cross-reactivity). This situation is reminiscent of the results of allotype analysis in long-term chimeras described in Section IV, A. In the anti-(T,G)-A-L response, distribution of *H-2* restricted and cross-reactive helper T cells may have been different between EA chimeras and irradiation chimeras (Bechtol *et al.*, 1974; Press and McDevitt, 1977).

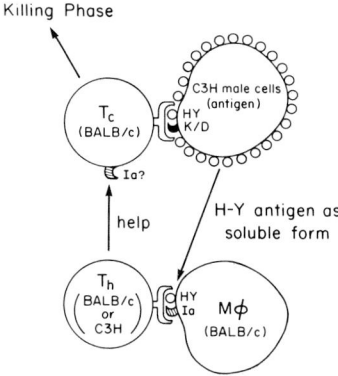

Fig. 4. In BALB/c↔C3H chimeras, cytotoxic T cell precursors (T_c), helper T cells (T_h) and antigen-presenting cells or macrophages (Mφ) all include cell populations of two different *H-2* types. C3H male cells used for immunization can be recognized directly by anti H-Y T_c precursors of either BALB/c or C3H type, but helper T cells can only recognize H-Y antigen which has been processed by BALB/c macrophages (Mφ) that bear associative Ia antigens. This introduces an *Ia* restriction to the helper T cells, which in turn could explain the *H-2* type of the anti H-Y cytotoxic T cells generated in these chimeras.

D. A New Dispute: Are T Cells Selected by Thymus MHC?

The thymus is an organ where T lymphocytes grow and differentiate into mature and functional cells. In adult mammals, T cells are drawn from a stem cell pool in bone marrow, pass through the thymus and migrate to peripheral immune tissues.

Zinkernagel *et al.* (1978) and Bevan and Fink (1978) observed that (A × B)F_1→A irradiation chimeras, when they were infected with virus or immunized with minor histocompatibility antigens on F_1 cells, could generate T_c responses restricted *only to A strain H-2* type. In this combination, both T cells and antigen-presenting macrophages are of F_1 origin, i.e., they have the *H-2* type of both A and B strain. Therefore, the failure to find B strain restricted responses cannot be attributed to antigen presentation. It was proposed that the MHC of the radioresistant portion of host thymus [e.g., thymus epithelium, A type in (A × B)F_1→A chimeras] may select or "educate" T cells to acquire capability for MHC-restricted recognition.

Erb *et al.* (1979) examined irradiation chimeras and A↔B EA chimeras for the interaction of T_h and antigen-presenting macrophages *in vitro*. Primary T_h induction restricted to either parental *H-2* but not to unrelated *H-2* types was always found in EA chimeras where both parental *H-2* types are expressed in thymus epithelium. In A + B→(A × B)F_1 irradiation chimeras, but not in A↔B EA chimeras, priming with antigen was necessary to demonstrate induction of T_h restricted to allogeneic (other parental type) *H-2*. In (A × B)F_1→A irradiation chimeras, helper T cells were induced only by A strain macrophages, in agreement with the idea of thymus MHC selection.

As was discussed previously, MHC restriction and immune response gene (*Ir* gene) phenomena represent opposite sides of the same coin. It follows that, if T cell selection by thymus MHC is correct, this step constitutes an important requirement for the *Ir* gene action to be realized. This expectation is generally met. If nonresponder type T cells are selected in a responder-type thymus, they acquire the status of responder (Matsunaga and Simpson, 1978; von Boehmer *et al.*, 1978).

On the other hand, several experiments had shown that the influence of thymus MHC on T cell responsiveness and restriction specificity may not be "black and white" effects, but rather that the selected T cells are biased toward the reactivity of thymus MHC.

A→B irradiation chimeras, when immunized repeatedly with minor histocompatibility antigen on A type *H-2* cells, could generate T_c responses restricted to A (Matzinger and Mirkwood, 1978). As A→B chimeras do not have A type MHC in the thymus, the original claim of thymus MHC influence was questioned. Similar results that seem at odds with the concept of thymus MHC learning have also been obtained with (A × B)F_1→A chimeras (Blanden and Andrew, 1979). At a relatively early stage in this debate, it seemed clear that the effect of thymus MHC is merely quantitative. In other words, they may always be *minor* T cell populations that are restricted to other MHC not present in the thymus.

More seriously, however, even this concept was recently challenged. It is possible to estimate the frequency of precursors of cytotoxic T cells using limited dilution culture. Using A→(A × B)F_1 irradiation chimeras that were thymectomized and grafted with B strain thymus (i.e., A type T cells learned B type H-2), Stockinger *et al.* (1980) found anti-Sendai virus T_c precursor frequency to have a ratio of about 3 B restricted to 1 A restricted. When such chimeras are immunized with virus-infected cells of either parental strain, T_c cells of either restriction specificity could be readily generated, depending on the *H-2* type of the immunizing cells used. Furthermore, when precursor frequency in normal mice was studied, the ratio of syngeneic to allogeneic *H-2* was 6 to 1 in spleen (Stockinger *et al.*, 1980), and between 3 to 1 and 10 to 1 in the thymus (Stockinger *et al.*, 1981). The precursor frequency for T_c seems to be biased toward thymus MHC, but the difference is not impressive.

Thus, facing these experimental challenges, the original interpretation (Zinkernagle *et al.*, 1978) for the role of thymus MHC might have to be reconsidered. Suppressor T cells have, indeed, been proposed to explain the restriction preference observed in (A × B)F_1→A irradiation chimeras (Blanden and Andrew, 1979).

Several years before the finding of *H-2* restriction in the virus-specific T_c system, Jerne had proposed a theory that implies a notion of MHC restriction

(Jerne, 1970). He argued that specificity of antibody v domains that are encoded by germ-line *v* genes are directed against MHC antigens of a species. Elimination of anti-self MHC clones occurs in the thymus through the self tolerance mechanism, but mutant clones would escape such elimination and migrate to the periphery where they prepare for anti-foreign responses.

Modification of this original idea was attempted by von Boehmer *et al.*, (1978) based on two-receptor models, but at that time the models were concerned with antibody molecules rather than with T cell receptors. Indeed, recent findings by Klinman's group (Wylie *et al.*, 1982) show that a large proportion of antibody clones against influenza virus in mice are *H-2* restricted. As antibody molecules of a clone have identical combining sites, this and other results (see below) indicate that two receptors are not necessary for restrictive recognition. MHC restriction makes more sense for T cells because they have to interact with other cells where MHC antigens are used as specific markers. However, it is also possible that we see only MHC-*restricted* T cells and MHC-*unrestricted* antibodies in most of our immunological assay systems.

On the other hand, T cell tolerance seems also to be induced outside the thymus (Besedovsky *et al.*, 1979). The question whether the thymus plays an important role for tolerance and selection (Jerne, 1970) with regard to MHC antigens is still largely open.

Several recent experiments argue in favor of a one-receptor altered-self model: (1) helper T cell "double" hybridomas that originated from parental cells different from each other with regard to both antigen X specificity and restriction *H-2 I* specificity did not show "recombinant type" antigen X plus *H-2* (Kappler *et al.*, 1981); (2) cross-reactions exhibited by many cloned T cell lines. For example, $H-2D^b$ restricted anti-H-Y T_c cells recognize $H-2D^d$ antigens (von Boehmer *et al.*, 1979). There is a T_c line with antigen X (minor histocompatibility antigen) plus A *H-2*, which cross-reacts with determinants created by antigen Y plus B *H-2* (Hunig and Bevan, 1982). (3) Anti-H-2 T_c clones seem to recognize conformational change of the molecules rather than epitopes determined by primary sequence (Sherman, 1982).

E. Chimeras and Idiotype Network

The immensity of immune diversity creates another dimension of interaction that the immune system seems to employ. V domains of antibody molecules, and probably of T cell receptors, carry a set of epitopes that can be recognized by other antibodies or T cells. These determinants are called idiotypes (Oudin and Michel, 1969). Thus, if a single mouse expresses an antibody repertoire of the order of tens of millions (Klinman *et al.*, 1976), i.e., tens of millions of different antibody clones, idiotype heterogeneity could be expected to be of similar magnitude. As foreseen by Jerne (1974), this would generate a network of idiotype and anti-idiotype interactions and could be a driving force for some aspects of immunological regulation and recognition. For example, when foreign epitopes are introduced into the immune system, the first anti-foreign antibodies would induce anti-idiotypic responses. This anti-idiotypic set might mimic epitopes on foreign antigens (Fig. 5, internal image) (Cazenave et al., 1977). Thus, the anti-idiotypic sets will in turn keep stimulating the first anti-foreign antibodies, and external epitopes are no longer needed. Immunological memory might be maintained for long periods of time by this kind of mechanism (Jerne, 1974). Helper T cells and suppressor T cells may also use idiotypic determinants of cells with which they interact (Eichmann, 1975; Eichmann et al., 1978). Such a mechanism would make helping or suppression more efficient. Although we do not know whether the immune system would work without an idiotype network, it is conceivable that evolutionary expansion of immune diversity created such a network as a by-product (Matsunaga and Ohno, 1981). As idiotype determinants are all self-antigens, an anti-idiotypic response is a kind of autoimmune response that happens to be beneficial to the immune system.

The idiotype network may also be important in the selection of clones involved in the development of the immune repertoire. The antibody response against phosphorylcholine (PC) determinants in BALB/c mice is characterized by a dominant clone or clones that have a *TEPC-15* idiotypic type. In antigen-primed BALB/c mice, the *TEPC-15* idiotype is also expressed by helper T cells. Genetically, the *TEPC-15* expression is controlled by the *Vh* gene cluster of *Igh* linkage group, and in some mouse strains *TEPC-15* is not expressed. EA chimeras of BALB/c↔C3H constitution (C3H is *TEPC-15* negative) were immunized with PC antigens; lymph node cells of several chimeras were then pooled and further stimulated by antigens in tissue culture in the presence of (BALB/c × C3H)F_1 macrophages (Augustin et al., 1980). After a few weeks of culture, more than 90% of cells were T cells, and alloreactive *H-2* cells had died

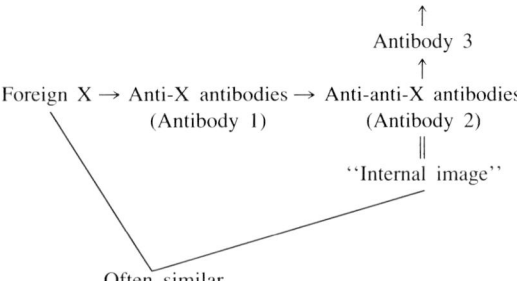

Fig. 5. Idiotype–anti-idiotype network elicited in immune response. Foreign antigen X may share similar epitopes with the antibody 2 (anti-anti-X antibodies). As antibody 2 remains in the animals, such epitopes (internal image of foreign antigen X) may keep stimulating anti-X antibodies (antibody 1), thus maintaining immunological memory.

IX. The Use of Chimeric Mice in Immunology

out. In the same protocol, anti-PC helper T cells from *normal BALB/c mice* exhibit *TEPC-15* and I^d restriction, whereas cells from *normal C3H mice* show I^k-restricted PC-specific helper function, but no *TEPC-15* expression. However, results from EA chimeras were surprising as well as informative. BALB/c cells from EA chimeras (C3H cells were removed by anti-H-2 plus complement) had two populations restricted to either I^d (BALB/c) or I^k (C3H), and *both* populations expressed *TEPC-15* idiotype to the same extent (Augustin *et al.*, 1980). C3H cells from EA chimeras also showed two restriction specificities but, most unexpectedly, they were both positive with the *TEPC-15* idiotype indistinguishable from the idiotype of BALB/c cells.

Although the origin of the anti-PC antibodies was not examined in these EA chimeras, the above result strongly indicates the presence of some selective force in the BALB/c environment leading to *TEPC-15* expression. It is likely that a set of *Vh* gene products rather than the presence or absence of particular *V* genes determines the expression of an available clone through an idiotypic network. Paradoxically, in BALB/c→BALB/c irradiated and bone marrow reconstituted mice, such interaction operates in the direction of *TEPC-15* suppression (Augustin *et al.*, 1977). The finding that the *TEPC-15* idiotype was expressed equally by T helper cells of both BALB/c and C3H origin in BALB/c↔C3H EA chimeras, in spite of the fact that T cells of each genotype included both I^d and I^k restricted cells, makes the notion of dual recognition more attractive but does not disprove the one receptor hypothesis.

Note Added in Proof

Recent reports (Groves and Singer, 1983; Lamb and Feldman, 1984) have appeared that indicate that self-tolerance is *H-2* restricted (Section III, C).

ACKNOWLEDGMENTS

I thank Dr. S. Ohno for many stimulating discussions, Dr. E. Simpson for kindly correcting my manuscript, Drs. J. Terz and J. Benfield for support, and Mrs. J. Flanagan for typing the manuscript.

REFERENCES

Anderson, B., Billingham, R. E., Lampkin, G. H., and Medawar, P. B. (1951). Heredity **5**, 379.
Augustin, A. A., Julius, M. H., and Cosenza, H. (1977). "Regulation of the Immune System," p. 195. Academic Press, New York.
Augustin, A. A., Julius, M. H. Cosenza, H., and Matsunaga, T. (1980). *In* "Regulatory T Lymphocytes" (B. Pernis and H. J. Vogel, eds.), p. 171. Academic Press, New York.
Bach, F. H. (1970). *Science* **168**, 1170.

Barnes, R. D., and Graham, C. G. (1976). *Cell. Immunol.* **21,** 146.
Bechtol. K. B., and McDevitt, H. O. (1976). *J. Exp. Med.* **144,** 123.
Bechtol, K. B., Freed, J. H., Herzenberg, L. A., and McDevitt, H. O. (1974). *J. Exp. Med.* **140,** 1660.
Benacerraf, B. (1978). *J. Immunol.* **120,** 1809.
Benacerraf, B., and McDevitt, H. D. (1972). *Science* **175,** 273.
Besedovsky, H. D., del Ray, A., and Sorkin, E. (1979). *J. Exp. Med.* **150,** 1351.
Bevan, M. J. (1975). *Nature (London)* **256,** 419.
Bevan, M. J., and Fink, P. J. (1978). *Immunol. Rev.* **42,** 4.
Billingham, R. E., Brent, L., and Medawar, P. B. (1953). *Nature (London)* **172,** 603.
Blanden, R. V., and Andrew, M. E. (1979). *J. Exp. Med.* **149,** 535.
Brent, L., Rayfield, L. S., Chandler, P., Fierz, W. Medawar, P. B., and Simpson, E. (1981). *Nature (London)* **290,** 508.
Burnet, F. M. (1959). "The Clonal Selection Theory of Acquired Immunity." Vanderbilt Univ. Press, Nashville, Tennessee.
Burnet, F. M., and Fenner, F. (1949). "The Production of Antibodies." Macmillan, New York.
Cantor, H., McVay-Bondreau, L., Hugenberger, J., Naidorf, K., Shen, F. W., and Gershon, R. K. (1978). *J. Exp. Med.* **147,** 1116.
Cazenave, P. A., Juy, D., and Bona, C. (1977). *Cold Spring Harbor Symp. Quant. Biol.* **41,** 307.
Cohn, M. (1968). "Nucleic Acid in Immunology," p. 671. Springer-Verlag, New York.
de la Chapelle, A., Schroder, J., Rantanen, P., Thomasson, B., Niemi, M., Tiilikainen, A., Sanger, R., and Robson, E. B. (1974). *Ann. Hum. Genet.* **38,** 63.
Dorsch, S., and Rosen, B. (1977). *J. Exp. Med.* **145,** 1144.
Ehrlich, P. (1900). *Proc. R. Soc. London, Ser. B* **66,** 424.
Eichmann, K. (1975), *Eur. J. Immunol.* **5,** 511.
Eichmann, K., Falk, I., and Rajewsky, K. (1978), *Eur. J. Immunol.* **8,** 853.
Erb, P. M., and Feldmann, M. (1975). *J. Exp. Med.* **142,** 460.
Erb, P. M. Meier, B., Matsunaga, T., and Feldmann, M. (1979). *J. Exp. Med.* **149,** 686.
Festenstein, H., Huber, B., Abbasi, K., and Barnes, R. D. (1975). *J. Immunogenet.* **2,** 351.
Gershon, R. K. (1974). *Contemp. Top. Immunobiol.* **3,** 1.
Gershon, R. K., and Kondo, K. (1970). *Immunology* **18,** 723.
Gorczynski, R. M., and Steele, E. J. (1980). *Proc. Natl. Acad. Sci. U.S.A* **77,** 2871.
Gordon, R. D., Simpson, E., and Samelson, L. E. (1975). *J. Exp. Med.* **142,** 1108.
Gorer, P. A., Lyman, S., and Snell, G. D. (1948). *Proc. R. Soc. London, Ser. B.* **135,** 499.
Groves, E. S., and Singer, A. (1983). *J. Exp. Med.* **158,** 1413.
Hasek, M. (1953). *Cesk. Biol.* **2,** 25.
Hunig, T., and Bevan, M. J. (1981). *Nature (London)* **294,** 460.
Hurme, H. M., Hetherington, C. M., Chandler, P. R., and Simpson, E. (1978). *J. Exp. Med.* **147,** 758.
Ishii, N., Zoltan, A., and Klein, J. (1982). *Nature (London)* **295,** 531.
Jerne, N. K. (1970). *Eur. J. Immunol.* **1,** 1.
Jerne, N. K. (1974). *Ann. Immunol. Pasteur, Paris* **125C,**373.
Kappler, J. W., Skidmore, B., White, P., and Marrack, P. (1981). *J. Exp. Med.* **153,** 1198.
Katz, D. H. (1977). "Lymphocyte Differentiation, Recognition, and Regulation." Academic Press, New York.
Katz, D. H., Hamaoka, T., and Benacerraf, B. (1973). *J. Exp. Med.* **137,** 1405.
Katz, D. H., Dorf, M. E., and Benacerraf, B. (1974). *J. Exp. Med.* **141,** 1717.
Kilshaw, P. J., Brent, L., and Pinto, M. (1975). *Nature (London)* **255,** 489.
Kindred, B. (1979). *Prog. Allergy* **26,** 137.

Kindred, B., and Schreffler, D. C. (1972). *J. Immunol.* **109**, 940.
Klein, J. (1979). *Science* **203**, 516.
Klinman, N. R., Sigal, N. H., Metcalf, E. S., Pierce, S. K., and Gearhart, P. J. (1976). *Cold Spring Harbor Symp. Quant. Biol.* **41**, 165.
Kreth, H. W., and Williamson, A. R. (1973). *Eur. J. Immunol.* **3**, 141.
Lamb, J. R., and Feldman, M. (1984). *Nature (London)* **308**, 72.
Langhorne, J., and Lindahl, K. F. (1982). *Eur. J. Immunol.* **12**, 101.
Liew, F. Y., and Simpson, E. (1980). *Immunogenetics* **11**, 255.
Loveland, B. E., Hogarth, P. M., Ceredig, R., and McKenzie, I. F. C. (1981). *J. Exp. Med.* **153**, 1044.
McDevitt, H. O., Deak, B. D., Schreffler, D. C., Klein, J., Stimpfling, J. H., and Snell, G. D. (1972). *J. Exp. Med.* **135**, 1259.
McLaren, A. (1976). "Mammalian Chimeras." Cambridge Univ. Press, London and New York.
McLaren, A., and Bowman, P. (1969). *Nature (London)* **224**, 238.
McLaren, A., Chandler, P., Buehr, M., Fienz, W., and Simpson, E. (1981). *Nature (London)* **290**, 513.
Matsunaga, T. (1973). Ph. D. Dissertation, University of Kyoto, Kyoto, Japan.
Matsunaga, T., and Ohno, S. (1981). "The Immune System," Vol. 1, p. 76. Karger, Basel.
Matsunaga, T., and Simpson, E. (1978). *Proc. Natl. Acad. Sci. U.S.A.* **75**, 6207.
Matsunaga, T., Brenan, M., Benjamin, D., and Simpson, E. (1979). *In* "T and B Lymphocytes: Recognition and Function" (F. Bach, B. Bonavida, E. S. Vitetta, and C. F. Fox, eds.), p. 551. Academic Press, New York.
Matsunaga, T., Simpson, E., and Meo, T. (1980). *Transplantation* **30**, 34.
Matzinger, P. (1981). *Nature (London)* **292**, 497.
Matzinger, P., and Bevan, M. J. (1977). *Cell. Immunol.* **29**, 1.
Matzinger, P., and Mirkwood, G. (1978). *J. Exp. Med.* **148**, 84.
Meo, T., David, C. S., Rijnbeck, A., Nabholz, M., Miggiano, V. C., and Schreffler, D. C. (1975). *Transplant. Proc.* **7**, 127.
Miller, J. F. A. P., and Mitchell, G. F. (1968). *J. Exp. Med.* **128**, 801.
Mintz, B. (1967). *Proc. Natl. Acad. Sci. U.S.A.* **58**, 344.
Mintz, B. (1974). *Annu. Rev. Genet.* **8**, 411.
Mintz, B., and Palm, J. (1969). *J. Exp. Med.* **129**, 1013.
Mintz, B., and Silvers, W. K. (1967). *Science* **158**, 1484.
Mintz, B., and Silvers, W. K. (1970). *Transplantation* **9**, 497.
Nabholz, M., Vives, J., Young, H. M., Meo, T., Miggiano, V. C., Rijnbeck, A., and Schreffler, D. C. (1974). *Eur. J. Immunol.* **4**, 378.
Ohno, S., Epplen, J. T., Matsunaga, T., and Hozumi, T. (1981). *Prog. Allergy* **28**, 8.
Oudin, J., and Michel, M. (1969). *J. Exp. Med.* **130**, 545.
Owen, R. D. (1945). *Science* **102**, 400.
Phillips, S. M., and Wegmann, T. G. (1973). *J. Exp. Med.* **137**, 291.
Phillips, S. M., Martin, W. J., Shaw, A. R., and Wegmann, T. G. (1971) *Nature (London)* **234**, 146.
Press, J. L., and McDevitt, H. O. (1977). *J. Exp. Med.* **146**, 1815.
Ranlet, D. H., and Bevan, M. J. (1981). *J. Exp. Med.* **155**, 1766.
Schwartz, R. H. (1978). *Scand. J. Immunol.* **7**, 3.
Shearer, G. M., Rehn, T. G., and Garbarino, C. A. (1975). *J. Exp Med.* **151**, 1348.
Sherman, L. A. (1982). *Nature (London)* **297**, 511.
Shevach, E. M., and Rosenthal, A. S. (1973). *J. Exp. Med.* **138**, 1213.
Simpson, E., and Gordon, R. D. (1977). *Immunol. Rev.* **35**, 59.

Simpson, E., and Matsunaga, T. (1979). *Transplantation* **27**, 295.
Sprent, J. (1978). *J. Exp. Med.* **148**, 1142.
Sprent, J., and von Boehmer, H. (1976). *J. Exp. Med.* **144**, 617.
Stockinger, H., Pfizenmaier, K., Hardt, C., Rodt, H., Rollinghoff, M., and Wagner, H. (1980). *Procl Natl. Acad. Sci. U.S.A.* **77**, 7390.
Stockinger, H., Bartlett, R., Pfizenmaier, K., Rollinghoff, M., and Wagner, H. (1981). *J. Exp. Med.* **153**, 1629.
Tarkowski, A. K. (1961). *Nature (London)* **190**, 857.
Tonegawa, S. (1976). *Proc. Natl. Acad. Sci. U.S.A.* **73**, 203.
von Boehmer, H., and Haas, W. (1979). *J. Exp. Med.* **150**, 1134.
von Boehmer, H., Sprent, J., and Nabholz, M. (1975a). *J. Exp. Med.* **141**, 332.
von Boehmer, H., Hudson, L., and Sprent, J. (1975b). *J. Exp. Med.* **142**, 987.
von Boehmer, H., Hass, W., and Jerne, N. K. (1978). *Proc. Natl. Acad. Sci. U.S.A.* **75**, 2439.
von Boehmer, H., Hengartner, H., Nabholtz, M., Lernhardt, W., Schreier, M. G., and Haas, W. (1979). *Eur. J. Immunol.* **9**, 592.
Waldman, H., Pope, H., and Munro, A. J. (1976). *Nature (London)* **258**, 728.
Warner, C. M., Graves, R. M., Tollefson, C. M., Schmerr, M. J. F., Stephens, T. J., Merryman, C. F., and Maurer, P. H. (1976). *Immunogenetics* **3**, 337.
Warner, C. M., McIvor, J. L., Mauer, P. H., and Merryman, C. F. (1977). *J. Exp. Med.* **145**, 766.
Wegmann, T. G., Hellström, I., and Hellström, K. E. (1971). *Proc. Natl. Acad. Sci. U.S.A.* **68**, 1644.
Weigle, W. O. (1980). *Adv. Immunol.* **30**, 159.
Wylie, D. E., Shermann, L. A., and Klinman, N. R. (1982). *J. Exp. Med.* **155**, 403.
Zinkernagel, R. M., and Doherty, P. C. (1974a). *Nature (London)* **248**, 701.
Zinkernagel, R. M., and Doherty, P. C. (1974b). *Nature (London)* **251**, 547.
Zinkernagel, R. M., and Doherty, P. C. (1975). *J. Exp. Med.* **141**, 1427.
Zinkernagel, R. M., and Doherty, P. C. (1979). *Adv. Immunol.* **27**, 51.
Zinkernagel, R. M., Callahan, G. N., Althage, A., Cooper, S. Streilein, J. W., Klein, P. A., and Klein, J. (1978). *J. Exp. Med.* **147**, 882.

4

Muscle and Skeleton

CHAPTER **X**

The Use of Chimeras in Analyses of Craniofacial Development

DREW M. NODEN

Department of Anatomy
New York State College of Veterinary Medicine
Cornell University
Ithaca, New York

I. Introduction 241
II. Defining Embryonic Origins of Cephalic Tissues 244
    A. Skeletal and Connective Tissues 244
    B. Craniofacial and Cervical Muscles 250
    C. Otic and Periotic Structures 253
    D. Periocular Tissues 258
III. Experimental Analyses of Craniofacial Development 261
    A. Establishment of Definitive Tissue Relationships 261
    B. Differentiation and Patterning of Craniofacial Connective Tissues 262
    C. Patterning of Voluntary Muscles 267
IV. Extrapolation to Mammals 269
V. Methodological Considerations 271
    A. Transplantation Artifacts 271
    B. Problems Encountered in Analyzing Chimeras 273
VI. Summary and Perspectives 275
    References 277

I. INTRODUCTION

Since the pioneering experiments of Born (1894, 1897) and Harrison (1898, 1903), the use of surgically created chimeric embryos has been a major tool in the arsenal of developmental biologists. Spemann (1901, 1921) was the first to introduce the term chimera into the embryological literature and to envision the great potential of this method in analyzing the mechansims of craniofacial development. Furthermore, it was he who first undertook such analyses (Spemann, 1921), taking advantage of inherent species differences in the number of intracellular pigment granules in amphibian embryos (Harrison, 1903; Spemann, 1918, 1938). These experiments culminated in the so-called organizer experiment of Spemann and Mangold (1924).

Later, using the same markers and also exploiting differences in the shapes or types of tissues found in various species of amphibians, several investigators examined the development of craniofacial pigment cells (Twitty, 1936), skeletal tissues (Andres, 1946, 1949; Wagner, 1949, 1959), and teeth (Raven, 1935). These analyses, which have been reviewed by Baltzer (1952) and Owen (1959), complemented the many extirpation and vital staining experiments performed during the same decades. However, because the intrinsic markers available in amphibians become depleted as tissues mature or are otherwise unstable, the usefulness of these chimeras was limited to early developing structures.

The discovery and popularization of the quail nucleolar marker by Le Douarin and co-workers (Le Douarin, 1969, 1973) opened a new era in experimental

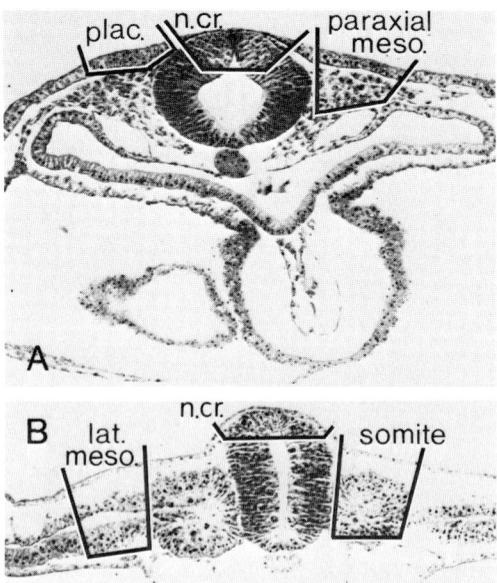

Fig. 1. Transverse sections through a stage 10⁻ (9 somite) quail embryo illustrating the tissues excised for transplantation. (A) at the level of somitomere 6, immediately rostral to the center of the otic placode. (B) At the level of somite 3. lat. meso., lateral mesoderm, n. cr., neural crest primordium; plac., otic placodal thickening.

X. The Use of Chimeras in Analyses of Craniofacial Development

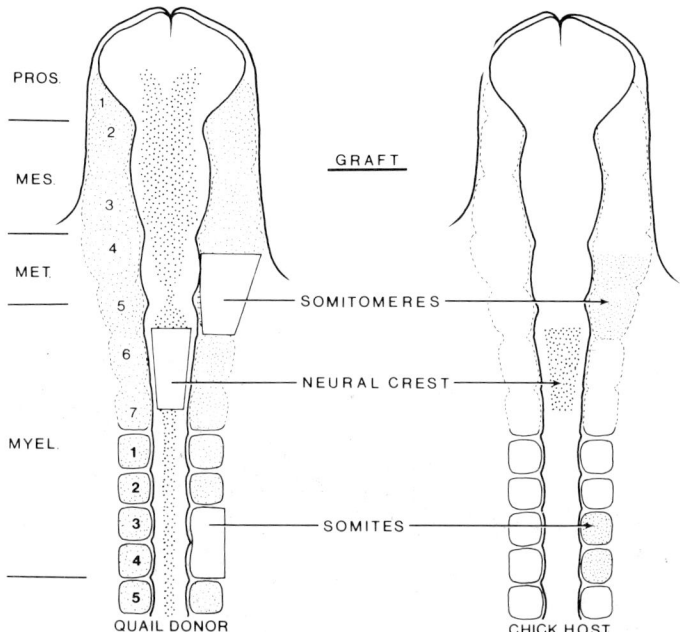

Fig. 2. Experimental design of transplantations used to trace the origins of craniofacial tissues. All levels of prospective craniofacial neural crest, paraxial mesoderm, and surface ectoderm (not shown) have been grafted. PROS., prosencephalon; MES., mesencephalon; MET., metencephalon; MYEL., myelencephalon. (From Noden, 1983b.)

analysis of craniofacial development, providing both a stable, replicating marker and the opportunity to perform analyses on an amniote embryo.

The principal embryonic tissues described and analyzed in this chapter are the avian neural crest and paraxial mesoderm, with brief supplementary comments concerning lateral mesoderm and surface ectoderm. Figures 1 and 2 show the locations of these primordia at the time of transplantation and the design utilized in control (orthotopic transplant) experiments.

The primordia of neural crest cells located between the midprosencephalon and somites number 5 and 6 have been grafted at Hamburger-Hamilton (1951) stages 9^- (4 somites) through 10^- (9 somites), with the older stages being used for the more caudal levels. To produce the data on neural crest development discussed in this chapter, only the dorsal margin of the closed neural tube was excised and transplanted (Noden, 1978b,c; 1983a). Experiments in which the entire neural tube in a defined region have been transplanted (Le Lièvre, 1974, 1978; Le Lièvre and Le Douarin, 1975; Le Douarin, 1975) yield comparable results.

Cephalic paraxial mesoderm includes the occipital (first through fifth) somites and the seven somitomeres, which are incompletely segmented populations of mesoderm located rostral to the first somite (Meier, 1979, 1981). These tissues are transplanted at stages 9–10, well in advance of the emigration of crest cells (Noden, 1982b, 1983b). To maintain the integrity of the tissue during transfer to the host, small areas of surface ectoderm are usually left attached to the mesoderm. Possible contamination by cells derived from this superficial eptihelium can be identified based on experiments in which only the surface ectoderm has been transplanted (D'Amico-Martel and Noden, 1983). Variations of these orthotopic transplantations will be described as needed throughout the chapter.

The objectives of this chapter are first to describe the principal results of research on craniofacial development performed using amphibian and avian chimeras. Emphasis will be on the developmental mechanisms and concepts elucidated using these methods rather than on specific anatomic details, most of which have been reviewed extensively (Le Douarin, 1975, 1980, 1982; Noden, 1978a, 1980, 1982a,b, 1984). Subsequently, a brief discussion of the methodological as well as interpretive problems encountered using quail ↔ chick chimeras will be presented.

II. DEFINING EMBRYONIC ORIGINS OF CEPHALIC TISSUES

A. Skeletal and Connective Tissues

Part of the head skeleton is, although unsegmented, fundamentally similar to the vertebral column in its structure and development. Included in this category would be the occipital, basisphenoid, and petrous regions of the lower braincase. These structures arise from occipital somites and several caudal somitomeres (Fig. 3B and 4). Unique to the head is the formation of large intramembranous (dermal) bones. These are derived from the same somitomeres, but differentiate lateral and dorsal to the brain and form the parietal and most of the frontal elements of the calvaria.

However, the most strikingly different feature of the head region is the formation of a large, ectodermally derived population of mesenchymal cells having skeletogenic and connective tissue-producing capabilities. This is the cephalic neural crest population. Some of these cephalic crest cells share the ability to form sensory and visceral efferent neurons, glia, melanocytes, and Schwann cells with crest cells originating at postcranial axial levels (see Chapter XII by Le Douarin, Teillet, and Fontaine-Perus; also Noden, 1978c; D'Amico-Martel and Noden, 1983). However, others form many of the types of tissues traditionally attributed to mesoderm. Most of the skeletal structures derived from the neural

X. The Use of Chimeras in Analyses of Craniofacial Development

crest are indicated in Fig. 4, with examples of grafted quail crest cells forming skeletal tissues shown in Fig. 5, 14, and 27.

In addition to documenting the precise origin of each skeletal structure in the head and the exact axial level from which the precursors are derived, the use of quail↔chick chimeras provides valuable information concerning the mechanisms of growth and secondary fusion of embryonic skeletal tissues (reviewed in Noden, 1978a, 1982b). Growth of a particular structure may bring it into an area occupied by heterologous mesenchyme. However, in every situation examined, each skeletal structure remains composed of cells derived exclusively from one source. This indicates that once condensation of chondroblasts (or osteoblasts) has occurred and the perichondrium (periosteum) is established, no additional mesenchymal cells are recruited into the structure.

In later embryonic stages many skeletal structures are found to contain both neural crest and mesodermal cells. These arise by the secondary fusion of separate primordia. The avian basisphenoid develops in this fashion, with the lateral wing derived from the neural crest (Fig. 3A) and the central part from mesoderm. Other examples include the otic capsule (Fig. 14), cartilaginous sclera, columella (Fig. 16), and frontal bone. In most cases there is little integration of crest and mesodermal cell populations at the zone of fusion, which is recognizable only in chimeras.

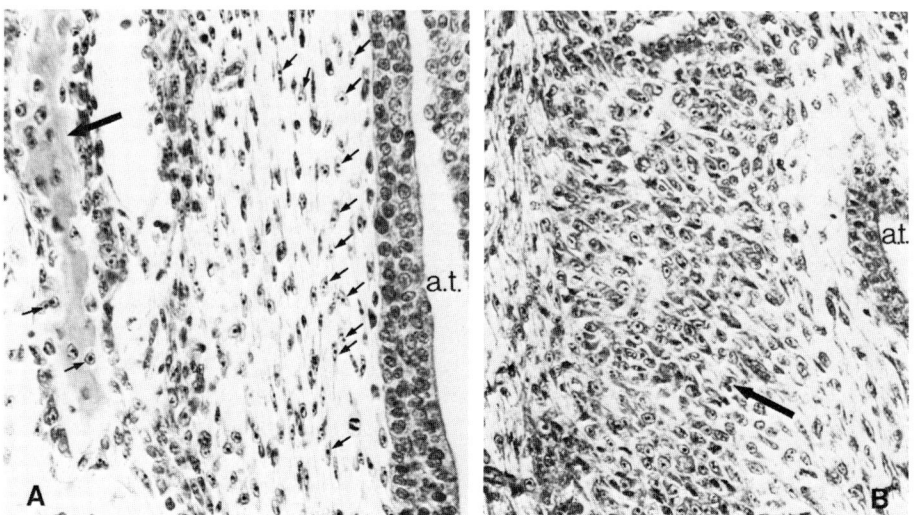

Fig. 3. The lateral (A) and medial (B) areas of the basisphenoid (large arrows) bone from chimeric hosts in which rostral myelencephalic neural crest (A) or postotic paraxial mesoderm (B) from a quail donor embryo had been transplanted into a chick host. Small arrows indicate quail cells in which the nucleolar marker is most evident. a.t., auditory tube.

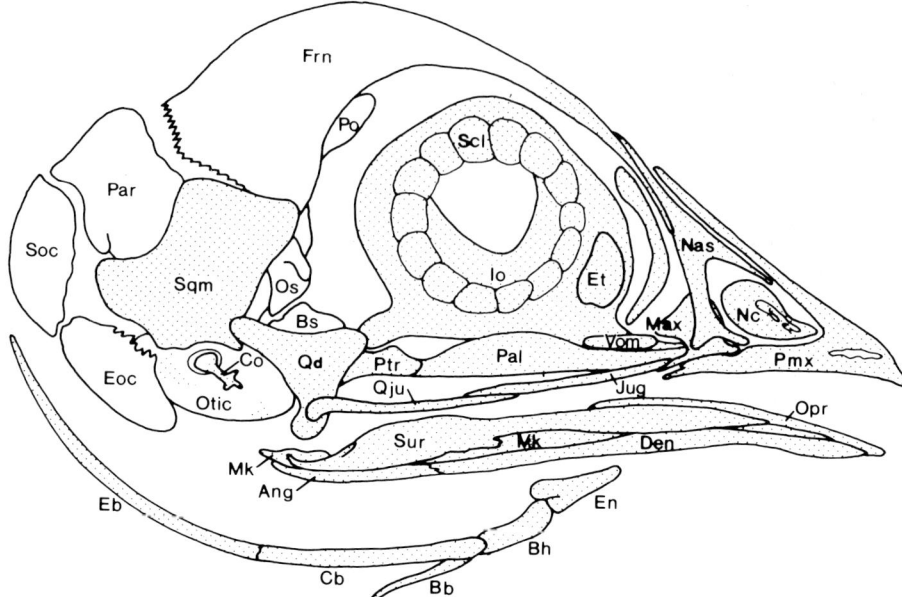

Fig. 4. Schematic drawing of the skull of a chick embryo at approximately 18 days of incubation. Shaded structures or regions are of neural crest origin; the remaining are derived from paraxial mesoderm. Cartilages and endochondral bones: Bb, basibranchial; Bh, basihyoid; Bs, basisphenoid; Cb, ceratobranchial; Co, columella; Eb, epibranchial; En, entoglossum; Eoc, exoccipital; Et, ethmoid; Io, interorbital; Mk, Meckel's; Nc, nasal capsule; Os, orbitosphenoid; Po, postorbital; Qd, quadrate; Soc, supra-occipital. Intramembranous (dermal) bones: Ang, angular; Den, dentary; Frn, frontal; Jug, jugal; Max, maxillary; Nas, nasal; Opr, opercular; Pal, palatine; Par, parietal; Pfr, prefrontal; Pmx, premaxilla; Ptr, pterygoid; Qju, quadratojugal; Scl, scleral ossicles; Sqm, squamosal (temporal); Sur, surangular; Vom, vomer.

The embryonic origins of each skeletal structure in the head are most easily understood by reconstructing the interface between mesodermal and neural crest populations, after the latter have occupied their definitive, postmigratory positions. This interface, shown schematically in Fig. 6, is established 1–2 days after the initial emigration of crest cells from the dorsal midline and demarcates skeletal and connective tissues of neural crest origin from those formed by mesoderm.

With the few exceptions mentioned below, every connective tissue and most perivascular and subcutaneous smooth muscles that develop ventral to the interface are of neural crest origin. This tremendous breadth of cell types formed by the cephalic neural crest, illustrated in Fig. 7 and 8, was not appreciated until the availability of the quail marker (Le Lièvre and Le Douarin, 1975; Noden, 1978b, 1983a; reviewed by Le Douarin, 1982).

However, the interface is not impenetrable. Outgrowing peripheral nerves and

X. The Use of Chimeras in Analyses of Craniofacial Development

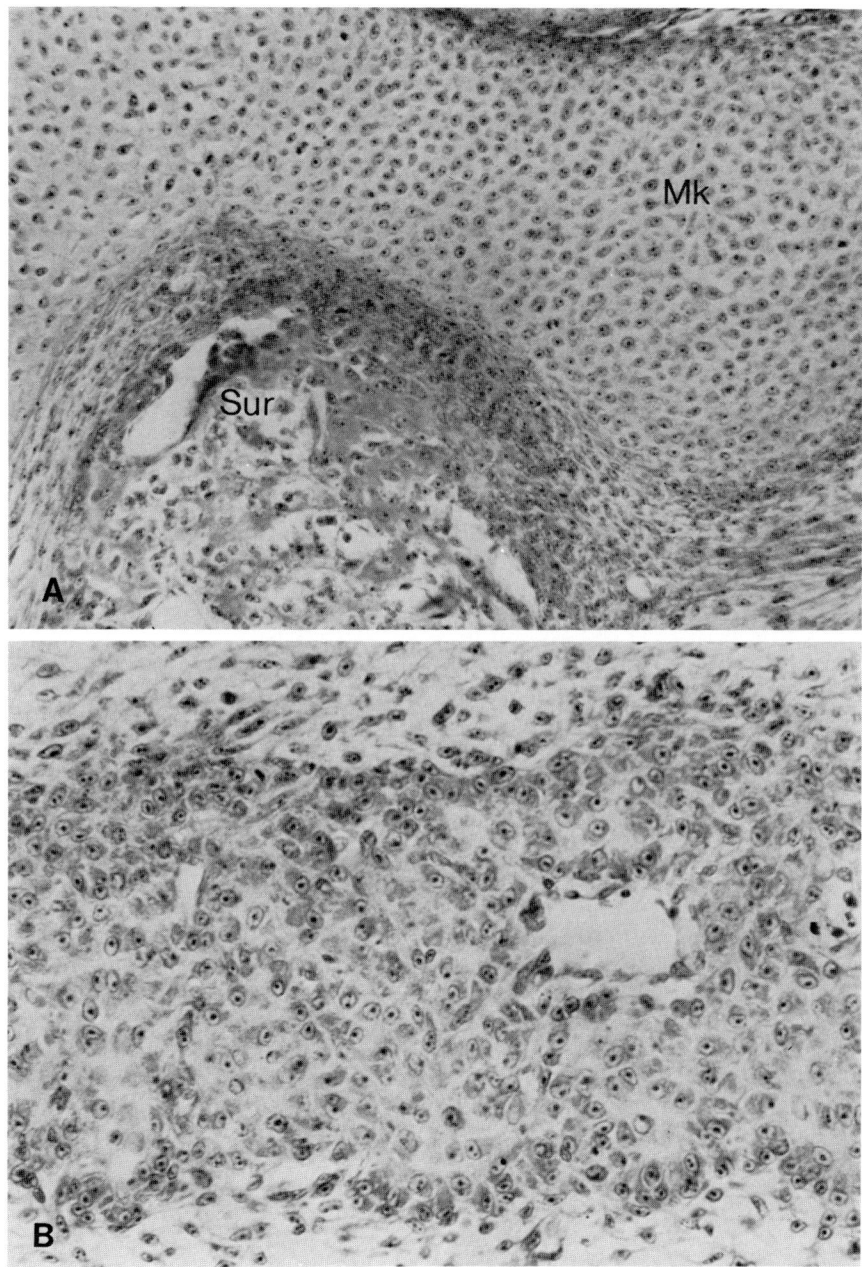

Fig. 5. Skeletal tissues fromed by transplanted quail neural crest cells. (A) Meckel's cartilage and surangular bone, 7.5-day host. (B) Palatine bone, 8-day host All chondrocytes and osteocytes are derived from the grafted tissue.

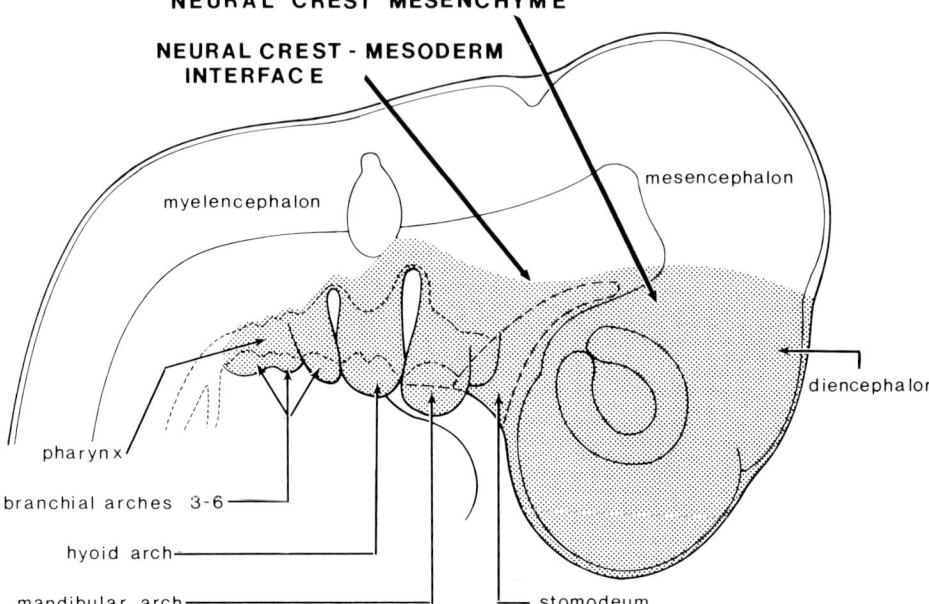

Fig. 6. The interface between postmigratory neural crest and paraxial mesoderm mesenchymal populations. (From Noden, 1983c.)

accompanying Schwann cell precursors cross it, so, too, do arterial angiogenic cords that presumably originate in cardiovascular splanchnic mesoderm (Rosenquist, 1970) and invade the mesenchyme of the head. Finally, as will be discussed in the following section, the anlagen of voluntary muscles cross the mesoderm–neural crest interface.

Differential growth of the head, in particular the flexures of the brain and secondary expansion of branchial arch and facial processes, causes the contours of the interface to change dramatically. Despite these extensive alterations in the absolute position and plane of the interface, the relationships established early in development do not change. The crest-derived mesenchymal populations associated with the adenohypophysis, salivary gland, and vascular endothelial tissues remain in intimate contact with these epithelia. Thus, for example, crest cells initially surrounding the fourth and sixth aortic arches later come to reside in the aortico-pulmonary septum, and their presence there is essential for the normal separation of systemic from pulmonary arterial channels (Le Lièvre and Le Douarin, 1975; Kirby, *et al.*, 1983).

Fig. 8. Sections from quail-chick chimeras showing that adenohypophyseal connective tissues (A) and telencephalic leptomeningeal tissues (B) are derived from transplanted neural crest cells. (From Noden, 1978b.)

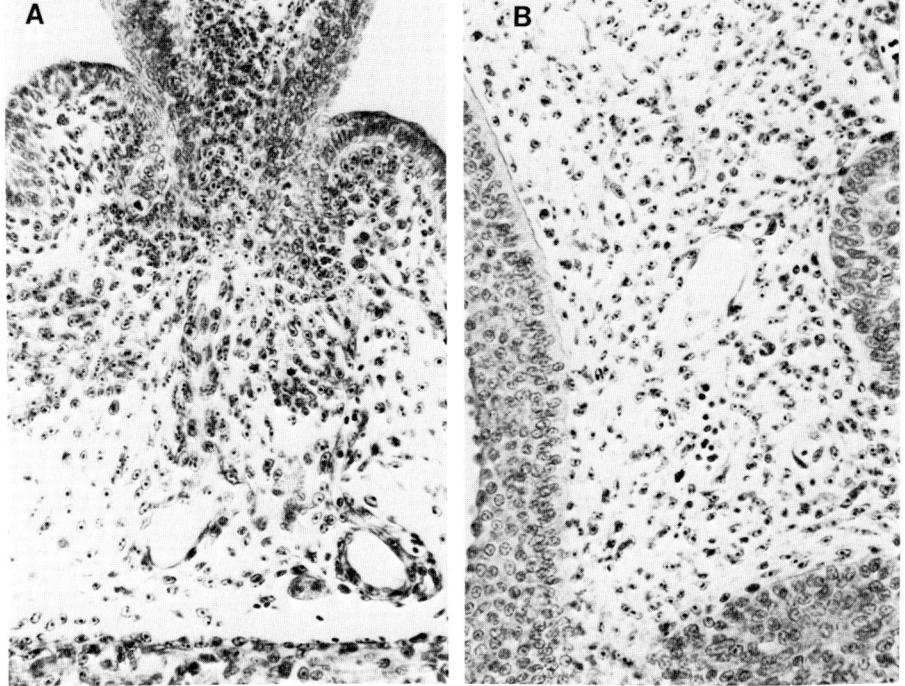

Fig. 7. Connective tissues of (quail) neural crest origin in chimeric hosts. (A) Feather and dermis. (B) Tongue mesenchyme. [(A) from Noden, 1982b.]

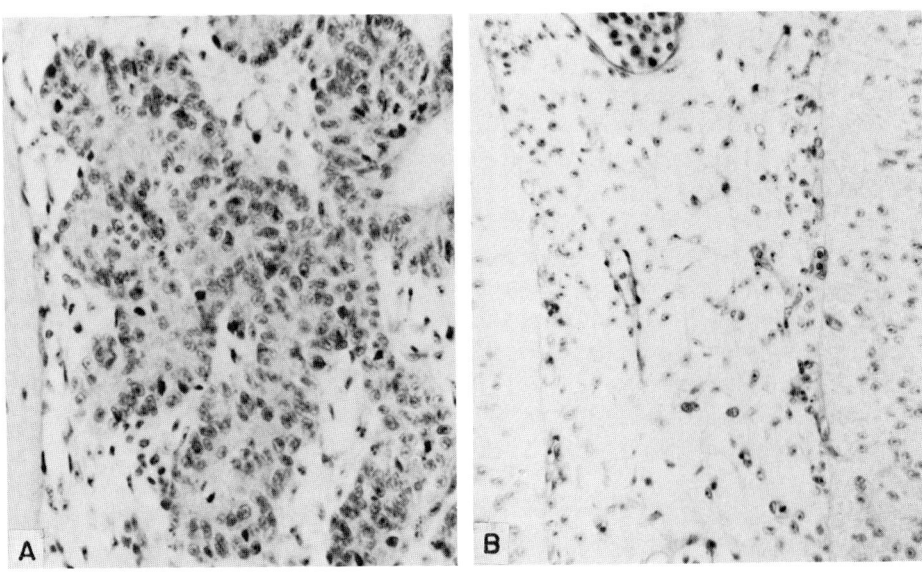

B. Craniofacial and Cervical Muscles

The voluntary muscles of the head are traditionally grouped into several categories. First are the muscles concerned with head posture and movement. These attach cranially to the skull and caudally to the vertebrae. In birds this set includes the biventer, complexus, splenius, rectus capitis, and longus colli muscle complexes. All are innervated by cervical nerves and are presumed to be derived from somitic myotomes. The intrinsic muscles of the tongue constitute a second group. Both extirpation and transplantation experiments (Piatt, 1938; Deuchar, 1958; Hammond, 1965; Hazelton, 1970) have shown that these arise from myotomal cells of occipital somites that coalesce to form the hypoglossal cord. This mesenchymal aggregate can be followed ventrally into the base of the tongue.

The third group includes the extrinsic ocular muscles, a set of six major and a variable number of ancillary muscles that has been very conservative throughout vertebrate evolution. Their embryonic antecedents have been described in all vertebrate classes as three bilateral pairs of preotic condensations (Neal, 1918; Slonaker, 1921; Adelmann, 1927; Edgeworth, 1935; Gilbert, 1957).

The fourth group is the most problematic. It includes the so-called branchiomeric muscles associated with the jaw and hyobranchial skeleton. The embryonic origins and evolutionary changes of these muscles and their associated skeletal tissues have been crucial features of most hypotheses concerning the emergence and evolution of vertebrates (reviewed by deBeer, 1937; Nelson, 1969; Romer, 1972; Jarvik, 1980; Moore, 1981; Northcutt and Gans, 1983; Gans and Northcutt, 1983; Noden, 1982b, 1983b,c). Based largely on comparative descriptions (Balfour, 1878; Scammon, 1911; Damas, 1944; Holmgren, 1940; Bjerring, 1968), it has been postulated that these muscles arise embryonically from lateral plate mesoderm present in branchial arches prior to the arrival of neural crest cells.

The last group includes the intrinsic and extrinsic muscles of the larynx and trachea. Some experimental data suggest that they are derived at least in part from somites (Hazelton, 1970), although owing to their innervation they are often included in the branchiomeric category.

There are two reasons to be skeptical about this entire categorization of muscle groups, especially those associated with the face and jaws. First, Meier and his co-workers have shown that the paraxial mesoderm rostral to the first somite is organized into seven incompletely segregated mesenchymal populations called somitomeres in mammals (Meier and Tam, 1982), birds (Meier, 1979, 1981), reptiles (Meier and Packard, 1984), and amphibians (S. Meier, personal communication). This suggests that the organization of mesoderm in the head might share more features with the somites of the trunk than heretofore suspected.

Second, recent investigations using guail↔chick chimeras have demonstrated that appendicular muscles are derived from somitic myotomes, and not from

X. The Use of Chimeras in Analyses of Craniofacial Development

lateral plate mesoderm as previously suspected (Chevallier, 1979; Chevallier *et al.*, 1977, 1978; Christ *et al.*, 1977, 1979; see Chapter XI by Gumpel-Pinot and review by Christ *et al.*, 1983). This leaves the branchiomerics as the only voluntary muscles not derived from paraxial mesoderm.

To investigate the origins of these muscles, segments of cephalic paraxial mesoderm equal in length to two somites were grafted from quail into chick embryos. The results are unequivocal; all voluntary muscles in the head of the bird are derived from paraxial mesoderm, either from somites or somitomeres depending upon the location. Figure 9 summarizes these data, representative examples of which are presented in Fig. 10. A complete account of these results is presented in Noden (1983b).

Although all voluntary muscles share comparable embryonic ancestry, they differ in one essential respect, that is the origin of their connective tissues. Any myogenic population that enters a neural crest-derived mesenchyme will derive its connective tissue elements from that ectodermal population. This includes all the muscles derived from somitomeres, and some, primarily those associated with the tongue, that originate in occipital somites. Other muscles, such as the intrinsic and caudal extrinsic laryngeals, derive their connective tissues from lateral plate mesoderm, while in the remainder both myogenic and connective tissue components originate as somitic mesoderm.

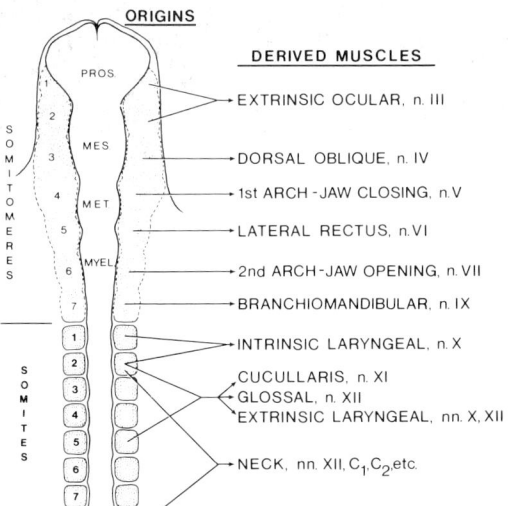

Fig. 9. Summary of the results of cephalic paraxial mesoderm transplants indicating the origins of myocytes in craniofacial voluntary muscles. Because the boundaries between somitomeres are not visible in living tissue, it is possible that somitomeres adjacent to those indicated may also contribute to particular muscle groups. PROS., prosoncephalon; MES., mesencephalon; MET., metencephalon; MYEL., myelencephalon.

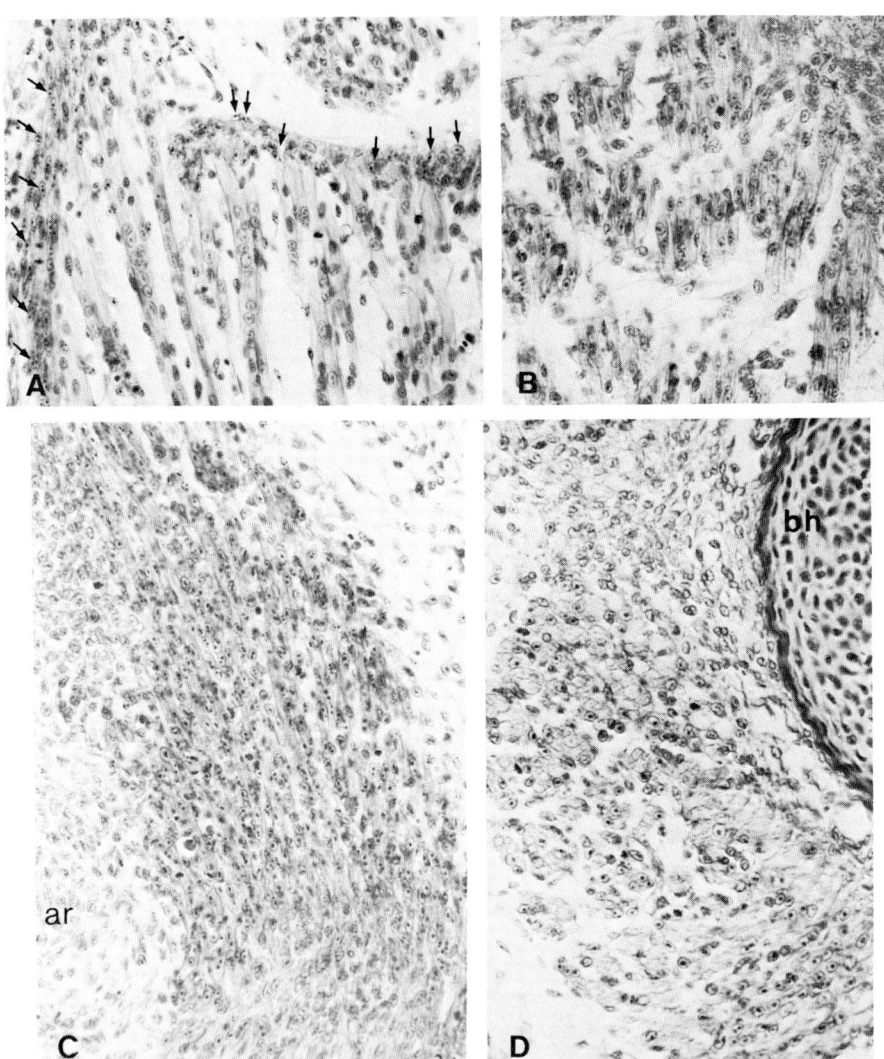

Fig. 10. Representative results of quail neural crest (A) or paraxial mesoderm (B)–(D) transplantation. (A) and (B) Mandibular depressor muscle showing the complementary pattern of labeling after neural crest or somitomere transplantation. Neural crest cells form fascia (small arrows) but not myocytes; mesoderm forms myocytes but not connective tissues. (C) Dilator glottidis and (D) rostral hyoglossal muscles following grafting of somites 1 and 2 and somites 3 and 4, respectively. ar, arytenoid cartilage; bh, basihyoid. (From Noden, 1983b.)

In the preceding analyses no labeled cells were found associated with the esophagus, trachea, or pharynx, nor were laryngeal connective tissues labeled. Lateral plate mesoderm is a logical condidate to form some, possibly all of these endoderm-associated tissues. Although a thorough investigation of the derivatives of cephalic lateral mesoderm has not been completed, some regions have been mapped. Figure 11 illustrates the results of transplanting lateral plate mesoderm, both somatic and splanchnic layers, at the levels of somites 3 and 4. The smooth muscle and connective tissues surrounding the esophagus are labeled.

Surprisingly, quail cells were found along the entire length of the esophagus and crop in these chimeric hosts. This indicates that a very short longitudinal strip of lateral mesoderm, equivalent to approximately two somites in length, elongates along the entire intrathoracic length of gut muscle. Presumably, the underlying endoderm expands as well, as suggested by the work of Le Douarin (1964), Rosenquist (1971), and Fukuda-Taira (1981).

Thus, the ability to transplant marked cells between species has resolved the problem of the embryonic origins of voluntary and some involuntary muscles of the head and neck. Precise definition of these origins, which has eluded anatomists and embryologists alike for over a century, is a necessary prerequisite both to establishing true homologies among muscles found in representatives of different vertebrate classes and to analyzing how muscles ontogenetically develop specific attachments and innervations (see Section III, C).

C. Otic and Periotic Structures

The vertebrate ear presents an unusually complex set of developmental problems. This is because many disparate structures, representing tissues of several different embryonic origins, participate directly or indirectly in the development of the inner and middle ear regions.

Both the membranous labyrinth and most neurons in ganglion VIII arise from a small area of surface ectoderm originally located immediately beside the midmyelencephalon (Fig. 12). The location and boundaries of the anlage of this otic placode have been defined using the quail-chick transplantation methods described previously.

As shown in Fig. 13, the membranous labyrinth in these hosts was often chimeric, indicating that the entire primordium was not included in the transplant. By comparing the presence and position of labeled neurons in the vestibular and acoustic ganglia with the patterns of labeling in the membranous labyrinth, it has been possible to define the site of origin of these sensory neurons. They are derived from the same otic epithelium as will later form the medial wall of the utriculus. Thus, neurons arise from a different part of the membranous labyrinth from that which forms most of their peripheral targets, including the cristae, maculae, cochlear duct, and lagena. The neurogenic foci

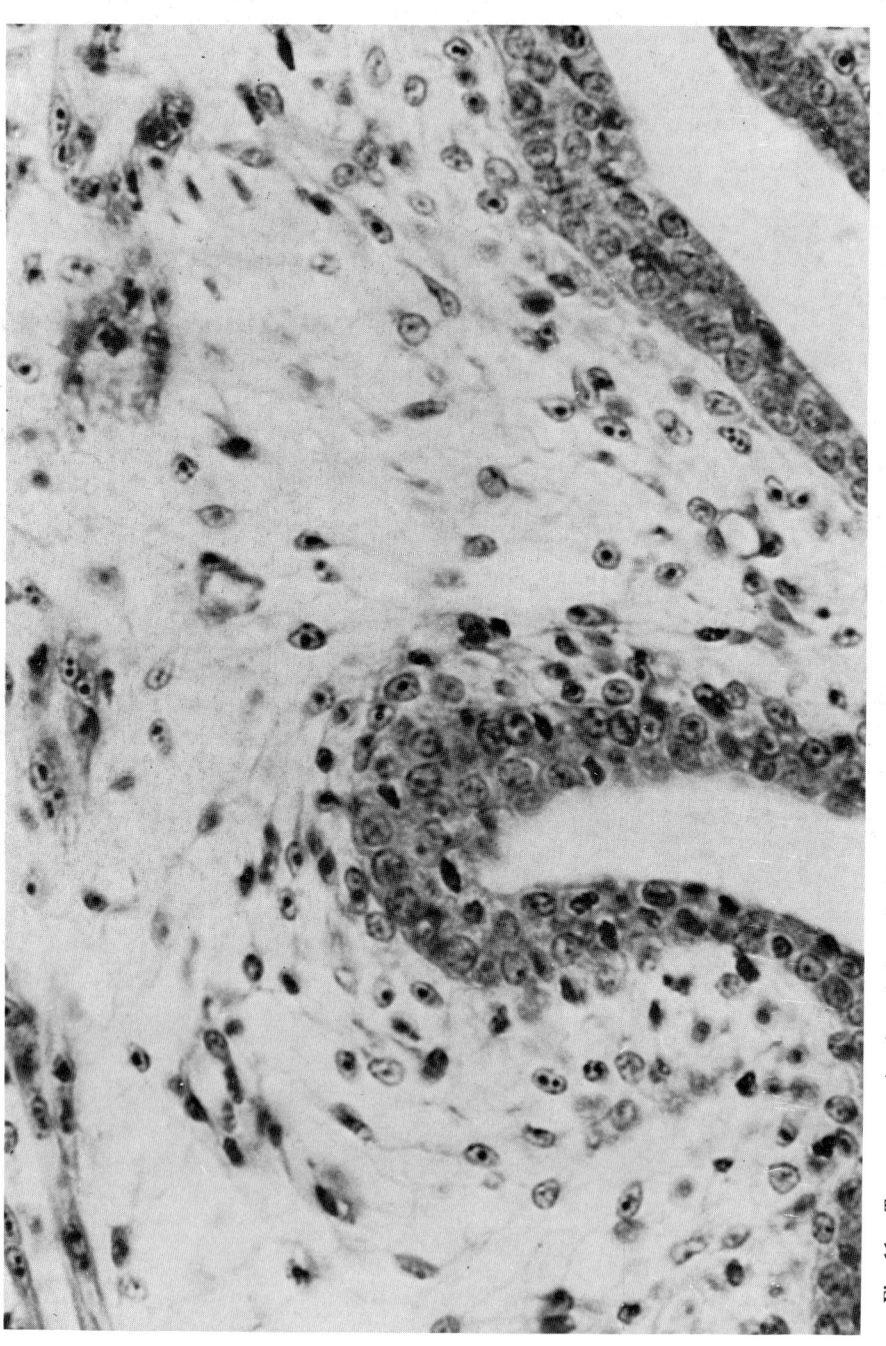

Fig. 11. Transverse section through the wall of the esophagus from a host chick that received a transplant of quail lateral plate mesoderm at the level of somites 2 and 3. Both smooth muscle and connective tissues are labeled.

X. The Use of Chimeras in Analyses of Craniofacial Development

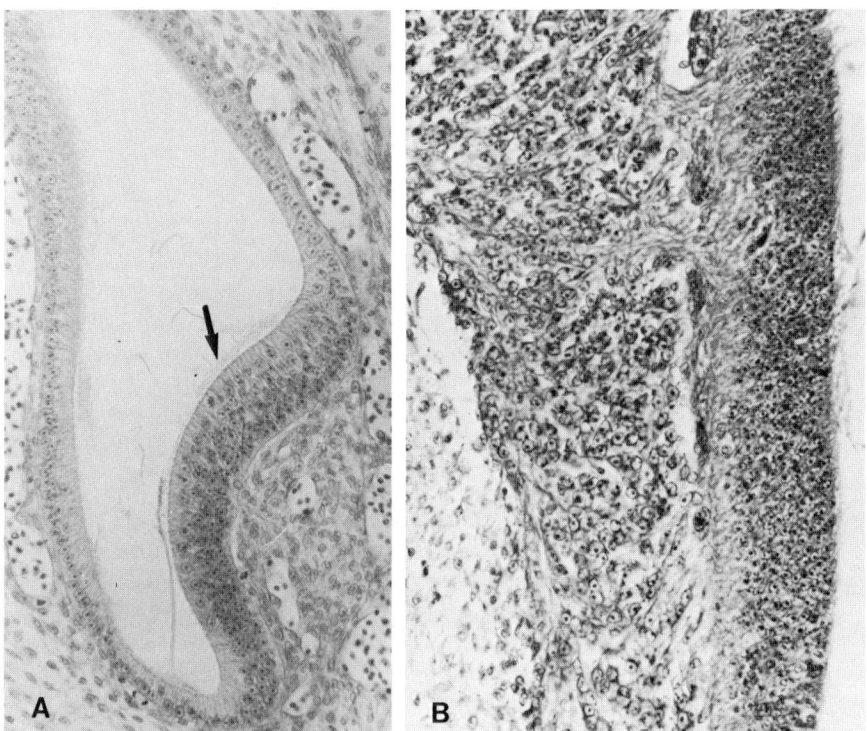

Fig. 12. Transplantation of the primordium of the otic placode results in labeling of the membranous labyrinth (A) but not associated connective tissues or Schwann cells beneath a christa (arrow), and the cochlear duct (B) and associated acoustic ganglion cells.

producing vestibular and auditory neurons are close together and are probably contiguous or slightly overlapping.

The membranous labyrinth is surrounded by mesenchymal cells that form the cartilaginous otic capsule and other periotic connective tissues. Le Lièvre (1974) was the first to note that neural crest cells made a contribution to this capsule. This occurs at three particular locations. Crest-derived chondrogenic cells are found in a column extending laterally from the caudal ampulla to the surface of the capsule (Fig. 14), and immediately dorsal to the medial part of the middle ear cavity at the level of attachment of the otic process of the quadrate, as illustrated schematically in Fig. 15. Neural crest cells also form the opisthotic process.

Developmentally, the first of these three sites of neural crest contribution is quite different from the others. As dorsal and ventral condensations of pre-chondrogenic paraxial mesoderm expand laterally around the rapidly expanding membranous labyrinth, they surround a population of crest mesenchymal cells. The mesodermal condensations grow toward each other lateral to the caudal

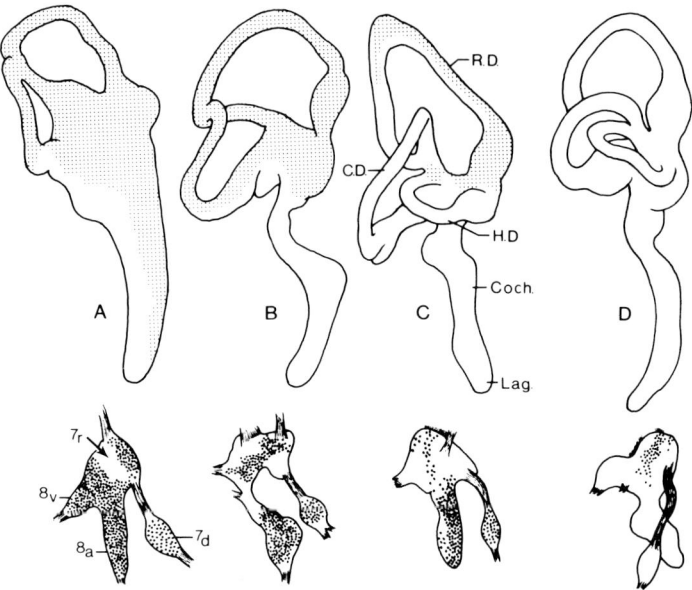

Fig. 13. (A)–(D) illustrate the different patterns of labeling in reconstructed membranous labyrinths following grafting of small pieces of quail dorsolateral surface ectoderm (A)–(C) or the neural crest (D) at the midmyelencephalic level. Stippling indicates quail epithelium. Below are reconstructions of ganglia associated with the seventh and eighth cranial nerves in each of these four hosts. Dots indicate quail neurons. These reconstructions are used to map the origins both of each part of the membranous labyrinth and of vestibular and acoustic neurons. R. D., H. D., and C. D. indicate rostral, horizontal, and caudal semicircular ducts, respectively; Coch. and Lag. are the cochlear duct and lagena, respectively; 7_r and 7_d are the proximal (root) and distal (geniculate) ganglia, respectively; 8_a and 8_v are the acoustic and vestibular ganglia, respectively.

ampulla, but do not fuse together. Instead, crest mesenchymal cells form a seam between them (Fig. 14), and subsequently undergo chondrogenesis to bridge the gap.

In the other two areas of crest contribution to the otic capsule, the mesodermal and crest-derived mesenchymal populations are contiguous prior to the onset of chondrogenesis, and both differentiate simultaneously. These two sites of neural crest contribution may perhaps represent vestigial dorsal hyobranchial skeletal tissues.

Most of the connective tissues associated with the middle ear are also of neural crest origin. These cells encompass the auditory tube (Eustachian tube; Fig. 3B) and most of the middle ear cavity. They surround the tympanic membrane and contribute to the sparse mesenchymal population located between its outer, ectodermal and inner, endodermal layers (Fig. 16A), and form the shaft and distal

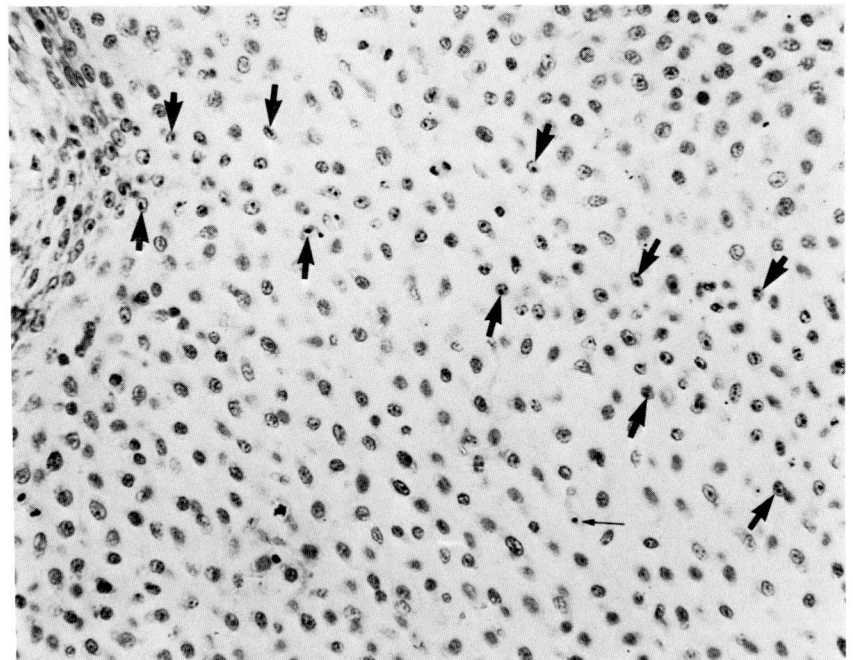

Fig. 14. A seam of quail neural crest-derived chondrocytes passing from the area of an ampulla (upper left) laterally toward the squamosal bone. With one exception (small arrow), all chondrocytes on either side of the seam in the otic capsule are of mesodermal origin.

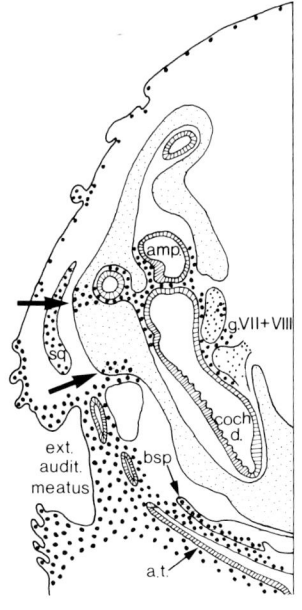

Fig. 15. Schematic indication of the contributions of neural crest mesenchymal cells to periotic structures. Arrows indicate two of the areas of crest contribution to the otic capsule. coch d., cochlear duct; ext. audit. meatus, external auditory meatus. (Revised from Noden, 1983a.)

parts of the columella. The proximal part of the columella, the footplate, is of mesodermal origin (Noden, 1982b; Maderson *et al.*, 1982).

Here again, neither these origins nor the precise relations of populations that often share similar fates but have greatly disparate embryonic histories could have been predicted from descriptive or comparative studies.

D. Periocular Tissues

Neural crest cells form shortly after the initial appearance of the optic evagination. Subsequently, they surround the optic vesicle except along its caudal (fu-

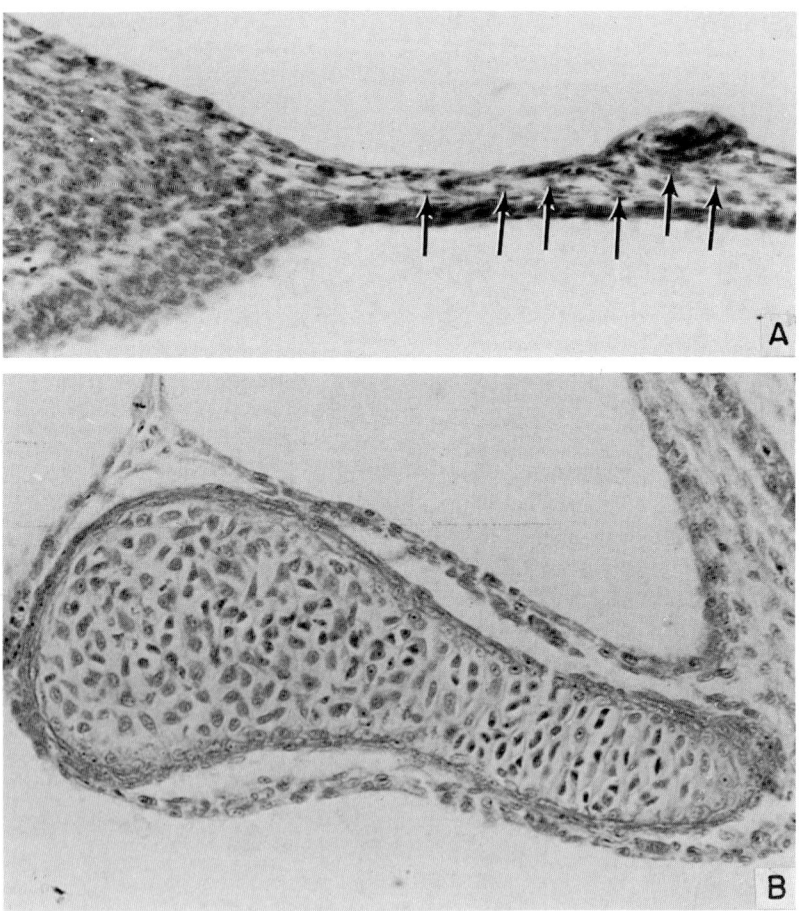

Fig. 16. Contributions of grafted quail neural crest cells (arrows) to the tympanic membrane (A) and columellar shaft (B). The presence of both labeled and unlabeled chondrocytes in the columella is explained in the text.

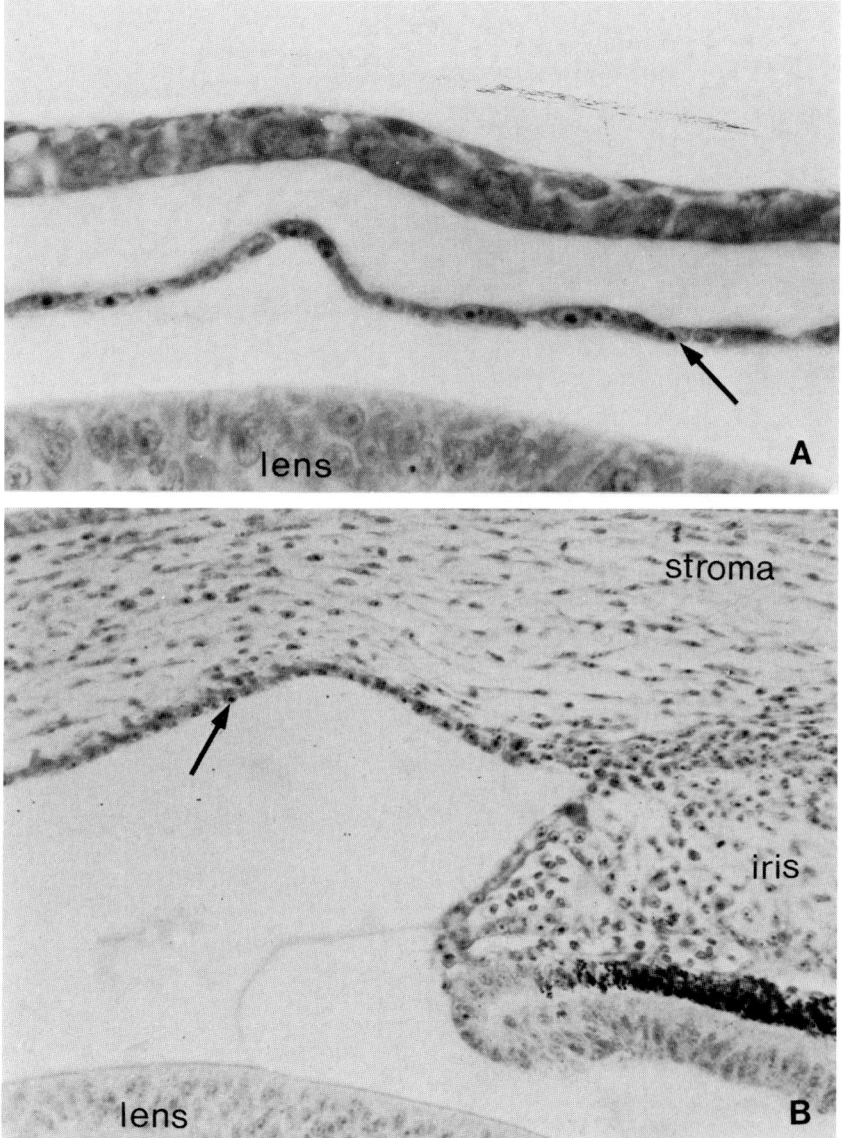

Fig. 17. Quail neural crest cells form the corneal endothelium (arrows) and, later, the keratocytes of the stroma and mesenchymal populations of the iris (B). (From Noden, 1978b.)

ture dorsotemporal) border, where it is contacted by paraxial mesoderm (Noden, 1978b, 1982a; Johnston *et al.*, 1979; Meier, 1980, 1982). Greatest attention has been focused on the development of anterior periocular structures, especially the cornea. In quail↔chick chimeras it is possible to show that the posterior epithelium (corneal endothelium) is derived from the neural crest, as are the mesenchymal cells that invade the acellular stroma and form keratocyes (Fig. 17).

Crest cells form all mesenchymal components of the iris and, surprisingly, are the source of involuntary ciliary muscles (Fig. 18). In birds this muscle group is striated, and it is the only such muscle formed by the neural crest (but see Le Lièvre and Le Douarin, 1975). The mesenchyme of the nictitating membrane is entirely of neural crest origin but, as illustrated in Fig. 29, the eyelids are chimeric in that myocytes of the palpebral muscles are derived from somitomeric mesoderm originally located beside the metencephalon (Noden, 1983b).

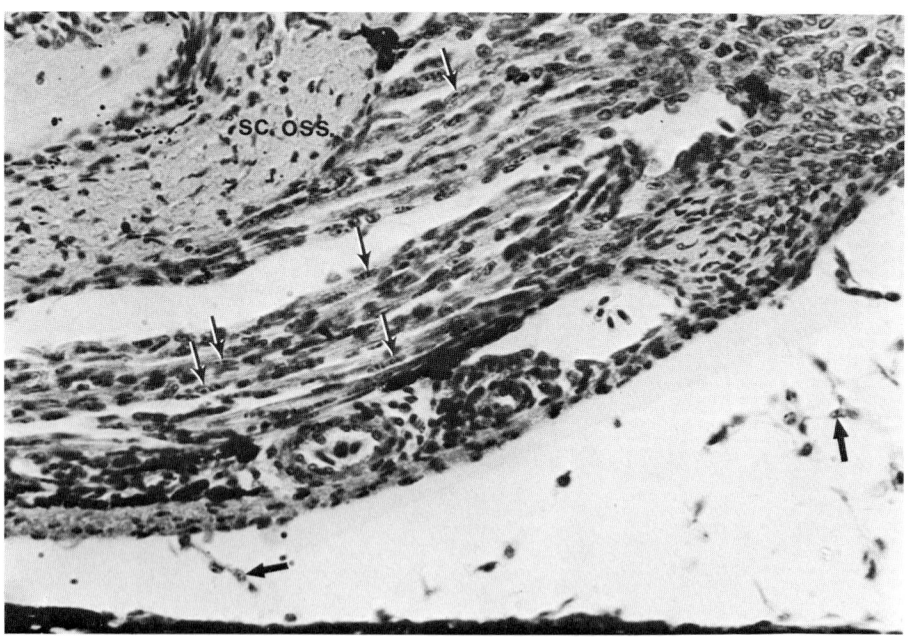

Fig. 18. The ciliary region from an 18-day host embryo that received a graft of quail mesencephalic neural crest cells. The pigmented retina is at the bottom, the lens is to the upper right. Labeled cells are visible in the ciliary muscle (black and white arrows) and fibers of the pectinate ligament (solid black arrows). The marker is not visible in osteocytes in the scleral ossicle (sc. oss.) due to shrinkage of the nuclei in this mature tissue. (From Noden, 1982a.)

III. EXPERIMENTAL ANALYSES OF CRANIOFACIAL DEVELOPMENT

A. Establishment of Definitive Tissue Relationships

The ability of neural crest cells to emigrate from their place of origin and, using precise pathways, to penetrate many more ventral and often deeper parts of the embryo has long been considered one of the most unusual and distinctive features of this population (reviewed by Weston, 1970, 1982; Noden, 1978a). Many of the biochemical components of the migratory routes have been defined (Pratt *et al.*, 1975; Pintar, 1978; Derby, 1978; Newgreen and Thiery, 1980; Thiery *et al.*, 1982) and their roles in affecting crest cell movements investigated (Noden, 1973, 1975; Weston, 1980, 1982). In most parts of the head neural crest cell populations enter a hyaluronate-rich, fibronectin-lined area bounded by the surface ectoderm superficially and paraxial mesoderm beneath (Bolender *et al.*, 1980; Duband and Thiery, 1982). Biochemical or surgical disruption of these components alters and sometimes eliminates the migratory pathways used by crest cells (Anderson and Meier, 1982), indicating that the integrity of the extracellular milieu is a necessary prerequisite for dispersal of the crest population.

In all these studies, the basic assumption has been that during the major dorsal-to-ventral migrations of crest populations the surrounding tissues are relatively static and not moving. However, the results of independently transplanting the overlying surface ectoderm and underlying paraxial mesoderm have challenged this assumption (Noden, 1984). As illustrated schematically in Fig. 19, both of these populations are undergoing dorsoventral displacements or expansions at the same time as crest cells are migrating. Thus the neural crest is not moving through a static environment. Rather, the entire complement of superficial cephalic tissues is expanding and shifting ventrally during a 12–20 hour period in avian development.

In view of the results of lateral splanchnic mesoderm grafting described previously, it is likely that a similar scenario may account in part for the distribution of enteric neuron precursors. Crest cells originating at the levels of somites 1–6 might become associated with gut endoderm and surrounding mesoderm located at the same axial levels. Subsequent elongation of the gut would result in a passive caudal displacement of these crest cells, although this cannot completely account for the final distribution of enteric neurons derived from the occipital region (Le Douarin and Teillet, 1974; Le Douarin, 1977, 1980).

Thus, none of these data negate the necessity for both cephalic and trunk crest cells to migrate actively relative to neighboring tissues. They do, however, suggest that active migration is but one component of a complex, integrated set of morphogenetic movements and tissue expansions occurring throughout the embryo during the late neurula stages of embryogenesis.

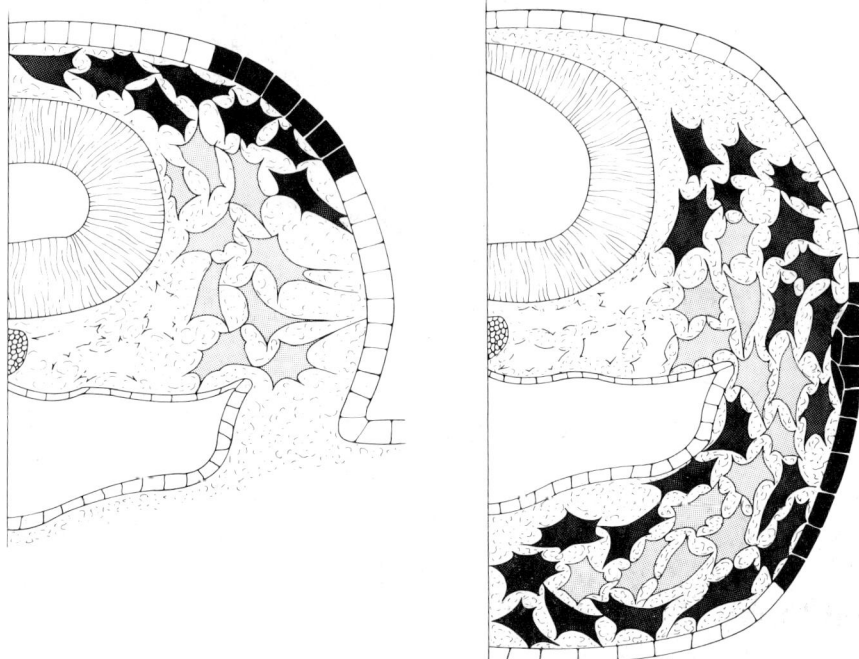

Fig. 19. This schematically illustrates the simultaneous morphogenetic movements and expansions of surface ectoderm (solid black), neural crest (dark gray) and paraxial mesoderm cells (light gray). These movements occur between stages 10 (10 somites) and 14 (20 somites) in avian embryos. (Redrawn from Noden, 1984.)

B. Differentiation and Patterning of Craniofacial Connective Tissues

Neural crest cells transplanted from one region of the head to another will form the types of tissues appropriate to their new location, even when this involves the formation of tissues not normally in their repertoire (but see Nakamura, 1982; Nakamura and Ayer-Le Lièvre, 1982). The same is generally true of trunk neural crest cells (Le Douarin and Teillet, 1974).

Many of the tissue interactions that promote skeletogenesis by cephalic crest cells have been defined, with the surface ectoderm (Bee and Thorogood, 1980; Tyler, 1978; Tyler and McCobb, 1980; reviewed by Hall, 1978, 1982), pigmented retinal epithelium (Newsome, 1972, 1976; Stewart and McCallion, 1975), pharyngeal endoderm (in amphibians, Corsin, 1975; Drews et al., 1972; Epperlein, 1974; Holtfreter, 1968; Okada, 1955), and the neural tube (Benoit and Schowing, 1970) all implicated in these processes.

X. The Use of Chimeras in Analyses of Craniofacial Development

The use of chimeras has been especially valuable in analyzing the interactions underlying odontogenesis (Osman and Ruch, 1981; Thesleff and Humerinta, 1981). Heterotypic combinations of avian oral mesenchyme (derived from the neural crest) or epithelium with murine oral epithelium or mesenchyme has clearly shown that epithelium taken from avian oral region will promote odontogenesis by mouse oral mesenchyme (Hata and Slavkin, 1978; Cummings et al., 1981). The reciprocal effects upon avian epithelium are more controversial. In an elegant series of intraocular heterotypic recombinations, Kollar and Fisher (1980) found that mouse oral mesenchyme could induce avian orofacial epithelium to form ameloblasts capable of synthesizing enamel matrix. However, these results cannot be duplicated in organ culture systems (Cummings et al., 1981).

While the interactions and mechanisms necessary to promote cyto differentiation and histogenesis of many craniofacial tissues have been well explored, less attention has been paid to the broader problem of patterning. Patterning is the process whereby each tissue forms in the correct location, develops its unique, characteristic shape, and establishes precise relations with other tissues. It is this process that distinguishes a developing embryo from a teratoma or an anidean monster.

With the large body of data described above all suggesting that extrinsic factors control the development of crest mesenchyme, most current investigators have viewed the neural crest population as a passive participant in the patterning process. In contrast, as illustrated in Fig. 20, xenoplastic (cross-species) grafts of amphibian cephalic crest primordia have clearly shown that species-specific aspects of patterning are inherent within the neural crest mesenchyme (Andres, 1946, 1949; Wagner, 1949, 1959; Baltzer, 1952; Henzen, 1957). However, these clues regarding inherent patterning capabilities have not been integrated into most contemporary analyses of neural crest development. Similarly, the formation of aberrant skeletal structures and ectopic teeth following the switching of amphibian jaw- and gill-producing neural crest primordia (Hörstadius and Sellman, 1946; Sellman, 1946) has been largely ignored.

To test directly the possible role of the neural crest in the patterning of skeletal structures, chick presumptive second or third branchial-arch crest cells were removed and replaced with quail presumptive first-arch populations (Noden, 1983a). In 28 out of 29 surviving hosts first-arch skeletal structures formed within the second (or third) branchial arch (Fig. 21). On each side of the head in these hosts were found supernumerary squamosal, pterygoid, and angular bones and quadrate and Meckel's cartilages, plus several duplicate distal jaw structures (Fig. 22). The possibilities that these results could be due to species incompatibility or differences in the initial size of the donor and host crest populations have been tested and negated.

The explanation proposed for the formation of this super-numerary first-arch

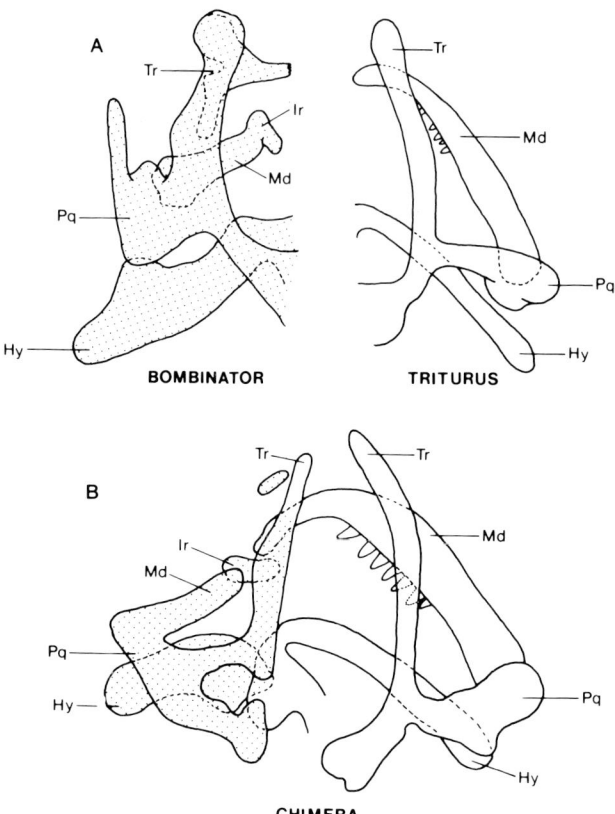

Fig. 20. Simplified reconstruction of Wagner's (1949, 1959) amphibian chimera experiments. (A) shows in dorsal view the rostral visceroskeleton of the left side in the frog *Bombinator* (stippled) and the right side in *Triturus*, a salamander. (B) is a chimera produced after unilateral neural crest xenoplastic transplantation. The tissue types and general morphology of the donor-type prevailed. Hy, hyoid; Ir, infrarostial; Md, mandible; Pq, palatoquadrate; Tr, rostial tratecula.

skeletal array is that the transplanted neural crest population contains inherent patterning capabilities. When this population enters a branchial arch and is surrounded by chondrogenic and osteogenic influences, the pattern of its response is determined by the axial level at which the crest cells initially formed. It is well known that the neural plate, from which crest cells originate, is polarized along the rostral–caudal axis of the embryo, and that by the neurula stage discrete regions within this epithelium are fixed (Corner, 1963, 1964). The neural crest populations that enter branchial arch regions carry this positional specification with them, and it determines the pattern of their subsequent development.

X. The Use of Chimeras in Analyses of Craniofacial Development

What is most striking about these data is that they parallel the results of studies using other mesenchymal systems. The similarities between the results of Hörstadius (1950), which parallel those described above, and data on patterning of the developing limb were first explored by Zwilling (1972). For example, it is well documented that when the mesenchyme of a wing is replaced with that of a leg or that from a different species, the pattern of development of the forelimb is determined by the donor mesenchyme (Harrison, 1921; Saunders et al., 1959; Zwilling, 1955, 1961). This demonstrates that regional as well as species-specific limb morphogenetic specification occurs within lateral somatic mesoderm (Chevallier and Kieny, 1982; Wachtler et al., 1982). Through the study of

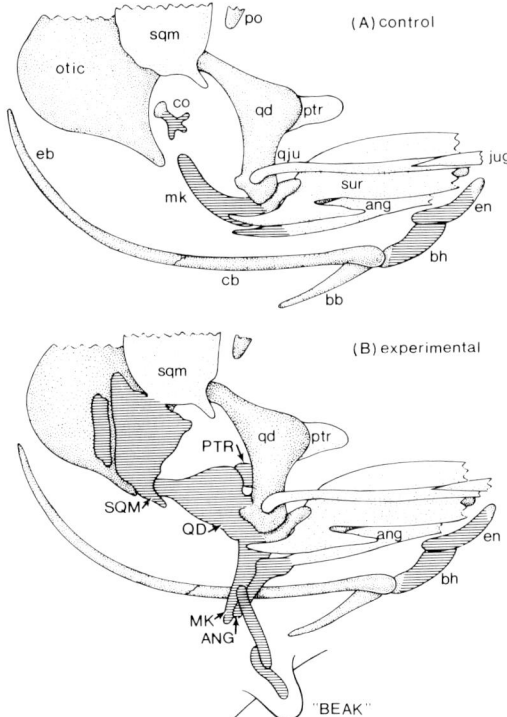

Fig. 21. Reconstructions of chimeric, quail↔-chick host embryos showing (dark horizontal lines) skeletal tissues derived from grafted neural crest cells. In (A) the rostral myelencephalic crest population was grafted orthotopically. In (B) this population was excised and replaced with met- and mesencephalic crest cells. In the latter series the transplanted crest population formed a nearly complete first arch (jaw) skeleton in the region of the second (hyoid) arch. Lower case labels indicate normal tissues, supernumerary skeletal structures are indicated with upper case labels. (Redrawn from Noden, 1983a.)

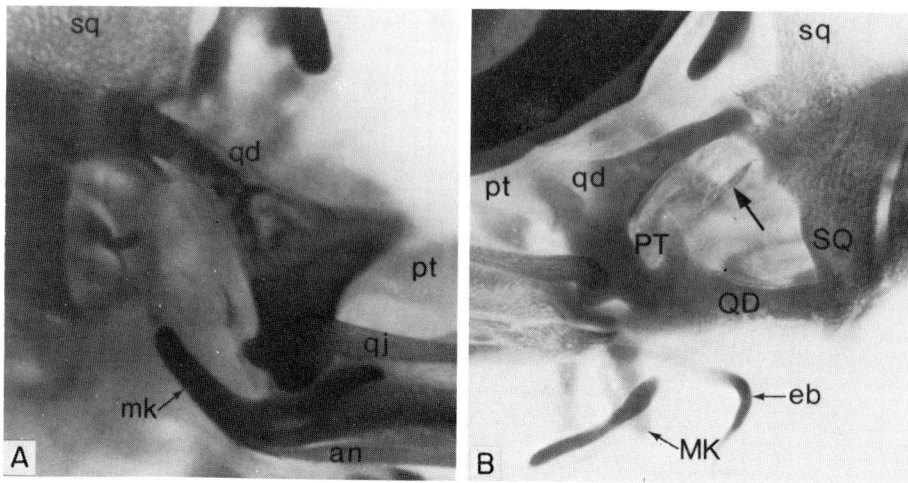

Fig. 22. Two 13.5-day embryos stained *in toto* for cartilage and bone and then cleared. (A) is a normal chick embryo viewed from the right. (B) is a host chick viewed from the left that received heterotopic neural crest grafts as in Fig. 21B, but from a *chick* donor. This was done to ensure that the duplication of skeletal structures was not the result of a species incompatibility. Capital letters indicate ectopic tissues. Unlabeled arrow indicates a skeletal fragment that could not be identified. (From Noden, 1983a.)

experimental chimeras, we now know that regional specification also occurs in the cephalic neural crest.

This patterning is not limited to crest-derived structures. Most of the host embryos that received heterotopic neural crest grafts developed bilateral projections from the ventrolateral surface of the neck (Fig. 23). Histological examination showed that these projections have an hypertrophy of the distal peridermal layer that is unique to the normal beak region of birds. The beak (Kato and Hayashi, 1963; Tonegawa, 1973; Zwilling, 1961) normally develops in response to inductive influences emanating from underlying (crest-derived) mesenchyme. The same is true for teeth (Henzen, 1957; Kollar and Baird, 1970; Platt, 1891; Wagner, 1949, 1959) and other specializations of cephalic integument (Lawrence, 1963; McLoughlin, 1961, 1963). Furthermore, many hosts had an ectopic external auditory meatus located caudal to the normal external ear. The formation of both of these ectopic specializations of the surface ectoderm of the first branchial arch was caused by the underlying transplanted neural crest-derived mesenchyme.

Thus, crest-dependent, as well as crest-derived, structures are patterned according to the original axial position of the neural crest. This population is behaving in a manner identical to mesenchymal populations found in other parts

X. The Use of Chimeras in Analyses of Craniofacial Development

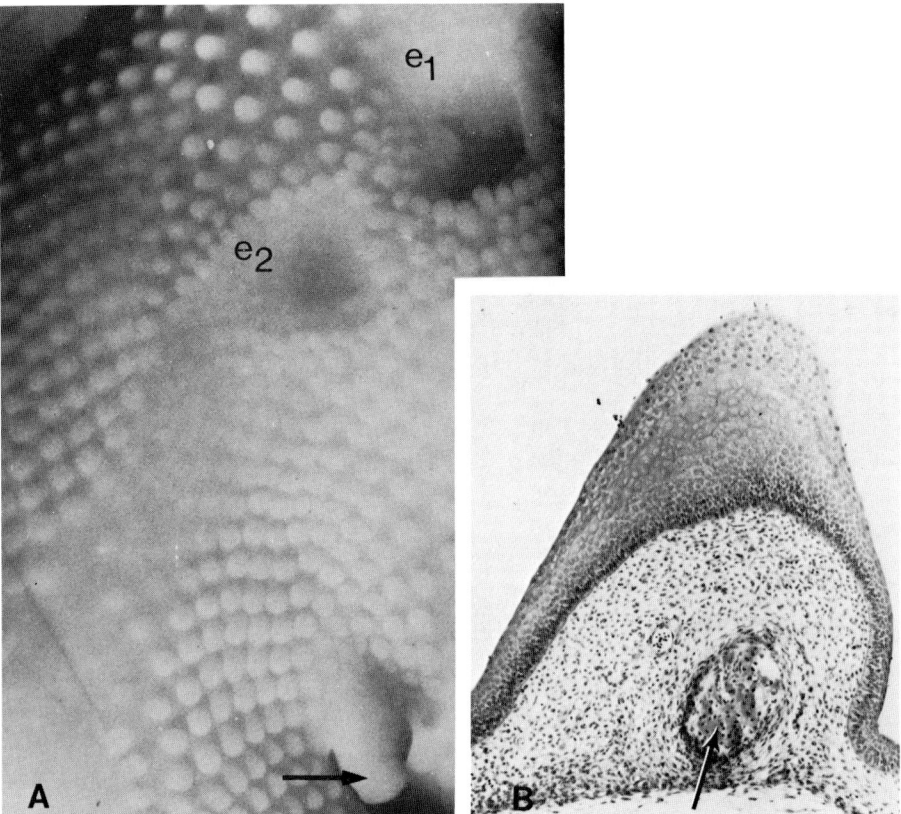

Fig. 23. (A) External view of an embryo from the same series of heterotopic neural crest grafts as in Fig. 21 and 22. There is a supernumerary external auditory meatus (e_2) located immediately caudal to the normal one (e_1), and an ectopic beak-like projection from the ventrolateral surface of the neck. (B) shows the internal structure of this ectopic beak, including the hypertrophied, cornified surface ectoderm, which is characteristic of the avian beak, and a small osseous nodule (arrow) derived from grafted crest mesenchyme. (From Noden, 1983a.)

of the body, including limb mesenchyme (lateral plate mesoderm) and sclerotomal mesenchyme (paraxial mesoderm).

C. Patterning of Voluntary Muscles

The results of experimental analyses of limb development have shown that the pattern of appendicular muscle formation is changed in harmony with alterations in the pattern of connective tissue development (Chevallier *et al.*, 1977; Javois and Iten, 1982). However, despite the widespread use of craniofacial muscle arrangements in determining taxonomic relations among vertebrate species, the embryonic basis for patterning of these muscles has never been investigated.

The branchial arch muscles in host embryos that developed with supernumerary jaw skeletal structures have been examined. Voluntary muscles form attachments and establish orientations appropriate to the skeletal structures present in the arch, even if these structures are different from those normally found there. Thus, associated with the supernumerary jaw skeletons are well-developed mandibular depressor, mandibular epibranchial, and, in some cases, intermandibular muscles. These results suggest that myoblasts entering a branchial arch will form muscles in response to a pattern dictated by the crest-derived mesenchyme.

To test this more directly, midbrain-level somitomeres were replaced with either presumptive or definitive thoracic somites (Noden, 1981, 1982b). As illustrated in Fig. 24, some of the myotomal cells from these transplants broke away from the body of the graft and invaded crest-derived mesenchyme, which is a type of mesenchyme that they normally would never encounter. Subsequently, these thoracic myotomal cells formed normal extrinsic ocular or mandibular adductor muscles, which were innervated by the correct nerves.

These results clearly indicate that the location and identity of voluntary muscles in the head are not inherent properties of the mesoderm from which they arise. Rather, such characteristics are acquired as a result of interactions with connective tissue-forming mesenchyme and the extracellular matrix (Chiquet *et*

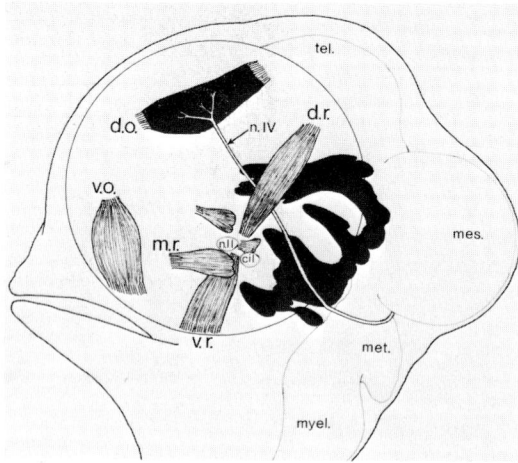

Fig. 24. Thoracic myotomal cells grafted into the head region differentiate into striated muscle masses (solid black). Some of these are associated with neural crest-derived connective tissues and form morphologically normal, innervated muscles, in this case the dorsal oblique (d.o.). Others remain isolated from head mesenchyme at the site of implantation. (Redrawn from Noden, 1982b.)

X. The Use of Chimeras in Analyses of Craniofacial Development

al., 1981; Turner *et al.*, 1983) into which myogenic primordia migrate. While a systematic analysis of efferent nerve projections has not yet been performed in these experimental animals, preliminary examination indicates that these peripheral nerves, like their target tissues, follow patterning cues located within the neural crest mesenchyme.

IV. EXTRAPOLATION TO MAMMALS

Millions of years have elapsed since the ancestors of birds and mammals separated into two divergent groups. During these millenia there have appeared many anatomical features of the head unique to each class, in addition to numerous family and species-specific characteristics. Given this diversity, it is essential to question carefully the extent to which avian data may be used in explaining mammalian cephalogenesis.

Cephalic paraxial mesoderm in mammals is initially identical to that described in birds (Meier and Tam, 1982). As illustrated in Fig. 25, cephalic neural crest cells appear prior to neural fold closure in humans (Bartelmez and Evans, 1926), laboratory animals (Bartelmez, 1962; Verwoerd and van Oostrom, 1979; Nic-

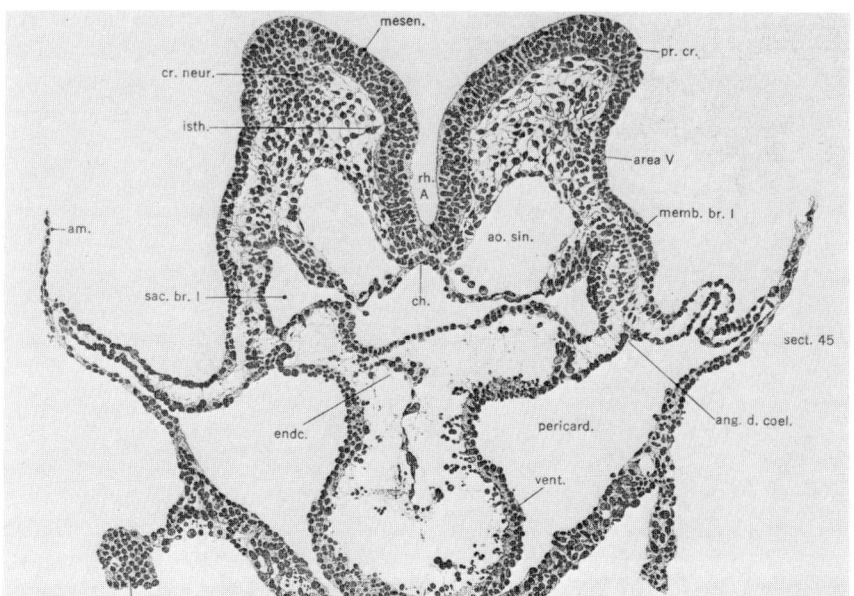

Fig. 25. Drawing of the mesencephalic neural folds and neural crest (cr. neur.) from an 8-somite human embryo. (From Bartelmez and Evans, 1926.)

hols, 1981) and domestic mammals (Halley, 1955; Noden and deLahunta, 1984), and after closure in birds. However, based on somite number and stage of cardiogenesis, these mammalian embryos are at a comparable stage of development when crest cells form. The two groups differ in the relative position at which the first cephalic crest cells form. As described by Nichols (1981), these crest cells leave the neural fold epithelium at the midlateral part of the head, not the dorsal midline as in chicks.

The critical issue is to determine precisely where in the mammalian head the mesoderm–neural crest interface subsequently forms. It seems likely that the rostral ossification centers in the mammalian frontal bone are of neural crest origin, as shown in birds. Indirect evidence for this comes from the study of aspirin-induced exencephaly in the rat. In these animals, the frontal bones are widely separated dorsally. However, rimming the rostral border of the schisis is a single, dorsomedian bone, clearly separate from nasal or maxillary elements. This appears to be a bone whose mesenchymal primordia were located near the dorsal midline and did not, as is the case of paraxial mesoderm, have to grow secondarily toward the midline.

The pre- and basisphenoid relations are less clear. This is due to presence in most birds and reptiles of a greater number of embryonic sphenoid ossification centers than is found in mammals and an inability to establish homologies between them (Jollie, 1971; Jarvik, 1981; Moore, 1981).

In the ear region, the use of avian chimeras has confirmed that the neural crest forms most, but not all, of the columella as well as the retroarticular and quadrate elements that give rise to the malleus and incus in mammals. What has been most surprising is that while the retroarticular process is, by position and associations, clearly a first arch skeletal structure, it is not derived from the same crest population as forms other jaw skeletal tissues. Rather, the retroarticular is derived from myelencephalic neural crest cells, most of which form second arch tissues (Le Lièvre, 1978).

Finally, it is unclear whether the caudal position of the mesoderm–neural crest interface is comparable in birds and mammals. Most birds have two laryngeal cartilages, a cricoid and arytenoid, and lack any known homologue to the mammalian thyroid cartilage. The interface lies rostral to the larynx; thus, neural crest cells make no contribution to these skeletal structures in birds (Le Lièvre and Le Douarin, 1975; Noden, 1983a). Presumably they arise from lateral plate mesoderm, but direct evidence for this is as yet unavailable. Based on supposed homology with branchial arch elements, it is usually stated that mammalian laryngeal skeletal structures are of neural crest derivation. While the avian data do not prove that the interpretation of descriptive accounts of mammalian laryngeal development are incorrect, they do provide support for an alternate hypothesis.

V. METHODOLOGICAL CONSIDERATIONS

During the course of the experiments described above, I have examined serial sections from over 300 quail↔chick chimeric embryos, which represents a postoperative survival rate of approximately 60%. Many situations have been encountered in which the data from one animal or series were equivocal, artifactual, or incomplete. The purpose of this concluding discussion is to describe the appearance of these compromising results and provide guidance for others in recognizing and, whenever possible, avoiding them in future studies using avian chimeras.

A. Transplantation Artifacts

Some of the problems encountered are related to the microsurgical procedures used, regardless of whether or not the tissues are from different species. Initial survival of embryos operated upon before the development of extraembryonic vasculature is most critically affected by two factors: evaporation and toxic chemicals. Fluid loss across the vitelline membrane causes the area opaca to become dehydrated. This prevents yolk sac blood islands from coalescing to form blood vessels. As the intraembryonic circulatory system develops, the embryo becomes edematous and remains colorless due to absence of hemoglobin-containing hematocytoblasts. This lethal situation is prevented by moistening embryos with saline and albumin before and again 1 hour after surgery (Silver, 1960), and then keeping the eggs in a saturated humidity environment for at least 1 day after surgery. Another frequent sign of insufficient moisture is the presence of spina bifida in the thoracic or lumbar region of the embryos.

The most frequent cause of immediate death of embryos, recognized by a loss of tissue integrity within 1 hour of surgery, is toxicity caused by vital staining. Neutral red is a photoreactive chemical and becomes toxic when exposed to high intensity illumination. Using low light intensity eliminates this problem. Some investigators prefer to use weak (0.05%) nile blue sulfate or to inject small amounts of a carbon–gelatin mixture into the subgerminal cavity to provide a higher contrast background.

The most difficult problems in surgical transplants on young embryos, and the ones that most frequently compromise experiments of this type, relate to the proper identification and clean isolation of the tissues to be transplanted. Proper identification requires a thorough appreciation of when and where the desired tissue appears, and of the grossly visible landmarks demarcating it from possible contaminating tissues. Given the proximity and lack of morphological distinctions among many embryonic populations, the investigator must recognize that not all populations are equally obtainable.

Isolation of primordia is best accomplished through careful dissection. However, there is often a risk of contamination by adjacent, adherent cells that may, a few days after transplantation, have formed as many progeny as the tissue under investigation. Treatment with collagenase or a chelating agent destroys or separates the basal lamina from epithelial tissues, allowing them to be readily separated from mesenchymal populations. However, it is now known that spatial information is stabilized and/or expressed in basement membrane components (Bernfield, 1981; Bernfield and Banerjee, 1982; Burrage and Lentz, 1982; Sanes, 1983), which argues against the use of these agents in some situations.

Trypsin is the most commonly used method of separating embryonic tissues prior to transplantation (see, for example, Le Douarin and Teillet, 1974; Noden, 1978c). However, by disrupting the integrity of the cell membrane there is potentially a risk of altering the properties being investigated. This has been clearly demonstrated, for example, in analyses of otocyst development *in vitro* (Orr, 1975; Friedman and Hodges, 1975; Saver and Van De Water, 1983).

A third essential aspect of the design of any transplantation experiment is to match correctly the size and developmental stage of the cell population to be grafted with that being replaced. Figure 26 illustrates how mismatching donor and host neural crest populations can alter the patterns of dispersion of the grafted cells. The same phenomena have been described for melanoblast dis-

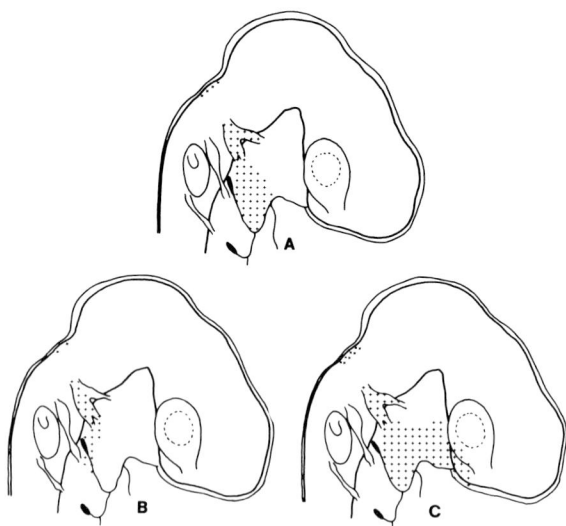

Fig. 26. (A) shows the normal pattern of distribution of neural crest cells (stipple) originating from the avian metencephalon. (B) illustrates the altered distribution if fewer than normal crest cells are grafted, or if the graft is implanted after host crest cell migration has begun. (C) is the converse, with a greater than normal number of crest cells or those ready to emigrate grafted into a younger host. (From Noden, 1980.)

X. The Use of Chimeras in Analyses of Craniofacial Development   273

tribution (MacMillan, 1976). This is an especially important and difficult parameter to account for in performing heterotopic, heterochronic, and/or xenoplastic transplants, and the results of some experimental protocols, in particular those in which a small population is to be substituted for a large one, are extremely difficult to interpret.

B. Problems Encountered in Analyzing Chimeras

As described in earlier chapters, the quail marker is remarkably free of artifacts. Of the cell types found in the head, only sensory neurons, in which the large, single heterochromatin condensation breaks apart into several smaller ones as the cells mature, and dense connective tissues, especially bone, in which the nuclei are highly compacted, present difficulties in identifying donor and host populations.

Quail and chick embryos are morphologically similar for the first 4 days of incubation. After this the development of most quail tissues becomes increasingly precocious in comparison with the chick. Figure 27 shows a section of the otic capsule from a chimeric embryo. Here, as in all other connective tissues examined, the differentiating quail cells are further developed than their chick counterparts.

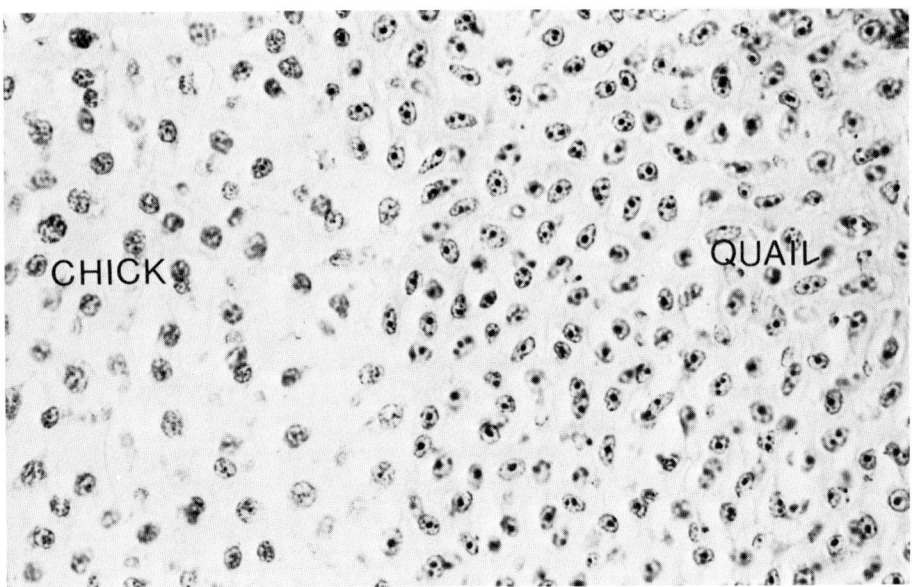

Fig. 27. Chondrocytes of the otic capsule in a 9-day chimeric host. This illustrates that quail skeletal tissues in chimeric host mature at their intrinsically programmed rate, i.e., in advance of chick tissues.

The potential benefits as well as problems associated with combining tissues from animals having disparate growth rates have been well documented by Harrison (1969). While quail tissues differentiate more rapidly, they generally form smaller skeletal structures than do chick tissues. The expression of these differences in growth rates in quail-chick chimeras are evident in Fig. 28 in which prospective upper or lower jaw neural crest cells were excised from the White Leghorn host embryos and replaced with homologous quail crest populations.

Despite these differences, most chimeric hosts develop normally. The only cephalic skeletal structure that is frequently abnormal in these chimeras is the columella. The crest-derived shaft and mesodermal footplate often fail to fuse together, although each component may be otherwise normal. In addition, the normally triradiate extracolumella often develops as one or several separate nodules located close to the retroarticular process of Meckel's cartilage. These problems are not unique to xenoplastic transplants, but are found more frequently in quail-chick hosts than in control, chick-chick embryos.

Undoubtedly the greatest difficulty in analyzing data from chimeric embryos is interpreting donor–host interfaces, and, in particular, identifying the exact source of host cells located adjacent to those of the donor. Many tissues are composites of cells having several disparate origins. For example, in the lower eyelid (Fig. 29) the connective tissues and Schwann cells arise from different regions of the neural crest, myocytes of the palpebral muscles are derived from the fourth and/or fifth somitomeres, and vascular endothelial tissues invade from primordial angiogenic cords in the lateral splanchnic mesoderm. Only when all

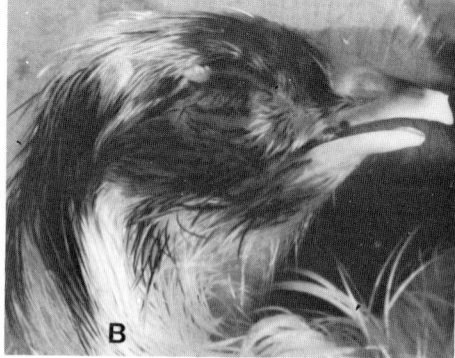

Fig. 28. Two host White Leghorn chick embryos that received quail neural crest grafts at the level of (A) the rostral midbrain and (B) the caudal midbrain and metencephalic. Regions of the skull derived from donor crest cells are evident by superior (A) and inferior (B) brachygnatha. The distribution of melanocytes derived from the graft in (B) is greater than that of the quail mesenchymal cells. (From Noden, 1980.)

X. The Use of Chimeras in Analyses of Craniofacial Development

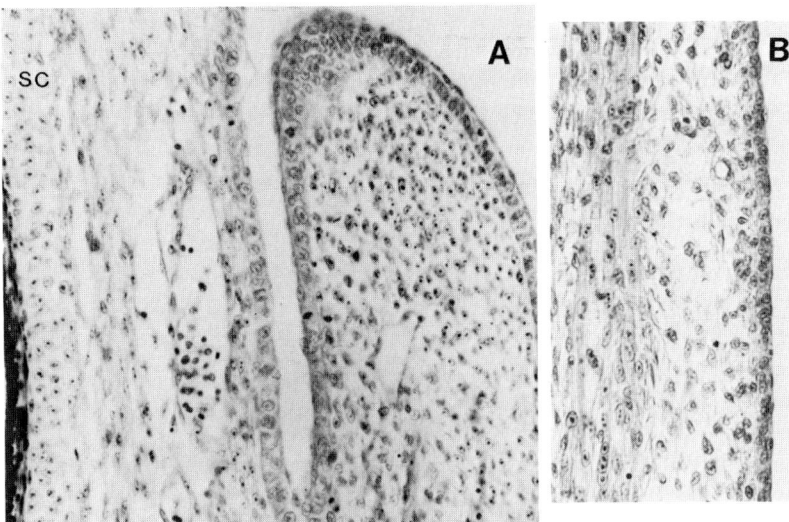

Fig. 29. Sections through anterior periocular tissues in host embryos that received quail (A) neural crest or (B) paraxial mesoderm (B) grafts. In (A) the mesenchyme and sclera (sc) are labeled; in (B) the ventral palpebral muscle is labeled. Vascular endothelial cells are not labeled in either series.

components have been labeled can the true meaning of a quail-chick interface or a chimeric tissue be understood.

The final note of caution concerns the thoroughness with which host embryos are examined. By way of example, in the past decade four different laboratories have transplanted neural crest cells at the level of the rostral somites. Each of these investigations has discovered a derivative of this crest population that escaped notice by the others. This is in large part due to an underestimation of the extent to which migrations and differential growth of neural crest and other primordia distribute them into widely disparate regions of the embryo. While not arguing against the scientific value of having several investigators independently working on similar problems, the need to examine with an unbiased eye a wide range of sections from chimeric host embryos is critical in order to prevent incomplete and erroneous interpretations. After succeeding in difficult surgery and nurturing the host embryo postoperatively, it is indeed a shame to ignore much of the data available.

VI. SUMMARY AND PERSPECTIVES

Many of the results described above dramatically alter the way in which development of the head and face are considered. The role of the neural crest in this process has been greatly expanded, from one of passive contribution to that of primary patterning influence, at least with respect to branchial arch skeletal and connective tissues. This property is inherent within the crest population before the onset of emigration, and differs depending upon the axial level at which the crest cells are formed. Furthermore, the spatial development of many tissues associated with crest-derived mesenchyme, including myogenic anlagen, efferent nerves, and surface ectoderm, is directed by the neural crest population. Whether the same is true for central and peripheral afferent nerve connections and angiogenic populations remains to be tested.

Finally, analysis of avian chimeras has revealed that all voluntary muscles of the head arise from paraxial mesoderm, although the precise distribution of somitomeric mesenchyme from which the more rostral facial and periocular muscles arise has not been defined. None of these conclusions would have been possible without the availability of a stable cell marker.

The complete spatial patterning of facial tissues is, of course, the result of interactions between neural crest-derived mesenchyme and many other tissues of the head. Rostrally, the growth of the prosencephalon, optic vesicles, and nasal invaginations alter the initial arrangement of the mesenchyme. Laterally, pharyngeal pouches expand and punctuate the cephalic crest population. Dorsally, crest mesenchymal cells contact mesodermal mesenchyme along a broad interface. Neither the extent of communication across this interface nor the basis for patterning of mesodermal skeletogenic populations has been experimentally analyzed.

While most of the "discoveries" facilitated through the use of avian chimeras had been suggested during the study of surgically created amphibian chimeras and other methods of experimental embryology, it is surprising how many of these suggestions had been overlooked. As a result, incorrect statements concerning craniofacial patterning and muscle origins are presented dogmatically in most textbooks. These recent data will necessitate not only that descriptive accounts be revised, but also that the etiology of many craniofacial dysmorphologies be reconsidered (see, for example, Couly and Ayer-Le Lièvre, 1983). Furthermore, they serve to remind developmental biologists that there are many secrets still locked within the embryo, and that problems associated with morphogenetic patterning and even defining embryonic cell lineages and tissue origins are unresolved and in need of vigorous scientific study.

ACKNOWLEDGMENTS

Original research reported in this chapter has been supported by The National Institutes of Health (NINCDS) Grant No. NS16408. The technical assistance provided by Dr. A. D'Amico-Martel, C. A. Wojcicki, D. Brummer, and A. McVey, and secretarial assistance of M. E. Lay are gratefully acknowledged.

X. The Use of Chimeras in Analyses of Craniofacial Development

REFERENCES

Adelmann, H. B. (1927). *J. Morphol. Physiol.* **44,** 29–87.
Anderson, C. B., and Meier, S. (1982). *J. Exp. Zool.* **221,** 329–335.
Andres, G. (1946). *Rev. Suisse Zool.* **53,** 502–510.
Andres, G. (1949). *Genetics* **24.** 387–534.
Balfour, F. M. (1878). "A Monograph on the Development of Elasmobranch Fishes." Macmillan, London.
Baltzer, V. F. (1952). *Experientia* **8,** 285–324.
Bartelmez, G. W. (1962). *Contrib. Embryol. Carnegie Inst.* **37,** 1–12.
Bartelmez, G. W., and Evans, H. M. (1926). *Contrib. Embryol. Carnegie Inst.* **85,** 1–67.
Bee, J., and Thorogood, P. (1980). *Dev. Biol.* **78,** 47–62.
Benoit, J., and Schowing, J. (1970). *In* "Tissue Interactions During Organogenesis" (E. Wolff, ed.), pp. 105–130. Gordon & Breach, New York.
Bernfield, M. R. (1981). *In* "Morphogenesis and Pattern Formation: Implications for Normal and Abnormal Development" (L. Brinkley, B. M. Carlson, and G. Connelly, eds.), pp. 139–162. Raven Press, New York.
Bernfield, M. R., and Banerjee, S. D. (1982). *Dev. Biol.* **90,** 291–305.
Bjerring, H. C. (1968). *Nobel Symp.* **4,** 345–357.
Bolender, D. L., Seliger, W. G., and Markwald, R. R. (1980). *Anat. Rec.* **196,** 401–412.
Born, G. (1894). *Jahresber. Schles. Ges. Vater. Kult.* pp. 1–11.
Born, G. (1897). *Arch. Entwicklungsmech. Org.* **4,** 349–365.
Burrage, T. G., and Lentz, T. L. (1982). *Dev. Neurosci.* **5,** 533–545.
Chevallier, A. (1979). *J. Embryol. Exp. Morphol.* **49,** 73–88.
Chevallier, A., and Kieny, M. (1982). *Arch. Entwicklungsmech. Org.* **191,** 277–280.
Chevallier, A., Kieny, M., and Mauger, A. (1977). *J. Embryol. Exp. Morphol.* **41,** 245–258.
Chevallier, A., Kieny, M., and Mauger, A. (1978). *J. Embryol. Exp. Morphol.* **43,** 263–278.
Chiquet, M., Eppenberger, H. M., and Turner, D. C. (1981). *Dev. Biol.* **88,** 220–235.
Christ, B., Jacob, H. J., and Jacob, M. (1977). *Anat. Embryol.* **150,** 171–186.
Christ, B., Jacob, H. J., and Jacob, M. (1979). *Experientia* **35,** 1376–1378.
Christ, B., Jacob, H. J., and Wachtler, F. (1983). *In* "Limb Development and Regeneration" (R. O. Kelley, P. F. Goetinck, and J. A. MacCabe, eds.), Part B, pp. 281–291. A. R. Liss, Inc., New York.
Corner, M. (1963). *J. Exp. Zool.* **153,** 301–312.
Corner, M. (1964). *J. Comp. Neurol.* **123,** 243–256.
Corsin, J. (1975). *J. Embryol. Exp. Morphol.* **33,** 335–342.
Couly, G., and Ayer-Le Lièvre, C. (1983). *Rev. Pediatr.* **19,** 5–21.
Cummings, E. G., Bringas, P., Jr., Grodin, M. S., and Slavkin, H. C. (1981). *Differentiation* **20,** 1–9.
Damas, H. (1944). *Arch. Biol.* **55,** 1–284.
D'Amico-Martel, A., and Noden, D. M. (1983). *Am. J. Anat.* **166,** 445–468.
deBeer, G. R. (1937). "The Development of the Vertebrate Skull." Oxford Univ. Press, London and New York.
Derby, M. (1978). *Dev. Biol.* **66:** 321–336.
Deuchar, E. M. (1958). *J. Embryol. Exp. Morphol.* **6,** 527–529.
Drews, U., Kocher-Becker, U., and Drews, U. (1972). *Arch. Entwicklungsmech. Org.* **171,** 17–37.
Duband, J. L., and Thiery, J.-P. (1982). *Dev. Biol.* **93,** 308–323.
Edgeworth, F. H. (1935). "The Cranial Muscles of Vertebrates." Cambridge Univ. Press, London and New York.
Epperlein, H. H. (1974). *Differentiation* **2,** 151–168.

Friedman, I., and Hodges, G. M. (1975). *Acta Oto-Laryngol.* **79,** 197–212.
Fukuda-Taira, S. (1981). *J. Embryol. Exp. Morphol.* **64,** 73–85.
Gans, C., and Northcutt, R. G. (1983). *Science* **220,** 268–274.
Gilbert, P. W. (1957). *Contrib. Embryol. Carnegie Inst.* **36,** 59–78.
Hall, B. K. (1978). *Arch. Oral Biol.* **23,** 1157–1161.
Hall, B. K. (1982). *J. Embryol. Exp. Morphol.* **68,** 127–136.
Halley, G. (1955). *J. Anat.* **89,** 133–152.
Hamburger, V., and Hamilton, H. L. (1951). *J. Morphol.* **88,** 49–92.
Hammond, W. S. (1965). *Anat. Rec.* **151,** 547–548.
Harrison, R. G. (1898). *Arch. Entwicklungsmech. Org.* **7,** 430–485.
Harrison, R. G. (1903). *Arch. Mikrosk. Anat.* **63,** 35–149.
Harrison, R. G. (1921). *J. Exp. Zool.* **32,** 1–136.
Harrison, R. G. (1969). "Organization and Development of the Embryo." (S. Wilens, ed.). Yale Univ Press, New Haven, Connecticut.
Hata, R.-I., and Slavkin, H. C. (1978). *Proc. Natl. Acad. Sci. U.S.A.* **75,** 2790–2794.
Hazelton, R. D. (1970). *J. Embryol. Exp. Morphol.* **24,** 455–466.
Henzen, W. (1957). *Arch. Entwicklungsmech. Org.* **149,** 387–442.
Holmgren, N. (1940). *Acta Zool. (Stockholm)* **21,** 51–267.
Holtfreter, J. (1968). In "Epithelial-Mesenchymal Interactions" (R. Fleischmaier and R. E. Billingham, eds.), pp. 1–30. Williams & Wilkins, Baltimore, Maryland.
Hörstadius, S. (1950). "The Neural Crest." Oxford Univ. Press, London and New York.
Hörstadius, S., and Sellman, S. (1946). *Nova Acta Regiae Soc. Sci. Up. [4]*. **13,** 1–170.
Jarvik, E. (1981). "Basic Structure and Evolution of Vertebrates," Vol. 2. Academic Press, New York.
Javois, L. C., and Iten, L. E. (1982). *Dev. Biol.* **90,** 127–143.
Johnston, M. C., Noden, D. M., Hazelton, R. D., Coulombre, J. L., and Coulombre, A. J. (1979). *Exp. Eye Res.* **29,** 27–45.
Jollie, M. (1971). *Acta Zool. (Stockholm)* **52,** 85–96.
Kato, Y., and Hayashi, Y. (1963). *Exp. Cell Res.* **31,** 599–602.
Kirby, M. L., Gale, T. F., and Stewart, D. E. (1983). *Science* **220,** 1059–1061.
Kollar, E. J., and Baird, G. (1970). *J. Embryol. Exp. Morphol.* **24,** 173–186.
Kollar, E. J., and Fisher, C. (1980). *Science* **207,** 993–995.
Lawrence, I. E., Jr. (1963). *J. Exp. Zool.* **152,** 205–217.
Le Douarin, N. M. (1964). *Bull. Biol. Fr. Belg.* **98,** 543–676.
Le Douarin, N. M. (1969). *Bull. Biol. Fr. Belg.* **103,** 435–442.
Le Douarin, N. M. (1973). *Dev. Biol.* **30,** 217–222.
Le Douarin, N. M. (1975). *Birth Defects, Orig. Artic. Ser.* **9,** 191–250.
Le Douarin, N. M. (1977). In "Cell Interactions in Differentiation" (M. Karkinen-Jääskelainen, L. Saxen, and L. Weiss, eds.), pp. 171–191. Academic Press, New York.
Le Douarin, N. M. (1980). *Curr. Top. Dev. Biol.* **116,** 32–86.
Le Douarin, N. M. (1982). "The Neural Crest." Cambridge Univ. Press, London and New York.
Le Douarin, N. M., and Teillet, M. A. (1974). *Dev. Biol.* **41,** 162–184.
Le Lièvre, C. S. (1974). *J. Embryol. Exp. Morphol.* **31,** 453–477.
Le Lièvre, C. S. (1978). *J. Embryol. Exp. Morphol.* **47,** 17–37.
Le Lièvre, C. S., and Le Douarin, N. (1975). *J. Embryol. Exp. Morphol.* **34,** 125–154.
McLoughlin, C. B. (1961). *J. Embryol. Exp. Morphol.* **9,** 385–409.
McLoughlin, C. B. (1963). *Symp. Soc. Exp. Biol.* **17,** 359–388.
MacMillan, G. J. (1976). *J. Embryol. Exp. Morphol.* **35,** 463–484.
Maderson, P. F. A., Noden, D. M., and Bank, S. (1982). *Anat. Rec.* **22,** 117A.
Meier, S. (1979). *Dev. Biol.* **73,** 25–45.
Meier, S. (1980). Original article series of the Society of Craniofacial Genetics.

Meier, S. (1981). *Dev. Biol.* **83**, 49–61.
Meier, S. (1982). *In* "Clinical, Structural and Biochemical Advances in Hereditary Eye Disorders" (D. L. Daentl, ed.), pp. 1–16. A. R. Liss, Inc., New York.
Meier, S., and Packard, P. (1984). In press.
Meier, S., and Tam, P. P. L. (1982). *Differentiation* **21**, 95–108.
Moore, W. J. (1981). "The Mammalian Skull." Cambridge Univ. Press, London and New York.
Nakamura, H. (1982). *Arch. Histol. Jpn.* **45**, 127–138.
Nakamura, H., and Ayer-Le Lièvre, C. (1982). *J. Embryol. Exp. Morphol.* **70**, 1–18.
Neal, H. V. (1918). *J. Morphol.* **30**, 433–453.
Nelson, G. J. (1969). *Bull. Am. Mus. Nat. Hist.* **141**, 475–552.
Newgreen, D., and Thiery, J.-P. (1980). *Cell Tissue Res.* **211**, 269–291.
Newsome, D. A. (1972). *Dev. Biol.* **27**, 575–579.
Newsome, D. A. (1976). *Dev. Biol.* **49**, 496–507.
Nichols, D. H. (1981). *J. Embryol. Exp. Morphol.* **64**, 105–120.
Noden, D. M. (1973). 9–35. *DHEW Publ. (NIH) (U.S.)* **NIH 73–546.**
Noden, D. M. (1975). *Dev. Biol.* **42**, 106–130.
Noden, D. M. (1978a) *In* "The Specificity of Embryological Interactions" (D. Garrod, ed.), pp. 4–49. Chapman & Hall, London.
Noden, D. M. (1978b). *Dev. Biol.* **67**, 296–312.
Noden, D. M. (1978c). *Dev. Biol.* **67**, 313–329.
Noden, D. M. (1980). *In* "Current Research Trends in Prenatal Craniofacial Development" (R. M. Pratt and R. L. Christiansen, eds.), pp. 3–25. Elsevier/North-Holland, New York.
Noden, D. M. (1981). *Am. Zool.* **21**, 19A.
Noden, D. M. (1982a). *In* "Biomedical Foundation of Ophthalmology" (F. A. Jacobiec, ed.), Sect. 3, pp. 1–23. Harper & Row, New York.
Noden, D. M. (1982b). *In* "Factors and Mechanisms Influencing Bone Growth" (A. D. Dixon and B. Sarnat, eds.), pp. 167–203. Alan R. Liss, Inc., New York.
Noden, D. M. (1983a). *Dev. Biol.* **96**, 144–165.
Noden, D. M. (1983b). *Am. J. Anat.* **168**, 257–276.
Noden, D. M. (1984). *Anat. Rec.* **208**, 1–13.
Noden, D. M., and deLahunta, A. (1984). "The Embryology of Domestic Animals: Developmental Mechanisms and Congenital Malformations." Williams & Wilkins, Baltimore, Maryland.
Northcutt, R. G., and Gans, C. (1983). *Q. Rev. Biol.* **58**, 1–28.
Okada, E. W. (1955). *Mem. Coll. Sci. Univ. Kyoto, Ser. B* **22**, 23–28.
Orr, M. F. (1975). *Dev. Biol.* **47**, 325–340.
Osman, M., and Ruch, J. V. (1981). *J. Dent. Res.* **60**, 1015–1027.
Owen, R. D. (1959). *J. Med. Educ.* **34**, 366–383.
Piatt, J. (1938). *J. Morphol.* **63**, 531–587.
Pintar, J. (1978). *Dev. Biol.* **67**, 444–464.
Platt, J. B. (1891). *J. Morphol.* **5**, 79–106.
Pratt, R. M., Larson, M. A., and Johnston, M. C. (1975). *Dev. Biol.* **44**, 298–305.
Raven, C. P. (1935). *Arch. Entwicklungsmech. Org.* **132**, 509–575.
Romer, A. S. (1972). *Evol. Biol.* **6**, 121–156.
Rosenquist, G. C. (1970). *Dev. Biol.* **22**, 461–475.
Rosenquist, G. C. (1971). *J. Embryol. Exp. Morphol.* **25**, 97–113.
Sanes, J. F. (1983). *Annu. Rev. Physiol.* **45**, 581–600.
Saunders, J. W., Jr., Gasseling, M. T., and Cairns, J. M. (1959). *Dev. Biol.* **1**, 281–301.
Saver, J. L., and Van De Water, T. R. (1984). *J. Exp. Zool.* **230**, 53–61.
Scammon, R. E. (1911). *Normen Taf. Entwicklungsgesch. Wirbelt.* **12**, 1–140.
Sellman, S. (1946). *Odontol. Tidskr.* **54**, 1–128.
Silver, P. H. S. (1960). *J. Embryol. Exp. Morphol.* **8**, 369–375.

Slonaker, J. R. (1921). *J. Morphol.* **35,** 263–357.
Spemann, J. (1901). *Arch. Entwicklungsmech. Org.* **12,** 224–264.
Spemann, J. (1918). *Arch. Entwicklungsmech. Org.* **43,** 448–555.
Spemann, J. (1921). *Arch. Entwicklungsmech. Org.* **48,** 533–570.
Spemann, J. (1938). "Embryonic Development and Induction." Yale Univ. Press, New Haven, Connecticut.
Spemann, J., and Mangold, H. (1924). *Arch. Mikrost. Anat. Entwicklungsmech.* **100,** 599–638.
Stewart, P. A., and McCallion, D. J. (1975). *Dev. Biol.* **46,** 383–389.
Thesleff, I., and Hurmerinta, K. (1981). *Differentiation* **18,** 75–88.
Thiery, J.-P., Duband, J. L., and Delouvée, A. (1982). *Dev. Biol.* **93,** 324–343.
Tonegawa, Y. (1973). *Dev., Growth Differ.* **15,** 57–71.
Turner, D. C., Lawton, J., Dollenmeier, P., Ehrismann, R., and Chiquet, M. (1983). *Dev. Biol.* **95,** 497–504.
Twitty, V. C. (1936). *J. Exp. Zool.* **74,** 239–302.
Tyler, M. S. (1978). *Anat. Rec.* **192,** 225–233.
Tyler, M. S., and McCobb, D. P. (1980). *J. Embryol. Exp. Morphol.* **56,** 269–281.
Verwoerd, C. D. A., and van Oostrom, C. G. (1979). *Adv. Anat., Embryol. Cell Biol.* **58,** 1–75.
Wachtler, F., Christ, B., and Jacob, H. J. (1982). *Anat. Embryol.* **164,** 369–378.
Wagner, G. (1949). *Rev. Suisse Zool.* **56,** 519–620.
Wagner, G. (1959). *Arch. Entwicklungsmech. Org.* **151,** 136–158.
Weston, J. A. (1970). *Adv. Morpho.* **8,** 41–114.
Weston, J. A. (1980). *In* "Current Research Trends in Prenatal Craniofacial Development" (R. M. Pratt and R. L. Chistiansen, eds.), pp. 27–46. Elsevier/North-Holland, New York.
Weston, J. A. (1982). *In* "Cell Behavior" (R. Bellairs, A. Curtis, and G. Dunn, eds.), pp. 429–470. Cambridge Univ. Press, London and New York.
Zwilling, E. (1955). *J. Exp. Zool.* **128,** 423–441.
Zwilling, E. (1961). *Adv. Morphog.* **1,** 301–338.
Zwilling, E. (1972). *Dev. Biol.* **28,** 12–17.

CHAPTER XI

Muscle and Skeleton of Limbs and Body Wall

MADELEINE GUMPEL-PINOT

Laboratoire de Neurochimie
INSERM
Hôpital de la Salpêtrière
Paris, France

I.	Introduction	281
II.	Birds	282
	A. Limbs	282
	B. Ribs, Sternum, Girdles, and Muscles of the Body Wall	300
III.	Mammals	302
	A. Morphological Markers	303
	B. Biochemical Markers: Muscle Formation	306
IV.	Concluding Remarks	307
	References	308

I. INTRODUCTION

Chimeric experiments involving the skeleton and muscles of the limbs and the body wall have been carried out mainly on two classes of vertebrates: birds and mammals. For technical reasons, primarily accessibility of the embryo to *in situ* experimental analysis and specificity of the markers used, the types of problem that can be elucidated by means of chimeric techniques are different in the two classes. Most of the chimeric experiments I shall describe here have been performed on bird embryos. The reasons are, first, that the bird embryo, particularly its axial and appendicular structures, is accessible to *in ovo* experimentation during a long period of incubation; second, that the quail nuclear marker is particularly useful, valid for practically all tissues, permanent (not modified by cell division), and easily revealed by a routine histological staining. In mammals, very early cleavage stages are open to *in vivo* manipulation in the sense that it is possible to produce chimeric embryos by the aggregation of blastocysts of genetically different strains of mice (see Chapter I). These chimeric embryos, surgically implanted in the uteri of foster mothers, develop to birth. Each newborn animal is thus the product of at least 2 pairs of parents from strains differing by a certain number of genetic factors that may be expressed phenotypically by morphological or biochemical characters, used as markers.

Chimeric grafts have also been done in amphibians. Limb regeneration, which consists of regrowth of a part that has been lost or removed, is very rare in vertebrates. However, it occurs in amphibians, especially in urodeles. The regeneration process is initiated by trauma, such as amputation, ligature, sectioning of a nerve, or grafting foreign tissue. The trauma is followed by healing, during which the epidermis covers the injured area. Then a phase of dedifferentiation proceeds, followed by regeneration.

One of the main problems in the regeneration process is the origin of the cells that form the new structures. During the dedifferentiation/redifferentiation process, the fate of the cells may be stable, a cartilage-derived cell producing only cartilage. In contrast, the redifferentiation can be a metaplasia, with change in the cell type.

To try to resolve this problem, fragments of tissue formed by triploid cells characterized by the presence of three nucleoli have been implanted in amputated diploid limbs of Urodele amphibians (Steen, 1968). These experiments were not as conclusive as later nonchimeric grafts, but the use of chromosome number as a permanent marker offers an interesting approach for tracing cell lineages.

II. BIRDS

A. Limbs

The experimental embryology of the bird limb has been dominated for years by considerations of mesoderm–ectoderm interactions (for review, see Zwilling, 1961).

1. MESODERM–ECTODERM INTERACTIONS: MESODERMAL CONTROL OF LIMB TYPE

Ectodermal caps of the limb can be separated from the mesodermal core using trypsin or other dissociating agents. The two components can be reassociated to form a complete limb, which is able to develop when grafted *in ovo*.

a. Wing-Leg Chimeras. When ectoderm and mesoderm of the chick wing and leg are cross-reassociated, the nature of the chimeric limb that develops depends on the mesoderm, i.e., a leg or a wing is obtained, according to the origin of the mesoderm (Hampé, 1959; Zwilling, 1955).

XI. Muscle and Skeleton of Limbs and Body Wall

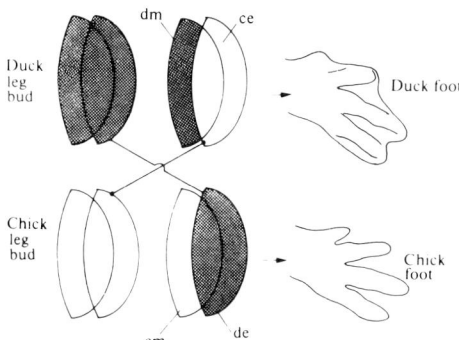

Fig. 1. Limb type is controlled by the mesoderm. Chimeric duck↔chick hindlimb buds develop according to the species that contributes the mesoderm. c, chick; d, duck; e, ectoderm; m, mesoderm. The duck foot is webbed, but not that of the chick. (After Pautou, 1968, from Hinchliffe and Johnson, 1980.)

b. Duck-Chick Chimeras. Heteroplastic exchanges have been carried out on different birds with specific morphological characters. For instance, the duck is web-footed, whereas the chick leg has separated digits. Chimeric legs develop according to the origin of the mesoderm; they are thus web-footed if the mesoderm is of duck origin (Hampé, 1959; Pautou, 1968, 1973) (Fig. 1).

c. Chick-Mammal Chimeras. Jorquera and Pujin (1971) constructed chimeric limbs from chick mesoderm and rat ectoderm grafted *in situ*. They obtained the development of a chick limb covered by hair germs. This experiment made two important points. First, even in xenoplastic exchanges, the mesoderm is the component responsible for the quality of the differentiation, i.e., the authors obtained chick legs or chick wings according to the origin of the mesoderm. Second, since limb development is the result of interactions between ectoderm and mesoderm (Zwilling, 1961), these xenoplastic experiments show that such interactions are possible between tissues belonging to different classes of vertebrates. In all interactions, the kind of differentiation (leg, wing, hair germs) depends on the origin of the reactive tissues, and the differentiation signal, even from tissues of a different species, is correctly interpreted.

2. ZONE OF POLARIZING ACTIVITY (ZPA)

In recent years, the mesoderm–ectoderm interaction hypothesis has been extended by the discovery of another element essential for the development of the limb: the zone of polarizing activity (ZPA). In grafts, the ZPA has been demonstrated to control the anteroposterior differentiation of the limb structures (for review, see Hinchliffe and Johnson, 1980; Hinchliffe and Gumpel-Pinot, 1983).

The discovery of the ZPA and its effect on limb bud development has initiated a great deal of work. The idea of the existence of such a polarizing zone was first suggested by the experiments of Saunders *et al.* (1958) and Amprino and Camosso (1959). In these studies, the rotation of the wing tip through 180° resulted in duplication of the wing tip skeleton. Moreover, grafting a small piece of mesoderm cut from the posterior margin (and only from the posterior margin) of the wing bud into a preaxial site produced the same effect (Saunders and Gasseling, 1968). Using the same type of transplantation experiments, the area of the wing bud that possesses the polarizing activity has been carefully mapped by McCabe *et al.* (1973) (Fig. 2).

The grafting experiments show that the role of the ZPA is to control the anteroposterior axis of the additional wing tip that has been initiated by its presence. Whether the ZPA controls the anteroposterior differentiation of the limb bud in normal development has been discussed for several years (for review, see Saunders, 1977; Hinchliffe and Johnson, 1980). Strong evidence in favour of this role has been obtained by the experiments of Summerbell (1979), Hinchliffe and Gumpel-Pinot (1981), and Hinchliffe *et al.* (1981).

The mode of action of the ZPA is still hypothetical. Chimeric grafts have been

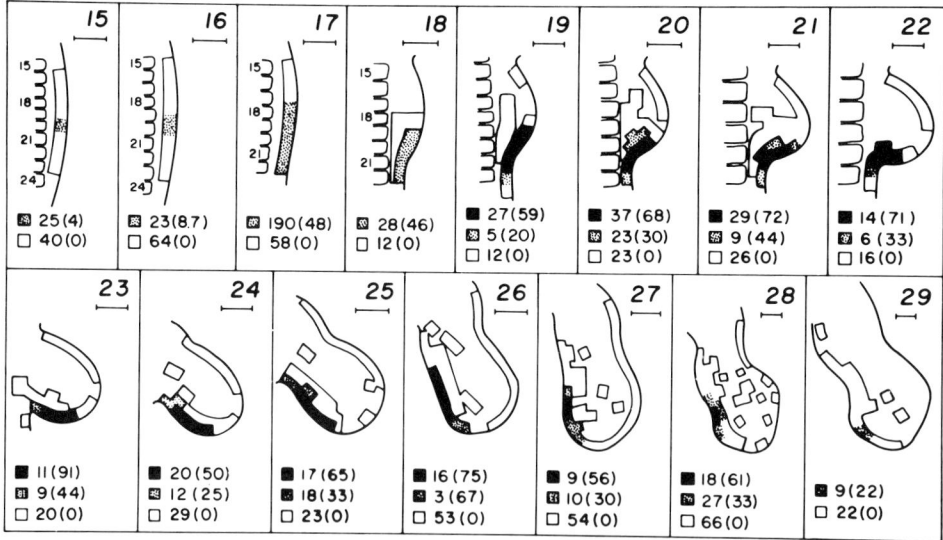

Fig. 2. Scheme showing the distribution of the zone of polarizing activity in the chick wing bud of stages 15–29. Dark shading represents areas of strong activity; at least 50% of the grafts taken from these areas induced duplications, and at least 50% of the duplications were major ones. Stippled areas are zones of weaker activity in which either less than 50% of the grafts produced duplications or fewer than 50% of the duplications were major ones. Unshaded outlined areas were tested, but showed no activity. The numbers following the small squares represent the numbers of successful grafts obtained from that area, and the numbers in parentheses indicate the percentage of grafts that induced duplications. The bar at the upper right hand corner of each rectangle corresponds to 0.5 mm. (After McCabe *et al.*, 1973.)

XI. Muscle and Skeleton of Limbs and Body Wall

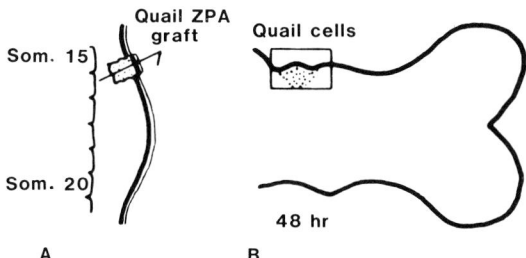

Fig. 3. Quail ZPA graft. (A) Operation on stage 19 wing bud. (B) Limb outline 48 hours later (stage 27), showing location of grafted quail cell and duplication of the limb tip. (After Summerbell and Honig, 1982.)

made to try to resolve this point. It has been demonstrated that the leg possesses a ZPA (Summerbell and Tickle, 1977) that produces duplications when grafted in the wing preaxial area. The skeletal duplications appear to be always of host origin. Fallon and Crosby (1977) demonstrated the presence of a ZPA in mouse, pig, human, ferret, and two species of turtles by grafting presumptive ZPA onto chick limbs. These results suggest that all amniote limbs have a polarizing zone and demonstrate interclass xenoplastic activity of the ZPA between phylogenetically very widely divergent amniotes (see p. 283). They are consistent with the concept of a signal receptor for limb induction and polarization that has been evolutionarily conserved.

Quail-chick chimeric grafting experiments have shown that the ZPA provokes duplication without itself making any cellular contribution to limb formation (Smith, 1979; Summerbell and Honig, 1982) (Fig. 3).

At present the most attractive model (Tickle *et al.*, 1975) is that the ZPA is the source of a morphogenetic signal interpreted by the limb bud mesenchyme. Chimeric interamniote experiments are in agreement with Wolpert's view (1981) that signal mechanisms may be universal in different embryonic fields, although the response varies according to the genome and developmental history of the responding tissue.

3. MAPPING OF THE BIRD LIMB BUD

The first attempts to map the *proximodistal* sequence of the different segments in the chick limb bud were made by insertion of carbon particles at different stages and at different levels (for review, see Hampé, 1959). By barrier insertion, Summerbell (1979) attempted to define skeleton position along the *anteroposterior axis*.

Chimeric grafting was used to determine precisely the position of the skeletal elements at different stages of development. Stark and Searls (1973), by following the fate of autoradiographically labelled fragments grafted into the wing bud, determined the position in the wing of the cells that give rise to each of the wing bones (Fig. 4). It seemed, from the results of recent deletion experiments (Hinchliffe and Gumpel-Pinot, 1981), that the previous maps had assigned the prospective skeletal areas to too narrow a territory on each side of proximodistal midline of the wing bud. An experiment was therefore designed (Hinchliffe *et al.*, 1981; Hinchliffe *et al.*, 1984) to map the different skeletal elements along the anteroposterior axis only. Using the chick-quail system, the authors replaced the excised chick wing bud tissue (anterior or posterior half, anterior or posterior third or two-thirds) by the corresponding part of a quail bud of similar stage. Most of these chimeric limbs developed normally, and histological examination enabled an accurate skeletal fate map to be constructed along the axis. This map (see Fig. 7) differs from the others in that it extends the presumptive skeleton ($2\frac{1}{2}$ somites instead of $1\frac{1}{2}$ or $1\frac{1}{3}$).

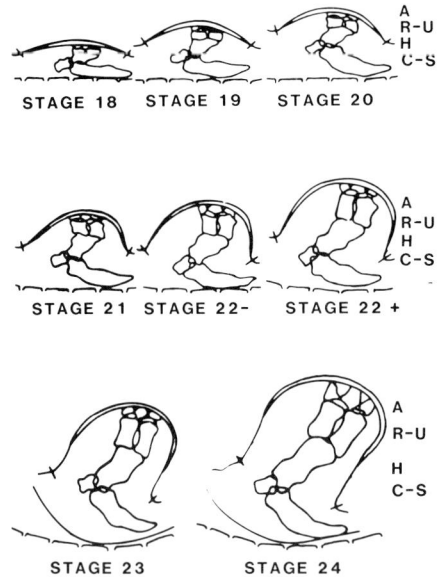

Fig. 4. Maps of the prospective bone-forming areas of the developing wing bud. A, autopod; C–S, coracoid–scapula; H, humerus; R–U, radius–ulna. Blocks of limb mesoderm from stage 19 to 22 embryos that had been labelled with [^{3}H]thymidine were implanted in wings of embryos from stage 18 to stage 24. After 2 to 3 days of growth, the host wings were fixed, sectioned, and autoradiographs were prepared. Every twentieth section through the wing was drawn, showing the location of cartilage and of the labelled cells. The host wing was reconstructed. The position of cells whose progeny were to form each of the wing cartilages at the time of the operation was determined by comparing the location of the labelled cells with respect to the cartilages in the reconstructed wing, with the location of the labelled cells at the time they were implanted. The hatch marks at the cranial and caudal base of the wing were determined by staining the ectoderm with 2% Nile blue sulphate in agar *in vivo;* the hatch marks indicate the limits of the stain. (After Stark and Searls, 1973.)

XI. Muscle and Skeleton of Limbs and Body Wall

These maps have been the basis of regulation experiments along either the promixodistal or the anteroposterior axis.

4. REGULATION IN THE LIMB BUD

During their development, organ rudiments generally pass through a labile phase during which regulation is possible. As differentiation progresses, this regulatory ability is lost.

a. Limb Regulation along the Proximodistal Axis. The first type of regulation to be studied was that of the leg bud along the proximodistal axis (for review, see Hinchliffe and Johnson, 1980).

If the distal part (with the ectoderm) is present, at least up to stages 22–25 (Hamburger and Hamilton, 1951), limb material can be removed from or added to the central portion and the limb is able to regulate and develop normally. If one-third of the leg bud is taken away, a leg develops that is frequently lacking a fibula (Hampé, 1959). Addition of the whole leg bud at stage 18 to a host leg bud at stage 21–22 whose distal tip has been removed, generally results in a leg with an enlarged fibula. The regulation of excess or deficiency seems to be less extensive in wing than in leg (Kieny, 1964a,b). For the regulatory process to take place, a number of conditions appear to be necessary (Kieny, 1977). The most important is that the distal part of the limb must be in an undifferentiated state. If the fate of the distal tip is already determined, serial repetitions of the skeletal elements appear in the composite limb (Summerbell *et al.*, 1973; Wolpert *et al.*, 1975).

In cases where regulation takes place, the main problem is the origin of the newly formed structures from either the added or the original material.

The first attempt to understand the origin of the structures on the suture line at the level of the graft was the construction of heterotopic wing-leg, leg-wing chimeric buds. The morphological characteristics of the bones, particularly of the zeugopod (tibia and fibula in the leg or radius and ulna in the wing), gave an idea of the origin of the cells forming these structures in the chimeric limbs (Kieny, 1964a,b). These regulatory mechanisms have been studied recently in more detail and analysed with greater precision by means of heterospecific chick-quail recombinations.

For studying *excess regulation,* a whole quail bud (leg or wing) was grafted to a chick wrist stump (Kieny and Pautou, 1976; Kieny, 1977). When perfect regulation occurred, three segments (stylopod, zeugopod, autopod) were obtained from five presumptive segments (stylopod, zeugopod + stylopod, zeugopod, autopod). They comprised an upper arm, a lower arm, and a hand in the case of a wing bud graft or a heteromorphic upper arm, a lower leg, and a foot in the case of a leg bud graft.

When the regulatory process was completed, histological analysis of the

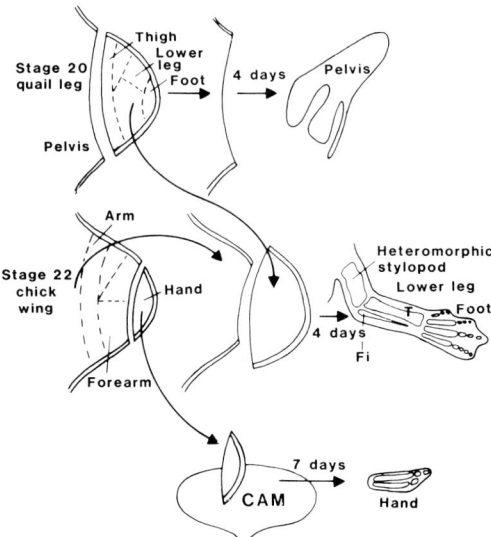

Fig. 5. Regulation of excess mesenchyme in an experimental limb bud in which both prospective stylopod and zeugopod are duplicated. Use of quail graft cells with their nucleolar marker enables the contribution of host and donor tissue to be analysed. Prospective chick zeugopod, now placed proximal to prospective quail zeugopod material, has its prospective fate shifted to the formation of more proximal stylopod structures. CAM, chorioallantoic membrane; Fi, fibula; T, tibia. (After Sengel, 1975, from Hinchliffe and Johnson, 1980.)

chimeric limbs showed that the three proximal presumptive segments had given rise to a composite quail-chick stylopodial element. Thus it is clear that the fate of the distal part of the stump tissues had been shifted to produce structures more proximal than normal (Fig. 5). In *deficiency* chimeric chick-quail experiments (Kieny, 1977), a two-segment constructed limb (stylopod + autopod) resulted in poor intermediate regulation when the quail autopod was grafted on a chick stump. Better regulation was obtained when a chick autopod was grafted on a quail stump, probably because the quail embryo differentiates more rapidly.

If intermediate regulation was obtained, both stump and graft but mainly graft tissues participated in the formation of the intercalate zeugopod. Thus in deficiency regulation too, distal cells (grafted autopod) have their presumptive positional fate shifted to more proximal values (Fig. 6).

b. Limb Regulation along the anteroposterior Axis. This type of regulation was neglected until recently. Barrier experiments by Warren (1934) seemed to exclude the presence of regulatory properties in either the anterior or posterior half of the limb bud, suggesting a mosaic development.

XI. Muscle and Skeleton of Limbs and Body Wall

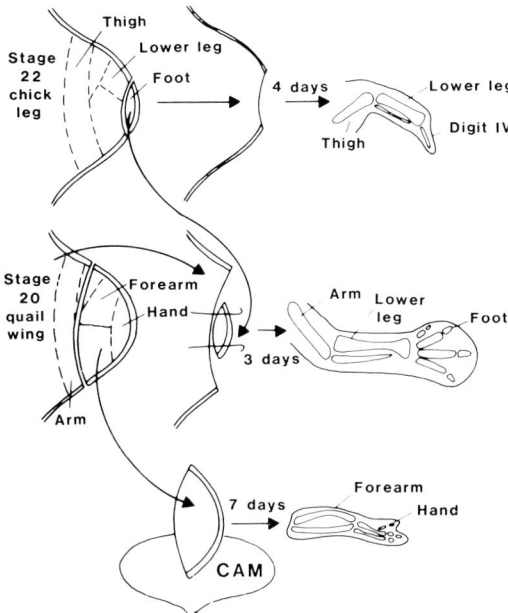

Fig. 6. Regulation of depletion of mesenchyme in an experimental limb bud in which the prospective zeugopod has been removed. As in Fig. 5, "labelled" quail cells are used in a heterospecific graft. Prospective chick autopod (normally forming foot paddle elements only) regulates and, in addition to foot paddle, forms also the intercalated zeugopod elements; thus, the prospective fate of some of these cells is shifted to development of more proximal structures. (After Sengel, 1975, from Hinchliffe and Johnson, 1980.)

Recent work on the ZPA (p. 283–285) gave prominence to the idea that the development of the anteroposterior axis was controlled by the existence of a morphogen produced by the ZPA, located at the posterior margin of the limb bud. This area, grafted preaxially at stages 18–22, was able to provoke the formation of extra digits from the anterior part, suggesting that the preaxial territories are not definitively determined at these stages. Moreover, after amputation of the posterior half, the anterior half develops less than predicted by its prospective fate. However, if amputation was performed in such a way as to preserve at least a part of the ZPA, perfect regulation was observed (Hinchliffe and Gumpel-Pinot 1981). Thus a certain degree of anteroposterior regulation is possible.

The problem has been studied by Yallup and Hinchliffe (1983), by chimeric quail-chick experiments (Fig. 7). The authors followed the anteroposterior regulation of chimeric wings formed by one-third anterior quail and one-third posterior chick (deficiency) or two-thirds anterior quail and two-thirds posterior chick (excess). Good regulation of pattern and size was seen at stages 19–22.

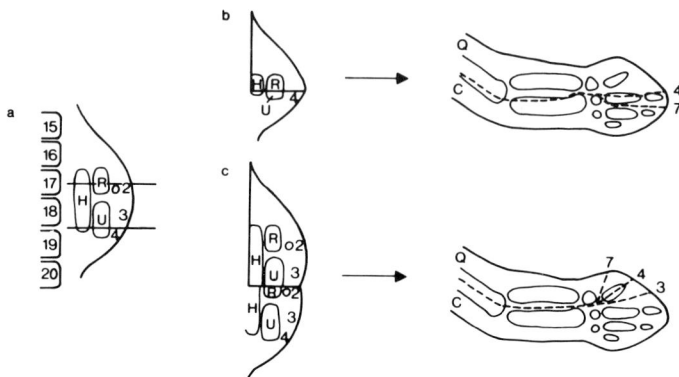

Fig. 7. (a) Fate map for a stage 20 wing bud (Hinchliffe *et al.*, 1981), the lines representing anterior and posterior cutting levels. Diagram of graft compositions and the chimaeric results for (b) deficiency operations and (c) excess operations. The dashed line represents the chick-quail boundary, (C, chick; Q, quail), and the numbers outside the limb skeleton drawing represent the number of replicates for each type of result. (After Yallup and Hinchliffe, 1983.)

Size regulation decreased gradually with increasing stage at operation and was poor when pattern regulation ceased at stage 24. Histological studies of the chimeric limbs showed that contribution to the intercalate regulatory structures occurred on both sides of the host–graft interface. The boundary was almost similar in both cases, the posterior contribution being somewhat greater, especially in the case of deficiency.

This capacity for regulation along the anteroposterior axis seems to conflict with the previous reports of mosaic development of the isolated posterior half of a limb bud with an intact ZPA. Anterior and posterior tissues can both be shifted either to more central or to more external positional values. According to the authors, "it is possible that for regulation to occur, the wing bud needs, in addition to the ZPA, a second anterior reference point which could be the anterior mesenchyme or apical ectodermal ridge" (Yallup and Hinchliffe, 1983).

5. ORIGIN OF THE CELL LINEAGES: SOMITIC CONTRIBUTION TO LIMB BUD MUSCULATURE

For a long time, the limb bud mesenchyme was considered to consist of a homogeneous population of pluripotential somatopleural cells. This view was supported by the work of Saunders (1948) who claimed, on the basis of carbon particle marking, that somites do not contribute to limb bud formation. However, Fischel (1895) described a diffuse migration of somite cells into the wing of birds, and, much more recently, Grim (1970) concluded from electron microscopy studies and observations of semithin sections that a few undifferentiated cells were released from the ventral edge of the dermomyotome at the level of the

XI. Muscle and Skeleton of Limbs and Body Wall

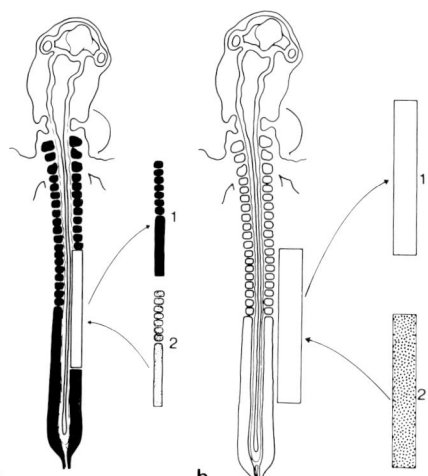

Fig. 8. (a) First experimental series, showing the unilateral extirpation of brachial somitic mesoderm (1) from a chick embryo at stage 13 (Hamburger and Hamilton, 1951) and the substitution by an equal mesodermal fragment of a quail (2). (b) Second experimental series. Removal of the somatic layer of lateral plate mesoderm together with the overlying ectoderm from a chick embryo stage 13 (Hamburger and Hamilton, 1951) at the level of the prospective wing bud (1). Transplantation of an equal graft (2) previously isolated from a quail embryo of the same stage of development. (After Christ *et al.*, 1977b.)

wing in the chick embryo. These cells were then integrated into the adjacent limb mesenchyme, where they could no longer be distinguished from the somatopleural cells, "their potential migration into the wing (being) neither confirmed nor excluded by (the) investigation" (Grim, 1970). Thus the idea arose that two subpopulations of mesodermal cells existed in the early limb bud. The cellular contribution of the somites to limb bud formation was established by Gumpel-Pinot (1974). After associating the limb territory with somites, either labelled with tritiated thymidine or using the quail-chick system, Gumpel-Pinot demonstrated that somitic cells do indeed migrate into the developing limb bud. This work, however, did not provide any information on the developmental fate of these somitic cells within the wing.

The problem of the developmental fate of the two cell lines forming the presumptive limb mesenchyme has been systematically studied by two groups, in France and West Germany, mainly using heterospecific quail-chick recombination and very similar basic experimental procedure, i.e., exchange of somites or portions of the somatopleural lateral plate at different levels *in ovo* (Fig. 8).

a. Somitic Origin of the Limb Striated Muscles—Developmental Fate of the Two Cell Lineages. Interspecific grafts of somites or parts of somatopleural mesoderm were made between quail and chick at stages 12–14, at the prospective wing level. Very rapidly after the graft, cells of somitic origin that had been integrated in the limb territory were limited to the prospective *myogenic areas*. Later on, when the muscles were well differentiated, the muscle cells were found to be of somitic origin, while the muscle connective tissue cells were derived from the somatopleural mesoderm. In contrast, if the presumptive limb territory isolated from the somites (stage 13–14) was grafted into the coelomic cavity, skeletal elements developed but no striated muscle differentiated (Christ *et al.*, 1974a, 1977a,b, 1979). Chevallier *et al.* (1977a,b) described very similar results. Moreover, by replacing the somitic mesoderm adjacent to the wing level by a piece of 9-day chick embryonic midgut, or by destroying, through local X irradiation, at least three somites or presumptive somites anterior and posterior to the wing level, Chevallier *et al.* (1978) obtained the development of wings in which the striated muscles were in most cases entirely missing.

The absence of a given muscle was accompanied by the absence of the corresponding *tendons*. However, Christ *et al.* (1977b) found that tendons in chimeric limbs were of somatopleural origin, and Kieny and Chevallier (1979) concluded from destruction by X ray of somites of the wing level that tendons start to develop autonomously from the muscle bulk. However, for their maintenance and further development, they require connexion to a muscle.

Smooth muscles, which lie under the epidermis and are implicated in feather movements, are of somatopleural origin. This was demonstrated by Christ *et al.* (1979) using coelomic quail-to-chick grafts: quail limb territories grafted in the coelom of chick embryos showed no development of striated muscles, but clear differentiation of smooth muscle fibres occurred. Kieny *et al.* (1979) reached the same conclusion after grafting quail somites at the level of the wing in the chick embryo: in the chimeric limbs, the smooth muscles were of chick origin.

Thus, in normal development, the fate of the two mesodermal cell lineages that form the wing bud is as follows: the striated skeletal muscles are of somitic origin, while the cartilage elements, the connective tissue, the tendons, and the smooth muscles derive from the somatopleure.

b. *Stage of the Migration from the Somite to the Limb Bud.* Christ *et al.* (1974a,b, 1976, 1977a,b) determined the stage at which migration occurs by means of quail-chick coelomic grafts of wing territories at different stages, the differentiation of striated muscles inside the grafts being considered as proof of the onset of migration. In this way, Christ *et al.* deduced that the migration starts at stage 13–14 at the level of the *wing*. The stage of the migration has also been studied by histological or ultrastructural examination of the normal relationships between somites and lateral plate by Christ *et al.* (1977a) and Jacob *et al.* (1979). These authors demonstrated that the migration starts at the level of the *leg,* at stage 16, from the newly formed somites 26–28. The migration lasts until stages 19–20. Chevallier (1978), by observation of the relationships between somites

XI. Muscle and Skeleton of Limbs and Body Wall

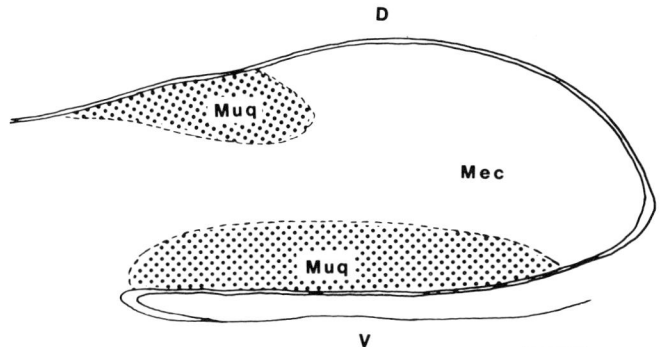

Fig. 9. Histological results of orthotopic implantation, at the level of the wing. The graft represents the homologous somitic mesoderm removed from a quail embryo. Host, 16 somites; donor, 17 somites. Fixed 48 hours after operation. The longitudinal section of the wing bud shows the localization of the two quail premuscular blastema (Muq) on each side of the bud axis which is still formed by chick mesenchyme (Mec.). D, dorsal; V, ventral. (After Chevallier, 1978.)

and lateral plate and by implantation of quail- or [^{3}H] thymidine-labelled somites in the chick embryo, obtained very precise information on the timing of the migration at the *wing* level. He concluded that the migration starts at the level of somite 15 when the embryo reaches 20 to 22 somites and begins at the level of somite 20 at the 23- to 25-somite stage. The migration is complete at the level of somite 15 at the 28-somite stage, and at the level of somite 20 at the 36-somite stage.

The migration thus does not depend on the state of differentiation of the somitic mesoderm. Whereas the brachial somites (15–20) are completely segmented when the migration starts, a large part of the para-axial mesoderm is still unsegmented at the onset of migration at the level of the leg (somites 26–32).

The first cells to penetrate the wing lie in the ventral part of the bud near the coelomic epithelium (Gumpel-Pinot, 1974; Chevallier, 1978). By virtue of their high mitotic activity (reflected by a fall in the intensity of the labelling), they invade the outgrowing limb bud. Later, myogenic cells gather into dorsal and ventral premuscular masses that are the sites of considerable mitotic activity (Fig. 9). Then, when the individual muscular masses became separated, the somitic cells are found in the muscular component (myocytes) of the muscle.

c. Regionalisation of the Muscle Derivatives of the Somitic Mesoderm. If the somites at the limb level normally give rise to limb musculature, one may ask whether this property is restricted to the somitic mesoderm at the corresponding limb levels. In experimental chick-quail exchange of somites from different levels (neck, wing, flank, leg) Chevallier *et al.* (1977a) and Christ *et al.* (1978) demonstrated that the organogenesis of the limb musculature did not depend on the cephalocaudal level from which the myogenic somite cells emigrated. The somitic mesoderm of the wing territory could be replaced by somitic mesoderm from any cephalocaudal level without disturbing or modifying limb muscle organogenesis (Fig. 10). Thus the myogenic cell lineage, originating from the somites, does not determine the muscular pattern; the somatopleure is responsible for this organisation (Jacob and Christ, 1980).

Several somites normally contribute to the wing musculature. The spatial distribution of myogenic somitic cells from these different somites in the intrinsic wing musculature has been studied. By replacing a single somite or two consecutive somites at the wing level of a chick embryo by the corresponding somites from a quail embryo, Kieny and Chevallier (1980) deduced that there is no strict relationship between a given somite and a particular intrinsic wing

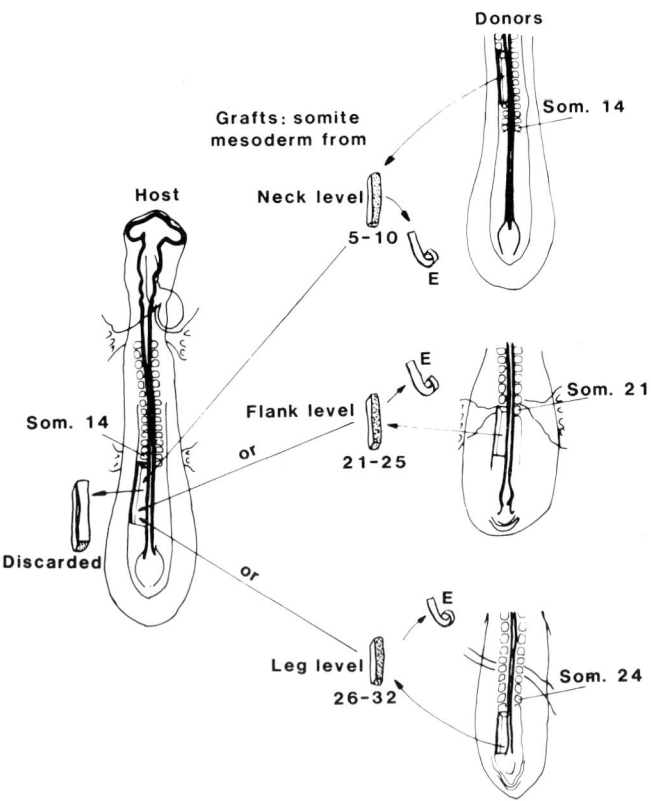

Fig. 10. Experimental scheme of heterotopic replacement of the somitic mesoderm from the wing level by heterospecific somitic mesoderm from the neck, flank, and leg levels. Combinations between chick host and quail donor and vice versa. E, discarded ectoderm. (After Chevallier et al., 1977a.)

muscle or muscle group. However, relatively little participation of the anterior somites in the formation of dorsal muscles was observed in a few cases.

d. Observations on the Mode of Migration of the Somitic Cells. The mode of migration of the somitic cells from the somite to the limb territory has been studied in detail using histological, histochemical, and ultrastructural (scanning and transmission) techniques (Christ *et al.*, 1977b; Jacob *et al.*, 1978, 1979).

First, cells at the ventrolateral border of the somites, distinguishable by their greater affinity for alcian blue, elongate from the somites in the direction of the limb territory. The cells move through a matrix consisting of collagen fibrils, hyaluronate, and proteoglycans. This matrix constitutes a fibrillar network and the cells contact the fibrils via a visible cell coat containing glycosaminoglycan granules. Migrating cells have filopodia at their leading end and contain microfilaments that indicate the ability to contract. Plaque-like contacts are made at the trailing end when the cell elongates and at the leading end to anchor it when the trailing end retracts. The Golgi apparatus is located in the trailing end.

e. Hypothesis as to Possible Causes of the Migration of Somitic Cells. The myogenic cells migrate from the somite to the lateral plate. This phase of the migration can be observed by light or electron microscopy. The cells migrate through a network of extracellular material. Jacob *et al.* (1979) found a correlation between the direction of cell migration and the preferred orientation of the collagen fibrils.

Cells from the premuscular mass of 4- to 5-day chick embryos, grafted in place of somites removed from the wing level, are still able to migrate in the wing bud and to participate in wing musculature (Mauger and Kieny, 1980). This finding suggests the existence of an attractive field, e.g., a gradient in the extracellular material, which could be controlled by the presence of ectodermal apical ridge (Gumpel-Pinot *et al.*, 1984). When migration starts, the limb bud primordium is still flat and seems undifferentiated, although it has been assumed that this territory is already committed to becoming a wing (Pinot, 1970).

However, the extracellular matrix situated between the somites and somatopleure is not essential for the migration of myogenic cells. When a piece of quail limb mesenchyme containing myogenic somitic cells was grafted in a chick wing bud, the myogenic cells present in the graft migrated to the apex of the limb (Gumpel-Pinot *et al.*, 1984). In contrast, cartilage differentiated only at the site at which the graft had been inserted. Wachtler *et al.* (1982) also observed the immobility of precartilage cells by grafting quail somatopleure that had not yet been invaded by somitic cells in the limb bud of a chick embryo. These results seem to indicate that the migration depends both on intrinsic migratory properties of the cells and on their environment. Another argument is provided by the fact that premuscular tissues of stage 25–26 quail limb bud grafted within a younger chick limb bud are able to migrate to the limb tip, just as they do when grafted in place of the somites (Mauger and Kieny, 1980).

From the nonregionalisation of the somitic mesoderm with respect to the muscle derivatives, it is obvious that the myogenic cell lineage does not determine the muscular pattern of the limb.

Neural tube and neural crest were excised or destroyed in order to obtain nerveless limbs (Jacob and Christ, 1980). These experiments clearly showed that the nerve exerts no influence on the migration of the cells and on the organisation of the muscle pattern in the limb. Moreover, when the leg somatopleure of a chick embryo is replaced by flank somatopleure of a quail embryo (Jacob and Christ, 1980), the pattern of the striated muscles formed in the somatopleure corresponded to the source of the graft (flank) and not to the position within the host (leg). This means that *connective somatopleural tissue* probably in relationship with skeleton determines muscle pattern.

The conditions encountered by the muscular cells during their migration between somite and limb bud and within the somatopleural connective tissue are thought to be responsible for the oriented migration of the cells and establishment of the muscular pattern, but we are still far from understanding the mechanism of this process.

f. Determination of Cell Type. It is clear from the experiments described above that the limb bud is composed of two mesodermal lineages: the somitic lineage, from which the skeletal striated muscles develop, and the somatopleural primary lineage, which gives rise to all the other types of tissues of mesodermal origin. The existence of a double lineage during normal development is now very widely accepted.

However, the question remains as to whether the limb bud mesenchymal cells are already definitively determined by their somitic or somatopleural origin or whether their fate can be experimentally modified by extrinsic factors, such as their position within the limb. Zwilling (1966) divided the proximal part of the limb into prospectively chondrogenic fragments and fragments regarded as myogenic. After grafting, both types of fragments developed cartilage in the majority of cases, until stage 25. This experiment suggests that the stabilisation of limb tissue type is a relatively late event.

In situ, up to stage 22 the mesenchyme of the limb appears to be homogeneous, but soon afterwards biochemical differences (increased condensation and synthesis of chondroitin sulphate) appear in the prechondrogenic areas at the level of the zeugopod. At stage 25, chondrogenic areas are easily recognisable by their metachromatically staining matrix.

In experiments carried out by Searls and Janners (1969), [^{3}H] thymidine-labelled prospective fragment of a chondrogenic region, transplanted *in ovo* to a nonchrondrogenic area of a limb bud, developed appropriately to its new position

XI. Muscle and Skeleton of Limbs and Body Wall

up to stage 24. Searls and Janners assumed that the fragment of implanted tissue maintained its integrity and that there was no migration from host to graft or vice versa. Comparable results were presented by Caplan and Koutroupas (1973), who claimed that the position of cartilage and muscle in the limb bud is due to differential vascularisation. Nathanson and Hay (1980) concluded from tissue culture experiments that fetal rat myoblasts could, under favourable conditions, differentiate into chondrocytes.

Recently, two parallel experiments by different authors gave unexpected opposing results. Both experiments consisted in grafting, into the coelom, presumptive wing territory that had not yet been invaded by somitic cells. Both made use of the quail-chick system. Both groups considered that the migration started at stage 13–14. McLachlan and Hornbruck (1979) found striated muscles in 14 out of 19 grafts from donors at stage 10–13. Christ *et al.* (1979) examined

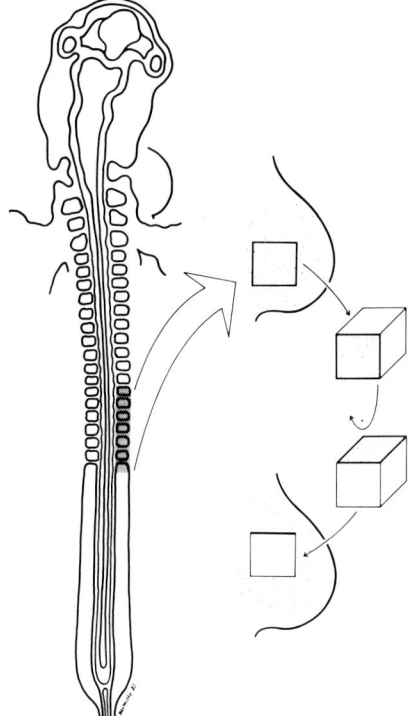

Fig. 11. Diagram illustrating the two-step operation described in the text. (a) Replacement of brachial somites. (b) Grafting of chick wing bud mesenchyme containing somitic cells of quail origin to another chick limb bud. These pieces of mesenchyme were oriented in such a way that prospective myogenic areas of the graft would come to lie in prospective chondrogenic regions of the host. (After Wachtler *et al.*, 1981.)

100 grafts made under similar conditions; in none of them was differentiated striated muscle present 2–12 days after the operation.

A possible explanation for these divergent results is that Christ *et al.* took their donors just before migration at stages 12–13. At stages 10–11, it is not yet possible to separate cleanly somitic and lateral plate mesoderm at the prospective wing level, and somitic mesoderm cannot be considered to be definitively eliminated from the graft (Christ and Jacob, 1980). However, the difference in the results when the grafts derived from embryos at stages 12–13 is not yet explained.

Christ *et al.* (1979) concluded that the developmental option was between fibroblasts and chondrocytes, and that the myoblasts were determined by their somitic origin. This idea was supported by the work of Wachtler *et al.* (1981), who performed a two-step experiment (Fig. 11). In the first step, chick brachial somites were replaced by quail somites to obtain a chimeric limb with quail

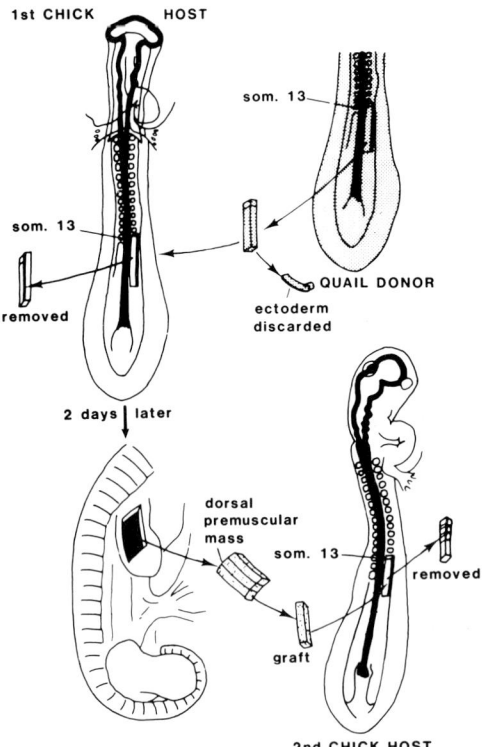

Fig. 12. Two-step replacement of chick somitic mesoderm of the wing level by a fragment of dorsal bispecific premuscular mass in which the myogenic cells are quail cells. First step of the experiment. (After Mauger and Kieny, 1980.)

XI. Muscle and Skeleton of Limbs and Body Wall

somitic cells. In the second step, a fragment of the chimeric wing containing chick somitic cells was grafted in the prospective chondrogenic area of a chick bud. The results of this experiment seemed to be quite clear, and the conclusions of Wachtler *et al.* were that, from stage 19 onward, there are two determined subpopulations in the wing bud: somitic cells yield striated muscle, and cells of somatopleural origin give rise to cartilage and connective tissue. Neither of these two populations can replace the other. Wing bud from precartilage area is determined to become cartilage at stage 20. Wing bud mesenchyme from soft, non-cartilage-forming regions (connective tissue) retains the option to form cartilage up to at least stage 26.

However another two-step experiment from the French group (Kieny, 1980; Mauger and Kieny, 1980) seems to challenge this view. In the first step (Fig. 12), the authors constructed a chimeric wing bud by replacement of the chick somites by corresponding quail somites at the wing level. After a few days of development, they took this chimeric wing bud, discarded the ectoderm, dissociated the cells, and reassociated them in another limb ectodermal jacket (Fig. 13). They grafted this associated limb in place of a previously removed 3- to 4-day chick embryo wing bud. All types of cells became randomly reassociated to form

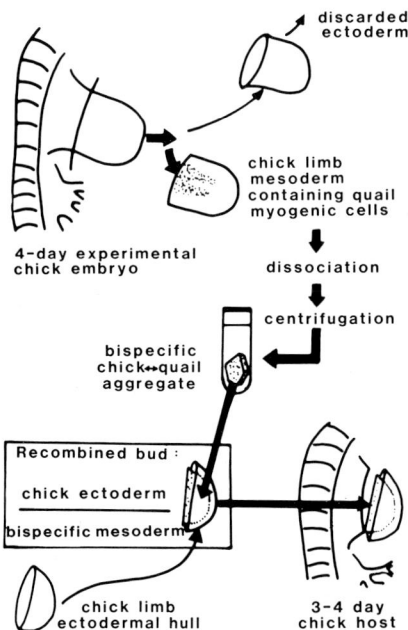

Fig. 13. Diagram illustrating the second step of the construction of a recombined limb bud, whose quail myogenic cells have been haphazardly redistributed among the chick limb cells (after Kieny, 1980).

a new bud. After development of this bud, quail cells of somitic origin were found in cartilage elements formed in the reassociated limb. Thus the possibility, under very extreme experimental conditions, of a developmental plasticity among the myogenic somitic cell lineage is perhaps not excluded. The question is still open.

g. *The Concept of Double-Lineage Limb Formation in Vertebrates.* The participation of the somites in the formation of the limb musculature has long been known in other classes of vertebrates. In reptiles, somitic outgrowths penetrate the limb bud very deeply, and the number of somites involved is related to the development or regression of the limb (for review, see Raynaud 1977). In mammals, in addition to numerous early observations, Milaire (1976) showed that somitic cells participate in leg development in the mouse embryo. Houben (1976), studying the ultrastructural features of these cells, concluded that they were myogenic. Christ and Jacob (1980) observed cells migrating between somites and limb buds in the human embryo. In birds, the evidence from chimeric experiments summarised above strongly supports these observations from other vertebrate classes. Thus it seems that there are no basic differences between embryos of higher vertebrates with regard to the origin of the limb cells.

B. Ribs, Sternum, Girdles, and Muscles of the Body Wall

The origin of the structures forming the body wall and the mechanisms of their morphogenesis were studied by several authors using different techniques before the introduction of chimeric chick-quail experiments; the latter confirmed definitively a number of points and demonstrated some others.

The somatopleural origin of the sternum was shown by Fell in the budgerigar as early as 1939. *In vitro* culture of fragments of lateral plate demonstrated that the sternum develops from two somatopleural cartilaginous anlagen that subsequently fuse.

The problem of the origin of the ribs and the muscles of the body wall was more difficult to resolve. Up to 1953, it was generally held that ribs and muscles of the body wall, including abdominal muscles, were derived at least partly from the somites. This was the consequence of work by Fischel (1895) who described in birds a diffuse migration of somitic cells that became integrated in the body wall structures. Strauss and Rawles (1955) reached different conclusions: after insertion of carbon particles along the somites at the moment of the formation of the ribs and coelomic grafts of portions of lateral plate taken at the level of the ribs, Strauss and Rawles claimed that only one-third of the body wall was of somitic origin, the ventral part being of somatopleural origin. At the level of the boundary between the two cell lines, a strip of tissue appeared to be formed by mixture of the two populations.

XI. Muscle and Skeleton of Limbs and Body Wall

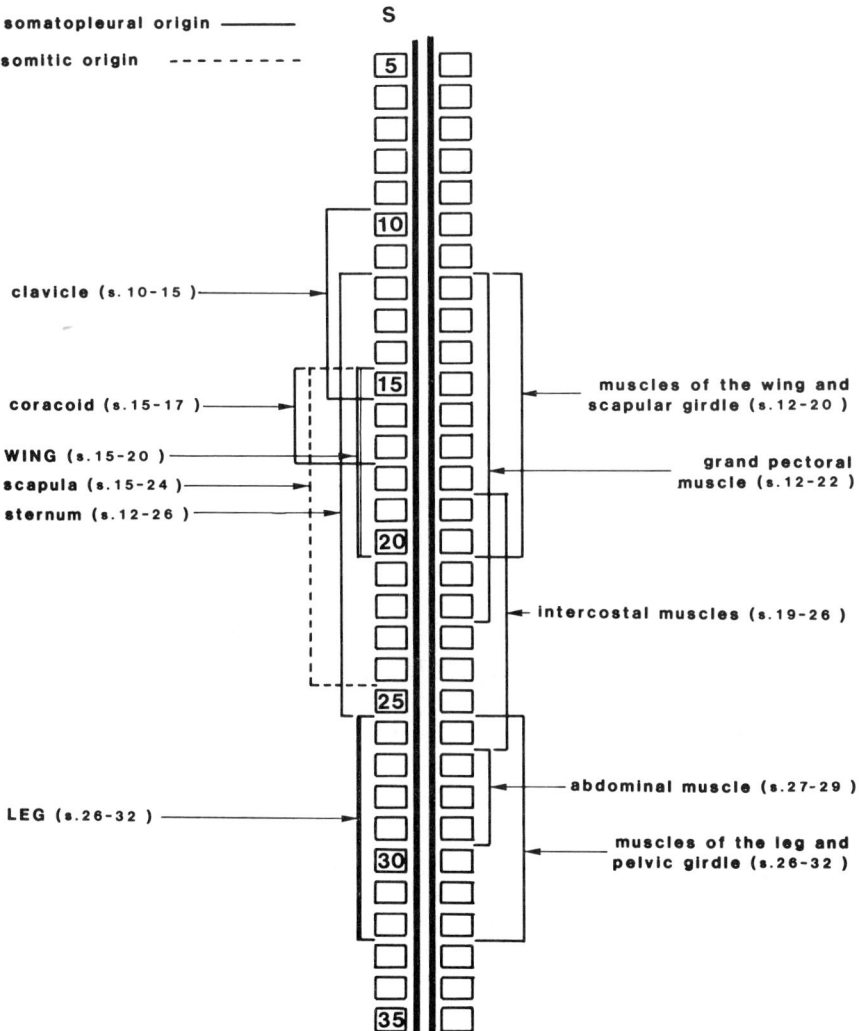

Fig. 14. Somitic level of the muscle and skeletal formations of the limbs and body wall in the chick embryo. All the striated muscles derive from somitic mesoderm. All the skeletal formations originate from the somatopleural mesoderm except the scapula considered to be of somitic origin. (Drawn from Chevallier, 1974, 1976, 1977.)

Seno (1961a,b), Murillo-Ferrol (1963), Pinot (1969), and Sweeney and Watterson (1969a,b) suggested that the somites gave rise to the whole rib structure and to all the lateroventral part of the body wall (intercostal and abdominal muscles), with the exception of the pectoral muscle, considered to be of somatopleural origin. According to these authors, the role of the lateral plate was to form the connective tissue into which somitic cells migrate and to give rise to sternum and pectoral musculature. However, some doubts remained concerning the sternal segment of the ribs, which did not develop from somitic material grafted in the coelom and did not develop *in vivo* after destruction by X irradiation of the rib-forming somites.

These problems, difficult to approach using classical methodology, remained unsolved until the advent of quail-chick chimeras.

Christ *et al.* (1974a,b, 1976), by replacement of somites in the quail-chick system, demonstrated that the entire rib, including the sternal segment, was of somitic origin. They assumed that the muscles of the body wall, including the pectoral muscle, also derived from somites, and they confirmed that the connective tissues and the sternum were of somatopleural origin. Beresford *et al.* (1978) reached similar conclusions. Chevallier (1975, 1977, 1978, 1979), using chick-quail recombinations of somites and lateral plate from different levels, as well as local destruction by X irradiation of definite portions of the somites, obtained similar results regarding the origin of the different structures. Moreover, he was able to localise with precision the somitic level from which the cells derived: pectoral muscle, somites 12–22; intercostal muscles, somites 19–26; abdominal muscles, somites 27–29; dorsal and intervertebral muscles, metameric level-specific origin. Chevallier (1977), using the same chimeric techniques associated with local destruction by X rays, studied the origin and the localisation of the girdles and associated muscles. He showed that the scapula derives from somitic mesoderm, while the clavicle, coracoid, sternum, and pelvic girdle originate from somatopleural mesoderm. The somitic level for these structures is as follows: scapula, somites 15–24; clavicle, somites 10–15; coracoid, somites 15–17; intrinsic and extrinsic muscles of the wing and scapular girdle, somites 12–20; sternum, somites 12–26; pelvic girdle, somites 26–32; intrinsic and extrinsic muscles of the leg and pelvic girdle, somites 26–32 (see Fig. 14).

III. MAMMALS

Because mammalian embryos are inaccessible to microsurgical intervention after implantation, chimerism is produced during the preimplantation stage (see Chapter I). No equivalent of the chick-quail nuclear difference is known in mammals, but various morphological and biochemical characteristics have been used as markers (see also Chapter II).

XI. Muscle and Skeleton of Limbs and Body Wall

A. Morphological Markers

1. BODY SIZE

The determination of body size, including skeletal dimensions, has been studied using chimeras between strains of mice selected for large (L) and small (S) body weight (Roberts et al., 1976; Falconer et al., 1981). To quantify the degree of chimerism, the authors used the classical technique of marking the constituent strains by contrasting coat colours, either albino or coloured. The strains also differed with respect to a genetically determined enzyme marker, so that the proportions of the two constituent cell types could be ascertained in a large number of different organs and tissues, including muscle. Body weight was found to be more closely correlated with the cellular composition of the coat of the chimeras than with the composition of any other organ or tissue examined. As melanocytes cannot be considered as determinants of size, the observed correlation must reflect some relationship between the proportion of melanocytes in the coat and the proportions of L and S cells in whatever tissue(s) regulates growth. If a single tissue is indeed involved in growth regulation, it cannot be any of those examined in this experiment.

2. VERTEBRAL COLUMN

Moore and Mintz (1972) constructed chimeric mice from two strains (C57BL and C3H) differing morphologically in the shape of their vertebrae. The authors were able to assess the likely parental origin of each half of each vertebra since some of the vertebrae studied showed a marked left–right asymmetry, with the lateral halves judged to have different strain phenotypes. Since each half-vertebra is formed by one anterior half-somite and one posterior half-somite, an asymmetry can also be observed between the anterior and the posterior part of

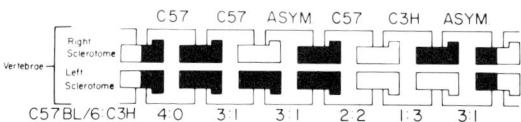

Fig. 15. Diagram illustrating the clonal model of vertebral development and how it would provide a basis for vertebral phenotypes observed in C57BL/6 C3H allophenic mice. Each vertebra (subtended by an upper and lower bracket) appears to be derived intersegmentally from four clones: the caudal and cranial sclerotomes of neighboring pairs of somites. The cranial and caudal sclerotome derivatives of one somite may be of the same or of different genotypes. Each caudal element is shown with a projection, to indicate that it gives rise to the transverse processes. Solid sclerotome elements are of C57BL/6 genotype, outlined sclerotome elements of C3H genotype. Some of the possible genotypic clonal compositions of individual vertebrae are shown at the bottom, in fourths C57BL/6:C3H, and the resultant most probable morphological phenotype of each is given at the top. The caudal sclerotomites influence the vertebral phenotype more than do the cranial elements. Some genotypically mosaic vertebrae could also be of indeterminate strain-type. (After Moore and Mintz, 1972.)

each vertebra. The different possible phenotypic expressions are summarised in Fig. 15. From this study, the authors hypothesise that each vertebra is formed by a minimum of 4 clones. This is a minimum estimate, since each element might contain several clones. Since each half vertebra is formed by an anterior half-somite and a posterior half-somite, it follows that each somite may also be formed by a minimum of four clones, at least for the vertebra-forming part of the somite.

3. SHORT EAR AND VESTIGIAL TAIL

McLaren and Bowman (1969) and Grüneberg and McLaren (1972) investigated chimeric animals in which the parental strains differed by nine genetic factors, including two genes producing skeletal abnormalities, short ear (*se*) and vestigial tail (*vt*). They found that the chimeras, with respect to their skeletal characteristics closely resembled heterozygotes between the two component strains. For example, the ears and tails of the chimeras closely resembled those of *se*/+ *vt*/+ animals and did not show the marked reductions characteristic of *se*/*se* *vt*/*vt* homozygotes.

The short ear gene is associated with a widespread defect of cartilage formation in the embryo so that the short ear homozygotes (*se*/*se*) show several

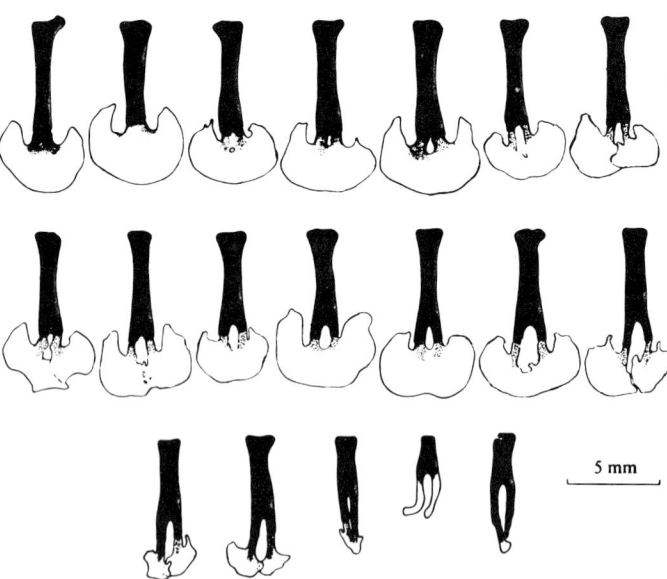

Fig. 16. The xiphisterna of nineteen chimeras arranged in ascending order of abnormality. In the first row, one normal and six with a varying degree of involvement of the xiphoid cartilage (white) and its calcified granules (stippled). The second row has mild and the third more marked involvements of the osseous xiphisternum (black). (After Grüneberg and McLaren, 1972.)

XI. Muscle and Skeleton of Limbs and Body Wall

Thus each somite appears to be formed by a number of precursor cells, as discussed by Mintz (1972). The myotomes of the somites that give rise to muscle are derived from at least two clones; the sclerotomes that form the vertebrae are derived from at least four precursor cells (see above). The dermatome probably comes from one cell responsible for a single hair follicle clone and perhaps others forming other dermal elements in the skin. If the somites in mammals as well as in birds are involved in the formation of the body wall, limb muscles, and ribs, the clonal origin of the somites could be more complicated.

With regard to muscle formation, a very interesting study from Peterson (1974) should be noted. Using electrophoretically distinguishable isozymes of malic enzyme as markers, this author studied chimeric mice formed from a normal strain and a strain characterised by dystrophic muscles. Although the muscles of the chimeric animals proved to contain genetically dystrophic as well as normal cells, histological examination showed them to be completely normal, with no phenotypic expression of dystrophy. This result suggests that the causal factors involved in the expression of muscular dystrophy lie outside the affected muscles, perhaps in the nervous system.

IV. CONCLUDING REMARKS

In birds, in which direct experimentation is possible at a stage when cell movements take place to achieve the definitive pattern of limb and the different structures forming the skeleton and muscles of the body wall, chimeric experiments have been particularly useful for identifying cell lineages. A number of problems could probably never have been solved without the discovery of the quail nuclear marker; species-related morphological markers are rather limited, and isotopic labelling is rapidly diluted by mitosis and cannot give a clear answer for phenomena that last more than a few hours. Other techniques used to study the embryological origin of tissues (barriers and culture of isolated fragments) have proved able to answer some specific questions, but present the major drawback of interfering with normal morphogenesis so that the results often need confirmation by chimeric quail-chick experiments. Some morphological problems arise from the fact that quail and chick embryos are not of exactly the same size, and do not have exactly the same developmental chronology. However, these problems are only pronounced during the last two-thirds of the incubation period and can be partially overcome by performing the quail-chick, chick-quail reciprocal experiments.

No cell marker comparable with the quail nuclear marker exists in mammals, nor is it possible to interfere microsurgically with the embryo during the postimplantation period when organogenesis occurs. Nevertheless, a number of developmental problems have been approached by chimeric experiments. Much information has been gathered on the distribution of precursor cells during development, but perhaps the most interesting possibility offered by the chimeric technique in mammals is to "track the genes" (Mintz, 1982). Chimeric experiments combining mutant and normal strains can provide a great deal of information concerning the mode of expression of the mutant genes, the role of the cellular environment, and the possibility of restoring normal function.

ACKNOWLEDGMENT

The author is most grateful to Dr. J. Smith for his help during the preparation of this manuscript.

REFERENCES

Amprino, R., and Camosso, M. (1959). *Arch. Anat. Microsc. Morphol. Exp.* **48,** 261–305.
Beresford, B., Le Lièvre, C., and Rathbone, M. P. (1978). *J. Exp. Zool.* **205,** 321–326.
Caplan, A. J., and Koutroupas, J. (1973). *J. Embryol. Exp. Morphol.* **29,** 571–583.
Chevallier, A. (1975). *J. Embryol. Exp. Morphol.* **33,** 291–311.
Chevallier, A. (1977). *J. Embryol. Exp. Morphol.* **42,** 275–292.
Chevallier, A. (1978). *Wilhelm Roux's Archi. Dev. Biol.* **184,** 57–73.
Chevallier, A. (1979). *J. Embryol. Exp. Morphol.* **49,** 73–88.
Chevallier, A., Kieny, M., and Mauger, A. (1977a). *J. Embryol. Exp. Morphol.* **41,** 245–258.
Chevallier, A., Kieny, M., Mauger, A., and Sengel, P. (1977b). *In* "Vertebrate Limb and Somite Morphogenesis" (D. A. Ede, J. R. Hinchliffe, and M. Balls, eds.), pp. 421–432. Cambridge Univ. Press, London and New York.
Chevallier, A., Kieny, M., and Mauger, A. (1978). *J. Embryol. Exp. Morphol.* **43,** 263–278.
Christ, B., and Jacob, H. J. (1980). *In* "Teratology of the Limbs" (H. J. Merker, H. Nau, and D. Neubert, eds.), pp. 67–77. de Gruyter, Berlin.
Christ, B., Jacob, H. J., and Jacob, M. (1974a). *Experientia* **30,** 1446–1448.
Christ, B., Jacob, H. J., and Jacob, M. (1974b). *Experientia* **30,** 1449–1451.
Christ, B., Jacob, H. J., and Jacob, M. (1976). *Verh. Anat. Ges.* **70,** 1007–1011.
Christ, B., Jacob, H. J., and Jacob, M. (1977a). *Verh. Anat. Ges.* **71,** 1231–1237.
Christ, B., Jacob, H. J., and Jacob, M. (1977b). *Anat Embryol.* **150,** 171–186.
Christ, B., Jacob, H. J., and Jacob, M. (1978). *Verh. Anat. Ges.* **72,**353–357.
Christ, B., Jacob, H. J., and Jacob, M. (1979). *Experientia* **35,** 1376–1378.
Falconer, D. S., Gauld, I. K., Roberts, R. C., and Williams, D. A. (1981). *Genet. Res.* **38,** 25–46.
Fallon, J. F., and Crosby, G. M. (1977). *In* "Vertebrate Limb and Somite Morphogenesis (D. A. Ede, J. R. Hinchliffe, and M. Balls, eds.), pp. 55–69. Cambridge Univ. Press, London and New York.
Fell, H. B. (1939). *Philos. Trans. R. Soc. London, Ser. B* **229,** 407–463.
Fischel, R. (1895). *Morphol. Jahrb.* **5,** (23), 544–561.
Gearhart, J. D., and Mintz, B. (1971). *Am. Zool.* **11,** 677–678.
Gearhart, J. D., and Mintz, B. (1972). *Dev. Biol.* **29,** 27–37.
Grim, M. (1970). *Z. Anat. Entwcklungs gesch.* **132,** 260–271.
Grüneberg, H., and McLaren, A. (1972). *Proc. R. Soc. London, Ser. B* **182,** 9–23.
Gumpel-Pinot, M. (1974). *C. R. Hebd. Seances Acad. Sci., Ser. D* **279,** 1305–1308.
Gumpel-Pinot, M., Ed., D. A., and Flint, O. P. (1984). *J. Embryol. Exp. Morphol.* **80,** 105–125.
Hamburger, V., and Hamilton, H. L. (1951). *J. Morphol.* **88,** 49–92.

XI. Muscle and Skeleton of Limbs and Body Wall

Hampé, A. (1959). *Arch. Anat. Microsc. Morphol. Exp.* **48,** 345–478.
Hinchliffe, J. R., and Gumpel-Pinot, M. (1981). *J. Embryol. Exp. Morphol.* **62,** 63–82.
Hinchliffe, J. R., and Gumpel-Pinot, M. (1983). *In* "Current Ornithology." (R.-F. Johnston, ed.), pp. 293–327. Plenum Publ. Corp.
Hinchliffe, J. R., and Johnson, D. R. (1980). *In* "The Development of the Vertebrate Limb," pp. 132–137. Oxford Univ. Press, London and New York.
Hinchliffe, J. R., Garcia-Porrero, J. A., and Gumpel-Pinot, M. (1981). *Histochem. J.* **13,** 643–658.
Hinchliffe, J. R., Gumpel-Pinot, M., Wilson, D. J. and Yallup, B. L. (1984). *In* "Matrices and Differentiation," pp. 453–470. Alan R. Liss Inc., New York.
Houben, J. J. G. (1976). *Arch. Biol.* **87,** 345–365.
Jacob, H. J., and Christ, B. (1980). *In* "Teratology of the limbs" (H. J. Merker, H. Nau, and D. Neubert, eds.), pp. 89–97. de Gruyter, Berlin.
Jacob, M., Christ, B., and Jacob, H. J. (1978). *Anat. Embryol.* **153,** 179–193.
Jacob, M., Christ, B., and Jacob, H. J. (1979). *Anat. Embryol.* **157,** 291–309.
Jorquera, B., and Pujin, E. (1971). *C. R. Hebd. Seances Acad. Sci.* **272,** 1522–1525.
Kieny, M. (1964a). *J. Embryol. Exp. Morphol.* **12,** 357–371.
Kieny, M. (1964b). *Arch. Anat. Microsc. Morphol. Exp.* **53,** 29–44.
Kieny, M. (1977). *In* "Vertebrate Limb and Somite Morphogenesis" (D. A. Ede, J. R. Hinchliffe, and M. Balls, eds.), pp. 87–103. Cambridge Univ. Press, London and New York.
Kieny, M. (1980). *In* "Teratology of the Limbs" (H.J. Merker, H. Nau, and D. Neubert, eds.), pp. 79–87, de Gruyter, Berlin.
Kieny, M., and Chevallier, A. (1979). *J. Embryol. Exp. Morphol.* **49,** 153–165.
Kieny, M., and Chevallier, A. (1980). *Arch. Anat. Microsc. Morphol. Exp.***69,** 35–46.
Kieny, M., and Pautou, M. P. (1976). *Wilhelm Roux's Arch. Dev. Biol.* **179,** 327–338.
Kieny, M., Mauger, A., Chevallier, A., and Sengel, P. (1979). *Arch. Anat. Microsc. Morphol. Exp.* **68,** 283–290.
McCabe, A. B., Gasseling, M. T., and Saunders, J. R. (1973). *Mech. Ageing Dev.* **2,** 1–12.
McLachlan, J. C., and Hornbruch, A. (1979). *J. Embryol. Exp. Morphol.* **54,** 209–217.
McLaren, A., and Bowman, P. (1969). *Nature (London)* **224,** 238–240.
Mauger, A., and Kieny, M. (1980). *Wilhelm Roux's Arch. Dev. Biol.* **189,** 123–134.
Milaire, J. (1976). *Arch. Biol.* **87,** 315–343.
Miller Sulik, K., and Atnip, R. L. (1978). *J. Embryol. Exp. Morphol.* **47,** 169–177.
Mintz, B. (1972). *In* "Molecular Genetics and Developmental Biology" (M. Sussman, ed.), pp. 455–474. Prentice-Hall, Englewood Cliffs, New Jersey.
Mintz, B. (1982). *Science* **215,** 45–47.
Mintz, B., and Baker, W. W. (1967). *Proc. Natl. Acad. Sci. U.S.A.* **58,**592–598.
Moore, J. W., and Mintz, B. (1972). *Dev. Biol.* **27,** 55–70.
Murillo-Ferrol, N. L. (1963). *Ann. Desarollo* **11,** 391–402.
Nathanson, M. A., and Hay, E. D. (1980). *Dev. Biol.* **78,** 301–331.
Pautou, M. P. (1968). *Arch. Anat. Microsc. Morphol. Exp.* **57,** 311–328.
Pautou, M. P. (1973). *J. Embryol. Exp. Morphol.* **29,** 175–196.
Peterson, A. C. (1974). *Nature (London)* **248,** 561–564.
Pinot, M. (1969). *J. Embryol. Exp. Morphol.* **21,** 149–164.
Pinot, M. (1970). *J. Embryol. Exp. Morphol.* **23,** 104–151.
Raynaud, A. (1977). *Colloq. Int. C. N. R. S.* **266,** 201–219.
Roberts, R. C., Falconer, D. S., Bowman, P., and Gauld, I. K. (1976). *Nature (London)* **260,** 244–245.
Saunders, J. W., Jr. (1948). *Anat. Rec.* **100,** 756.
Saunders, J. W., Jr. (1977). *In* "Vertebrate Limb and Somite Morphogenesis" (D. A. Ede, J. R. Hinchliffe, and M. Balls, eds.), pp. 1–24. Cambridge Univ. Press, London and New York.

Saunders, J. W., Jr., and Gasseling, M. T. (1968). *In* "Epithelial-Mesenchymal Interactions" (R. Fleischmaier and R. F., Billingham, eds.), pp. 78–97. Williams & Wilkins, Baltimore, Maryland.
Saunders, J. W., Jr., Gasseling, T., and Gfeller, M. D. (1958). *J. Exp. Zool.* **137,** 39–74.
Searls, L. R., and Janners, M. Y. (1969). *J. Exp. Zool.* **170,** 365–376.
Sengel, P. (1975). *Ciba Found. Sympo.* [N.S.] **29,** 119–121.
Seno, T. (1961a). *Acta Anat.* **45,** 60–82.
Seno, T. (1961b). *Anat. Anz.* **110,** 97–101.
Smith, J. C. (1979). *J. Embryol. Exp. Morphol.* **52,** 105–113.
Stark, R. J., and Searls, R. L. (1973). *Dev. Biol.* **33,** 138–153.
Steen, T. R. (1968). *J. Exp. Zool.* **167,** 49–78.
Strauss, W. L., and Rawles, M. E. (1955). *Am. J. Anat.* **92,** 471–509.
Summerbell, D. (1979). *J. Embryol. Exp. Morphol.* **50,** 217–233.
Summerbell, D., and Honig, J. (1982). *Am. Zool.* **22,** 105–116.
Summerbell, D., and Tickle, C. (1977). *In* "Vertebrate Limb and Somite Morphogenesis" (D. A. Ede, J. R. Hinchliffe, and M. Balls, eds.), pp. 42–53. Cambridge Univ. Press, London and New York.
Summerbell, D., Lewis, J. H., and Wolpert, L. (1973). *Nature (London)* **244,** 492–496.
Sweeney, R. M., and Watterson, R. L. (1969a). *Am. J. Anat.* **126,** 127–150.
Sweeney, R. M., and Watterson, R. L. (1969b). *Teratology* **2,** 199–219.
Tickle, C., Summerbell, D., and Wolpert, L. (1975). *Nature (London)* **254,** 199–202.
Wachtler, F., Christ, B., and Jacob, H. J. (1981). *Anat. Embryol.* **161,** 283–289.
Wachtler, F., Christ, B., and Jacob H. J. (1982). *Acta. Embryol.* (in press).
Warren, A. E. (1934). *Am. J. Anat.* **54,** 449–485.
Wolpert, L. (1981). *Philos. Trans. R. Soc. London, Ser. B* **295,** 441–450.
Wolpert, L., Lewis, J., and Summerbell, T. (1975). *Ciba Found. Symp.* [N.S.] **29,** 95–130.
Yallup, B. L., and Hinchliffe, J. R. (1983). *In* "Limb Development and Regeneration" (J. Fallon and A. Caplan, eds.), pp. 131–140. Alan R. Liss, Inc., New York.
Zwilling, E. (1955). *J. Exp. Zool.* **128,** 423–442.
Zwilling, E. (1961). *Adv. Morphog.* **1,** 301–330.
Zwilling, E. (1966). *Ann. Med. Exp. Biol. Fenn.* **44,** 134–139.

5

Nervous System

CHAPTER **XII**

Chimeras in the Study of the Peripheral Nervous System of Birds

NICOLE M. LE DOUARIN, M. A. TEILLET, AND J. FONTAINE-PERUS

*Institut d'Embryologie du CNRS
et du Collège de France
Nogent-sur-Marne, France*

I.	Neural Crest and Placodal Origin of the PNS................	318
	A. Tracing of Cells from Neural Crest and Ectodermal Placodes to Sensory Ganglia by the Quail-Chick Marker System..........	318
	B. Schwann Cells and Cutaneous Sensory Corpuscles	333
	C. Ontogeny of the ANS Ganglia and Paraganglia	336
II.	Distribution of Developing Potentialities along the Neural Crest	339
III.	Development of Peptidergic Neurons in Chimeras	341
IV.	All Embryonic PNS Ganglia Contain Greater Developmental Potentialities (Both Quantitatively and Qualitatively) Than are Expressed in Normal Development	344
	A. Back-Transplantation of PNS Ganglia in the Neural Crest Migratory Pathway	344
	B. The Dual Cell Line Segregation Model....................	346
V.	Quail↔Chick "Spinal Cord Chimeras" Can Hatch and Survive: Observations and Perspectives	348
	References	349

The ontogeny of the peripheral nervous system is one of the fields where the use of avian chimeras has been the most fruitful during the last 10 years.

The peripheral nervous system (PNS) includes the sensory ganglia of the spinal and cranial nerves and the autonomic nervous system (ANS) that innervates the cardiovascular system, the viscera, and the smooth muscles of the eyes and skin. All the cell bodies of adult vertebrate primary sensory neurons lie outside the central nervous system (CNS) with the exception of the primary sensory neurons of the visual and olfactory systems and the neurons of the mesencephalic nucleus of the trigeminal nerve. The nerve connections of the ANS with the viscera are generally distinct from those of the somatic nervous system. Visceral sensory endings, present in most visceral organs, reach the spinal cord and brain through various visceral nerves, one of the most important of which is the vagus nerve that innervates most of the gut. The efferent visceral nerve impulse is carried from the CNS to the target organ through a two-neuron chain. The cell body of the preganglionic neuron is located in the rachidian bulbus or in the spinal cord. Its typically myelinated axon enters an autonomic ganglion where the nerve impulse is relayed to a second, postganglionic neuron, the nonmyelinated axon of which ensures transmission to the end organ.

In higher vertebrates, the ANS is divided into sympathetic and parasympathetic systems (Gaskell, 1916; Kuntz, 1953). The sympathetic division includes preganglionic neurons located in the intermediolateral column of the spinal cord; they establish synapses with postganglionic neurons of the sympathetic chain ganglia and plexuses (Fig. 1).

Parasympathetic preganglionic fibres are present in cranial nerves III, VII, IX, and X as well as in sacral nerves. The parasympathetic neurons lie in ganglia situated close to or within the innervated organs. As a general rule, the preganglionic myelinated axons are therefore much longer in the parasympathetic than in the sympathetic system, while unmyelinated postganglionic fibres are shorter.

The enteric nervous system is sometimes considered as a separate division of the ANS (Langley, 1921; Gershon, 1981). In fact, many of the neurons of the gut do not have connections with the CNS, and intrinsic peristaltic reflexes can be produced *in vitro,* which means that the gut has sensory receptors, intrinsic primary afferent neurons, and integrative and motor neurons. To a certain extent, the intramural nervous system of the gut thus functions independently of the other divisions of the ANS and CNS (Fig. 1).

Classically, synaptic neurotransmission in the autonomic ganglia of both sympathetic and parasympathetic systems is mediated by the same transmitter, acetylcholine (ACh), while catecholamines (CA), predominantly norepinephrine (NE) (Norberg, 1964), are the transmitters synthesized by the sympathetic postganglionic neurons and ACh is that produced by the parasympathetic nerves. However, there are exceptions to this rule: cholinergic neurons have been demonstrated in sympathetic ganglia [for example, neurons innervating the sweat glands of the footpad in cat and rat (Sjöqvist, 1963a, b; Landis, 1981)], while a few adrenergic cells are present in the parasympathetic system of some species.

In addition, various kinds of neuropeptides also found in the brain have been discovered in neurons of sympathetic, parasympathetic, and enteric ganglia and plexuses. Therefore the classical concept of two kinds of neurons—cholinergic and adrenergic—has to be replaced by a much more complex system of cell

XII. Chimeras in the Study of the Peripheral Nervous System of Birds

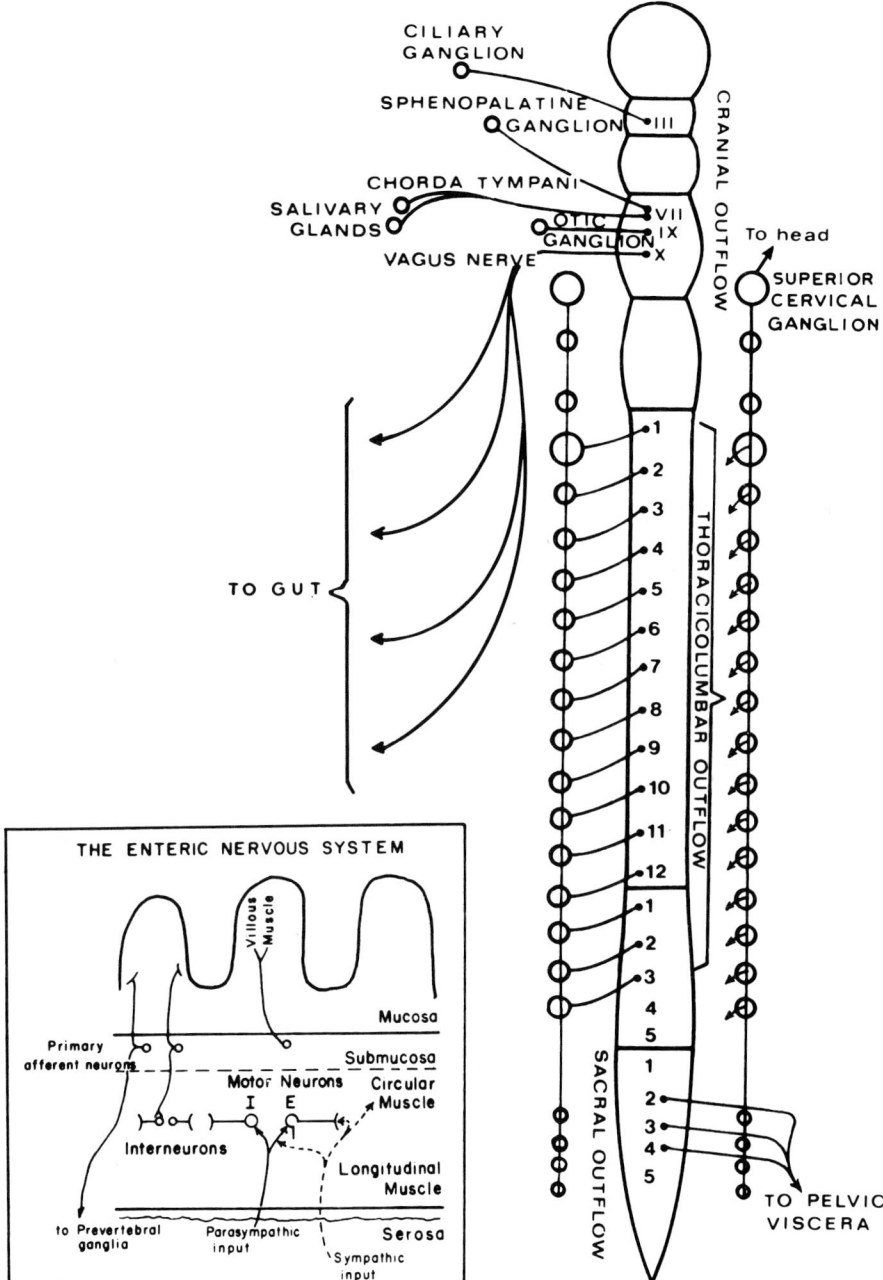

Fig. 1. Schematic representation of the autonomic nervous system. (From Gershon, in Burnstock *et al.*, 1979, with permission.)

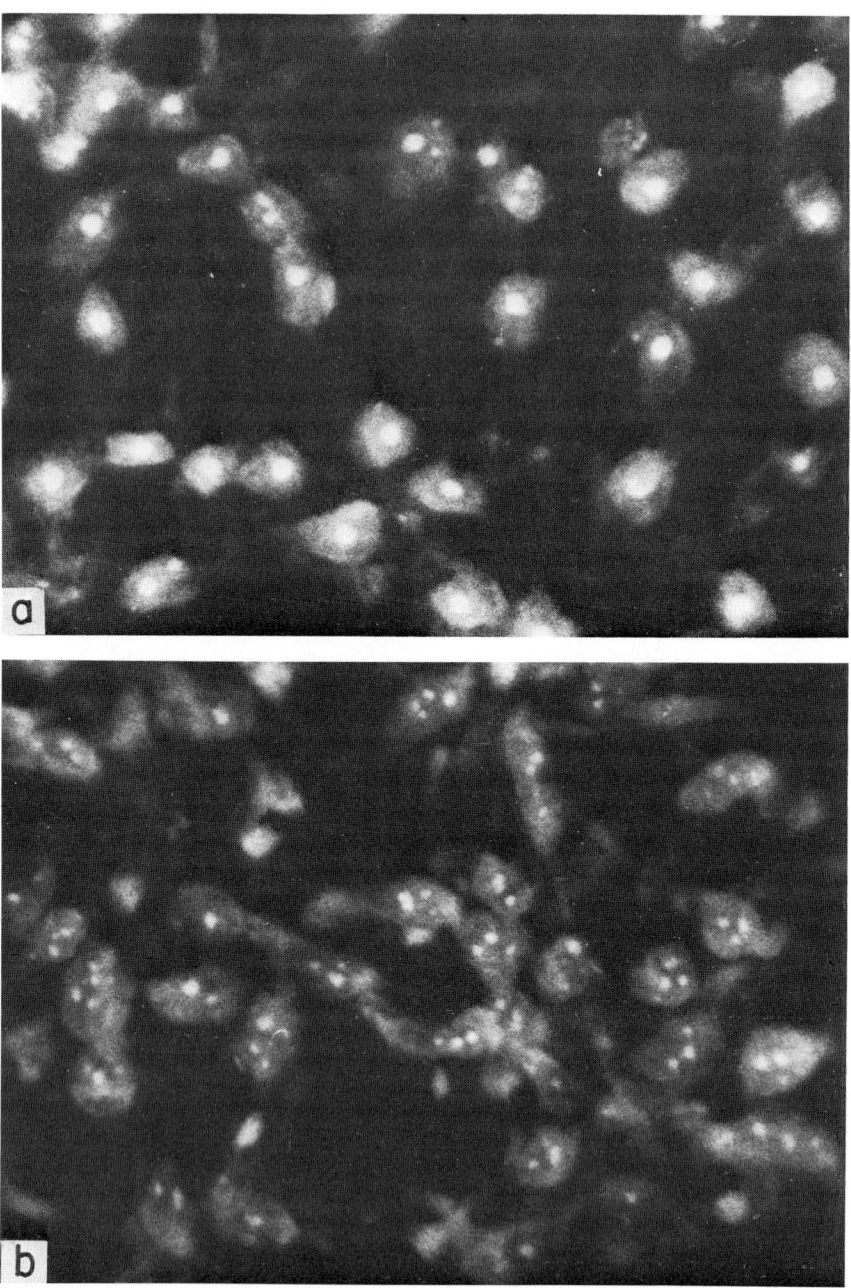

Fig. 2. Distinction of quail and chick cells. Acridine–orange staining of the neural tube in 4 day quail (a) and chick (b) embryos. In quail nuclei the heterochromatin stained in yellow appears as a large central nuclear mass, while in chick nuclei it is dispersed in several smaller chromocenters. × 1000. Ciliary neurons of 10-day quail (c) and chick (d) embryos. In the quail the nucleoli (arrows)

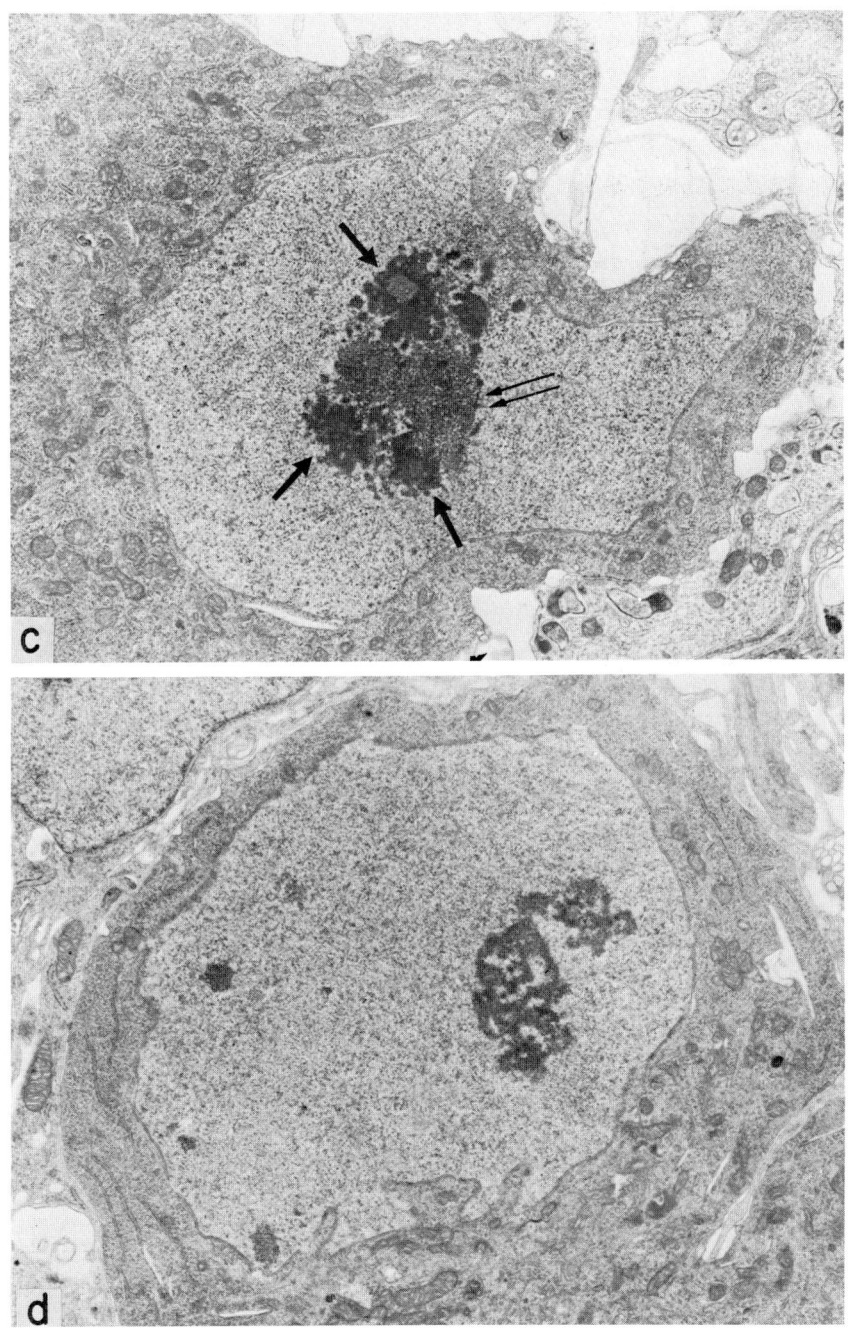

are associated with a large mass of heterochromatin (double arrow), which is lacking in the chick. × 8500.

types with a variety of neuractive substances, whose role is not completely elucidated (see Burnstock *et al.*, 1970; Hökfelt *et al.*, 1980; Furness and Costa, 1980; Gershon, 1981).

I. NEURAL CREST AND PLACODAL ORIGIN OF THE PNS

The role of the neural crest in PNS ontogeny was first demonstrated beyond doubt by means of various microsurgical techniques, including early removal or *in situ* destruction of parts of the neural primordium. However, these experiments did not yield clear-cut conclusions when attempts were made to determine in detail the level of origin of the PNS component cells. This resulted mainly from the regulation capacities of the young embryonic tissues, which tend to replace the missing area through proliferation and differentiation of neighbouring embryonic cells.

This type of analysis is more accurate if cell markers are used. Natural cell markers relying on genetic differences between closely related species or between strains of the same species were used by pioneer workers to carry out xenoplastic grafts of the neural fold in Amphibia. They were made between two species, whose cells could be distinguished either by their size or by their staining properties (Raven, 1936, 1937; Triplett, 1958).

Artificial labelling of nuclei with tritiated thymidine ($[^3H]TdR$) applied to embryos from which pieces of neural primordium were removed and grafted isotopically onto unlabelled embryos of the same species provided even more precise information both in amphibians (Chibon, 1964, 1966) and in birds (Weston, 1963; Johnston, 1966; Noden, 1975). However, the latter method is reliable only for a limited period of time due to the dilution of the isotopic labelling through rapid proliferation of the neural crest cells.

The quail-chick marker system already described in Chapter I was of great help in studying migration and differentiation of neural crest cells in bird embryos because the conspicuous differences existing between the cells of these two avian species are easy to identify in light and electron microscopy (Fig. 2) and remain stable through the whole developmental period. It has been extensively used by our group as well as in other laboratories.

A. Tracing of Cells from Neural Crest and Ectodermal Placodes to Sensory Ganglia by the Quail-Chick Marker System

Several kinds of labelling experiments have been performed to investigate the precise origin of the cell components of the PNS. They consisted of interspecific exchanges between quail and chick embryos at the same developmental stage.

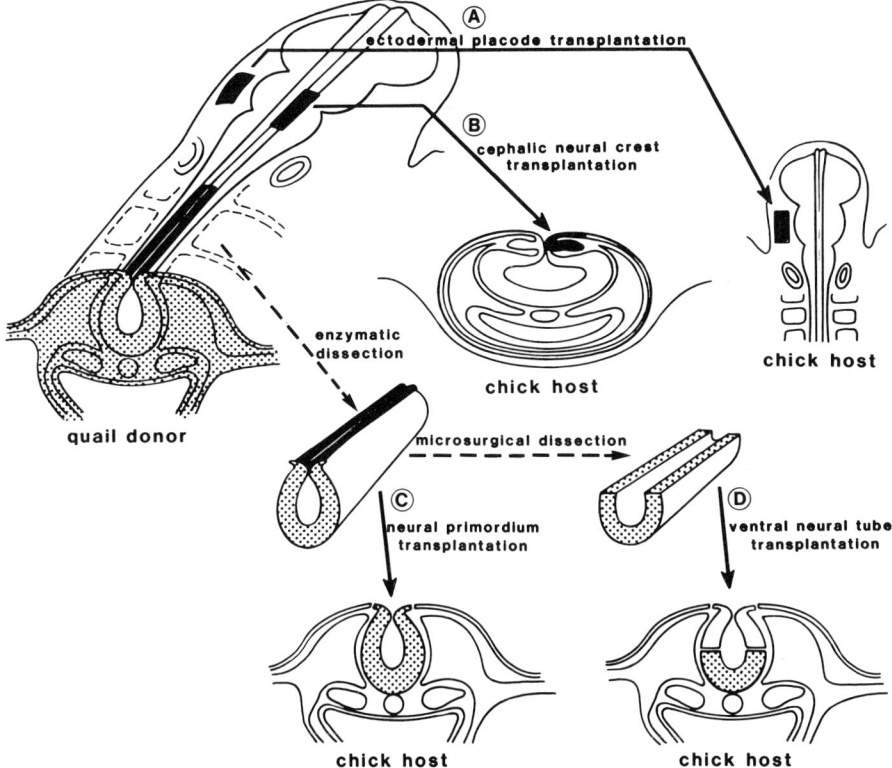

Fig. 3. Experimental designs used to trace the cells of neural crest and placodal origin during PNS ontogeny. Orthotopic and isochronic grafts of placodal ectoderm (A), cephalic neural crest (B), neural primordium composing the neural tube and the neural folds (C), or the ventral neural tube (D) between quail and chick embryos.

Exchanged fragments included either the entire neural primordium (i.e., neural tube and neural folds) or only the neural folds or the ventral part of the neural tube as indicated in Fig. 3 B–D. Exchanges of ectodermal placodes were also performed to study their role in the ontogeny of cranial sensory ganglia (Fig. 3A).

1. ONTOGENY OF THE CRANIAL SENSORY GANGLIA

The development of the sensory ganglia located along the cranial nerves involves the participation of the neural crest and, in some cases, of ectodermal placodes. In all ganglia, the glial cells are derived from the neural crest. In some, the crest also gives rise to the neuronal population; in others, the neurons have a placodal origin. Finally, in the trigeminal ganglion of cranial nerve V, both crest and placodes contribute to the neuronal population.

Cells leaving the neurogenic placodes to form ganglia in the head were already described by pioneer histologists at the turn of the century in all vertebrate classes (fishes: von Kuppfer, 1894; Landacre, 1910, 1912; amphibia: Landacre and McLellan, 1912; Coghill, 1916; birds: His, 1868; Goronowitsch, 1893; Goldby, 1928; mammals: Adelmann, 1925; Da Costa, 1931; Halley, 1955; Verwoerd and Van Oostrom, 1979). This origin was later confirmed by removing one of the two sources of ganglion cells (i.e., the neural crest or the placodes) (Van Campenhout, 1937, 1940; Yntema, 1942, 1943, 1944). By means of a detailed series of extirpation experiments performed in the chick embryo, Hamburger (1961) established that the trigeminal ganglion has a dual origin, large neurons located in the more distal part of the ganglion deriving from an ectodermal placode, and smaller ones, proximally located, from the neural crest.

Neural crest cell labelling with [^{3}H]TdR carried out by Johnston (1966), Johnston and Hazelton (1972), and Noden (1975) provided interesting information on the levels of the neural crest that participate in the formation of some cranial sensory ganglia; however dilution of the isotope precluded tracing further the differentiation of these cells.

Subsequently, the quail-chick marker system was used in the study of the formation of the trigeminal ganglion (Noden, 1978; D'Amico-Martel and Noden, 1983) and of the geniculate, superior, petrose, jugular, and nodose

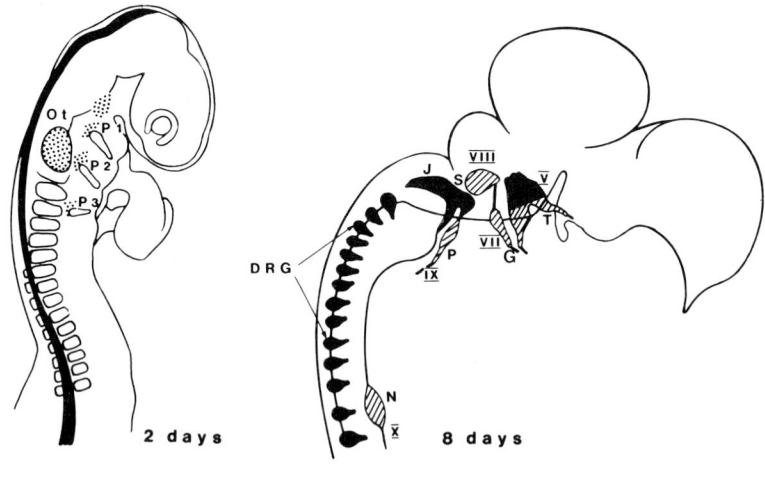

Fig. 4. (A) Distribution of placodes in a 2-day-old embryo. Ot, otic placode; P1, P2, P3, epibranchial placodes. (B) Sensory spinal (DRG) and cranial ganglia at 8 days, the neurons of which are either of crest (black) or placodal (strips) origin. V, VII, VIII, IX, and X indicate the number of cranial nerves bearing sensory ganglia. T, trigeminal; G, geniculate; J. and S, jugular and superior; P. and N., petrosal and nodose.

XII. Chimeras in the Study of the Peripheral Nervous System of Birds

ganglia (Narayanan and Narayanan, 1980; Ayer-Le Lièvre and Le Douarin, 1982; D'Amico Martel and Noden, 1983). Figure 4 shows the anatomical situation of these ganglia in chick and quail embryos around midincubation.

a. Ontogeny of Trigeminal Ganglion of Cranial Nerve V. The most rostral cranial sensory ganglion is the trigeminal associated with cranial nerve V. Its origin is especially complex since, as mentioned above, both the neural crest and placodal cells contribute to its neurons. Their respective role was elucidated by orthotopic transplants carried out by D'Amico-Martel and Noden (1983) as shown in Fig. 3A and B.

These authors demonstrated that the caudal mesencephalic and metencephalic neural crest is the source of the neurons located in the proximal region of the trigeminal ganglion. In contrast, the large distally located neurons in both the maxillomandibular and ophthalmic lobes are of placodal origin. They are labelled with the quail marker following transplants of quail surface ectoderm from

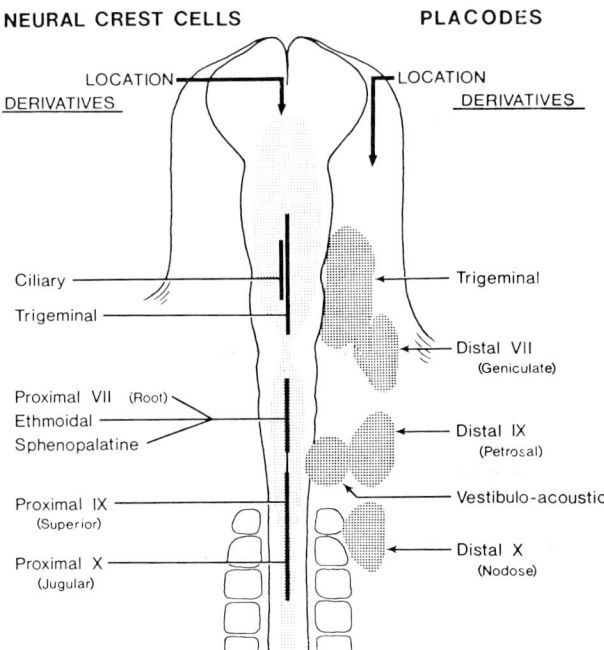

Fig. 5. Schematic drawing of a stage 9.5 chick embryo indicating positions of neural crest and placodal anlagen for cranial sensory and autonomic ganglia. The limits of each anlage were ascertained by comparing measurements taken at the time of surgery with the distribution of neurons containing the quail marker. Although the actual placodal neurogenic areas are undoubtedly smaller than drawn here, it is not possible to be more precise due to the regulative properties of surface ectoderm. (From D'Amico-Martel and Noden, 1983, with permission.)

the metencephalic level (Fig. 5). Occasionally however, isolated crest cells differentiate into neurons in the distal region of the ganglion. In all parts of the ganglion the glial cells are of crest origin. In the normal chick embryo, D'Amico-Martel and Noden (1983) describe the trigeminal placodal ectoderm as small foci of ectodermal cells starting to produce neuroblasts at the most rostral level at stage 11 of Hamburger and Hamilton (1951), with a peak of proliferative activity at stages 14–16 and cessation of neuroblast production at stage 21.

    b. Ontogeny of the Ganglion of Cranial Nerves VII and VIII. Cranial nerve VII (facial nerve) has a distal ganglion (geniculate) but no distinct proximal one, as is the case for nerves IX and X. In fact, clusters of neurons located within the vestibular ganglion are considered as the proximal (root) ganglion of nerve VII.

    The vestibular and acoustic ganglia are associated with cranial nerve VIII. Their neurons are formed by cells released from the otic placode located lateral to the myelencephalon (Fig. 5). The activity in the proximal placode occurs at about 60 hours of incubation (stage 17) (D'Amico-Martel and Noden, 1983).

    Grafting from quail to chick embryo of the otic placode, which forms slightly rostral to the transient first somite, resulted in labelling most neurons in the vestibular ganglion and all neurons in the acoustic ganglion. By contrast, after transplants of quail neural crest at the level of the rostral myelencephalon, Schwann and satellite cells associated with the ganglia of nerves VII and VIII were labelled. In addition, neurons of neural crest origin were present in the vestibular (but not in the acoustic) ganglion. They were often found as clusters at the rostroventral aspect of the vestibular ganglion along the path of cranial nerve VII and can be considered as the root ganglion of this nerve. A few crest-derived neurons with the quail nucleus were also commonly found along cranial nerve VII in the facial canal.

    In such experiments, no neurons were labelled within the geniculate ganglion unless the first epibranchial placode (at the level of the first branchial cleft) was composed of quail cells. This occurred when quail ectoderm of the midmetencephalon and rostral myelencephalon was exchanged between quail and chick (Fig. 5).

    c. Ontogeny of Ganglia of Nerves IX and X. Grafts of the entire neural primordium (Ayer-Le Lièvre and Le Douarin, 1982) or only the neural fold

Fig. 6. Transverse section of a 32-somite chimera at the level of the glossopharyngeal nerve (IX). A quail rhombencephalon (NT) has been implanted at 8-somite stage. The nerve (IX) is lined with quail neural crest cells which also form the primordium of the superior ganglion (S). Quail cells surround and penetrate (arrows) the placodal primordium of the petrosal ganglion (P). This primordium is elongated dorsally from the ectoderm (E) of the second branchial slit (2) corresponding to the second epibranchial placode (P2 of Fig. 4). Feulgen–Rossenbeck staining. × 196. (From Ayer-Le Lièvre and Le Douarin, 1982, with permission.)

XII. Chimeras in the Study of the Peripheral Nervous System of Birds

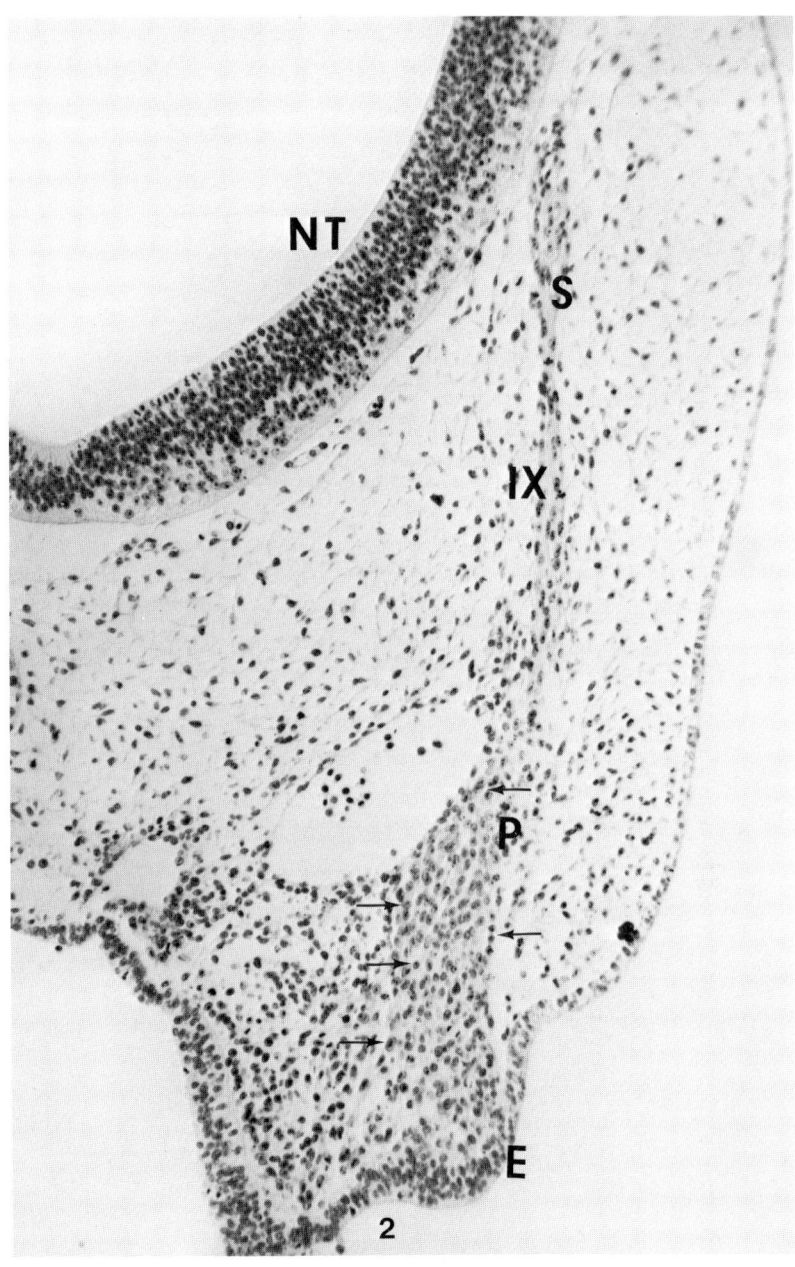

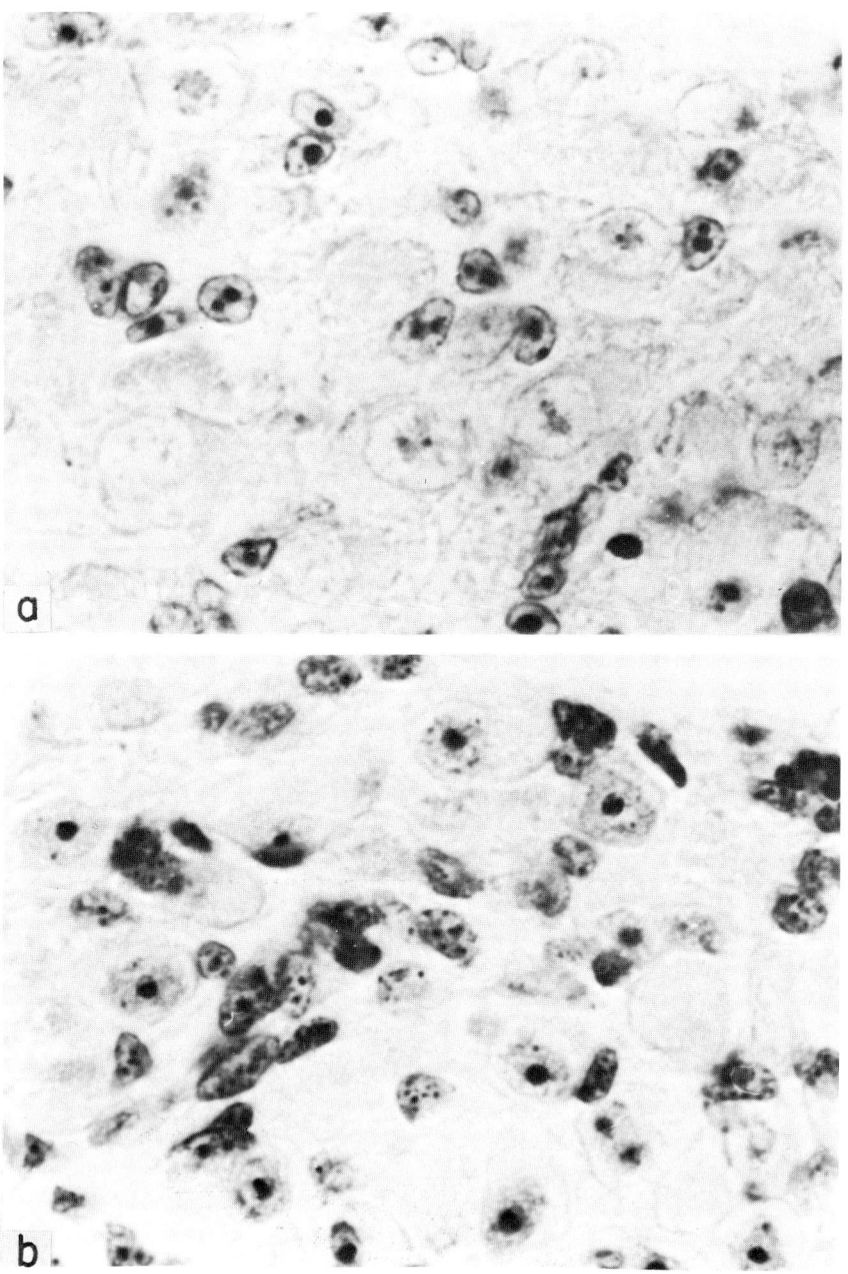

Fig. 7. Chimeric nodose ganglia resulting, respectively, of the graft of a quail rhombencephalon into a chick embryo (a) and vice versa (b). In (a) the neurons of placodal origin belong to the chick host species while the glia originates from the graft. In (b) it is the reverse situation. Note the large nuclei of the neurons with the quail marker. Feulgen–Rossenbeck staining. × 1000.

XII. Chimeras in the Study of the Peripheral Nervous System of Birds

(Narayanan and Narayanan, 1980; D'Amico-Martel and Noden, 1983) at the level of the myelencephalon consistently resulted in the labelling of neurons, satellite, and Schwann sheath cells of the proximal ganglia of nerves IX and X (i.e., superior and jugular).

Neurons of crest origin (i.e., with the quail nuclear marker) were practically never identified in the distal ganglia of nerves IX and X after neural crest grafting. Such experiments gave rise to chimeric nodose and petrosal ganglia in which the neurons derived from the host and the accessory cells from the graft. Figure 6 shows the ganglionic complex located along cranial nerve IX following the graft of a quail rhombencephalon on to a chick host embryo.

The reverse graft of a chick rhombencephalon into a quail host gave a complementary picture with nerve cells of quail type and satellite cells of chick (Fig. 7). In both cases it was clearly shown that blood vessel endothelia within the ganglia and a few connective cells are provided by the host embryo.

For quail neural transplants into chick embryos, analysis was carried out at successive stages of development so that the pathways followed by the crest cells could be identified. At the rhombencephalic level, the crest cells migrate mainly

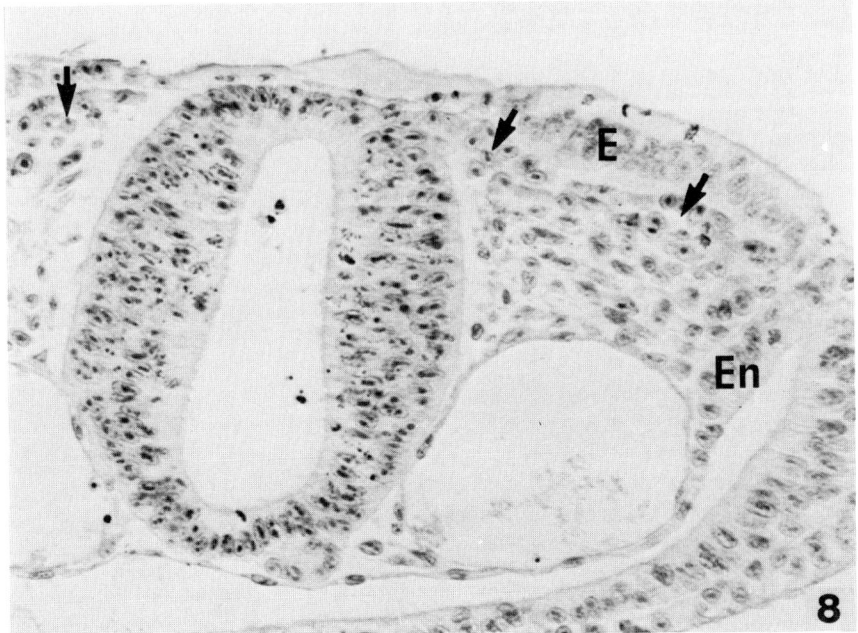

Fig. 8. Transverse section of a 15-somite quail-chick chimera, just posterior to the otic placodes. Quail neural crest cells (arrows) migrate under the ectoderm (E). The front of migration is reaching the lateral aspect of the pharyngeal endoderm (En). Feulgen-Rossenbeck staining. × 360. (From Ayer-Le Lièvre and Le Douarin, 1982, with permission.)

under the ectoderm, that is, between ectoderm and mesoderm (Fig. 8), and, early in the migration process, two streams of crest cells become individualized: a dorsolateral one whose cells aggregate dorsally to form the superior and jugular ganglia close to the CNS, and a more ventral one whose cells migrate further away and stop in the vicinity of the placodes at the same time as the latter release the neuronal precursors. This occurs at stage 18–19 for the second epibranchial placode, which contributes to the petrosal ganglion, and at stage 19–21 for the third, which provides the nodose ganglion neuronal supply. The role of the placodes in this process was clarified by D'Amico-Martel and Noden (1983), who transplanted from quail to chick superficial ectoderm located lateral to the myelencephalon and the three first somites (Fig. 5). Neurons of quail type never populated the proximal ganglionic complex of nerves IX and X, whereas they comprised practically the entire neuronal population of the petrosal and the nodose when the grafted epithelial area was large enough. These transplantation experiments also clearly established that no accessory and glial cells in these ganglia derived from the placodes.

2. ONTOGENY OF THE SPINAL GANGLIA

Spinal ganglia are metamerically distributed along the neural axis. It has been known since the first observations of His (1868) that they originate from the neural crest (sometimes called, for this reason, "ganglionic crest"). In experiments involving neural crest extirpation, the absence of spinal ganglia was often considered as proof of successful and complete crest excision at the level involved (Yntema and Hammond, 1945; Hammond and Yntema, 1947).

As early as 20 hours after orthotopic grafting of a piece of [^{3}H]TdR labelled neural primordium into an unlabelled chick embryo, Weston (1963) found labelled crest cells between the neural tube and the somites. Fifty hours after grafting, the spinal ganglia corresponding to the precise level of the graft, as well as the Schwann cells of the developing rachidian nerves, were found to be entirely of graft origin.

In our hands, application of the quail-chick marker system according to the experimental design represented in Fig. 3C, always resulted in the labelling of all the glial and the neuronal cells at the level of the operation (Le Douarin, 1969; Le Douarin et Teillet, 1970; Teillet, 1971). We have been interested in determining precisely the level of origin of the first dorsal root ganglion (DRG) at the cervical level. As shown in detail in Fig. 9, it appeared that the first spinal ganglion arises at the level of somite 6 in most embryos. Dorsal root ganglia corresponding to somites 6 and 7 are often smaller than the following cervical DRGs, and the first one is completely absent in some embryos (M. A. Teillet, unpublished data).

Both in the chick embryo (Hamburger and Levi-Montalcini, 1949; Pannese, 1974; Carr and Simpson, 1978) and in the quail (N. M. Le Douarin, M. A. Teillet, and J. Fontaine-Perus, personal observations), two populations of neu-

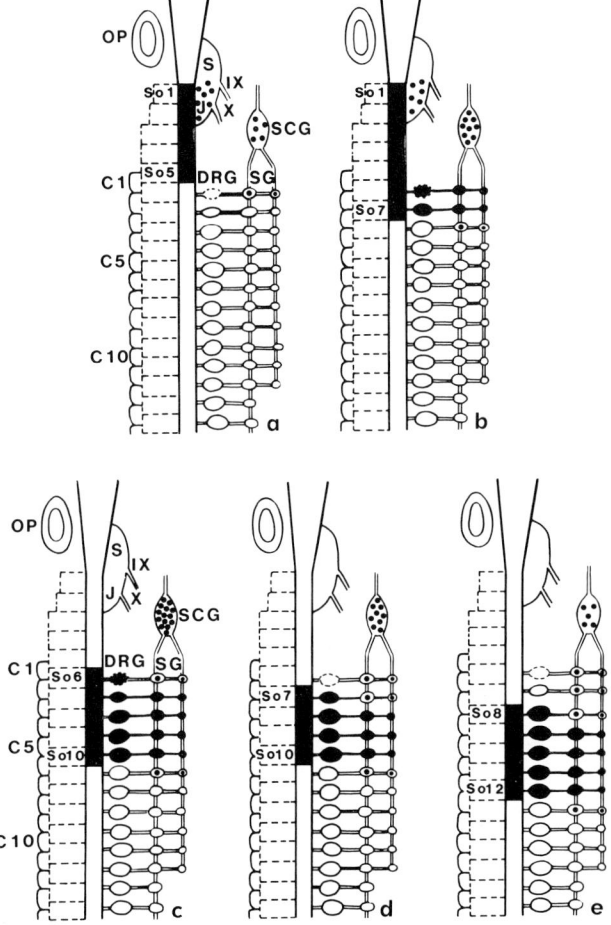

Fig. 9. Level of origin of PNS ganglion cells from the region of the neural crest corresponding to the first 12 somites. In black on the neural axis is represented the level at which the quail neural tube has been implanted. In the ganglia, black dots indicate a participation of donor cells to the ganglion. When the ganglion is totally black it means that it is completely made up of quail cells for neurons and glia. When the operation is made between somite 1 (So1) and somite 5 (So5) (a), a few neurons of the root ganglia of nerve IX and X [superior (S) and jugular (J)] but no DRG are labelled. Donor cells are also found in the superior cervical ganglia (SCG) and in the first ganglia of the sympathetic chains. When the quail neural primordium is grafted down to somite 7 (So7) DRG and sympathetic ganglia of two segments are made of quail cells and quail cells are more numerous in the SCG. When the graft is made between somite 6 (So6) and somite 10 (So10), all the corresponding DRG and spinal ganglia (SG) are made of quail cells, and the SCG cells are almost all of the donor type. When it is made from So7 or So8, one or two DRGs cranial to the anterior limit of the graft are of the host type, and SCG is made up of a mixture of host and donor cells. In conclusion, the first pair of DRG develops in front of So6 and from the neural crest cells of this level (it is lacking in some embryos). SCG arises from neural crest cells belonging to several segments comprised between So5 and So10. Note that the same crest level also provides several other sympathetic ganglia with neurons and glia. OP; otic placode; C1; cervical vertebra 1.

rons in the developing DRG can be distinguished by their rate of growth, their birthdate, and their differentiation. They are referred to, according to their spatial distribution, as lateroventral and mediodorsal. Both types of neuron, and their corresponding glial cells, originate from the neural crest. At the level of the graft the only host cells in the DRG were the blood vessel endothelial cells, a few connective cells, and the periganglionic connective capsule.

By following the early steps of neural crest cell migration in the chick embryo and in quail-chick chimeras, Thiery and co-workers (1982) have shown that the neural crest cells that migrate between the bulk of the somites and the neural tube (i.e., in the intersomite-neural tube pathway) aggregate to form the primordium of the sensory ganglia, whilst the cells migrating ventrally through the pathway located between two consecutive somites, do not contribute to the DRG but reach the para-aortic and paranotochordal area, where they form the autonomic sympathetic ganglia anlagen (Fig. 10). In the experiments of Thiery and co-workers, the crest cell migration pathways were visualised by means of a fluorescent antibody directed against fibronectin, a molecule of the extracellular matrix in which the migrating crest cells are embedded and whose role in migration seems critical (Rovasio *et al.*, 1983).

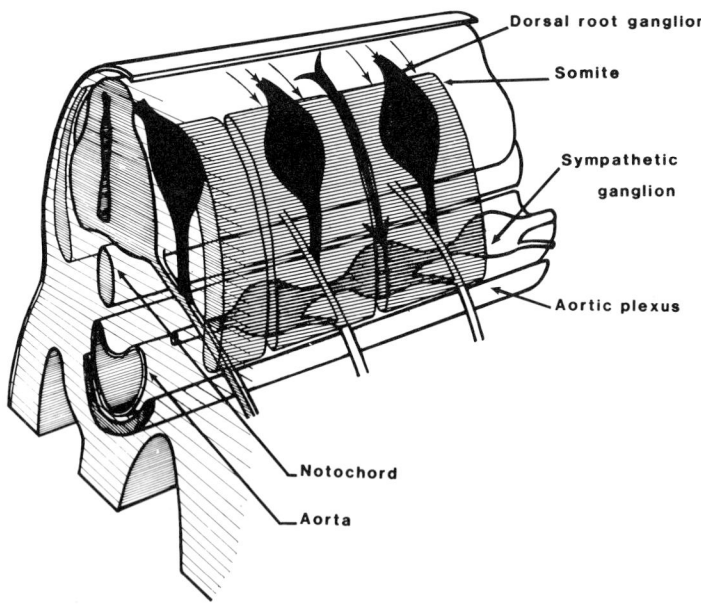

Fig. 10. Diagram showing the final distribution of the trunk neural crest derivatives with respect to the pathway of migration taken by neural crest cells. Large arrow, intersomitic pathway through which most of the cells which give rise to the sympathetic chain ganglia and plexuses migrate. Small arrows; migratory pathway located between somite and neural tube through which the cells which form the DRG migrate.

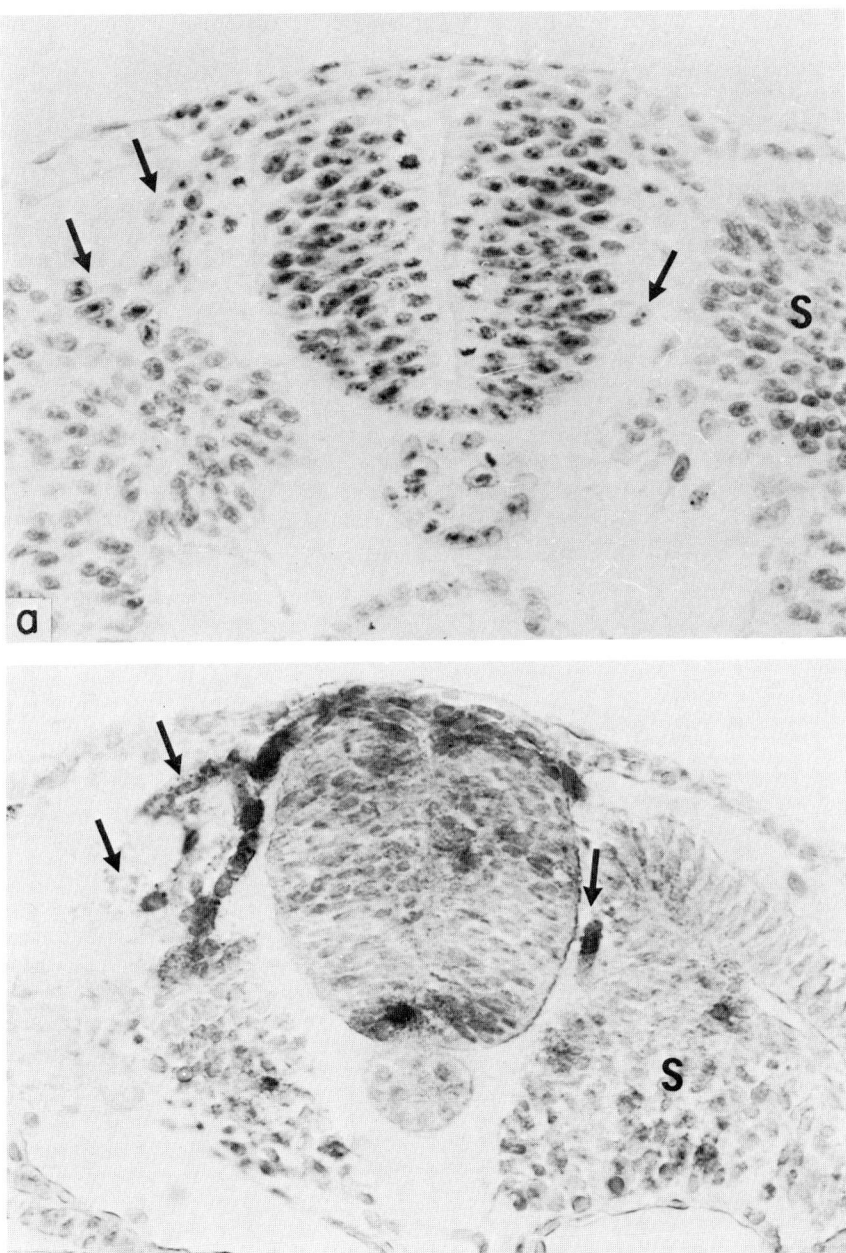

Fig. 11. Transverse sections of a 2-day quail-chick chimera (a) and of an intact chick embryo (b) at the trunk level. On the left side, neural crest cells (arrows) labelled either by the quail marker (a) or expressing AChE activity (b) are migrating between two somites. Since the section is not strictly transverse but somewhat oblique one can see on the right side of the figure crest cells migrating in the inter neural tube–somite pathway. S, somite. (a) Feulgen-Rossenbeck staining. × 460. (b) Karnovsky and Roots technique. × 480.

The existence both of the "intersomitic" and the "intersomite–neural tube" pathways is conspicuous in quail-chick chimeras (Fig. 11a); it can also be visualised by means of a natural marker of neural crest cells at the time of their early migration, the enzyme acetylcholinesterase (AChE) (Cochard and Coltey, 1983) (Fig. 11b). Pictures taken after grafting a quail neural primordium into a chick embryo or in the normal embryo after AChE labelling of the crest cells or immunostaining of fibronectin in the extracellular matrix are absolutely concordant.

3. SUBSTANCE P IN CRANIAL AND SENSORY GANGLIA

A systematic investigation has been carried out on the development of substance P (SP)-containing neurons in sensory ganglia of normal quail and chick embryos and of quail-chick chimeras. Substance P-containing cells were found in spinal ganglia at all levels (cervical, thoracic, and lumbar) of the trunk. In quails, they appear on day 4, in chickens on day 5 as weakly immunostained nerve cells dispersed throughout the ganglia. When the mediodorsal and lateroventral neurons become distinguishable, the SP-positive cells are restricted to the mediodorsal region where practically all the neurons exhibit immunofluorescent granules clearly concentrated in the perinuclear space (Fig. 12). As hatching time approaches, mediodorsal neurons all display SP immunoreactivity, although of variable intensity.

In quail-chick chimeras resulting from orthotopic and isochronic grafting of a quail neural primordium into chick embryos, SP nerve cells also developed in the mediodorsal region of the DRGs with the temporal pattern characteristic of the donor rather than the host species.

It was particularly interesting to study the differentiation of SP-containing neurons in the cranial sensory ganglia of quail-chick chimeras, since their localization paralleled strikingly that of neural crest-derived sensory neurons. In the trigeminal ganglion, SP was restricted to the small proximal neurons whose neural crest origin has been well established through the quail-chick marker system by D'Amico-Martel and Noden (1983) and our own unpublished observations. In contrast, large distal neurons of the maxillomandibular and ophthalmic lobes, which are of placodal origin, were found to be SP negative (Fig. 13).

In the proximal ganglia (superior and jugular) of cranial nerves IX and X, virtually all neurons were SP positive (Fig. 14), while in the petrosal ganglion we

Fig. 12. (a) Transverse section in the trunk of a 10-day quail embryo showing the DRG with small neurons located in the mediodorsal (MD) area and large neurons concentrated in the lateroventral region (LV). Toluidine blue staining. × 280. (b) Immunocytochemical localization of substance P. Frontal section in the trunk of a 10-day chick embryo. Substance P is present in the small nerve cell bodies of the mediodorsal area (MD). The large neurons of the lateroventral ganglionic portion (LV) is devoid of substance P. × 260.

XII. Chimeras in the Study of the Peripheral Nervous System of Birds

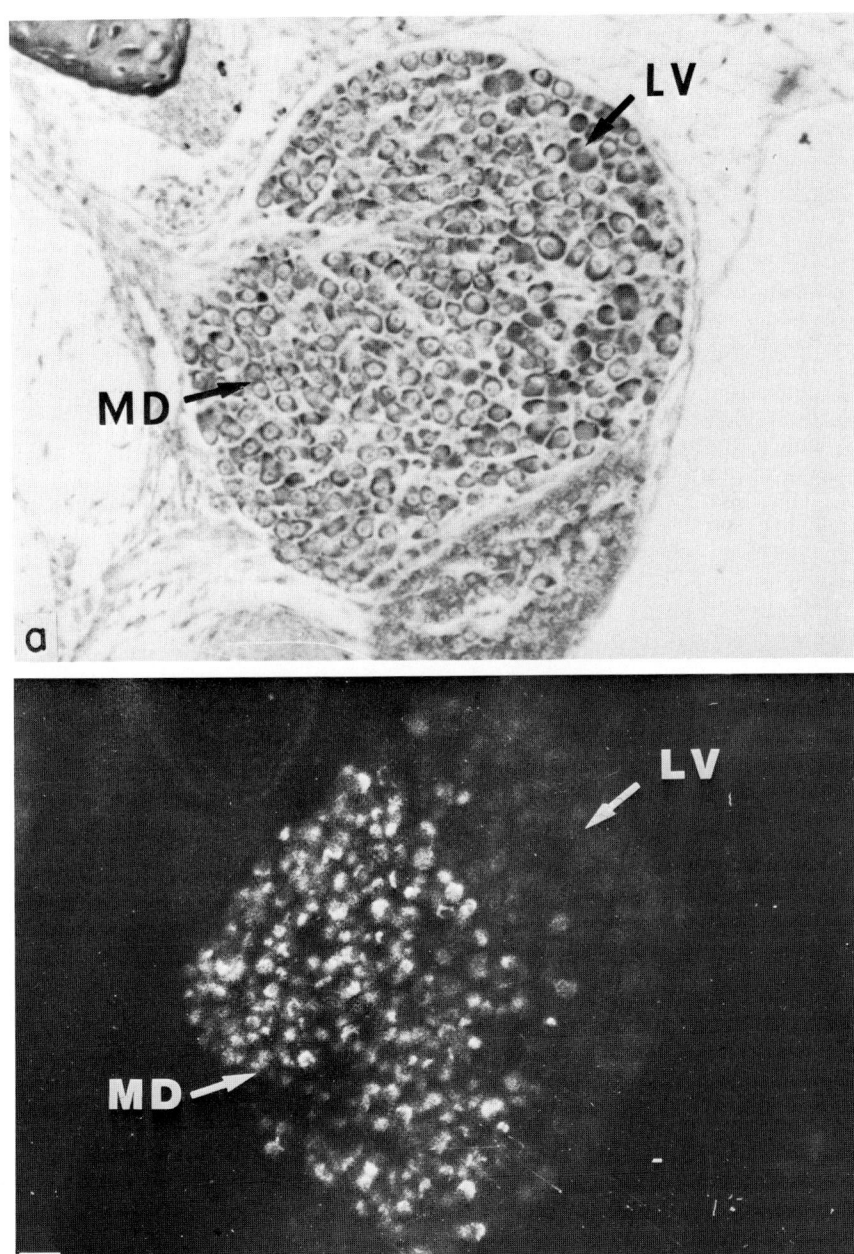

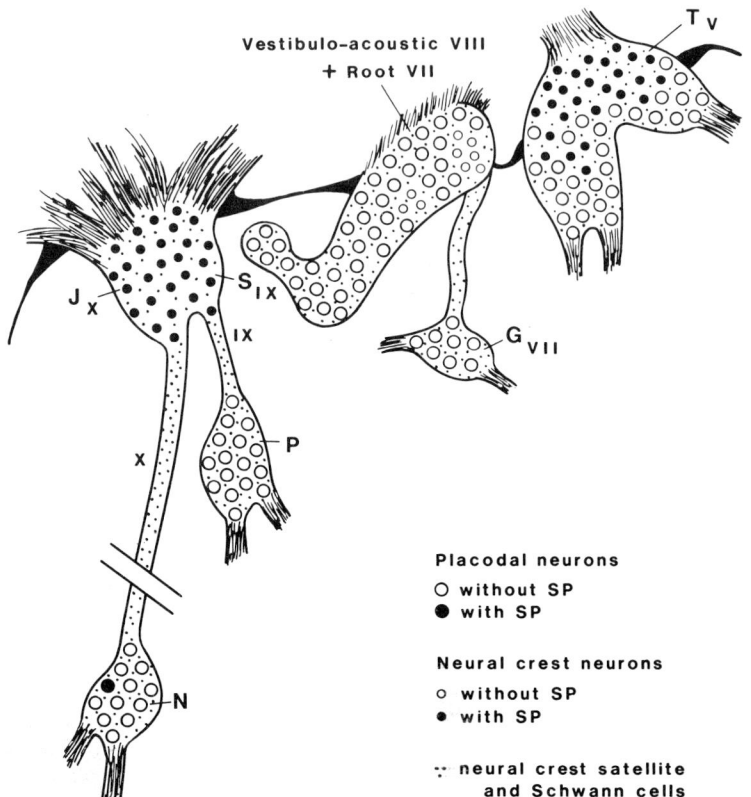

Fig. 13. Schematic representation (adapted from D'Amico-Martel and Noden, 1983) of the sensory ganglia of cranial nerves V, VII, VIII, IX, and X (N, nodose; P, perosal; J, jugular; S superior; G, geniculate; T, trigeminal). Distribution of substance P-containing neurons and of the placodal or neural crest origin neurons of the various ganglia is indicated. It appears that at the exception of a few placodal neurons of the nodose ganglion that show SP immunoreactivity, the placodal neurons are basically SP negative as are the LV neurons of DRG. In contrast the crest-derived neurons of the proximal cranial sensory ganglia are SP positive as are the MD neurons of the DRG. The small number of crest-derived neurons described by D'Amico-Martel and Noden (1983) in the vestibulo-acoustic complex (see root VII, small open circles) have not been found until now with SP (Fontaine-Pérus et al., 1984).

could not detect any SP-positive neuron; only a few were observed in the nodose ganglion, which, like the petrosal, contains neurons of placodal origin. Special efforts were devoted to determining whether the SP-positive nodose neurons observed both in chick and quail were the result of a minor contribution of neural crest to the neuronal population of this ganglion. After grafting quail rhombencephalon into chick embryos as described earlier (Ayer-Le Lièvre and Le Douarin, 1982), we treated sections of the chimeric nodose ganglia successively with

XII. Chimeras in the Study of the Peripheral Nervous System of Birds

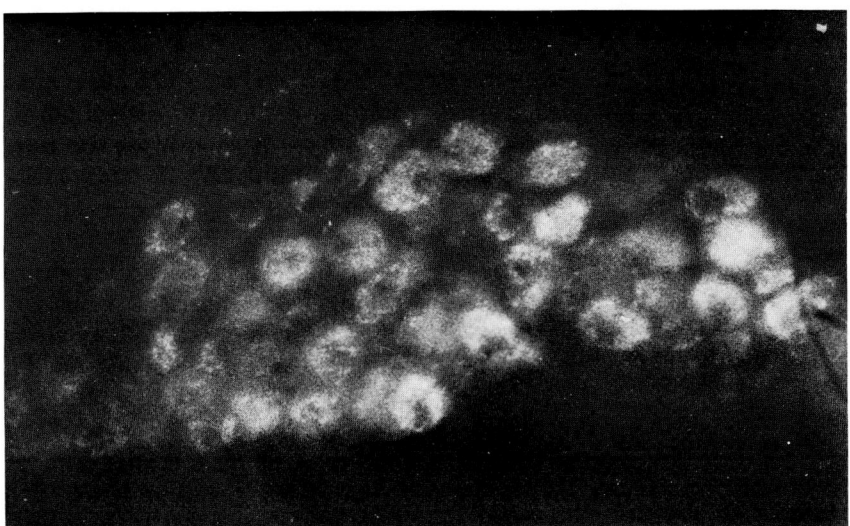

Fig. 14. Jugular–superior sensory complex of the IX and X cranial nerves in a quail at 14 days of incubation. Numerous nerve cell bodies (apparently all nerve cells of the ganglion) contain substance P. × 800.

an antibody against SP and with acridine orange. The latter technique enables quail and chick cells to be distinguished readily, but, unlike the Feulgen method, it can be carried out after peptide immunocytochemistry. In all cases, the SP-positive neurons of the chimeric nodose ganglion were found to be of placodal (chick) origin, as were all the SP-negative neurons (Fig. 13) (J. C. Fontaine-Pérus et al., 1984).

To summarise, as a general rule, the large placode-derived neurons that differentiate in the cranial sensory ganglia do not contain detectable SP; the few exceptions concern certain nerve cells of the nodose ganglion. In the root ganglia, by contrast, the neurons are both SP positive and neural crest derived. At the spinal cord level, the neural crest provides the DRGs with two kinds of sensory neurons, smaller SP-positive ones, located in the mediodorsal area of the ganglion and larger SP-negative ones, located more distally and ventrally.

B. Schwann Cells and Cutaneous Sensory Corpuscles

1. SCHWANN CELLS OF PERIPHERAL NERVES

Although it is unanimously recognised that most Schwann cells lining the peripheral nerves originate from the neural crest, some authors (Raven, 1937; Weston, 1963) suggested that certain sheath cells might originate from the neural tube itself. To clarify this question, either the dorsal part (with the crest) or the ventral part (without crest) of the chick neural tube was replaced by its quail counterpart (Fig. 3C and D). In all the embryos observed (at 6 days of incubation), the sheath cells of both ventral and dorsal nerve roots and of the rachidian nerves were derived from the neural crest; no Schwann cells were ever seen to emerge from the ventral neural tube (M. A. Teillet, unpublished observations).

2. RESPECTIVE ROLES OF DERMAL AND NEURAL CELLS IN THE ONTOGENY OF CUTANEOUS SENSORY CORPUSCLES

The ontogeny of cutaneous sensory corpuscles of vertebrates has been the subject of active research concerning the relationships between the nerves and their peripheral target organs. Experimental analysis and ultrastructural studies have attempted to analyse their morphogenesis, but no conclusive results were obtained since the origin of their component cells was uncertain. A number of

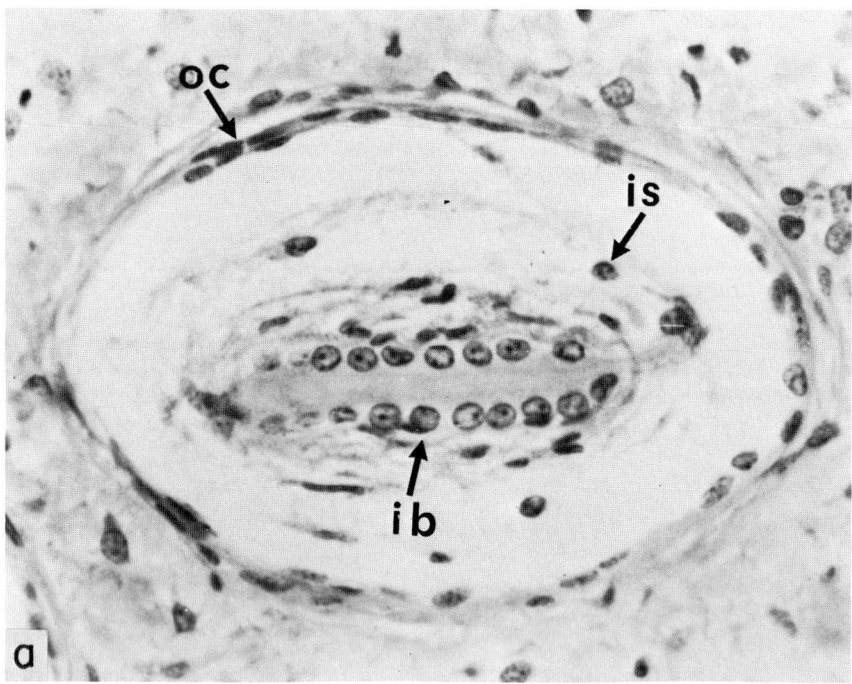

Fig. 15. (a) Herbst corpuscle of the beak of a 15-day posthatching duck. Longitudinal section, Feulgen staining. ib, inner bulb; is, inner space; oc, outer capsule. × 700. (b) Chimeric Herbst corpuscle obtained in a 5-day quail Gasserian ganglion–5-day duck frontal bud association. The inner bulb is composed almost entirely, if not totally, of quail cells (arrows). Most inner space cells and all capsule and dermal cells lack the quail-type nuclear granule. Longitudinal section. × 700. (c) Chimeric Herbst corpuscle obtained in 6-day chick Gasserian ganglion–5-day quail frontal bud association. The cells of the outer capsule and of the inner space contain the characteristic quail type nuclear granule. Most inner bulb cells are devoid of it (arrows). Oblique section. × 800. (From Saxod, 1973, with permission.)

XII. Chimeras in the Study of the Peripheral Nervous System of Birds

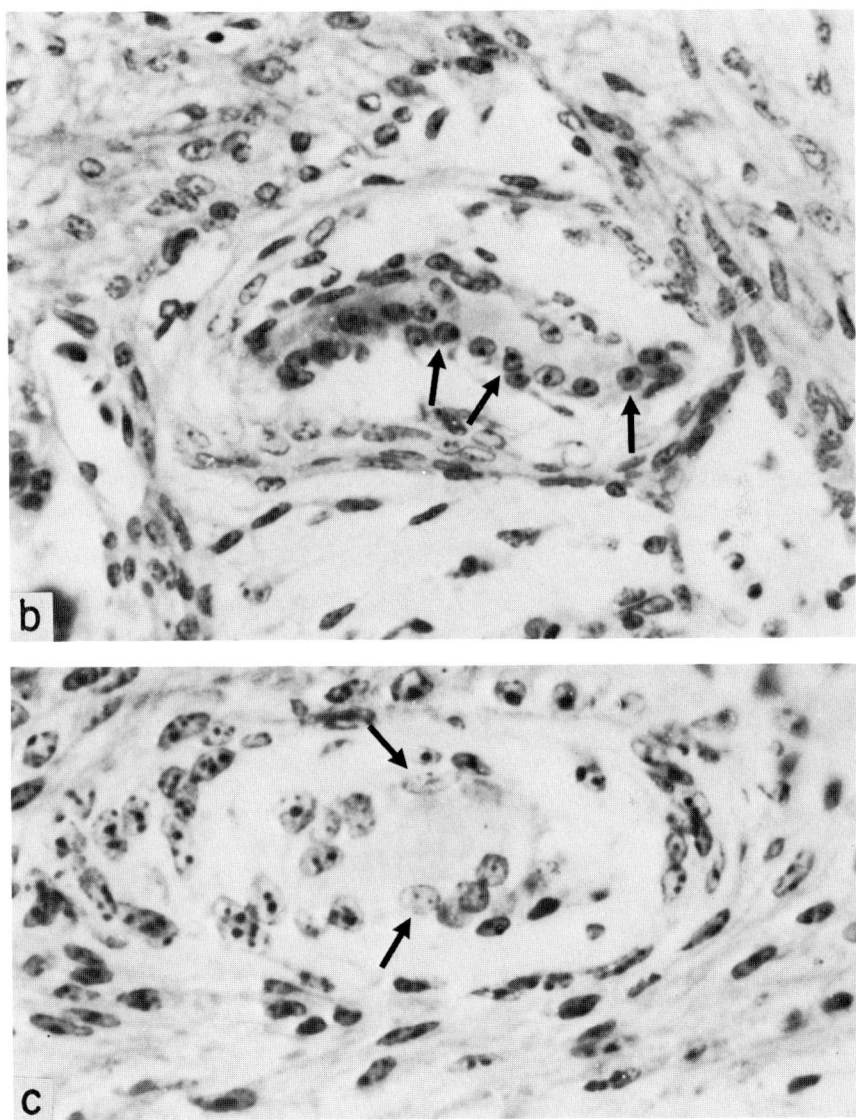

hypotheses have been formulated concerning, for example, Herbst corpuscles of birds and Pacinian corpuscles of mammals, which have a very similar organisation (see review of Shanta and Bourne, 1968). Three main cell types can be distinguished in the Herbst corpuscle: the cells of the inner bulb along the axial nerve ending, the cells of the inner space, and the capsular cells (Fig. 15).

Different opinions about their origins either from the dermal mesoderm or from Schwann cells prevailed (see Saxod, 1973), until Saxod used heterospecific combinations between chick and quail, or duck* and quail, to approach this problem. The design of the experiments was to associate the frontal bud that gives rise to the beak—a structure rich in Herbst corpuscles—with sensory ganglia of another species carrying nuclei with a different structure. The following combinations were performed: (1) grafts of a quail frontal bud on a duck or chick wing bud (2) coelomic grafts of frontal buds associated with a sensory ganglion (spinal or Gasserian). From the chimerism analysis of the Herbst corpuscles that developed, it appeared that the vast majority of the cells of the inner bulb, located along the axial nerve endings and some cells of the inner space, are derived from neural crest cells that accompany the nerve during its outgrowth. These cells are present in the spinal and Gasserian ganglia. All the other cell types of the corpuscle (capsular cells and most cells of the inner space) are provided by the dermal mesenchyme (Fig. 15) (Saxod, 1973).

In another interesting study by the same author (Saxod, 1972), it was shown that the onset of histogenesis of Herbst's and Grandry's corpuscles is entirely dependent on the presence of a somatosensory nerve ending. The nerve ending is also necessary for the maintenance of the structural integrity of the previously differentiated corpuscles.

Recently, Iwanaga *et al.* (1982) reported the presence of S100 protein (a protein specific to glial and Schwann cells) in the inner bulbs of Meissner and Pacinian corpuscles of man and rat. Capsular cells, in contrast, do not show immunoreactivity to S100 antiserum, thus confirming the dual origin of the sensory corpuscles.

C. Ontogeny of the ANS Ganglia and Paraganglia

It has long been recognised that the ANS is entirely derived from the neural crest. However, the precise level of origin of its component cells and the pathways they follow remained controversial, mainly because of biases introduced by the various experimental methods used (for references, see Le Douarin, 1980, 1982). We undertook a systematic study of the origin of the cells forming the ANS ganglia and paraganglia using the quail-chick chimera system (Le Douarin and Teillet, 1971a,b, 1973). Grafts of small fragments (corresponding to a length of four to six somites) of quail neural primordium into chick embryos (and vice versa as controls) were systematically carried out along the whole length of the neural axis. The developmental stages of host and donor embryos were identical and varied according to the level selected for the operation, in order to ensure

---

*The nuclear structure of duck cells stained for DNA is close to that of the chick and very different from that of the quail.

that crest cell migration had not started at the time of the intervention. The migrating neural crest cells were thereafter observed on serial sections of the trunk of the host (for the sympathetic chain and the adrenomedulla) and of its viscera (for the enteric and intravisceral ganglia). A correspondence could be established between the level of the graft and the definitive location of the ganglion cells, as a result of the stability of the labelling provided by the quail-chick cell association. A fate map of the neural crest was thus established (Fig. 16).

In recent experiments, Teillet has shown that the superior cervical ganglion (SCG) is formed by crest cells originating from the level of somites 5 to 10 (Fig. 9). When the grafted quail neural primordium (neural tube + neural fold) occupies this entire territory, the host SCG is completely made up of quail cells. When the graft is done either cranial to the fifth somite or caudal to the tenth

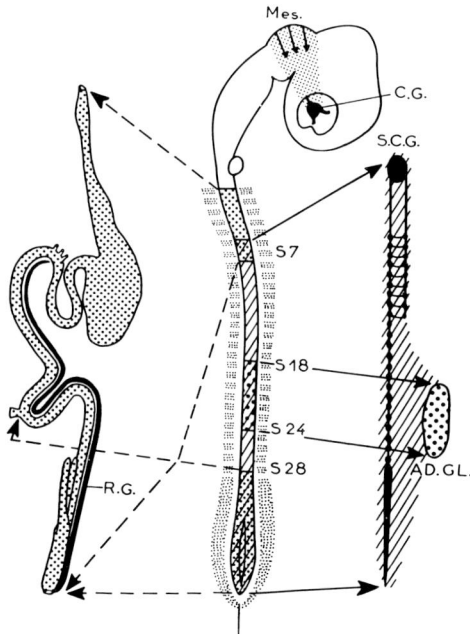

Fig. 16. The origin of adrenomedullary cells and autonomic ganglion cells as deduced from neural primordium transplantations between quail and chick embryos. The spinal neural crest caudal to the level of the fifth somite gives rise to the ganglia of the orthosympathetic chain. The adrenomedullary cells originate from the spinal neural crest between the level of somites 18 and 24. The vagal neural crest (somites 1–7) gives rise to the parasympathetic enteric ganglia of the preumbilical region, the ganglia of the postumbilical gut originating from both the vagal and lumbosacral neural crest. The ganglion of Remak (R.G.) is derived from the lumbosacral neural crest (posterior to the level of somite 28). The ciliary ganglion (C.G.) is derived from the mesencephalic crest (Mes). AD.GL., adrenal gland; S.C.G., superior cervical ganglion.

somite, then no quail cells are present in the SCG. Interestingly, along with the SCG, the 5- to 10-somite neural crest territory also provides the first five paravertebral ganglia with neurons and glial cells. The cells of the following paravertebral ganglia originate in their overwhelming majority from the crest posterior to somite 10.

It is clear that, in the area of the trunk between somites 7 and 28, neural crest cell migration is strictly confined to the dorsal mesenchyme derived from the somites, since the only crest derivatives in this area are the sensory and sympathetic chain ganglia, the aortic and adrenal plexuses, and the adrenomedullary cords. No crest cells penetrate the dorsal mesentery except for the Schwann cells that follow the nerve bundles to the periphery.

In contrast, orthotopic grafts carried out at the vagal and lumbosacral levels of the neural primordium resulted in colonisation of both the dorsal mesenchyme, the mesentery and the splanchnopleure. The migration of the lumbosacral neural crest gives rise, on the one hand, to the caudal part of the sympathetic chain and to the coeliac and pelvic plexuses and, on the other hand, to the ganglion of Remak and some of the intramural neurons of the postumbilical gut. The largest contribution to the myenteric plexuses is made by the crest at the level of somites 1–7, whereas that of the lumbosacral crest is comparatively small.

At the vagal level, the cells migrate lateroventrally under the superficial ectoderm. They reach the lateral wall of the pharynx and become incorporated into its mesodermal wall. Thereafter, they expand craniocaudally by migrating and dividing actively. They reach the level of the pancreatic ducts at about stage 20 and the umbilicus at about 5 days of incubation. The colorectum is not fully colonised before 8 days. This is true not only for the cells that arise in the vagal neural crest but also for those originating in the lumbosacral region of the neuraxis. Teillet (1978) actually demonstrated that the lumbosacral crest cells, which normally participate in enteric ganglion formation, stop in their ventral migration within the anlage of the ganglion of Remak, where they remain for a while before invading the gut itself.

Although it has been observed both in grafts of quail neural primordium into chick embryo, and vice versa, the contribution of the lumbosacral region of the crest to the postumbilical intestine is small; even in the hindgut, both the myenteric (Auerbach's) and the submucosal (Meissner's) plexuses arise mainly from vagal crest cells.

The main parasympathetic derivative of the lumbosacral crest is the ganglion of Remak, a complex structure, peculiar to birds, which develops in the dorsal mesentery. The ganglioblasts arising from the neural crest posterior to the level of somite 28 accumulate first in the mesorectum at stage 24 of Hamburger and Hamilton in the chick and at stage 18 of Zacchei (1961) in the quail. They subsequently migrate cranially, along the ileum and jejunum, and reach the level of the hepatic and pancreatic ducts. In addition to groups of ganglion cells distributed throughout its length, the ganglion of Remak appears to be the route followed by descending and ascending nerve fibers. At the level of the cloaca, it is in close relationship with the pelvic plexus, which also originates from the lumbosacral crest (Teillet, 1978).

Mapping of ANS derivatives in the neural crest reveals that the levels of origin of the various supplies of enteric, parasympathetic, and sympathetic ganglia roughly parallel the areas of the intermediate motor column from which the preganglionic fibers of the parasympathetic (vagal and sacral) and sympathetic (thoracolumbar) systems arise.

II. DISTRIBUTION OF DEVELOPING POTENTIALITIES ALONG THE NEURAL CREST

The heterogeneity of the neural crest with respect to its fate as cholinergic parasympathetic and adrenergic sympathetic autonomic derivatives (Fig. 16) suggested that, at each level of the crest, the cells were already committed before migration to a given type of differentiation.

By changing the initial position of the crest cells along the neural axis before they started migrating, it could be shown that this was not the case. Both the cephalic and vagal neural crest, transplanted at the level of somites 18–24 ("adrenomedullary" level of the crest), provided adrenomedullary-like cells for the suprarenal paraganglia. Conversely, the cervicotruncal neural crest, grafted into the vagal region, colonised the gut and gave rise to cholinergic enteric ganglia (Le Douarin and Teillet, 1974; Le Douarin et al., 1975).

The latter finding was confirmed in the following experimental system. Culture of the hindgut, taken from an embryo before receiving the neural crest cells, on the chorioallantoic membrane (CAM) for 7–10 days resulted in apparently normal muscular development but a total absence of ganglia (Smith et al., 1977). When a fragment of a neural crest was associated with the aneural hindgut before culture, enteric plexuses appeared, irrespective of whether the presumptive fate of the crest cells was to give rise to enteric ganglia or to sympathoblasts and adrenomedullary cells. The cholinergic nature of the ganglia formed in the culture was attested to by the presence of choline acetyltransferase (CAT) activity and high levels of acetylcholinesterase (AChE), while neither tyrosine hydroxylase, the key enzyme for CA synthesis, nor formol-induced fluorescence (FIF) of CA was detected.

Initial pluripotency of premigratory neural crest cells was also shown by Noden (1978) in a study involving heterotopic transplantation of different regions of the cranial neural crest. When the forebrain crest, which normally never gives rise to neural elements, was grafted at the mesencephalic–metencephalic region, crest cells emigrated from their new position and responded to the mid-hindbrain environment by forming normal ciliary and trigeminal ganglia. These ganglia were, however, absent when the reverse transplantation (i.e., replacement of diencephalic by mid-hindbrain crest) was performed.

On the other hand, experiments in which fragments of the quail neural crest were grafted within the crest cell migration pathway of chick embryos (Fig. 17)

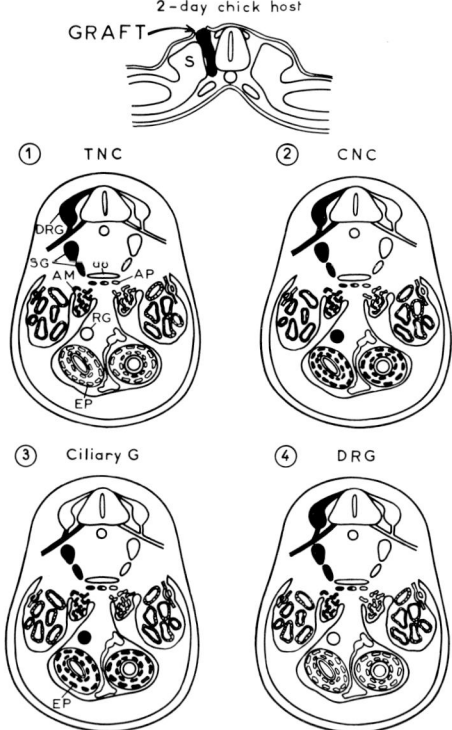

Fig. 17. Selective distribution of quail cells in the host following the transplantation of neural crest cells or of developing PNS ganglia of quail embryos (i.e., ''back-transplantation experiments'') into the neural crest migration pathway of 2-day chicks at the level of somites 18–24. Several days after the graft, the quail cells (in black) are exclusively distributed in various neural crest-derived structures of the host. (1) TNC: graft of trunk neural crest. Quail cells are abundant in the DRG, the sympathetic ganglia (SG), aortic plexus (AP), and adrenal medulla (AM) on the operated side. No cells migrated into the Remak ganglion (RG) or the enteric plexuses (EP). (2) CNC: graft of cephalic neural crest. Same distribution as in TNC plus contribution of the graft to the RG and EP. (3)–(4) Ciliary ganglion and DRG grafts from 4 to 6 day quail. In (3), no quail cells are found in the host DRG, but they are numerous in the gut structures. The reverse situation is found in (4). In both cases quail cells contribute to adrenergic sympathetic ganglia and to the adrenal medulla.

revealed that all PNS derivatives (spinal sympathetic, enteric ganglion cells, and adrenal paraganglion) can arise from any levels of the neuraxis, including the prosencephalon (Le Douarin et al., 1979; Le Lièvre et al., 1980). This does not necessarily imply that all the regions of the crest are strictly equivalent. Differences exist between crest cell populations of the different neuraxis levels as revealed by transplantation of neural crest fragments as supernumerary structures in the chick crest cell migration pathway: (1) the number of spinal sensory neurons and glial cells of graft origin are less numerous when cephalic rather than trunk neural crest cells are transplanted. In this respect a craniocaudal gradient seems to exist, inasmuch as the number of graft-derived cells in the host DRGs increases from prosencephalic to truncal levels of the crest. With the latter level it is not rare to find host DRGs almost entirely composed of quail cells, while only a few DRG cells arise from prosencephalic crest grafts. (2) The migration capabilities of cells at the various levels of the neural crest are different. As mentioned earlier, crest cells at the trunk level (somites 8 to 28) do not extend their migration to the dorsal mesentery and the intestinal wall but remain localized in the dorsal structures of the trunk where they form DRGs, sympathetic chains, and plexuses plus the adrenergic paraganglia. This is so both after isotopic, isochronic grafts and in supernumerary transplantation. In contrast, grafts of cephalic (mainly mesencephalic, rhombencephalic, and vagal) crest at the trunk level result in the colonisation of the gut by a significant number of cells that participate in enteric ganglia (Le Douarin and Teillet, 1974; Schweizer, 1980; Le Lièvre et al., 1980) (Fig. 17). When transplanting supernumerary crest cells, efforts were made to graft roughly the same number of cells, regardless of the level of origin of the implant. Our provisional interpretation of these results is that cephalic crest cells have a higher proliferation capacity than trunk crest cells, on the assumption that cell proliferation is an important component of the migration process.

III. DEVELOPMENT OF PEPTIDERGIC NEURONS IN CHIMERAS

In birds, as in other vertebrates, the gut is the site of differentiation of cells containing a variety of regulatory peptides (for general reviews, see Hökfelt et al., 1980; Schultzberg et al., 1980). So far, distribution of four peptides has been analysed in the adult avian gut in some detail: enkephalins (Ek), vasoactive intestinal peptide (VIP), substance P (SP), and somatostatin (S) (Polak et al., 1974; Sundler et al., 1977, 1979; Alumets et al., 1977, 1978; Vaillant et al., 1980; Brodin et al., 1981; Epstein et al., 1981; Reinecke et al., 1981).

The development of peptidergic neurons was studied in chick and quail embryos by Fontaine-Pérus et al. (1981) and Saffrey et al. (1982). As a general rule, immunoreactive cells develop in the enteric ganglia chronologically according to a craniocaudal gradient. On the other hand, their differentiation takes place 1 or 2 days earlier in the quail than in the chick for every cell type and every level of the embryo considered. For example, immunoreactive fibres for VIP and SP appear first at day 9 in the quail, and at day 10 in the chick gut. Soon afterward, cell bodies become detectable with antibodies against VIP, while SP-positive neurons appear only toward days 13 and 14 in quail and chick, respectively. Enkephalin-positive neurons follow the same pattern of differentiation as VIP and SP, except that they never appear in the oesophagus and start differentiating from the stomach downward around day 9 in quail and 10 in chick. In addition to the above-mentioned peptides, Saffrey et al. (1982) also distinguished neurotensin-positive nerve cells in the chick embryo gizzard and proventriculus from day 11 of incubation.

Besides the gut, sympathetic ganglia of mammals and birds contain cells with VIP and S immunoreactivity (Lundberg et al., 1979, for mammals; Fontaine-Pérus et al., 1982, for birds), and adrenomedullary cells contain Ek and S (our own unpublished observations in birds). Vasoactive intestinal peptide can be first detected in sympathetic cell bodies in quail and chick embryos only from about day 9–10 of incubation. In contrast, S appears earlier, i.e., as early as 3.5 days in the most anterior sympathetic ganglia of the quail, meaning that immunoreactivity develops at about the same time as CA fluorescence.

In order to find out whether peptidergic neurons differentiated normally in chimeric embryos in which quail neural crest cells had been implanted into chick at an early developmental stage, we looked for the appearance of peptide-containing nerve perikarya in the following situations: (1) when the quail neural primordium had been grafted orthotopically and isochronically into the chick host either at the adrenomedullary (somites 18–24) or at the vagal (somites 1–7) levels of the neural axis; (2) when the quail adrenomedullary neural primordium had been heterotopically implanted at the vagal level of the chick host. Under all these conditions, VIP, SP, Ek, and S immunoreactivities were observed in a number of quail cells located either in the peripheral ganglia of the trunk and in the adrenal medulla at the level of the graft (in orthotopic grafts of the adrenomedullary neural primordium) or in the enteric ganglia of the chick gut (in the other types of grafts) (Fig. 18). The developmental stage at which the first neurons become detectable in the host conforms to the genetic characteristics of the effector cells, i.e., they differentiate at the same stage in normal quail neurons *in situ* and in quail neurons transplanted into chick hosts. In contrast, the distribution of the peptidergic neurons in the host depends on the tissue into which the neural crest cells migrate and not on their origin in the neural axis or their fate in normal development (Fontaine-Pérus et al., 1982).

XII. Chimeras in the Study of the Peripheral Nervous System of Birds

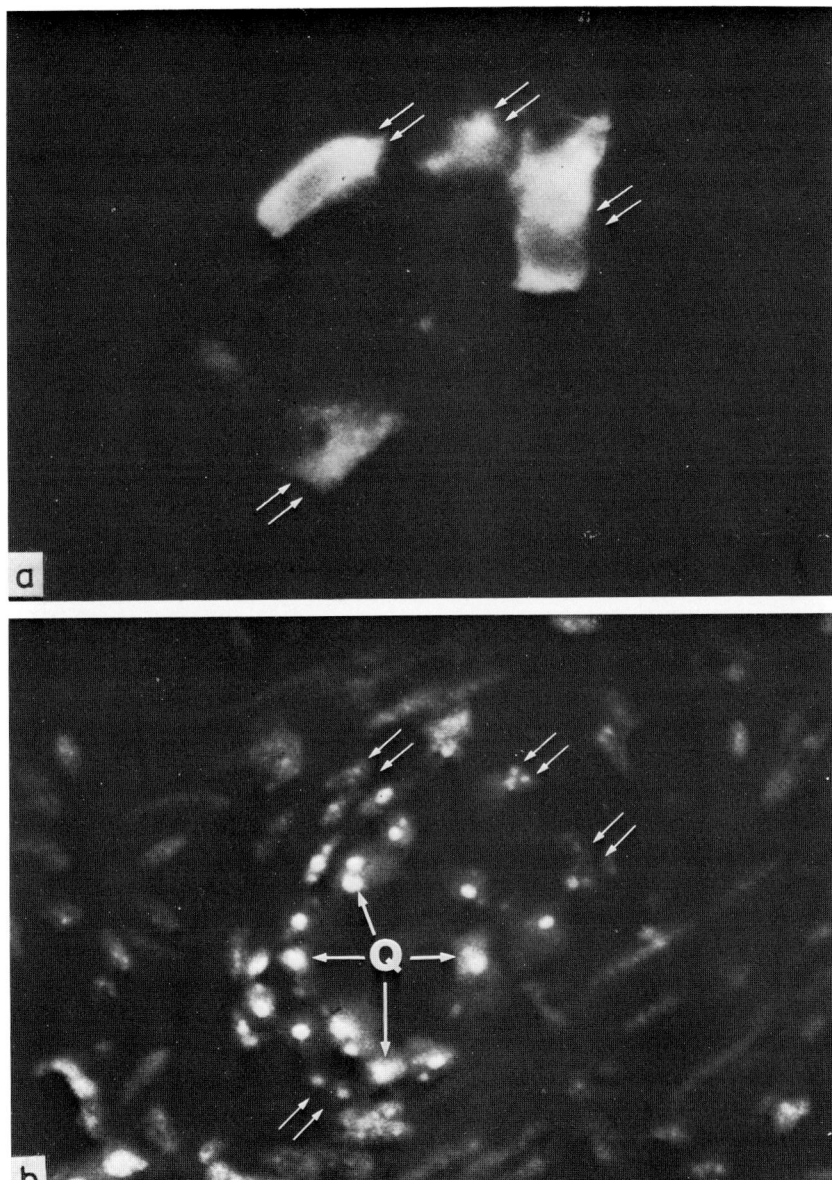

Fig. 18. Section in the oesophagus of a 15-day chick embryo which, at the 9-somite stage, has received the isotopic and isochronic graft of a quail vagal primordium at the level of somites 1–7. (a) Intense substance P immunoreactivity in nerve cell bodies (arrows) of myenteric plexus ganglia. (b) Same section post-stained with acridine orange. The ganglion is entirely composed of grafted quail cells (Q). × 1200.

IV. ALL EMBRYONIC PNS GANGLIA CONTAIN GREATER DEVELOPMENTAL POTENTIALITIES (BOTH QUANTITATIVELY AND QUALITATIVELY) THAN ARE EXPRESSED IN NORMAL DEVELOPMENT

A. Back-Transplantation of PNS Ganglia in the Neural Crest Migratory Pathway

Once neural crest cell migration has taken place and a contingent of cells has reached its destination and aggregated to form a particular ganglion, one would expect their fate rapidly to become irreversible even though, as reported above, their differentiation is regulated through cell–cell interactions with the environment. The fact that, in the developing PNS ganglia, most of the neurons become postmitotic soon after ganglion cells aggregate is in favour of such a view. For example, in chicken DRGs, birthdate studies have demonstrated that most of the neurons become postmitotic between 4.5 and 7.5 days of incubation, whereas the ganglia start to form during day 4 (Carr and Simpson, 1978). In the ciliary ganglion (Dupin, 1984), some neurons are already postmitotic at the very earliest stages of ganglion formation. This could mean that, early in development, the ganglion cell population divides into two lines, the neuronal line (autonomic or sensory) and the non-neuronal (glial) line from which the only cells to arise later are satellite and Schwann cells. In fact, back-transplantation experiments, in which developing quail PNS ganglia were implanted into the neural crest cell migration pathway of a younger chick host (2 days of incubation), revealed that this is not the case (Fig. 17).

From a series of experiments extending over several years (Le Douarin *et al.*, 1978, 1979; Ziller *et al.*, 1979; Le Lièvre *et al.*, 1980; Ayer-Le Lièvre and Le Douarin, 1982; Schweitzer *et al.*, 1983; Dupin, 1984), we can draw the following conclusions:

1. Differentiation of neurons can be evoked from the non-neuronal cell population of every type of embryonic PNS ganglion when it is grafted into the neural crest migration pathway of a 2-day-old embryo at the time of host crest cell migration.

2. The non-neuronal cells divide actively while migrating in the host and stop in the specific crest cell sites of arrest where they contribute to the PNS host structures. This results in a considerable increase in population size in all types of grafted ganglia. As an example, it was shown by Dupin (1983) that transplantation of about 6000 cells from a 4-day quail ciliary ganglion will, when the chick host has reached 8 days of incubation, yield about 1×10^6 cells with a diversity of differentiated phenotypes: neuronal, paraganglionic, satellite, and Schwann

XII. Chimeras in the Study of the Peripheral Nervous System of Birds

cells. In normal development, these 6000 cells from 4-day quail ciliary ganglion would have given rise only to about 15,000 cells over a similar period of development.

3. Only mitotic neuronal precursors can survive in the graft, while the neurons, postmitotic when transplanted, will die. Therefore, all the neurons arising from the graft are derived from undifferentiated precursor cells. Such precursors (or stem cells) still exist in the ganglia until at least the second half of the incubation period and, in the ciliary ganglion, even up to nearly hatching time (Ziller *et al.*, 1979).

4. Two categories of ganglia must be considered with respect to their developing potencies in this type of experiment. Those from which sensory neurons and DRG glia arise and those from which these cell types always fail to develop and which give rise only to the autonomic derivatives (sympathetic ganglia, paraganglia and enteric ganglion cells). In the experiments designed as represented in Fig. 17, sensory ganglion cells arise only from grafted DRGs and never from any kind of autonomic ganglia or from distal ganglia of sensory cranial nerves (such as the nodose ganglion, for example). In addition, grafted DRGs fulfil this function only if they are taken from the quail donor before day 7 of development, i.e., before the birthdate of all their constituent neurons (Schweizer *et al.*, 1983).

The experiments that led to the conclusion that postmitotic neurons do not survive the graft conditions will be described briefly. The nodose ganglion, whose mixed placodal and crest origin is well established, was particularly appropriate to study this problem because, as reported in a recent paper (Ayer-Le Lièvre and Le Douarin, 1982), it is possible to construct chimeric nodose ganglia in which either the neuronal or the non-neuronal cell population is selectively labelled by the quail nuclear marker. The fate of either component can then be followed after back-transplanting chimeric ganglia in chick hosts. It was found that no sensory neurons ever arose from the graft, and that the operation was rapidly followed by the death of the nodose primary sensory neurons. In contrast, a number of quail autonomic neurons developed in the host from the non-neuronal population of the grafted ganglion. They were distributed not only in sympathetic chains and plexuses but also in the adrenal medulla and enteric ganglia.

If postmitotic sensory neurons do not survive under the conditions of back-transplantation, one may wonder how postmitotic autonomic neurons behave. To investigate this point, Dupin (1982) labelled embryonic quail ciliary neurons during their last mitosis by injecting [^{3}H]TdR at 3 and 4 days of incubation. The ciliary ganglion was back-transplanted from a 9- to 11-day donor; at that time, only the neurons were still heavily labelled, whereas the isotope had become diluted in the dividing satellite cell population. Through this radioactive labelling, the quail neurons could be distinguished from the quail glial cells when transplanted into the chick embryo. It quickly became apparent that the neurons did not survive in grafts for more than 2 to 3 days, and that the transplanted cells exhibiting the quail nuclear marker were all derived, in this case also, from the non-neuronal cell population.

Our observations may be applicable to transplantation experiments involving all types of peripheral ganglia. In other words, ganglionic non-neuronal cells retain the potential to generate autonomic neurons, even after the ganglion neurons have become postmitotic. In spinal sensory, as well as in autonomic, ganglia, only ANS precursors remain after the neurons are postmitotic.

B. The Dual Cell Line Segregation Model

In view of these results, one of us proposed as a working hypothesis a model for cell-line divergence during the ontogeny of neural crest derivatives (Fig. 19) (Le Douarin, 1984).

Early in PNS ganglion development and maybe already in migrating neural crest cells, two populations of neuronal precursor cells are segregated; one will end up in DRG, as sensory neurons, whereas the other will give rise to autonomic neurons and paraganglion cells with all the diverse phenotypes implied.

Precursors for sensory neurons of crest origin develop as such only in root ganglia of sensory cranial nerves and in DRG. If they ever reach the sites where autonomic ganglia or distal ganglia of cranial nerves develop, they rapidly disappear. Such a view is suggested by the fact that even when back-transplanted very early, autonomic or distal cranial ganglia fail to give rise to sensory neurons in the host.

The common characteristic of DRG and root ganglia of cranial nerves, which differentiates them from the autonomic and the trunk cranial nerve ganglia, is their close proximity to the CNS. It therefore seems as though the cells of the "sensory line" have strict survival requirements that can be fulfilled *in vivo* only if they develop close to the CNS. DRG cells extend neurites rapidly toward the CNS, suggesting that specific growth factor(s) are produced by the neural tube, thus ensuring their survival and growth.

In addition, while neural crest cells from any part of the neural axis, transplanted in the crest migration pathway, develop into dorsal root ganglia sensory neurons (Le Douarin *et al.*, 1979; Le Lièvre *et al.*, 1980; Schweizer, 1980), the same crest, when cultured *in vitro* for 2 days, gives rise after transplantation to autonomic, but not sensory, neurons (Erickson *et al.*, 1980). This strongly suggests that the presumptive sensory neuron precursor cells die in culture, whereas the autonomic neuronal precursors survive and further develop in the graft.

The autonomic precursors have a large range of developmental options with respect to the transmitter that they will synthesize (e.g., CA, ACh, neuropeptides), and these options are largely under the control of environmental factors

XII. Chimeras in the Study of the Peripheral Nervous System of Birds

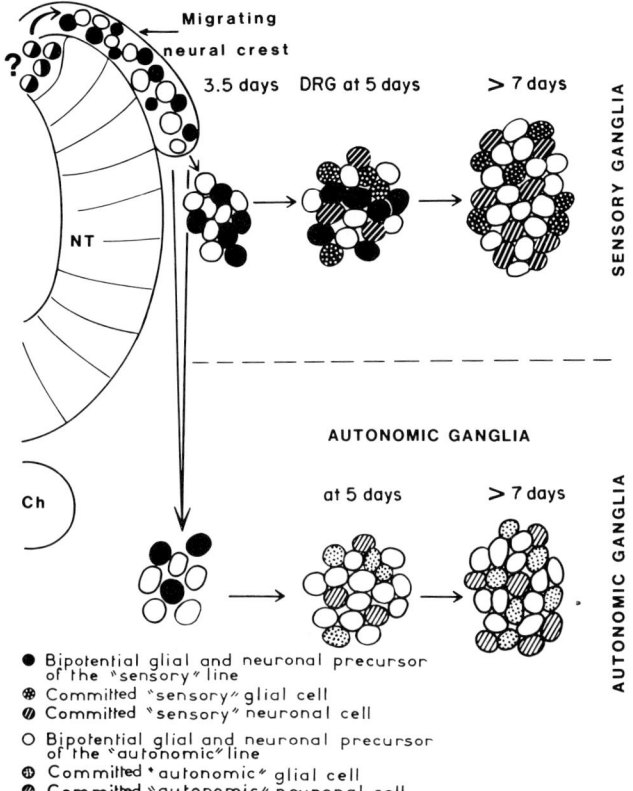

Fig. 19. Hypothetical model for segregation of "sensory" and "autonomic" cell lines in the neural crest and its derivatives at the trunk levels. Sensory precursors survive only in ganglia located in the vicinity of the neural tube (NT). Both sensory and autonomic precursors exist in DRG and autonomic ganglia. See text for more explanations. Ch, notochord.

produced by non-neuronal cells of the embryonic rudiment in which the crest cells differentiate (see Le Douarin, 1982; Fontaine-Pérus *et al.*, 1982).

Autonomic precursors are also present in all types of PNS ganglia, including cranial sensory ganglia such as the nodose. They survive there at least until day 10 of development, although they never express their differentiating capacities in the sensory ganglia environment. These can, however, be elicited if sensory ganglia are back-transplanted into the neural crest migration pathway of a younger host, and, as described above, various kinds of autonomic derivatives arise in the host from grafted sensory ganglia.

Whether there exists a single autonomic precursor from which various neuronal phenotypes (i.e., adrenergic, cholinergic, peptidergic) can arise or whether several specialized precursors are precommitted early, remain open questions.

However, the fact that both ACh and CA can be produced simultaneously by the same autonomic neuron (Furshpan *et al.*, 1976; Landis, 1976; Potter *et al.*, 1981) supports the view that, within a given committed autonomic neuroblast, both cholinergic and adrenergic developmental options coexist.

V. QUAIL↔CHICK "SPINAL CORD CHIMERAS" CAN HATCH AND SURVIVE: OBSERVATIONS AND PERSPECTIVES

Interspecific exchanges of neural primordium (pieces of neural tube together with their attached neural folds), as in Fig. 3C, between quail and chick embryos result in chimeras that can be called "spinal cord- chimeras", since their spinal cord is a mosaic of host and donor parts. The aim of this type of experiment was so far to identify the neural crest migration pathways and fate in normal development and to gain some insight into the mechanisms governing cell differentiation within the various lineages arising from the initial crest cell population. Lately we were tempted to find out whether spinal cord chimeras could hatch and survive and whether their heterospecific central connections and neuromuscular junctions would allow normal behavior.

Hatching is a difficult crisis for an embryo whose eggshell has been opened, an aliquot (albeit small) of albumin removed, and, in addition, has been subjected to surgery. However, recently we have successfully raised posthatching eight spinal cord chimeras, one of which survived for 20 days and three others for more than 2 months; the latter are still alive at the time of writing. All result from isochronic and isotopic grafts of quail neural tube at the level of somites 17 to 20 into chick embryos at the 20- to 24-somite stage (M. Kinutani and N. M. Le Douarin, unpublished).

As expected from previous observations on similarly operated embryos, the chimeras had a normal overall morphology with a transverse strip of quail-like pigmented feathers extending over the wings and the belly and the pattern of pigmentation in the quail is closely reproduced in the feathers of the white Leghorn chick host (Fig. 20).

The four chimeras that survived had no trouble in standing, walking and, at a later period, flying (see frontispiece photograph).

It remains to be seen whether the graft will be finally rejected through immune mechanisms, a highly probably occurrence. Even so, these interesting animals should provide a unique model for studying neural connectivity and certain histogenetic events, such as the relationships between the nerve cell processes and glia at a given transverse level of the neuraxis. For this purpose antibodies selectively recognizing quail (and not chick) nerve cells will be necessary in

XII. Chimeras in the Study of the Peripheral Nervous System of Birds

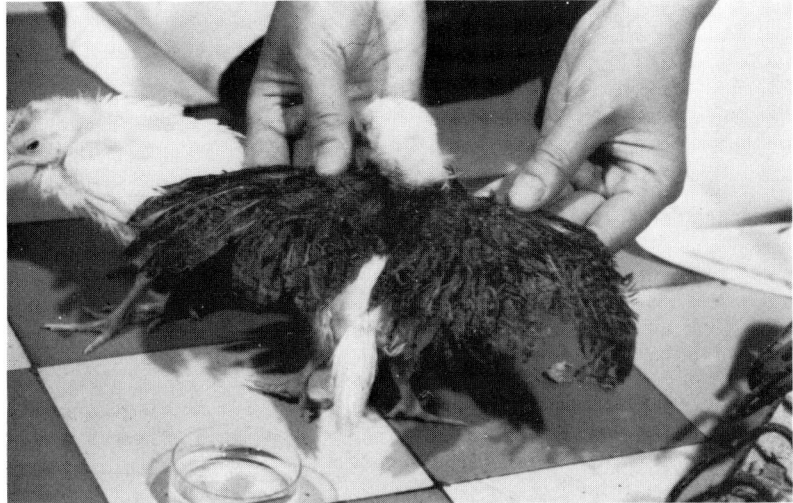

Fig. 20. Spinal cord quail-chick chimera showing a quail-type pigmentation on the feathers of the chick host 22 days after birth.

order to identify quail axons and nerve terminals at all levels of the CNS. Efforts are presently in progress to prepare antibodies with the required specificity.

In any case, the fact that the chimeric birds look quite normal shows most strikingly that all cell–cell interactions and cell recognition mechanisms involved in the building of a nervous system can operate perfectly well between cells of two different animal species.

ACKNOWLEDGMENTS

This work was supported by the Centre National de la Recherche Scientifique, the Délégation Générale à la Recherche Scientifique and by the NIH (Grant 2 RO1 DEO 4257-06).

REFERENCES

Adelmann, H. B. (1925). *J. Comp. Neurol.* **39,** 19–171.
Alumets, J., Sundler, F., and Hakanson, R. (1977). *Cell Tissue Res.* **185,** 465–479.
Alumets, J., Hakanson, R., Sundler, F., and Chang, K. J. (1978). *Histochemistry* **56,** 187–196.
Ayer-Le Lièvre, C., and Le Douarin, N. M. (1982). *Dev. Biol.* **94,** 291–310.
Brodin, E., Alumets, J., Hakanson, R., Leander, S., and Sundler, F. (1981). *Cell Tissue Res.* **216,** 455–469.

Burnstock, G., Campbell, G., Satchell, D., and Smythe, A. (1970). *Br. J. Pharmacol.* **40**, 668–688.
Burnstock, G., Hökfelt, T., Gershon, M. D., Iversen, L. L., Kosterlitz, H. W., and Szurszewski, J. H. (1979). *Neurosci. Res. Program Bull.* **17**, 377–519.
Carr, V. McM., and Simpson, S. B. (1978). *J. Comp. Neurol.* **182**, 727–740.
Chibon, P. (1964). *C. R. Hebd. Seances Acad. Sci.* **259**, 3624–3627.
Chibon, P. (1966). *Mm. Soc. Zool. Fr.* **36**, 1–107.
Cochard, P., and Coltey, P. (1983). *Dev. Biol.* **98**, 221–238.
Coghill, G. E. (1916). *J. Comp. Neurol.* **26**, 247–340.
Da Costa, A. C. (1931). *Arch. Biol.* **42**, 71–106.
D'Amico-Martel, A., and Noden, D. M. (1983). *Am. J. Anat.* **166**, 445–468.
Dupin, E. (1982). Thèse de 3e cycle, Université de Paris VI.
Dupin, E. (1984). *Dev. Biol.*, in press.
Epstein, M. L., Lindberg, I., and Dahl, J. L. (1981). *Peptides* **2**, 271–276.
Erickson, C. A., Tosney, K. W., and Weston, J. A. (1980). *Dev. Biol.* **77**, 142–156.
Fontaine-Pérus, J., Chanconie, M., Polak, J. M., and Le Douarin, N. M. (1981). *Histochemistry* **71**, 313–323.
Fontaine-Pérus, J. C., Chanconie, M., and Le Douarin, N. M. (1982). *Cell Differ.* **11**, 183–193.
Fontaine-Pérus, J. C., Chanconie, M., and Le Douarin, N. M. (1984). *Dev. Biol.*, in press.
Furness, J. B., and Costa, M. (1980). *Neuroscience* **5**, 1–20.
Furshpan, E. J., McLeish, P. R., O'Lague, P. H., and Potter, D. D. (1976). *Proc. Natl. Acad. Sci. U.S.A.* **73**, 4225–4229.
Gaskell, W. H. (1916). "The Involuntary Nervous System." Longmans, Green, London.
Gershon, M. D. (1981). *Annu. Rev. Neurosci.* **4**, 227–272.
Goldby, F. (1928). *Am. J. Anat.* **62**, 135–138.
Goronowitsch, N. (1893). *Morphol. Jahrb.* **20**, 187–259.
Halley, G. (1955). *J. Anat.* **89**, 133–152.
Hamburger, V. (1961). *J. Exp. Zool.* **148**, 91–124.
Hamburger, V., and Hamilton, H. L. (1951). *J. Morphol.* **88**, 49–92.
Hamburger, V., and Levi-Montalcini, R. (1949). *J. Exp. Zool.* **111**, 457–502.
Hammond, W. S., and Yntema, C. L. (1947). *J. Comp. Neurol.* **86**, 237–266.
His, W. (1868). "Untersuchungen uber die erste Anlage des Wirbeltierleibes. Die erste Entwicklung des Hühnchens im Ei." F. C. W. Vogel, Leipzig.
Hökfelt, T., Johansson, O., Ljungdahl, A., Lundberg, J. M., and Schulzberg, M. (1980). *Nature, (London)* **284**, 515–521.
Iwanaga, T., Fujita, T., Takahashi, Y., and Nakajima, T. (1982). *Neurosci. Lett.* **31**, 117–121.
Johnston, M. C. (1966). *Anat. Rec.* **156**, 143–156.
Johnston, M. C., and Hazelton, R. D. (1972). *In* "Third Symposium on Oral Sensation and Perception: The Mouth of the Infant" (J. B. Bosma, ed), pp. 76–97. Thomas, Springfield, Illinois.
Kuntz, A. (1953). "The Autonomic Nervous System," pp. 117–134. Baillière, London.
Landacre, F. L. (1910). *J. Comp. Neurol.* **20**, 309–411.
Landacre, F. L. (1912). *J. Comp. Neurol.* **22**, 1–55.
Landacre, F. L., and McLellan, M. (1912). *J. Comp. Neurol.* **22**, 461–486.
Landis, S. C. (1976). *Proc. Natl. Acad. Sci. U.S.A.* **73**, 4220–4224.
Landis, S. C. (1981). *Br. Soc. Dev. Biol. Symp.* **5**, 147–160.
Langley, J. N. (1921). "The Autonomic Nervous System," Part 1. Heffer, London.
Le Douarin, N. M. (1969). *Bull. Biol. Fr. Belg.* **103**, 435–452.
Le Douarin, N. M. (1980). *Curr. Top. Dev. Biol.* **16**, 31–85.
Le Douarin, N. M. (1982). "The Neural Crest." Cambridge Univ. Press, London and New York.

XII. Chimeras in the Study of the Peripheral Nervous System of Birds 351

Le Douarin, N. M. (1984). *In* "Cellular and Molecular Biology of Neuronal Development" (I. Black, ed.). Plenum, New York. pp. 3–28.
Le Douarin, N. M., and Teillet, M. A. (1970). *C. R. Seances Soc. Biol. Ses. Fil.* **164,** 390–397.
Le Douarin, N. M., and Teillet, M. A. (1971a). *C. R. Hebd. Seances Acad. Sci.* **273,** 1411–1415.
Le Douarin, N. M., and Teillet, M. A. (1971b). *C. R. Hebd. Seances Acad. Sci.* **272,** 481–484.
Le Douarin, N. M., and Teillet, M. A. (1973). *J. Embryol. Exp. Morphol.* **30,** 31–48.
Le Douarin, N. M., and Teillet, M. A. (1974). *Dev. Biol.* **41,** 162–184.
Le Douarin, N. M., Renaud, D., Teillet, M. A., and Le Douarin, G. H. (1975). *Proc. Natl. Acad. Sci. U.S.A.* **72,** 728–732.
Le Douarin, N. M., Teillet, M. A., Ziller, C., and Smith, J. (1978). *Proc. Natl. Acad. Sci. U.S.A.* **75,** 2030–2034.
Le Douarin, N. M., Le Lièvre, C. S., Schweizer, G., and Ziller, C. M. (1979). *In* "Cell Lineage, Stem Cells and Cell Determination" (N. Le Douarin, ed.), pp. 353–365. Elsevier/North-Holland, Amsterdam.
Le Lièvre, C. S., Schweizer, G. G., Ziller, C. M., and Le Douarin, N. M. (1980). *Dev. Biol.* **77,** 362–378.
Lundberg, J. M., Hökfelt, T., Schultzberg, M., Uvnäs-Wallensten, K., Köhler, C., and Said, S. I. (1979). *Neuroscience* **4,** 1539–1559.
Narayanan, C. H., and Narayanan, Y. (1980). *Anat. Rec.* **196,** 71–82.
Noden, D. M. (1975). *Dev. Biol.* **42,** 106–130.
Noden, D. M. (1978). *Dev. Biol.* **67,** 313–329.
Norberg, K. A. (1964). *Int. J. Neuropharmacol.* **3,** 379–382.
Pannese, E. (1974). *Adv. Anat. Cell Biol.* **47,** 1–97.
Polak, J. M., Pearse, A. G. E., Garaud, J. C., and Bloom, S. R. (1974). *J. Br. Soc. Gastroenterol.* **15,** 720–724.
Potter, D. D., Landis, S. C., and Furshpan, E. J. (1981). *Ciba Found. Symp. [N.S.]* **83,** 123–138.
Raven, C. P. (1936). *Wilhelm Roux' Arch. Entwicklungs mech. Org.* **134,** 122–145.
Raven, C. P. (1937). *J. Comp. Neurol.* **67,** 220–240.
Reinecke, M., Schlüter, P., Yanaihara, N., and Forssmann, W. G. (1981). *Peptides* **2,** Suppl. 2, 149–156.
Rovasio, R. A., Delouvée, A., Yamada, K. M., Timpl, R., and Thiery, J. P. (1983). *J. Cell Biol.* **96,** 462–473.
Saffrey, M. J., Polak, J. M., and Burnstock, G. (1982). *Neuroscience* **7,** 279.
Saxod, R. (1972). *J. Embryol. Exp. Morphol.* **27,** 277–300.
Saxod, R. (1973). *Dev. Biol.* **32,** 167–178.
Schultzberg, M., Hökfelt, T., Nilsson, G., Terenius, L., Rehfeld, J. F., Brown, M., Elde, R., Goldstein, M., and Said, S. (1980). *Neuroscience* **5,** 689–744.
Schweizer, G. (1980). Thèse de 3e cycle, Université de Paris VI.
Schweizer, G., Ayer-Lelièvre, C., and Le Douarin, N. M. (1983). *Cell Differ.* **13,** 191–200.
Shanta, T. R., and Bourne, G. H. (1968). *In* "The Structure and Function of Nervous Tissue" (G. H. Bourne, ed.), Vol. 1, pp. 379–459. Academic Press, New York.
Sjöqvist, F. (1963a). *Acta Physiol. Scand.* **57,** 339–351.
Sjöqvist, F. (1963b). *Acta Physiol. Scand.* **57,** 352–362.
Smith, J., Cochard, P., and Le Douarin, N. M. (1977). *Cell Differ.* **6,** 199–216.
Sundler, F., Hakanson, R., Larsson, L. J., Brodin, E., and Nilsson, G. (1977). *In* "Substance P" (H. S. von Euler and B. Pernow, eds.), pp. 59–65. Raven Press, New York.
Sundler, F., Alumets, J., Fahrenkrug, J., Hakanson, R., and Schakfalitzky de Muckadell, O. B. (1979). *Cell Tissue Res.* **196,** 193–201.
Teillet, M. A. (1971). *C. R. Assoc. Anat.* **152,** 734–743.
Teillet, M. A. (1978). *Wilhelm Roux's Arch. Dev. Biol.* **184,** 251–268.

Thiery, J. P., Duband, J. L., and Delouvée, A. (1982). *Dev. Biol.* **93,** 324–343.
Triplett, E. L. (1958). *J. Exp. Zool.* **138,** 283–312.
Vaillant, C., Dimaline, R., and Dockray, G. J. (1980). *Cell Tissue Res.* **211,** 511–523.
Van Campenhout, E. (1937). *Arch. Biol.* **48,** 611–666.
Van Campenhout, E. (1940). *C. R. Seances Soc. Biol. Ses Fil.* **134,** 112–113.
Verwoerd, C. D. A., and Van Oostrom, C. G. (1979). *Adv. Anat. Embryol. Cell Biol.* **58,** 1–71.
von Kuppfer, C. (1894). "Studien zur vergleichenden Entwicklungsgeschichte des Kopfes der Kranioten. II. Die Entwicklung des Kopfes von Ammocoetes Planeri." Lehmann, Munchen.
Weston, J. A. (1963). *Dev. Biol.* **6** 279–310.
Yntema, C. L. (1942). *Anat. Rec.* **82,** 455.
Yntema, C. L. (1943). *J. Exp. Zool.* **92,** 93–120.
Yntema, C. L. (1944). *J. Comp. Neurol.* **81,** 147–167.
Yntema, C. L., and Hammond, W. S. (1945). *J. Exp. Zool.* **100,** 237–263.
Zacchei, A. M. (1961). *Arch. Ital. Anat. Embriol.* **66,** 36–62.
Ziller, C., Smith, J., Fauquet, M., and Le Douarin, N. M. (1979). *Prog. Brain Res.* **51,** 59–74.

CHAPTER **XIII**

Ontogeny and Genetics of the Mammalian Nervous System

RICHARD J. MULLEN

Department of Anatomy
University of Utah School of Medicine
Salt Lake City, Utah

I.	The Normal Cerebellum	355
	A. Normal Development	355
	B. Cell Genotype Markers	355
	C. Purkinje Cell Lineage	357
II.	Cerebellar Mutants	359
	A. Purkinje Cell Degeneration	359
	B. Lurcher	360
	C. Staggerer	361
	D. Weaver	362
	E. Reeler	363
	F. Neuron–Glia Interactions in Cerebellar Mutants	364
	G. Miscellany	364
III.	Other Regions of the Nervous System	365
	A. Retinal Degeneration	365
	B. Retinal Dystrophy in Rats	366
	C. Myelination Mutants	366
IV.	Conclusion	367
	References	367

The nervous system is by far the most complex of all the body systems. Its development begins extremely early and is not complete until well after birth. It contains several types of glia and a far greater number of types of neurons. To build and maintain such a system requires an enormous amount of genetic information. It has been estimated that over half of the genome codes for proteins specific to the nervous system and that the young adult mouse brain contains about 140,000 mRNAs (Hahn *et al.,* 1982), and other RNAs are undoubtedly found in the rest of the nervous system and the developing nervous system. One approach to simplifying this system so that it will be more amenable to analysis is to use genetic mutations to perturb single events in development or function and then to analyze the consequences. This has been an extremely successful approach, but it too has problems when interpreting the results. Many mutants that at first appear to affect primarily a single cell type, on closer examination will reveal effects on other cell types. This raises the question of which effects are primary effects of the gene and which are secondary. Although this is a problem with any system, it is a particularly severe problem with the nervous system because of the multitude of cell types, the protracted period of development, and the fact that the whole design of the nervous system is for cell interaction that makes it susceptible to pre- and postsynaptic alterations. Thus, for these mutants to be most effectively utilized, it is necessary to determine where the genes are acting by constructing a mosaic system in which mutant and normal cells can interact to determine where the primary defect lies. Sidman (1982) has coined the expression "confrontation analysis" for "the many biological experimental designs that juxtapose cells with new neighbors to observe their interactions." This may be done by transplantation, by mixing cells in culture, or by producing chimeras, each of these methodologies having its own advantages and disadvantages. Although the term "confrontation" often implies a bold or hostile meeting, chimeras offer the most gentle and natural association of cells because the cells are together for virtually their entire life and no aberrant development is known to be caused by chimerism per se.

In this chapter, I will review the progress made during the past decade in utilizing chimeras to study the ontogeny and genetics of the mammalian nervous system. Included will be the use of chimeras in studying cell lineage in the nervous system. Lineage studies are not only important in their own right, but are also necessary for correct interpretation of those studies aimed at determining site of gene action. For an extreme example, if each half of the cerebellum were a clone, chimeras would be practically useless for studying the gene action because all of the interacting cells would be of the same genotype. Fortunately, that is not the case.

Before beginning, I would like to emphasize a few points. As we stated in an earlier review (Mullen and Herrup, 1979), "when we state that a particular gene is acting in a particular cell type, we mean only that and do not mean to imply that the gene is not acting anywhere else. Likewise, when we state that a mutant gene does not act in a particular cell type, we mean that if it is acting there, its action alone is not sufficient to elicit the mutant phenotype." Finally, at the cellular level, many of the mutants we study have many facets to their phenotype, and even when a gene is shown to act intrinsically to a particular cell type, that does not mean that all abnormalities of that cell type are due to intrinsic action of the gene.

XIII. Ontogeny and Genetics of the Mammalian Nervous System

I. THE NORMAL CEREBELLUM

Though large, the mammalian cerebellum is a relatively simple and highly organized structure (Fig. 1). The cerebellar cortex, a highly convoluted structure, consists of lobules and smaller folia. These in turn consist of three layers: an outer molecular layer with few cells, a single-cell layer of the large Purkinje cells that are the only cortical cells to send their axons out of the cortex to the deep cerebellar nuclei, and the internal granule cell layer that is packed with millions of small granule cells. The axons of the granule cell rise and bifurcate in the molecular layer where they will synapse on the tree-shaped (but planar) dendrites of the Purkinje cells. There are three other less prominent neuronal cell types in the cortex—stellate, basket, and Golgi II cells—but they are of minor importance to the topic at hand. The most prominent glial cell type is the Bergmann glia, whose soma lies in the Purkinje cell layer and whose processes traverse the molecular layer. In the following discussions, two functions of this glial cell are particularly germane: in developing cerebellum it serves as a guide for migrating granule cells and in the adult it ensheaths the Purkinje cell dendrite.

A. Normal Development

The cerebellum develops from the roof of the fourth ventricle. The Purkinje cells are generated at the ventricular surface on embryonic days 11–13 and migrate outward to take their place in the developing cortical plate. The granule cells, on the other hand, are derived from the subventricular zone in the rhombic lip and migrate across the surface of the developing cerebellum, continuing to divide as they migrate. When they have completed their final division, which is not until after birth, they first become bipolar cells as they send out processes in opposite directions parallel to the folia. These processes, called parallel fibers, synapse on the tertiary branchlet spines of the Purkinje cell dendrites. The nucleus then translocates in a third process descending across the molecular layer, past the Purkinje cells to its final position in the granular layer. During this inward migration, the granule cell migrates along the radial Bergmann glia fiber.

B. Cell Genotype Markers

The scarcity of good cell genotype markers has been the single most limiting factor in fully utilizing the potential of chimeras. The topic is more fully discussed by West in Chapter II, and here I will only briefly mention some of the markers that have been used in studies of the nervous system.

Although we have been limited by cell markers, we have also been quite fortunate. The histochemical visualization of the high- and low-activity alleles of β-glucuronidase first demonstrated in chimeric liver by Condamine *et al.* (1971) and later adapted to use for some cell types in the nervous system (Feder, 1976;

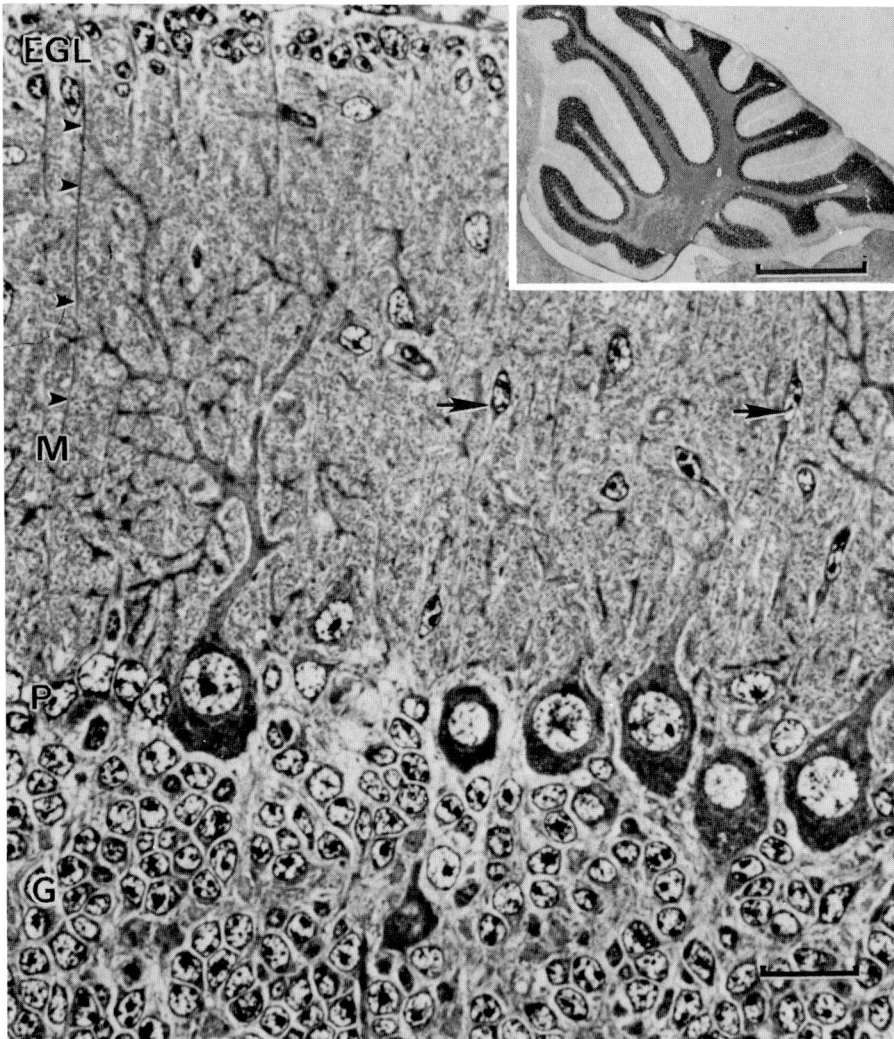

Fig. 1. A 1-μm section of cerebellum from normal mouse 13 days after birth showing several characteristic features of the cerebellar cortex. The mature cortex consists of the molecular layer (M), Purkinje cell layer (P), and granule cell layer (G). At this age, however, a few cells are still present in the external granular layer (EGL) but these will eventually migrate to the granule cell layer along Bergmann glial fibers (arrowheads). Migrating granule cells (arrows) can also be seen. The Purkinje cell to the left shows part of its expansive dendritic tree branching in the molecular layer. Scale, 20 μm. The inset is a 5 μm midsagittal section of a mature cerebellum. Scale, 1mm.

Mullen, 1977a) has proved invaluable in studies of the cerebellum. Although the enzyme appears to be present in a great variety of cell types, it is most obvious in larger neurons as opposed to granule cells, for example, which contain very little cytoplasm. However, in chimeras one does not usually see stained and unstained cells but rather stained and lightly stained cells. Several years ago, we showed that a similar phenomena in chimeric liver was due to transfer of enzyme rather than induction of enzyme in low activity cells (Herrup and Mullen, 1979c). Presumably this also happens in nervous tissue, but for some unknown reason does not happen to Purkinje cells. The majority of cerebellar mutants have some effect on Purkinje cells, and so, fortunately, the glucuronidase marker system has made them amenable to study. It can be used to mark other neuronal and glial cell types but must be used with a bit more caution.

Cerebellar granule cells also contain β-glucuronidase, but it is usually visualized as only a small spot of staining in each cell and, therefore, is not very reliable as a cell marker. Green *et al.* (1975) reported an abnormal distribution of heterochromatin in leukocytes from ichthyosis (*ic*) mutant mice. We examined the nervous system and found that certain neuronal cell types, such as granule cells, displayed a similar abnormal distribution of heterochromatin (Goldowitz and Mullen, 1982a). In sections of normal cerebellum, the heterochromatin is usually seen as a few small clumps along the nuclear membrane, whereas in *ic/ic* mice, the heterochromatin is usually seen as a single large mass in the center of the nucleus. Serial reconstructions revealed that virtually all *ic/ic* granule cells had the centrally placed mass, but in single sections only a certain percentage of the cells, depending on section thickness, would display it. In addition, a small percentage of normal granule cells have a similar phenotype. Thus, ichthyosis can be used to determine what proportion of granule cells in a chimera are *ic/ic*, but it is less useful for determining the genotype of individual cells.

Glucosephosphate isomerase (GPI) variants have been used in numerous studies to demonstrate tissue mosaicism by electrophoresis. However, it was not until Gearhart and Oster-Granite (1980; Oster-Granite and Gearhart, 1980) developed an antibody against the GPI-1B variant that it could be used as a cell genotype marker by the peroxidase–antiperoxidase (PAP) technique.

C. Purkinje Cell Lineage

Cerebellar Purkinje cells are the only cell type in the brain whose lineage has been studied to any significant extent with chimeras. The reasons for this include the fact that the available cell genotype markers work well on Purkinje cells, the cells are easily identified and in a single row in sections, and, finally, their numbers (a few hundred per section, less than 200,000 total) are manageable.

All laboratories that have examined Purkinje cell mosaicism in chimeras agree that the mosaicism is "fine-grained" (Dewey *et al.*, 1976; Mullen, 1977a, 1978;

Oster-Granite and Gearhart, 1981). In Purkinje cell degeneration (*pcd*) chimeras, which will be further discussed below, in which the mutant cells had degenerated, the surviving cells appeared to be near randomly distributed both in single sagittal sections and in a two-dimensional reconstruction of 25 serial sections (Mullen, 1977a). Since that distribution could conceivably have somehow been due to rearrangement of $+/+$ cells after degeneration of the *pcd* Purkinje cell population, a high–low β-glucuronidase chimera in which there was no cell loss was also examined. The average patch size of high glucuronidase cells, which comprised 17.5% of the population of 1937 cells in 5 sections, was 1.26, which was only slightly greater than what would be expected in a random linear array. However, it was also found that the proportion of the two genotypes varied significantly in different cerebellar regions, so it could not be said that the distribution throughout the entire cerebellum was random (Mullen, 1977a, 1978). Oster-Granite and Gearhart (1981) also observed regional variations in the proportions of the two genotypes in chimeras they have examined with the anti-GPI marker. However, they found that the average patch size was greater than would be expected in a random array with an estimated clone size of 1.81 to 4.13 cells compared with my estimate of 1.03. In addition to different cell markers, there were also a few other differences between the two studies. My analysis was based on single 8 μm sections, whereas Oster-Granite and Gearhart used serial reconstructions of three 5 μm sections. Also, different strains of mice were used. Regardless of whether the differences are real or due to different techniques, the two studies do demonstrate the mosaicism is indeed fine-grained and, if there are clones, they are extremely small and of variable size.

Another approach to Purkinje cell lineage has been reported by Wetts and Herrup (1982b) and the results are fascinating. They were counting the number of surviving Purkinje cells in lurcher (*Lc*) chimeras, which are similar to *pcd* chimeras in that the mutant Purkinje cells degenerate leaving only the genetically normal cells (Wetts and Herrup, 1982a). They noticed that the number of surviving cells in four chimeras, as well as the numbers in three controls in which there was no cell loss, were all integral multiples of 10,200, the lowest number of wild-type cells in half of the cerebellum of one of the chimeras. Allowing for a 3% counting error, the probability of obtaining the numbers they observed by chance alone was 0.0053. They proposed that 10,200 represented a single clone of Purkinje cells and, by comparing that with the total number of Purkinje cells, which varied from 82,000 to 114,000 per half-cerebellum depending on the mouse strain, concluded that the Purkinje cells in each half-cerebellum are derived from 8 to 11 progenitors. As pointed out by Wetts and Herrup (1982), it is not known whether these progenitors also give rise to other cell types. Interestingly, in the chimera in which only a single clone survived, the surviving cells were found throughout the half of the cerebellum examined.

In summary, Purkinje cells do have a clonal history, but some time during development the cells become dispersed so that in the adult a coherent clone would consist of, at most, a few cells. With regard to chimeras, I think the above studies are quite definitive and give a general picture of Purkinje cell mosaicism. However, from the developmental neurobiology point of view, they probably raise more questions than they answer. For example, when do the cells intermingle? Does the intermingling mean that the final position of a cell is not important, or is there some reason, such as its ultimate connectivity, for the cells to become dispersed? Oster-Granite and Gearhart (1981) have suggested that the small patches of Purkinje cells they observed may form the basis of functional units. Can this be tested with chimeras?

II. CEREBELLAR MUTANTS

There are now 18 independent loci known to cause abnormalities in the cerebellum (see Sidman, 1983, for review). The application of chimera technology to study sites of gene action in the cerebellum has been very successful over the past several years. I will describe the results of studies of five mutants: Purkinje cell degeneration (*pcd*), lurcher, (*Lc*), staggerer (*sg*), weaver (*wv*), and reeler (*rl*). Each of these mutants affects more than one cell type in the nervous system. Thus, with each there is the possibility that the most obviously affected cell type is not the primary site of gene action. I should point out that at the cellular level, a single mutant can have several "phenotypes", such as cell position, cell size, cell number, and it is important to define which phenotype you are looking at when interpreting the results. Thus, where a simple statement, such as "the gene does not act in Purkinje cells," might seem sufficient, an admittedly more awkward statement may appear such as "the ectopic positioning of Purkinje cells is not due to intrinsic gene action." Though awkward, the latter is a far more accurate statement of the phenotype being studied.

A. Purkinje Cell Degeneration

The *pcd* mutant is characterized by total degeneration of Purkinje cells as well as neurons in other regions of the central nervous system (Mullen *et al.*, 1976). This was the first cerebellar mutant to be examined with chimeras and is probably the simplest. The chimeras were constructed so that the *pcd* component had the high β-glucuronidase allele (*pcd/pcd Gus^b/Gus^b* ↔ *+/+ Gus^h/Gus^h*). In these chimeras, none of the surviving Purkinje cells was stained by glucuronidase histochemistry, indicating that they were all derived from the +/+ component. Thus, it was concluded that the death of the Purkinje cells resulted from intrinsic action of the *pcd* gene (or lack of action by the mutant gene). Can we also conclude that the Purkinje cell is the site of gene action of the wild-type allele at the *pcd* locus? Although that may be a relatively safe assumption, it is not an absolute corollary because some mutant phenotypes may be caused by a gene being expressed in a cell type where it normally would not be active.

The other neuronal cell types that degenerate in the *pcd* mutant (e.g., photoreceptor cells and mitral cells) also degenerate in chimeras, and the one male chimera examined also had the abnormal sperm characteristic of *pcd* males but we do not know if these defects are also intrinsic to these affected cell types.

B. Lurcher

Heterozygous lurcher, $Lc/+$, mutant mice are similar to *pcd* mutants in that all of the Purkinje cells degenerate, but differ in that 90% of the granule cells and 75% of the inferior olivary neurons, both of which synapse on Purkinje cells, also degenerate (Swisher and Wilson, 1977; Caddy and Biscoe, 1979). Caddy and Biscoe hypothesized that the Purkinje cell was the primary site of gene action and that the degeneration of the granule cells and inferior olivary neurons was a secondary event. Obviously, this was a good candidate for analysis with chimeras. Wetts and Herrup (1982a) produced $Lc/+$ chimeras and used β-glucuronidase as a cell genotype marker for the Purkinje cells. The results were similar to the *pcd* chimeras described above in that all of the surviving Purkinje cells were $+/+$ in genotype indicating that the Lc gene does act intrinsically to the Purkinje cells.

Wetts and Herrup also used the β-glucuronidase marker to examine the inferior olivary neurons, but here the analysis is somewhat more complex. Since only 75% of these cells degenerate in lurcher mutants, we would expect to see both normal and mutant cells in the chimeras regardless of where the gene was acting. In the chimeras, the size and number of cells in the inferior olivary nuclei were less than normal but larger than $Lc/+$. Thus, the mutant phenotype was being expressed in the chimeras. They reasoned that if only $Lc/+$ cells were degenerating, then there should be an increase in the proportion of $+/+$ cells, that is, the $+/+$ cells should be over-represented compared with other regions of the CNS. What they found with each of three chimeras was that the proportion of $+/+$ cells was lower than expected if only $Lc/+$ cells were degenerating and was similar to the proportion seen in other regions of the CNS. Therefore, they concluded that degeneration of these neurons was not due to intrinsic action of the Lc locus and was probably a secondary effect resulting from loss of their post synaptic target, the Purkinje cell.

To examine the granule cell defect in $Lc/+$ chimeras, Wetts and Herrup (1982c) used the ichthyosis cell marker system. When using ichthyosis as a cell marker it is important to remember that the proportion of cells expressing the ichthyosis phenotype in single sections is dependent on fixation, the strictness

XIII. Ontogeny and Genetics of the Mammalian Nervous System   361

with which the cells are scored, and, in particular, the thickness of the sections. Therefore, the numbers given by Goldowitz and Mullen (1982a) should not be used as control values but rather, each investigator must do their own controls as Wetts and Herrup have done. For the two $Lc/+$ $+/+ \leftrightarrow +/+$ ic/ic examined by Wetts and Herrup, it was calculated that 95% and 75%, respectively, of the granule cells were $Lc/+$ in genotype. They concluded that degeneration of the granule cells was not an intrinsic effect of the gene and was probably a result of the Purkinje cell defect.

If both pcd and Lc cause degeneration of all Purkinje cells and both act intrinsically to the Purkinje cells, why do the granule cells and inferior olivary neurons undergo secondary degeneration in Lc but not in pcd? The answer to this probably lies in timing of the degeneration of Purkinje cells. In pcd, the Purkinje cells do not begin to degenerate until relatively late, around 18–20 days after birth. By this time, most granule cells have completed their migration and formed synapses on the Purkinje cells. That these synapses are functionally normal can be deduced from the fact that the mutant mice are behaviorally normal until around 22–24 days of age, by which time some 50% of the Purkinje cells have degenerated. Since the granule cells had formed synapses with the Purkinje cells, they had, in a sense, justified their existence and hence survive (at least until the mice are much older). In $Lc/+$ mutants, however, the Purkinje cells are abnormal at a much earlier age, so most of the granule cells cannot form synapses and degenerate. In the $Lc/+$ chimeras, there are intermediate numbers of normal Purkinje cells and intermediate numbers of granule cells survive.

C. Staggerer

The most obvious defect in staggerer (sg) is the degeneration of granule cells after they have migrated to the internal granule cell layer. In addition, the Purkinje cells are small in size, aberrantly positioned, and lack branchlet spines, the postsynaptic target of the granule cell axon. We produced sg chimeras for examination with the β-glucuronidase marker (Herrup and Mullen, 1979b). In the chimeras, those Purkinje cells that were normal in size and in position were $+/+$ in genotype, while those that were small in size and/or out of position were sg/sg in genotype. Thus, both the abnormal size and abnormal position phenotypes were due to intrinsic action of gene in Purkinje cells.

The staggerer chimeras also displayed another abnormality. There were some regions of the cortex that looked like a pcd chimera in that, even allowing for the small size and abnormal positioning of cells, there appeared to be gaps where there were no Purkinje cells, and overall there appeared to be a reduction in the total number of Purkinje cells. This observation led to a reexamination of the staggerer mutant, and it was found that, indeed, the number of Purkinje cells in the mutant was reduced by about 75% (Herrup and Mullen, 1979a). In staggerer chimeras, the total number of Purkinje cells is reduced by the proportion that one would expect if the *sg* gene were acting intrinsically in determining the number of *sg* Purkinje cells (Herrup and Mullen, 1981). This result also suggests that the wild-type cells play out their own developmental program and do not compensate for the reduced number of *sg* cells. It is not known whether the reduced number of Purkinje cells in *sg* is due to reduced genesis or degeneration. Thus, all of the defects of *sg* Purkinje cells appear to be due to intrinsic action of the mutant gene.

Herrup (1983) has recently used staggerer↔ichthyosis chimeras to demonstrate that the degeneration of granule cells is secondary. The reason for secondary degeneration of granule cells in *sg* is probably similar, though not identical, to the degeneration of granule cells in *Lc*. In *Lc*, the Purkinje cells degenerate, whereas in *sg*, some Purkinje cells are present but they lack the branchlet spines on which the granule cells synapse. This suggests that it is the synapse per se that is necessary for granule cell survival and not, for example, some humoral factor from the Purkinje cells.

Finally, it should be noted that although neither *Lc* nor *sg* acts in granule cells, one does not see a normal number of granule cells in *Lc* or *sg* chimeras. This is most likely due to matching of the numbers of presynaptic cells (i.e., granule cells) to the number of available postsynaptic target cells (i.e., Purkinje cells), which is an important concept in understanding how cell numbers are controlled in the developing nervous system. These chimeras provide an excellent model for examining this phenomenon.

D. Weaver

Weaver (*wv*) is a very complex mutant. The most obvious defect in young adult *wv/wv* is the absence of granule cells. However, unlike *sg* and *Lc*, the granule cells degenerate in the external granule layer before they migrate inward (Sidman, 1968). In addition, there are reports of reduced numbers of Purkinje cells (Rezai and Yoon, 1972) and abnormal and reduced numbers of Bergmann glia (Rakic and Sidman, 1973). Interestingly, heterozygous weaver mice exhibit similar defects but to a much lesser extent. In *wv*/+ mice, many granule cells do migrate to the granule cell layer, but many remain in ectopic positions in the molecular layer. Two very plausible hypotheses have emerged regarding the primary defect in weaver. According to one, the Bergmann glia is the primary defect and the granule cell migration defect is secondary (Rakic and Sidman, 1973). However, Sotelo and Changeux (1974) argued that since some granule cells do migrate but still degenerate, the granule cells "are closer than the glial cells" to the primary defect.

We have examined heterozygous weaver chimeras using the ichthyosis mutant as a cell marker for granule cells (i.e., *wv*/− +/+ ↔ +/+ *ic/ic*) (Goldowitz and

Mullen, 1982b). The symbol "$wv/-$" is used because the chimeras were known to be at least heterozygous, but some may have been homozygous, though we think that is unlikely. Since we were using ichthyosis as a marker, it was necessary to score large numbers of cells in normal position in the granule cell layer and in ectopic positions in the molecular layer, and from that data calculate what proportion of cells in each area were $wv/-$. Regardless of the cell marker being used, a quantitative analysis was also necessary because even in $+/+$ mice, a few granule cells fail to complete migration and remain in the molecular layer. In four of the $wv/-$ chimeras we examined, the granule cell layer was a mixture of $wv/-$ and $+/+$ cells, as would be expected because granule cells can migrate and survive in heterozygous mutants. In these same chimeras, there were excessive numbers of ectopic granule cells in the molecular layer, and it was calculated that 100% of them were $wv/-$ in genotype. Thus, the migration defect of $wv/-$ granule cells is due, at least in part, to intrinsic action of the gene. Unfortunately, we were not able to determine the genotype of the Bergmann glia so we cannot say that they do not play any role in the migration defect. The possibility remains that both cell types must be mutant to impede migration. Also, we have not examined known homozygous weaver chimeras so we cannot say that the degeneration of granule cells is due to intrinsic gene action. Nevertheless, weaver is the only mutant examined to date that has been shown to act in granule cells.

Abnormally positioned Purkinje cells were also observed in the $wv/-$ chimeras. In two chimeras processed for β-glucuronidase histochemistry, the aberrantly positioned Purkinje cells included both $wv/-$ and $+/+$ cells, indicating that the position of Purkinje cells was not being determined by intrinsic action of the weaver gene.

E. Reeler

Reeler has been an intensely studied mutant because it affects the positioning of cells in almost all cortical regions of the brain (Caviness and Rakic, 1978). In the cerebellum there are very few folia, a greatly reduced granule cell population, and the majority of Purkinje cells lie in a mass beneath the granule cells, although a few are found in normal positions. In reeler chimeras, both rl/rl and $+/+$ Purkinje cells are found in ectopic positions, indicating that they are not being positioned according to their own genetic information (Mullen, 1977b; Mullen and Sidman, 1984). The cerebellar granule cells have not been examined in reeler chimeras, but there is a good chance that they represent not a secondary defect, but a tertiary defect. The primary defect is unknown, but must occur very early because cortical regions are already abnormal at embryonic day 13. The abnormal positioning of Purkinje cells probably represents a secondary defect. Because the Purkinje cells are so far removed from their normal position, the number of granule cells probably matches up with the very small number of Purkinje cells in normal position and hence may represent a tertiary defect.

F. Neuron–Glia Interactions in Cerebellar Mutants

Dr. Marilyn Fisher has developed an exciting new method for studying neuron–glia interactions using chimeric mice. Fisher and Kozak (1983) were examining the level of expression of the enzyme glycerol-3-phosphate dehydrogenase (GPDH) in neurological mutant mice by biochemical and immunocytochemical techniques. By immunocytochemical techniques, using an antibody prepared against GPDH, they had earlier shown that the enzyme was localized in Bergmann glia and oligodendrocytes (Fisher et al., 1981). They found reduced levels of the enzyme in several neurological mutants (Fisher and Kozak, 1983). Fisher recognized that many of the mutants with reduced levels were cerebellar mutants (e.g., Lc, pcd, and sg) with Purkinje cell defects and hypothesized that the reduced levels of enzyme were not due to the genes acting directly on the glia but rather were due to failure of the Purkinje cells to induce normal enzyme expression in the glia (Fisher, 1984). To test this hypothesis, Fisher has examined enzyme expression by immunocytochemistry in pcd and Lc chimeras. When normal cerebellum is stained by the PAP procedure using anti-GPDH, virtually the entire molecular layer is intensely stained because of the ubiquitous Bergmann glia processes and the ensheathments that they send out to envelop the Purkinje cell dendrites. In pcd and Lc chimeras in which the mutant Purkinje cells had degenerated leaving only the genetically normal cells, the immunocytochemistry gave a very different picture and was quite dramatic. The only regions of the molecular layer to show high levels of staining were those areas where there were surviving Purkinje cells (Fisher and Mullen, 1984). The other regions exhibited the low levels of staining characteristic of the mutants, although there were Bergmann glia present as demonstrated using anti-GFA (glial fibrillary acidic protein) antiserum by the PAP procedure on adjacent sections. These results support Fisher's hypothesis that the Purkinje cell is needed for normal expression of this glial enzyme. The experiments also demonstrate an exciting new application of chimera technology.

G. Miscellany

Most of the studies on cerebellar mutant chimeras have paid relatively little attention to the behavior of the chimeras, primarily because most of the investigators are more interested in what is happening at the cellular level than at the behavioral level. In general, when a chimera exhibits the ataxia that characterizes cerebellar mutants, the cerebellum of that chimera will be predominantly mutant in phenotype. There are several reasons why this may be so. Mice have a big advantage over us bipeds for they have four legs to stand on, and the lateral position of their legs gives them a wide stance compared to, for example, a horse. Thus, ataxia is sometimes difficult to detect even in some cerebellar mutant mice (not chimeras), such as nervous (nr) when it is maintained on a hybrid background. When these nervous mice and some chimeras stand up on their hindlegs, they will occasionally topple over. However, there may be somewhat more biological reasons why the chimeras seldom exhibit ataxia. It is known that different regions of the cerebellum have different functions and receive signals from, and send signals to, different parts of the body. Because of the intermingling of cells described above, in mutant chimeras you seldom see areas of any magnitude that are entirely mutant in phenotype. Thus, even in chimeras that are, for example, 60% mutant, there will be no large areas totally devoid of function. Another possibility is that the normal cells in the chimera somehow compensate for the defective cells by forming new or additional connections. There is, however, no evidence that such plasticity occurs in these chimeras.

III. OTHER REGIONS OF THE NERVOUS SYSTEM

Compared with the chimera studies of the cerebellum, studies on other regions of the nervous system have been meager, primarily because of inadequate cell genotype markers. A few studies have been informative and should be included in this chapter.

A. Retinal Degeneration

In mice with retinal degeneration (rd), the photoreceptor cells degenerate rapidly, beginning 10 days after birth. The retinal pigment epithelium is known to be intimately associated with the photoreceptor cells both anatomically and physiologically. Since the genotype of the pigment epithelial cells in chimeras can be readily determined using pigmentation as a cell genotype marker in pigmented↔albino chimeras, this was one of the first regions of the nervous system to be studied with chimeras. When retinal degeneration chimeras were examined using pigmentation to mark the pigment epithelial cells, no correlation was found between the genotype of the pigment epithelium and whether the underlying photoreceptors were degenerating or surviving (LaVail and Mullen, 1976). Thus, the pigment epithelium does not appear to be involved in the etiology of retinal degeneration.

Two other mutations in mice that cause degeneration of photoreceptor cells are the *pcd* and *nr* mutants. The retinas from chimeras of both of these mutants have also been examined, and the photoreceptor cells do degenerate in the chimeras.

However, in most of the chimeras we examined, both component strains were pigmented or the retina was predominantly of one genotype, so the role of the pigment epithelium could not be determined. Thus, the site of gene action is not known for any of these three mouse mutants.

B. Retinal Dystrophy in Rats

In rats with retinal dystrophy (*rdy*), the photoreceptor cells also degenerate but somewhat later than in *rd* mice. However, this mutant was known to have a pigment epithelial cell abnormality, a failure to phagocytize rod outer segment discs that are shed daily from rod photoreceptor cells (see LaVail, 1981, for review). The problems associated with making our first (and last) rat chimeras were made worthwhile when we examined the retinas of the chimeras. There was degeneration of photoreceptors but, unlike the mouse, the degeneration was occurring only under those regions where the pigment epithelial cells were *rdy/rdy* in genotype (Mullen and LaVail, 1976). Thus, the primary site of *rdy* gene action is the pigment epithelial cell.

C. Myelination Mutants

Shiverer (*shi*) is a mouse mutant with severe myelin defects in the central nervous system. Of particular significance is the absence of myelin basic protein (MBP) in *shi/shi* mice, demonstrated both biochemically and immunohistochemically using an antiserum against MBP. This abnormality could be either the result of an intrinsic defect in the oligodendrocyte, which forms the myelin, or failure by the axon to induce the glial cell (as with GPDH described above). To investigate this, Mikoshiba *et al.* (1982) have examined *shi* chimeras and, using the anti-MBP antiserum, found both MBP-positive and -negative fibers mixed in the white matter. Electron micrographs revealed that single axons were enveloped by both normal and shiverer-like myelin suggesting that the axon is not responsible for the abnormal myelination. Although it also suggests that the oligodendrocyte is the primary site of gene action, that could not be proved because the genotype of the oligodendrocytes was not demonstrated.

The negative staining with anti-MBP in *shi* is already finding use as a marker for Schwann cells (the myelinating cell type in the peripheral nervous system) in studies of other mutants such as muscular dystrophy (Peterson *et al.*, 1982).

Transplantation studies by Aguayo and his colleagues (1977) had implicated the Schwann cell as the defective cell type in the trembler (*Tr*) mouse mutant. Trembler chimeras appear similar to the *shi* chimeras described above, with both myelinated and hypomyelinated Schwann cells on individual axons (Rayburn *et al.*, 1980), further demonstrating that the axon is not responsible for the defect, and that even when axon and glia are together throughout development, the mutant phenotype will be expressed by the glia.

The same rationale and the same results have also been reported regarding the quaking (*qk*) mutant, namely, single axons with both normal and quaking-like myelin (Berarducci *et al.*, 1981).

IV. CONCLUSION

The studies reviewed above clearly demonstrate that we need no longer argue for the potential of chimeras in studying the nervous system: their worth is now evident. Future studies will undoubtedly define the sites of gene action of other mutant genes and a genetic architecture of the nervous system will begin to emerge. Once the site of a gene action is determined, the chimeras, as well as the mutants themselves, become more useful as tools for neurobiological investigation. Many areas of the nervous system, such as spinal cord and cerebral cortex, have barely been touched by the technology. With few exceptions (see West, Chapter II) the embryonic nervous system has not been touched except by retrospective analysis from observations made on the adult. When those laboratories working with younger and younger stages meet with those laboratories studying older and older stages (see Rossant, Chapter IV), some very exciting discoveries are inevitable.

REFERENCES

Aguayo, A. J., Attiwell, M., Trecarten, J., Perkins, S., and Bray, G. M. (1977). *Nature (London)* **265**, 73–75.
Berarducci, A., Peterson, A., Aguayo, A., and Tretjakoff, I. (1981). *Soc. Neurosci. Abstr.* **7**, 545.
Caddy, K. W. T., and Biscoe, T. J. (1979). *Philos. Trans. R. Soc. London. Ser. B* **287**, 167–201.
Caviness, V. S., Jr., and Rakic, P. (1978). *Annu. Rev. Neurosci.* **1**, 297–326.
Condamine, H., Custer, R. P., and Mintz, B. (1971). *Proc. Natl. Acad. Sci. U.S.A.* **68**, 2032–2036.
Dewey, M. J., Gervais, A. G., and Mintz, B. (1976). *Dev. Biol.* **50**, 68–81.
Feder, N. (1976). *Nature (London)* **263**, 67–69.
Fisher, M. (1984). *Proc. Natl. Acad. Sci. U.S.A.* In press.
Fisher, M., and Kozak, L. P. (1983). Personal communication.
Fisher, M., and Mullen R. J. (1984). Submitted for publication.
Fisher, M., Gapp, D. A., and Kozak, L. P. (1981). *Dev. Brain Res.* **1**, 341–354.
Gearhart, J., and Oster-Granite, M. L. (1980). *J. Histochem. Cytochem.* **28**, 245–249.
Goldowitz, D., and Mullen, R. J. (1982a). *Dev. Biol.* **89**, 261–267.
Goldowitz, D., and Mullen, R. J. (1982b). *J. Neurosci.* **2**, 1474–1485.
Green, M. C., Shultz, L. D., and Nedzi, L. A. (1975). *Transplantation* **20**, 172–175.
Hahn, W. E., Van Ness, J., and Chaudhari, N. (1982). *In* "Molecular Genetic Neuroscience" (F. O. Schmitt, S. J. Bird, and F. E. Bloom, eds.), pp. 323–334. Raven Press, New York.
Herrup, K. (1983). *Dev. Brain Res.* **11**, 267–274.
Herrup, K., and Mullen, R. J. (1979a). *Brain Res.* **172**, 1–12.
Herrup, K., and Mullen, R. J. (1979b). *Brain Res.* **178**, 443–457.
Herrup, K., and Mullen, R. J. (1979c). *J. Cell Sci.* **40**, 21–31.

Herrup, K., and Mullen, R. J. (1981). *Dev. Brain Res.* **1**, 475–485.
LaVail, M. M. (1981). *Invest. Ophthalmol. Visual Sci.* **21**, 638–657.
LaVail, M. M., and Mullen, R. J. (1976). *Exp. Eye Res.* **23**, 227–245.
Mikoshiba, K., Yokoyama, M., Inoue, Y., Takamatsu, K., Tsukada, Y., and Nomura, T. (1982). *Nature (London)* **299**, 357–359.
Mullen, R. J. (1977a). *Nature (London)* **270**, 245–247.
Mullen, R. J. (1977b). *Soc. Neurosci. Symp.* **2**, 47–65.
Mullen, R. J. (1978). *In* "The Clonal Basis of Development" (S. Subtelny and I. M. Sussex, eds.), pp. 83–101. Academic Press, New York.
Mullen, R. J., and Herrup, K. (1979). *In* "Neurogenetics: Genetic Approaches to the Nervous System" (X. O. Breakefield, ed.), pp. 173–196. Am. Elsevier, New York.
Mullen, R. J., and LaVail, M. M. (1976). *Science* **192**, 799–801.
Mullen, R. J., and Sidman, R. L. (1984). In preparation.
Mullen, R. J., Eicher, E. M., and Sidman, R. L. (1976). *Proc. Natl. Acad. Sci. U.S.A.* **73**, 208–212.
Oster-Granite, M. L., and Gearhart, J. (1980). *J. Histochem. Cytochem.* **28**, 250–254.
Oster-Granite, M. L., and Gearhart, J. (1981). *Dev. Biol.* **85**, 199–208.
Peterson, A., Bray, G., and Marler, J. (1982). *Soc. Neurosci. Abstr.* **8**, 1008.
Rakic, P., and Sidman, R. L. (1973). *Proc. Natl. Acad. Sci. U.S.A.* **70**, 240–244.
Rayburn, H. B., Peterson, A. C., and Aguayo, A. (1980). *Soc. Neurosci. Abstr.* **6**, 660.
Rezai, Z., and Yoon, C. H. (1972). *Dev. Biol.* **29**, 17–26.
Sidman, R. L. (1968). *In* "Physiological and Biochemical Aspects of Nervous Integration" (F. D. Carlson, ed.), pp. 163–193. Prentice-Hall, Englewood Cliffs, New Jersey.
Sidman, R. L. (1982). *In* "Molecular Genetic Neuroscience" (F. O. Schmitt, S. J. Bird, and F. E. Bloom, eds.), pp. 389–400. Raven Press, New York.
Sidman, R. L. (1983). *In* "Genetics of Neurological and Psychiatric Disorders" (S. S. Kety, L. P. Rowland, R. L. Sidman, and S. W. Matthysee, eds.), pp. 19–46. Raven Press, New York.
Sotelo, C., and Changeux, J. P. (1974). *Brain Res.* **77**, 484–491.
Swisher, D. A., and Wilson, D. B. (1977). *J. Comp. Neurol.* **172**, 205–218.
Wetts, R., and Herrup, K. (1982a). *J. Embryol. Exp. Morphol.* **68**, 87–98.
Wetts, R., and Herrup, K. (1982b). *J. Neurosci.* **2**, 1494–1498.
Wetts, R., and Herrup, K. (1982c). *Brain Res.* **250**, 358–362.

CHAPTER **XIV**

Mouse Chimeras in the Study of Behavior

MURIEL N. NESBITT

Department of Biology
University of California at San Diego
La Jolla, California

I.	The Background	369
	A. The Mice	370
	B. The Behaviors	370
II.	The Problem	372
III.	How Can Chimeras Provide Answers?	373
IV.	Methodology	374
	A. Chimera Production	374
	B. Analysis of Tissue Composition	374
	C. Behavior Testing	375
	D. Data Analysis	375
V.	Results: Covariation among Behaviors	375
VI.	Significance of the Data	377
	A. Covariance among Behaviors	377
	B. Covariance of Behavior Groups with Tissues	378
	References	379

I. BACKGROUND

Existing inbred strains of mice incorporate a sample of the genetic variability that was present in their wild ancestors. Thus inbred strains differ from one another in a wide variety of characteristics, including many behavioral traits. Behavioral differences among inbred strains have been recognized and studied for many years, but success in elucidating their fundamental physiological and genetic mechansims has been limited. Chimeric mice, incorporating cells from differently behaving strains, can be used to learn things about complex behavioral differences between strains, as this chapter will illustrate.

A. The Mice

The mouse strains chosen for this study were A/J and C57BL/6J. They were chosen because they differ in several behaviors that are relatively easy to measure and for which the strain differences are large. In addition, these two strains differ in other characteristics that are useful in tracing the distribution of cell types in chimeras. A/J mice are albino, and have the a electrophoretic form of the enzyme glucosephosphate isomerase (GPI). C57BL/6J mice are pigmented and have the b form of GPI. Because GPI is nearly ubiquitous in tissue distribution and is expressed at easily detectable levels, it provides a ready means for measuring the relative contributions of the two strains to particular tissues.

B. The Behaviors

1. ALCOHOL PREFERENCE

Mice of some strains will drink 10–15% ethanol in preference to plain water when the choice is available (McClearn and Rodgers, 1959). Mice of other strains will avoid the ethanol and drink the water in similar circumstances. When alcohol preference is expressed in terms of the ratio of ethanol solution to total fluid consumed, A/J mice score 0.18 ± 0.08, with C57BL/6J mice score 0.80 ± 0.11. Alcohol preference scores have been determined for many strains, and several studies have shown that the strain differences in this characteristic are inherited (e.g., McClearn, 1972), and are controlled by more than one gene (Fuller and Collins, 1972). Individual genes contributing in a major way to the control of alcohol preference have not been identified, but it has been shown that changes at the albino locus can influence alcohol preference (Henry and Schlesinger, 1967), and that the level of activity of liver alcohol dehydrogenase has a small effect on preference (Sheppard *et al.*, 1968). Although it has been demonstrated that damage to, electrical stimulation of, or administration of drugs active on the central nervous system can dramatically influence alcohol preference, the site of action of murine genes affecting alcohol preference has not been determined.

2. OPEN FIELD ACTIVITY

An "open field" is usually the bottom of a container such that the mouse being tested has approximately 1 m^2 at his disposal. In its crudest form, the field is partitioned by a gridwork of painted lines, and an observer counts the number of lines the mouse crosses in a given period of time. Our experiments were done in such an apparatus, with a 10-minute trial time. Mice that move actively around the field cross many lines and thus acheive a high score. In our hands, C57BL/6J mice score 308 ± 68. A/J mice, on the other hand, move rather little in these circumstances, scoring 55 ± 21. Open field activity scores are available for many strains (DeFries and Hegmann, 1970; McClearn, 1959; Thompson, 1953). Our tests were done under ordinary fluorescent room lighting.

Open field activity levels have been shown to be inherited in a complex fashion (DeFries and Hegmann, 1970; McClearn, 1959). Individual genes contributing to the control of open field activity levels have not been identified.

3. OPEN FIELD DEFECATION

Some kinds of mice are especially prone to respond to stressful situations by urinating and/or defecating. Some mice respond the same way to the open field test. It is a simple matter to quantify defecation during the open field test by counting the number of fecal boli produced. During our open field testing this has been $0.2 \pm .01$ for C57BL/6J and 4.6 ± 1.1 for A/J. Open field defecation scores exist for several strains, and there is some tendency for the more active strains to defecate less, but this is not an absolute correspondence (Bruell, 1963; DeFries and Hegmann, 1970). Again inheritance is complex, and individual genes and mechanisms are obscure (DeFries and Hegmann, 1970; McClearn, 1959).

4. CRICKET ATTACKING

In 1973, Karla Butler Thomas accidentally discovered that some mice will quickly attack and eat a cricket (*Achaeta domestica*) introduced into their cage. She pursued the matter and found that mice of different strains differ in their latency to attack (i.e., in the time that elapses between the introduction of the cricket and the attack by the mouse). C57BL/6J mice were found to be relatively slow among the attackers, and to show a sex difference in latency to attack (male: 251 ± 45 seconds, female: 533 ± 42 seconds). A/J mice, however, share with mice of the DBA/2J strain the distinction of being nonattackers. Mice of these strains only rarely attack crickets, be it within the 30-minute test period, or when left with the cricket overnight. Thomas (1973) showed that this characteristic is inherited, but it is not yet known how many genes underlie it.

5. ROPE CLIMBING

Karla Butler Thomas also discovered (personal communication) that mice of different strains differ in their latency to climb and climbing time when placed on a knot in a vertically stretched piece of rope 18 inches long. C57BL/6J mice move off the knot fairly quickly (76 ± 13 seconds), while A/J mice do so slowly (245 ± 32 seconds). C57BL/6J mice are also quick to finish climbing once they have started (4.3 ± 0.79 minutes), while A/J mice are slower (7.8 ± 1.5 minutes). Unfortunately this test discriminates least well between our two strains (there is more overlap in the distributions of rope climbing scores for the two strains than there is for the distributions of the other scores). No work has been done on the inheritance or physiology of rope climbing.

II. THE PROBLEM

The question of precisely how genes and their products bring about and modulate behaviors is a fascinating one. An important step along the way toward answering this question would be to know through what tissues specific genes act in generating behavioral characteristics. In working on mammalian behaviors, one too frequently encounters the prejudice that genes modulating behaviors must be expressed in the brain. While it seems intuitively obvious that the brain is involved in the generation of behaviors, such as alcohol preference or open field activity, and while (as mentioned above) a great many studies have shown that physical or pharmacological manipulation of the brain can certainly affect behaviors such as these (e.g., Hoffman, 1982; Tuomisto *et al.*, 1982; Vecsei *et al.*, 1982), nevertheless it is equally clear that organs other than the brain and nervous system participate in the generation of these behaviors. Using alcohol preference as an example, one might imagine a chain of events similar to the following in the generation of the behavior. A naive (with respect to ethanol) mouse finds he has two drinking bottles available, and drinks from both in the course of a few hours. Ethanol has an odor and a flavor different from water. The mouse will perceive these differences or not, according to the qualities of his peripheral sense organs and the relevant brain regions. The odor and/or flavor may be interpreted in the brain as pleasant or unpleasant. The ethanol solution consumed will travel down the esophagus to the stomach and gut. It will be absorbed into the bloodstream at a rate depending on the qualities of the stomach and gut. The rate of absorbtion and of metabolism of the ethanol and the rate of excretion of ethanol and its metabolites all depend on the qualities of various organs (gut, liver, kidneys), and together they determine what substances and what concentrations the brain and viscera will be exposed to via the blood. These exposures give rise in the brain to the sensations the mouse experiences in response to ethanol. The strength and qualities of these sensations depends upon the qualities of viscera and brain. The brain interprets sensations as pleasant or unpleasant, and associates them with the odor and flavor of ethanol, thus determining what behavior will be seen when the mouse next encounters alcohol. There are plenty of places in this scheme where genetic changes affecting the brain could affect the response of the mouse to ethanol, but similarly many steps at which genetic changes in other organs could affect the outcome. For example, a mouse whose metabolism and brain were such that he felt quite ill after consuming ethanol, but whose peripheral sensory organs were such that he could not distinguish the ethanol solution from water would be unlikely to avoid ethanol (or to prefer it). Two mice with equivalent brains, sensory organs, and metabolic capacities might differ in alcohol preference if one were to absorb it very slowly from the gut and the other very quickly, so that in the latter case there was a swift accumulation to high levels of ethanol and acetaldehyde in the

XIV. Mouse Chimeras in the Study of Behavior

blood, whereas in the former the rise of blood levels of these substances would rise slowly and perhaps stay low because slower absorption places a lesser burden on the system of excretion.

This discussion is meant to emphasize that the question of what tissues express the genes controlling behaviors is in fact an open and interesting one, and to point out that in delving into the mechanisms underlying a particular strain difference in a behavior one can only study the one or few steps in the generation of that behavior that happen to differ between the two strains. One cannot, in this way, study the behavior as a whole and has no handle on those steps of the process that do not differ between the strains being studied.

Another question of interest in the present study is that of the relatedness or lack thereof among the various characteristics being examined. At one extreme, it might be proposed that a single genetically generated physiological difference between A/J and C57BL/6J mice gives rise to all seven of the behavioral differences we measure. At the other extreme, one might imagine that each of the seven differences is independently genetically and physiologically determined. Various possibilities exist between the two extremes.

III. HOW CAN CHIMERAS PROVIDE ANSWERS?

Since chimeras made by the aggregation of morulae differ from one another in the relative proportions and in the distribution of the two cell types, the proportion of A/J cells in a given tissue in a series of A↔C57BL/6 chimeras will vary. The score for a given behavioral test, e.g., alcohol preference, should thus vary over a series of A↔C57BL/6 chimeras, and would be expected to reflect the cellular composition of the tissue(s) in which the gene(s) whose product(s) regulate(s) this behavior is(are) expressed. Any two cell populations in a chimera that are not intimately developmentally related could differ in their proportions of A and BL/6 cells. Thus two behavior traits controlled by different tissues or groups of tissues would vary independently over a series of chimeras, while two behaviors controlled by the same tissue or set of tissues would covary. Several behavior traits can be measured in the same series of chimeras, and the degree of covariation among them can be determined. Thus if any two or more behaviors are simply different reflections of the same underlying physiological mechanism, the group can be recognized by its covariance.

Since any behavior for which A and C57BL/6 mice differ should vary over a series of A↔C57BL/6 chimeras as the cellular composition of the relevant tissue(s) varies, and since it is possible to score the cellular composition of tissues as well as behaviors over the series of chimeras, it should be possible to identify the tissue(s) relevant to control of a particular behavior by the covariance of its cellular composition with the score for the behavior in question. Thus, for example, if the difference in alcohol preference between A and C57BL/6 were to reflect a difference in, say, liver between these strains, the alcohol preference scores should covary with the proportion of C57BL/6-derived cells in the liver over a series of chimeras.

In summary, then, a series of chimeras derived from two differently behaving strains can be used to relate variation in a particular behavior or group of behaviors to variation in a given tissue or group of tissues. Identifying the tissue in which a behavior difference is generated is the first step toward discovery of the physiological mechanism underlying the behavior difference.

IV. METHODOLOGY

A. Chimera Production

Aggregation of morulae was the technique used to produce the chimeras in this study (Nesbitt et al., 1979). The techniques used were minor modifications of those of Tarkowski (1961) and Mintz (1962). Our efforts at chimera production produced 47 individuals (Nesbitt et al., 1981) of which 12 subsequently proved to contain only one of the two cell types. Data presented here are thus drawn from 35 chimeric individuals.

B. Analysis of Tissue Composition

To determine the relative contributions of A and C57BL/6 cells to our chimeric tissues, we prepared a homogenate of each tissue in an equal volume of water, and carried out cellulose acetate gel electrophoresis on each homogenate (Zipzone system, Helena Laboratories). Electrophoresis was done for 30 minutes at 200 V using a buffer consisting of 3 g Tris + 14.4 g glycine per liter at pH 8.5. Gels were stained for GPI activity by the method of Carter and Parr (1967), by overlaying each gel with a total volume of 4 ml of stain containing 1% agar in a molten state. Since GPI is present in significant amounts in erythrocytes, it was necessary to perfuse each chimera with saline + heparin at the time of killing.

Standard mixtures, made from hemolysates containing known numbers of A and C57BL/6 cells, were run on every gel. Each tissue homogenate was run twice. The first time, each was run on the same gel with standards ranging from 10 to 90% C57BL/6, and, by comparison to the standards, assigned to a given 10% interval. Each homogenate was then run a second time on a gel containing standards ranging at 2% intervals over the relevant 10% interval. Finally each homogenate was given a specific score, between 0 and 100, representing the percentage of C57BL/6 contribution to its GPI pattern.

Since A mice are albino while C57BL/6 are pigmented, we could also esti-

XIV. Mouse Chimeras in the Study of Behavior

mate the composition of the coat melanocyte population. This was simply done by eye, and was no doubt less accurate than the estimates done via GPI.

C. Behavior Testing

Brief descriptions of the behavior tests were presented in Section IB above, and detailed descriptions can be found in Nesbitt *et al.* (1979).

D. Data Analysis

The data are gathered in two forms. The behavior tests yield numbers of boli, numbers of line traversed, seconds or minutes of latency to attack or climb, or climbing time, and ratio of ethanol to total fluid intake. The tissue composition analysis yields numbers between 0 and 1 reflecting the proportion of BL/6 contribution to the tissue. We need to look for covariance between behaviors, and between any behavior and a tissue or set of tissues. Therefore, we have calculated pairwise correlation coefficients among behaviors and between tissues and behaviors. In addition, we have used a technique called factor analysis to identify covarying groups of behaviors. The "factor loading" entities, as shown in Table I below, are regression coefficients for predicting individual behaviors from a given factor. The factors are determined by maximizing the correlation within, while minimizing correlation between, the clusters (of behaviors) being defined. A standard computer program BMDP4M (Dixon, 1975) was used for the factor analysis, specifying options for varimax rotation and the method of principal components. Only factors whose eigenvalues are at least 1 are accepted, because an eigenvalue less than 1 indicates that the factor contains no more information than one of the existing variables. Factor loadings less than 0.250 are rejected in the analysis, because they represent very little correlation between the factor and the behavior.

V. RESULTS: COVARIATION AMONG BEHAVIORS

Table I shows the outcome of factor analysis on the behavior data for the chimeras. Three factors were identified, meaning that our seven behaviors can be assigned to three groups. Members of a group covary with each other, but members of different groups do not covary. The negative signs in the table reflect the fact that in some behaviors C57BL/6-like performance leads to high scores (e.g., alcohol preference), while in other behaviors (e.g., open field defecation) it leads to low scores. Dashes in the table represent loadings less than 0.250.

TABLE I.

Rotated Factor Loadings

Behavior	Factor I	Factor II	Factor III
Alcohol preference	—	0.398	0.795
Defecation	—	—	−0.872
Open field activity	−0.765	—	0.307
Cricket I	0.889	—	—
Cricket II	0.834	—	—
Rope latency	—	0.885	—
Rope time	—	0.880	—

TABLE II.

Correlations between Tissues and Factor I

Tissue	r	$P(r = 0)$
Left brachial nerve	−0.568	0.0070
Right lower foreleg muscle	−0.590	0.0019
Left lower foreleg muscle	−0.542	0.0043
Left upper foreleg muscle	−0.583	0.0018
Left upper hindleg muscle	−0.557	0.0026
Left lower hindleg muscle	−0.605	0.0005
Right lower hindleg muscle	−0.646	0.0007
Left maxillary muscle	−0.559	0.0084
Right maxillary muscle	−0.543	0.0110
Heart	−0.520	0.0038
Brachial lymph nodes	−0.596	0.0056

TABLE III.

Correlations between Tissues and Factor III

Tissue	r	$P(r = 0)$
Coat melanocytes	0.646	0.0001
Eye fat pad	0.500	0.0070
Adrenal gland	0.583	0.0056
Bladder	0.508	0.008
Eyeball	0.476	0.009
Stomach	0.725	0.0002
Esophagus	0.704	0.0001
Total intestine	0.556	0.0038
Right lower hindleg muscle	0.534	0.0070

XIV. Mouse Chimeras in the Study of Behavior

TABLE IV.

Correlations between Brain and Factors

	Factor I		Factor II		Factor III	
Tissue	r	$P(r = 0)$	r	$P(r = 0)$	r	$P(r = 0)$
Total brain	−0.217	0.288	0.002	0.993	−0.086	0.675
Brainstem	−0.375	0.095	0.136	0.556	0.315	0.165
Cerebellum	−0.086	0.741	−0.026	0.920	0.136	0.603
Cerebrum	−0.248	0.254	−0.082	0.711	0.111	0.614
Olfactory lobes	−0.541	0.025	−0.012	0.344	−0.270	0.294
Optic nerves	−0.002	0.994	0.141	0.520	0.179	0.412

The data on covariance between behavior groups and tissues is most easily presented in terms of correlation between each factor (in a sense the "mean" of a covarying group of behaviors) and the various tissues. Table II shows the correlation coefficients for tissues with factor I. The tissues shown are those for which we have at least 20 observations, and for which the probability that the real correlation coeffecient (r) is zero is 0.01 or less, except for right maxillary muscle, which is included for discussion purposes.

Table III shows the tissues correlated with factor III, according to the same criteria as applied above. According to the same criteria, factor II was correlated only with one of the right lower foreleg muscles. Otherwise, the best correlations with factor II involved another right foreleg muscle ($p = 0.02$), ovaries ($p = 0.24$), eyeballs ($p = 0.19$), heart ($p = 0.065$), and submaxillary gland ($p = 0.105$). Of these correlations, only the one with submaxillary gland is negative, as would be expected of a biologically significant correlation (C57BL/6 mice score low in rope climbing, while tissues with much C57BL/6 contribution get a high score).

Neither brain nor brain parts appear in Table II or Table III. Since brain is of special interest in work involving behavior, the correlation data on brain parts with the factors is presented in Table IV. Only tissues where there are at least 20 observations are included.

VI. SIGNIFICANCE OF THE DATA

A. Covariance among Behaviors

The factor loadings shown in Table I indicate that the seven behaviors that we measure fall into three groups. Open field activity covaries with latency to attack crickets by naive (cricket I) and experienced (cricket II) mice. Rope climbing time covaries with latency to climb, and alcohol preference covaries with defecation in the open field. The covariances of alcohol preference with factor II, and of open field activity with factor III are of marginal statistical significance, and that of alcohol preference with factor II is of the wrong sign to be of biological significance.

When two or more behaviors (or traits of any kind) covary in a series of chimeras, it indicates that they are functions of the same group(s) of cells, or perhaps of two or more developmentally related groups of cells. Covariance in chimeras says nothing about common control at the genetic level.

B. Covariance of Behavior Groups with Tissues

Tables II and III give clues as to what tissues are in fact the sources of the A versus C57BL/6 differences in behavior. A wide array of muscles is correlated with open field activity and cricket attacking, and the sign of correlation is appropriate. (Factor I was derived in such a way that it correlates positively with A-like scores, negatively with C57BL/6-like.) In fact, all muscles that were scored, with the exception of diaphragm and one of the right lower foreleg muscles (which showed a higher but inappropriately signed correlation with factor II), were correlated more strongly with factor I than with any other factor. In all cases apart from the two mentioned above, the sign of correlation was appropriate, and the highest $P(r = 0)$ was 0.0602. Thus muscles not meeting criteria for inclusion in Table II were still well correlated with factor I, as illustrated by right maxillary muscle in the table. Compare the worst correlation between muscle and factor I, ($P = 0.0602$) to the P values appearing in Table IV. This consistency in pattern suggests that the difference(s) between A and C57BL/6 mice in open field activity and cricket attacking are related to one or more differences expressed in muscle. Our matrix of correlations (three factors by >100 tissues) is large enough so that we would expect to see a few spurious correlations reach the level of statistical significance by chance. Probably the one nerve and one pair of lymph nodes appearing in Table II represent examples of such spuriously significant correlations, along with the eye fat pad and the one muscle in Table III. If a gene or genes acting in adipocytes, for example, were genuinely involved in regulating alcohol preference and open field defecation, then every fat pad examined should show correlation with factor III (as muscles do with factor I). Thus cases where a single member of a family of tissues (one fat pad, one lymph node) has shown correlation with a behavior have not been taken seriously.

In the array of tissues correlated with factor III (Table III), our interest is becoming focused on neural crest (represented by coat melanocytes and adrenal gland) and gut (represented by stomach, esophagus, and intestine). These are the five best correlations in the table, and are interesting in being two families of related tissues. Unfortunately we did not dissect apart adrenal cortex from medulla before our assays.

Tables II–IV taken together indicate that none of the major contributors to A versus BL/6 differences in our particular behaviors is expressed in brain. If a small brain region, which we either did not take or took as part of a much larger segment, were to be a major determining factor in one of our behaviors, we would expect to see correlation with the brain parts we did study. Any one brain part should be correlated with others, due to their developmental relationship, as esophagus is correlated with stomach and melanocytes are correlated with adrenal gland.

Perhaps the major drawback of our technique is the fact that we cannot study separately tissues that we cannot physically separate. This kind of research needs an easily detectable, cell limited marker scorable at the cell level.

To summarize, our work has suggested that the A versus C57BL/6 differences in cricket attacking and open field activity is a reflection of a genetic difference or differences expressed in muscle, while the differences between A and C57BL/6 in alcohol preference and open field defecation are reflections of differences expressed in gut and neural crest.

ACKNOWLEDGMENT

Original research reported in this chapter has been supported by National Institutes of Health Grant CA 30082.

REFERENCES

Bruell, J. H. (1963). *Am. Psychol.* **17,** 445.
Carter, N. D., and Parr, C. W. (1967). *Nature (London)* **216,** 511.
DeFries, J. C., and Hegmann, J. P. (1970). In "Contributions to Behavior-Genetic Analysis: The Mouse as a Prototype" (G. Lindzey and D. Thiessen, eds.), p. 23. Appleton, New York.
Dixon, W. J. (1975). "BMPD Biomedical Computer Programs." Univ. of California Press, Berkeley.
Fuller, J. L., and Collins, R. L. (1972). *Ann. N.Y. Acad. Sci.* **197,** 42–48.
Henry, K. R., and Schlesinger, K. (1967). *J. Comp. Physiol. Psychol.* **63,** 320–323.
Hoffman, P. L. (1982). *Pharmacol. Biochem. Behav.* **17,** 685–690.
McClearn, G. E. (1959). *J. Comp. Physiol. Psychol.* **52,** 62–67.
McClearn, G. E. (1972). *Ann. N.Y. Acad. Sci.* **197,** 26–31.
McClearn, G. E., and Rodgers, D. A. (1959). *Q. J. Stud. Alcohol* **20,** 691–695.
Mintz, B. (1962). *Am. Zool.* **2,** 432.
Nesbitt, M. N., Spence, M. A., and Butler, K. (1979). *Behav. Genet.* **9,** 277–287.
Nesbitt, M. N., Guthrie, D., Spence, M. A., and Butler, K. (1981). In "Genetic Research Strategies for Psychobiology and Psychiatry" (E. S. Gershon, S. Matthysse, X. O. Breakefield, and R. D. Ciaranello, eds.), pp. 105–111. Boxwood Press, Pacific Grove, CA.

Sheppard, J. R., Albersheim, P., and McClearn, G. E. (1968). *Biochem. Genet.* **2,** 205–212.
Tarkowski, A. K. (1961). *Nature (London)* **190,** 857–860.
Thomas, K. B. (1973). *J. Comp. Physiol. Psychol.* **85,** 243–249.
Thompson, W. R. (1953). *Can. J. Psychol.* **7,** 145–155.
Tuomisto, L., Airaksinen, M. M., Peura, P., and Eriksson, C. J. P. (1982). *Pharmacol. Biochem. Behav.* **17,** 831–836.
Vecsei, L., Telegdy, G., Schally, A. V., and Coy, D. H. (1982). *Pharmacol. Biochem. Behav.* **17,** 633–637.

CHAPTER **XV**

Chimeras and Sexual Differentiation

ANNE McLAREN

MRC Mammalian Development Unit
University College London
London, England

I.		Hydra	382
II.		Nemertine Worms	383
III.		Amphibia and Birds	386
IV.		Mammals	387
	A.	Sex Ratio	387
	B.	Mechanisms of Sex Determination	392
	C.	Hermaphrodites	393
V.		Conclusions	397
		References	397

The distinction between males and females is the most common polymorphism seen in animals. What is its causal basis? In some species, exogenous factors play a major determining role. In the limpet *Crepidula,* for example, the sex of an individual changes during its lifetime, and is dependent in part on social interactions with other individuals, while in turtles and alligators the temperature at which the eggs are hatched determines the sex of the young. For other species, endogenous factors are more important: these are particularly hard to investigate.

In mammals, we know that there exists an endogenous chromosomal "sex-determining mechanism" such that males are almost always XY in sex chromosome constitution, and females are XX; and we know that once a testis or an ovary has formed, further sexual differentiation is under the control of sex hormones produced by the gonads (androgens and Müllerian regression factor by the testis and oestrogens by the ovary). But what is the causal chain linking chromosomal sex to gonadal sex? Is the direction of gonadal differentiation determined by the chromosome constitution of the body as a whole, or by that of the mesonephric tissue adjacent to the gonad primordium? Or is it the chromosomes of the gonadal cells themselves that play the determining role? And if so, is it the somatic cells of the gonad or the germ cells or both? Is there a testis-determining substance, controlled by the Y-chromosome? If so, is it one of the male-specific antigens, either the H-Y transplantation antigen or a serologically detected male antigen?

One obvious approach to such questions is to construct chimeras, animals made up of two distinct populations of cells, one male and one female, and then to study their sexual differentiation. As will become apparent in the course of this chapter, valuable insights have been obtained in this way, although the number of species in which the approach is technically feasible is limited, and a further constraint is that the relative proportions and distribution of the two cell populations is usually not under the control of the experimenter.

I. HYDRA

Parabiotic chimeras can be made in freshwater Hydra by slitting two individuals longitudinally and applying the cut edges together. Size regulation occurs, and the double complement of tentacles is rapidly reduced to the normal number, but experiments in which the two components were stained with different vital dyes suggest that cells from the two halves do not mingle extensively.

Brien (1963) used this technique to study sexual differentiation in *Hydra fusca,* a species in which no continuous germ cell lineage appears to exist. At 19°C, asexual reproduction occurs, but when the temperature of the water is lowered to 8°C, gametogenesis in either the male or the female direction is seen in some but not all strains. Brien made chimeras between individuals of the two types of strain at 19°C, then lowered the temperature to 8°C. He reports that gametogenesis was always induced in the asexual component, though with some delay. In 19 technically satisfactory chimeras, the sexual component developed in the female direction in 8 and in the male direction in 11; sexual differentiation in the asexual component was invariably in the same direction as in the sexual, suggesting that a genetic sex was either lacking in the asexual strain, or was overruled by the influence of the parabiotic partner. When chimeras were made between male and female individuals of the sexual strain, the female component became wholly or partly masculinized. Brien concludes that the effects on sexual differentiation were exerted by diffusible hormones.

Recent work by Littlefield (1984) establishes that the sexual phenotype of an individual Hydra is determined by the interstitial cells rather than the epithelial cells, but sheds little further light on whether the masculinization of the female component in male–female chimeras is due to male interstitial cells invading the female tissue and taking over gamete production, or to the same cells producing a diffusible substance that masculinizes the female interstitial cells.

XV. Chimeras and Sexual Differentiation

II. NEMERTINE WORMS

In species that can be propagated asexually, a state of chimerism once achieved may be propagated indefinitely. In the nemertine worm *Lineus sanguineus,* adult worms of contrasting colour (light brown and dark brown) can be cut in half sagitally and recombined by parabiosis; the resulting bilateral chimeras can then be divided transversely, into several segments, each of which will undergo complete regeneration. This process can be repeated at will, so as to yield a chimeric worm clone of any required size.

Sivaradjam and Bierne (1981) have used this technique to examine sexual

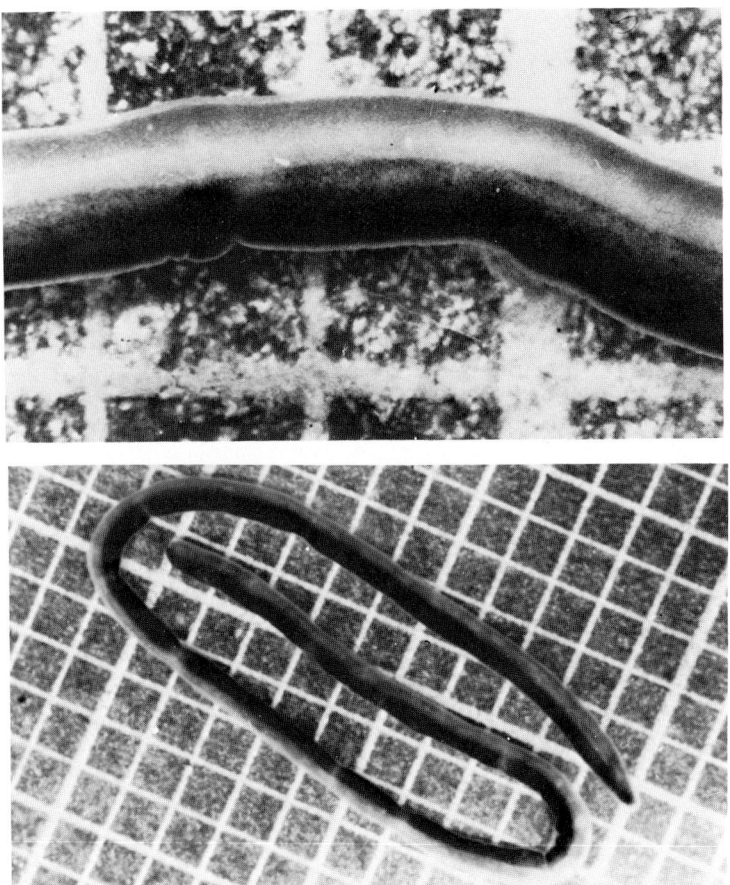

Fig. 1. Dorsal views of a chimeric nemertine worm propagated vegetatively. The dark and light components were taken from male and female worms, respectively: the difference in pigmentation persists indefinitely. The background paper is marked off in millimetres. (From Sivaradjam and Bierne, 1981.)

differentiation in heterosexual chimeras. In the example shown in Fig. 1, the dark component was from a clone of male worms, while the light tissue derived from a female clone. Individuals regenerated from any part (head, middle, or tail; large or small) of a sexually undeveloped worm went through a similar sexual cycle. This consisted first of a "primary gynandromorphous state" (Fig. 2a), in which the gonads developed autonomously, that is they became ovaries on the female and testes on the male side, and gametogenesis began; there followed a state of unilateral sex reversal, with ovarian development proceeding normally, but the testes gradually converting to ovotestes as groups of oocytes appeared among the spermatogenic cells; and finally a state of "secondary feminization" (Fig. 2b) was achieved, in which mature ovaries were seen on both sides of the animal and no testicular tissue remained, so that the chimera was indistinguishable from a normal fertile female apart from its contrasting bilateral pigmentation.

During the gynandromorphous state, sexual development as judged by the progress of gametogenesis was clearly more advanced on the female than on the male side of the chimeras. When the testes still contained only proliferating gonia, the contralateral ovaries contained growing oocytes. By the time that spermatogenesis had begun in the testes, the oocytes were already undergoing vitellogenesis. In contrast, when heterosexual chimeras were made in a similar fashion (though without a colour marker) in the closely related species *Lineus ruber*, although sex differentiation was again autonomous at the beginning of gonadal development, the testes showed more advanced development than did the contralateral ovaries. In this species it was the ovaries that underwent sex reversal, so that the sexually developed worms had mature active testes on both the male and the "female" side (Bierne, 1967, 1970).

Sivaradjam and Bierne (1981) speculate that the more precociously developing gonads may in each species bring about sex reversal of the contralateral gonads, either by means of directed migration of sexually differentiated cells from the dominant gonads, producing lysis or inhibition of the equivalent resident cells, or by some hormonal action emanating from the dominant gonads, exerting on both sides of the animal a feminizing action in *L. sanguineus* and a masculinizing action in *L. ruber*. Of possible relevance is the observation that the heterogametic sex (with unlike sex chromosomes) is the female in *L. sanguineus* and the male in *L. ruber*. Since no markers exist to distinguish between cells of the two chimera components within the gonads, it has so far not proved possible to decide between the cell migration and the hormonal hypothesis.

Fig. 2. Transverse sections of chimeric worms from the clone illustrated in Fig. 1. (From Sivaradjam and Bierne, 1981.) (a) Primary gynandromorphous state, with testis on left side and ovary on right. Scale, 50 μm. (b) Feminized worm, in which the left gonad, initially a testis, has been transformed into an ovary. No difference can be observed from a normal female control worm. Scale, 120 μm.

XV. Chimeras and Sexual Differentiation

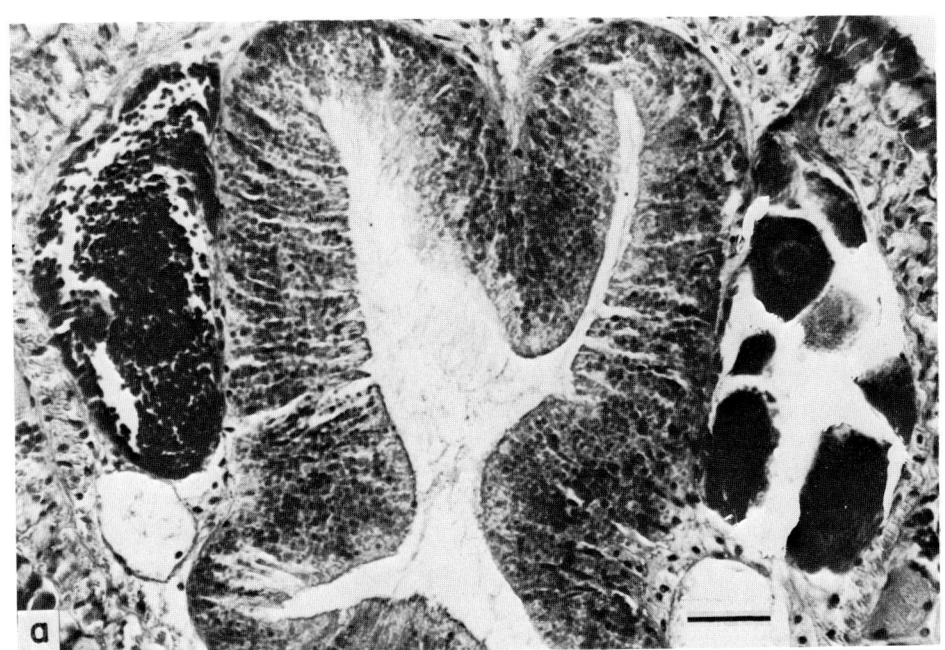

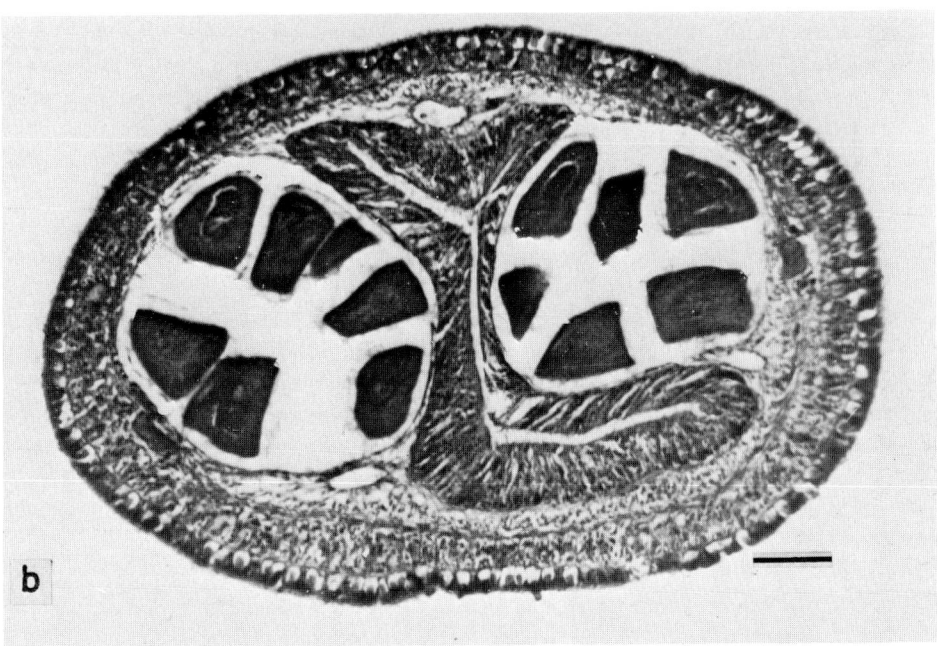

A further systemic influence on sexual differentiation in *Lineus* stems from the brain region. In regenerating chimeric fragments that temporarily lacked this region, gonadal development proceeded faster than when the brain was present, though the direction of differentiation was unaffected (Bierne, 1970; Sivaradjam and Bierne, 1981).

III. AMPHIBIA AND BIRDS

In lower vertebrates, chimeric individuals have been constructed by grafting pieces of one embryo into another. When Blackler (1965) transferred *Xenopus* neurula fragments that contained primordial germ cells into host neurulae from which some or all of the corresponding region had been excised, germ cells migrated out of the grafted tissues and colonized the developing host gonads. The chimeric embryos grew into adult toads, which could be shown by the presence of a nuclear marker to produce gametes of donor type. Since in *Xenopus* the male is known to be the homogametic (ZZ), and the female the heterogametic (ZW) sex, the fact that some of the female germ cell chimeras gave all-male donor progeny, and some of the male chimeras gave a 3 : 1 ratio of male to female donor progeny, proved that the chimeras were still fertile when the chromosomal sex of the graft differed from that of the host, and also showed that when a gonad of one genetic sex (e.g., ZW) was colonized by germ cells of the opposite sex (WW), it was the germ cells that underwent sex reversal (see Chapter V), while the differentiation of the somatic tissues of the gonad was unaffected. Thus in *Xenopus,* sexual differentiation appears to depend only on the chromosomal sex of the somatic component of the gonad and is not affected by the germ cells.

Similar grafting experiments have been carried out in birds. For example, Haffen (1968, 1975) made reciprocal grafts between the anterior part of one early chick embryo and the posterior part of another, and was able to rear the chimeric embryos, first *in vitro* and then in the coelomic cavity of a succession of host embryos, to a stage where their sexual differentiation could be studied. The primordial germ cells were located in the anterior germinal crescent at the time of grafting, while the posterior part contained the future gonadal region; the somatic component of the chimeric gonad and the germ cell population that colonized it would therefore have originated largely or entirely from different embryos. Haffen identified several reciprocal pairs in which one member developed an ovary and the other a testis, and argued that these must represent the chimeras that had been made between embryos of contrasting sex, in which the somatic component of the gonads was of one chromosomal sex and the germinal component of the other. There was no direct way of knowing the chromosomal sex of the different components (unlike the *Xenopus* experiment, the chimeric chick embryos could not be maintained long enough for breeding experiments to be carried out), but since the germ cells in the ovaries of such pairs degenerated, while those in ovary–ovary pairs underwent normal oogenesis, Haffen's results probably indicate that it is the somatic component that determines gonadal sex, in birds as in amphibia, and that the degenerating germ cells were chromosomally male.

Hajji *et al.* (1984) have made chick-chick and chick-quail embryo chimeras in which the sex of the embryos could be determined by the use of a sex-linked albino marker, and in this way have conclusively confirmed that the phenotypic sex of the gonad is determined by the chromosomal sex of the somatic component, both female (ZW) and male (ZZ), and not by the germ cells.

Thus, in both birds and amphibia, chimeric gonads can be produced in which the somatic and germinal components are largely or entirely of different genetic constitution. In both groups, the chromosomal sex of the somatic component has been shown to be responsible for the sexual differentiation of the gonad. (For a discussion of the fate of germ cells in a gonad of inappropriate sex, see Chapter V). No information is yet available, however, on the differentiation of a chimeric gonad containing a mixed population of chromosomally male and female somatic cells.

IV. MAMMALS

In mammalian chimeras, chromosomally male (XY) and female (XX) cells may coexist from early cleavage onward, closely intermingled in every organ of the body. This provides a particularly favourable opportunity to investigate the relationship between chromosomal and phenotypic sex.

A. Sex Ratio

At the time that the first experimental chimeras were made in mammals, little was known about the extent of cell mingling during development, and less about the cellular basis of sex determination. There was, therefore, no *a priori* expectation as to how XX↔XY mice would develop, whether as hermaphrodites, gynandromorphs, males or females. Moreover, there was initially no way of distinguishing mixed-sex (XX↔XY) from single-sex animals, since adequate procedures for chromosomal sexing of mitotic cells in mice (see Chapter II) had not yet been devised, and even now are tedious to apply. However, from the earliest report of the successful production and birth of chimeric mice (Tarkowski, 1961), and increasingly in later reports, it was clear that the sex ratio was biased in the male direction, and that intersexes were relatively uncommon, certainly less than the 50% of chimeras expected to be of mixed-sex constitution.

Only in "unbalanced" strain combinations (Mullen and Whitten, 1971), where one component strongly predominated over the other, was a normal 1 : 1 ratio of males to females seen. An earlier summary of published reports of aggregation chimeras from balanced strain combinations (McLaren, 1976, Table 4) shows that, of 260 animals for which independent evidence of chimerism was available, 168 were male, 84 were female, and 8 were hermaphrodites. If male and female embryos were produced in equal numbers and aggregated at random, so that an average of 50% of all chimeras were XX↔XY (see McLaren, 1976, Table 3), these figures would suggest that about 15% of XX↔XY chimeras developed as females, 6% as hermaphrodites, and the remainder, i.e., nearly 80%, as males. However, since "balanced" and "unbalanced" are relative rather than absolute categories, some heterogeneity can be expected, even among apparently balanced strain combinations.

A more direct estimate of the fate of XX↔XY chimeras, but inevitably based on much smaller numbers, can be derived from those studies in which chromosomal sexing was routinely performed. Of 41 known XX↔XY aggregation chimeras (Table I), 22% were females, 7% hermaphrodites, and 71% males.

When chimeras are made by injecting a single cell into a blastocyst, phenotypic sex usually follows the chromosomal sex of the host blastocyst, but again the XY cells tend to dominate. When an XX cell was injected into an XY blastocyst, all 21 chimeras were male; but when the donor cell was XY and the host blastocyst XX, 2 fertile males and a hermaphrodite developed, as well as 6 females (Table II).

Chimeras involving other species show a similar picture. Of 36 weaned *Mus musculus*↔*Mus caroli* chimeras, 27 were male and 9 female, a ratio consistent with all XX↔XY individuals developing as males (Rossant and Frels, 1980), while two sheep XX↔XY chimeras were both male (Tucker et al., 1974).

TABLE I.

Phenotypic Sex of XX ↔ XY Aggregation Chimeras from Balanced Strain Combinations

Reference	♂	⚥	♀	Total
Mystkowska and Tarkowski (1968)	2	1	0	3
Mystkowska and Tarkowski (1970))[a]	4	0	2	6
Milet et al. (1972)	2	1	1	4
Ford et al. (1974)	2	0	2	4
McLaren (1975)	7	1	0	8
Gearhart and Oster-Granite (1981)	8	0	4	12
A. McLaren and M. L. Buehr (unpublished)[a]	4	0	0	4
Total	29	3	9	41

[a] Fetal.

TABLE II.

Phenotypic Sex of XX ↔ XY Injection Chimeras[a,b]

Donor	Host	
	XX	XY
XX	18 fertile ♀, 1 sterile ♀	19 fertile ♂, 2 sterile ♂
XY	6 fertile ♀, 1 hermaphrodite, 2 fertile ♂	23 fertile ♂

[a] Single donor cells from blastocysts 3½ or 4 days postcoitum were injected into 3½-day host blastocysts.
[b] From Gardner and Lyon (1971), Gardner (1977), and R. L. Gardner and M. F. Lyon (personal communication).

Does the male bias in XX↔XY sexual differentiation reflect unequal numbers of XX and XY cells, perhaps due to greater proliferative ability of the latter, or do XY cells exert a masculinizing influence out of proportion to their numerical frequency? Figure 3 shows that all but one of the female XX↔XY aggregation chimeras had less than 40% of the XY component (about 25% on average), while about two-thirds of the males had more than 50% of XY cells. These estimates are based on coat colour or skin karyotyping in the adult, liver karyotyping in the fetus; karyotypes of lymphomyeloid tissues have not been included, as they are more variable and less correlated with phenotypic sex. None of these tissues are likely to be directly involved in sex differentiation, but at least three estimates suggest that no overall excess of XY cells is apparent, and different tissues within a chimera tend to be positively correlated with respect to the proportions of the two components that they contain.

The occasional XX↔XY intersexes that show a testis on one side and an ovary on the other, or an ovotestis on one side and an apparently normal ovary or testis on the other, suggest that the control of gonadal development is local rather than systemic. Experiments in which very early genital ridges, taken from mouse embryos at about $11\frac{1}{2}$ days postcoitum, have undergone sexual differentiation *in vitro* in accordance with their chromosomal sex when isolated from the rest of the embryo (Taketo and Koide, 1981), and even when isolated from the attached mesonephric region (Byskov and Grinsted, 1981; A. McLaren, unpublished observations), show that at least from this stage onward the control of sexual differentiation lies within the genital ridge itself. Much discussion has centred on the question of which gonadal tissue, germinal or somatic, determines gonadal sex (for references, see McLaren, 1976). On the germinal side, adult XX↔XY chimeras have been described with at least 95% of the somatic tissues of the

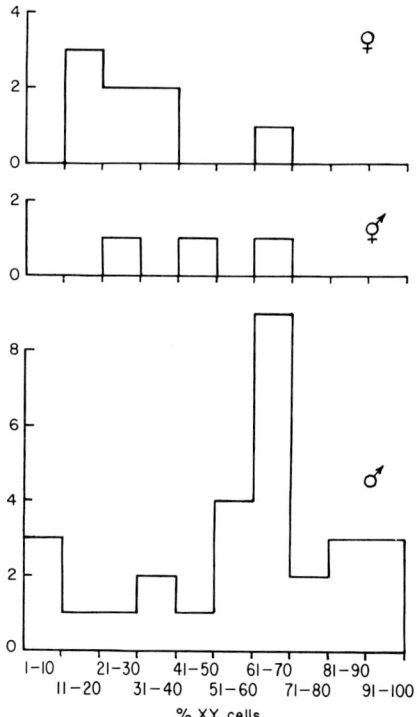

Fig. 3. The proportion of XY cells in individual XX↔XY aggregation chimeras (8 female, 3 hermaphrodites, 29 males) from balanced strain combinations. The animals are the same as those listed in Table I, except that no estimate is available for one female described by Milet *et al.* (1972); the estimates of the proportion of XY cells are based on visual inspection of the coat (Mystkowska and Tarkowski, 1968; Ford *et al.*, 1974; McLaren, 1975; Gearhart and Oster-Granite, 1981) or skin biopsy cultures (Milet *et al.*, 1972) for adult animals, and liver chromosomes for fetuses (Mystkowska and Tarkowski, 1970; A. McLaren and M. Buehr, unpublished).

gonad (testis or ovary) made up of cells of the inappropriate type; but the composition of the somatic cells of the adult gonad does not necessarily reflect that of the genital ridge at the critical stage of sexual differentiation. On the somatic side, gonadal differentiation proceeds normally, at least for a time, when the number of germ cells entering the genital ridges is greatly reduced, for genetic or other reasons; but perhaps even a few germ cells are enough to trigger the direction of sexual differentiation.

In a recent experiment (A. McLaren and M. L. Buehr, unpublished observations), we have made aggregation chimeras between embryos homozygous for *W* (white-spotting), in which very few germ cells reach the genital ridges, and normal (+/+) embryos. In some instances, normal testicular development occurred when the *W/W* component was XY and the +/+ component was XX, i.e., when the overwhelming majority of the germ cells in the testis were XX in chromosome constitution. A similar chimera was reported by Whitten (1975). This result appears to exonerate the germ cells from any responsibility for the direction of gonadal differentiation.

There is no general agreement as to whether the somatic cells of the gonad are derived from the coelomic epithelium, by ingrowth, or the mesonephros, by outgrowth, or whether they are of dual origin (for discussion, see Merchant-Larios, 1979; Wartenberg, 1981; Upadhyay *et al.*, 1979, 1981). Their origin may vary with the species or with the tissue within the gonad. The follicle cells of the ovary seem to be homologous to the Sertoli cells of the testis, and the thecal cells of the ovary to the Leydig cells of the testis. Ford *et al.* (1974) karyotyped follicle cells in a series of female chimeras, and found a strong correlation between right and left ovaries in the relative proportions of the two chimera components. This throws little light on the origin of the follicle cells, however, since both mesonephros and coelomic epithelium are mesodermally derived, so that for both tissues right and left sides would have a common origin within the primitive streak. Ford *et al.* (1974) also observed a female XX↔XY chimera in which 98% of the follicle cells were XY in chromosome constitution, a finding that might suggest a greater role for the Leydig/theca cells than for the Sertoli/follicle cells in the control of gonadal differentiation. However, as pointed out above, extrapolation from the adult to the embryo may be misleading, especially for the follicle cells, which undergo extensive proliferation in postnatal life.

The observation of XY follicle cells confirms that somatic cells within the gonad develop in accordance with the phenotypic sex of the gonad and not in accordance with their own chromosomal sex. The same is true of somatic cells in the secondary sex glands, as was elegantly shown for the epididymis and seminal vesicles by Mintz (1969) and Mintz *et al.* (1972). They identified one sterile XX↔XY chimera in which these glands appeared to consist entirely of the XX component, and another in which the seminal vesicle fluid contained seminal vesicle proteins characteristic of both component strains, XX as well as XY. Thus XX cells are capable of responding to an androgenic stimulus by forming morphologically normal seminal vesicles and secreting a specifically male protein, though the form of the protein is determined by the genotype of the XX cells themselves.

The relation between chromosomal sex and phenotypic sex in germ cells is discussed in Chapter V. It seems that most germ cells in XX↔XY chimeras start to differentiate in accordance with the phenotypic sex of the gonad, whatever their own chromosomal sex; XY germ cells in an ovary are capable of undergoing oogenesis, but XX cells in a testis degenerate soon after birth.

B. Mechanisms of Sex Determination

Since sex-specific structures differentiate earlier in the male gonad (testis cords) than in the female, the most economical hypothesis to explain the observations on gonadal differentiation in XX↔XY chimeras is to postulate that a critical proportion of XY cells is required in some somatic component of the genital ridge to ensure development of all somatic cells, XY and XX, in a male direction. If the proportion of XY cells is below the critical level at the moment of decision, then either an ovotestis forms (see below) or, more usually, all somatic cells, XY and XX, develop in a female direction, giving rise to an ovary.

How does the critical proportion of XY cells induce XX as well as XY cells to cooperate in producing a testis? Presumably XY cells secrete a chemical inducer that at a certain concentration acts on all cells, irrespective of their chromosome constitution. Two substances have so far been claimed to be secreted by XY but not by XX cells, the H-Y histocompatibility (transplantation) antigen, and the serologically detected male (SDM) antigen, also often termed H-Y antigen. No conclusive evidence has yet been obtained for or against either of these antigens playing a testis-inducing role (for reviews, see Wachtel and Koo, 1982; Silvers *et al.*, 1982; Simpson and Eicher, 1983). Most individuals with testes are positive for both H-Y and SDM, and most individuals lacking testes are negative for both antigens. A few exceptions are known. XY mice of certain genetic constitutions are H-Y positive, but develop as fertile females (for review, see Eicher, 1982), so if H-Y is the inducer they must lack appropriate receptors. Also a single H-Y negative male XO mouse has been reported (Melvold *et al.*, 1977). Female XO mice, though H-Y negative (Simpson *et al.*, 1982), are reported to be weakly SDM positive, so if SDM is the inducer, its concentration in these embryos must be below the threshold required for testicular differentiation. No relevant experiments have yet been done on H-Y histocompatibility antigen in XX↔XY chimeras, but our presumed mosaic T(X;16)16H/X *Sxr* females (see below) have been shown to express this antigen (Simpson *et al.*, 1984). For SDM, a report has appeared (Ohno *et al.*, 1978), claiming that in an XX↔XY male mouse, antigen was transferred from XY to XX Sertoli and Leydig cells in the testis, but was not transferred to XX cells in the spleen or epidermis. Unfortunately only one mouse was studied, and no chromosome analysis was done, so that the evidence for its XX↔XY status (other than the SDM antigen absorption tests) rested on the detection of only a single component in its germ line. Since as many as 50% of XY↔XY males may produce only a single type of progeny (McLaren, 1978; see Chapter V), this cannot be considered conclusive. The technical problems inherent in attempting to use absorption tests to quantitate SDM antigen levels have been discussed by Silvers *et al.* (1982). Further studies on male-specific antigens in XX↔XY chimeras are clearly required.

Serologically detected male antigen has also been suggested as the masculinizing agent in freemartinism. Female calves sharing a common blood circulation with male twins are usually sterile, with female external genitalia but poorly developed uterus and oviducts, some male secondary sex organs, and small sterile gonads that develop as ovaries initially, but show a variable degree of secondary masculinization, often containing some testicular tissue (Jost, 1970). The male twin is also affected, often becoming more or less sterile in later life. Because of the shared placental circulation, both twins are blood chimeras, yet blood chimerism in heterosexual twin pairs in man (see Chapter VII) and in the marmoset is not accompanied by any abnormalities of sexual development. The consequences of secondary chimerism for sexual development have been reviewed by Short (1970). Ohno et al. (1976), again using absorption tests to quantitate SDM antigen levels, claimed that SDM levels were as high in masculinized XX fetal gonads of bovine freemartins as in normal bull testes at a comparable developmental stage. They suggested that male cells colonizing the gonad of the female twin transmit antigen to the somatic cells of the gonad, which induces testicular development. It now seems, however, that the presence of XY cells in the female twin is not sufficient to masculinize the gonad, since Vigier et al. (1976) have succeeded in preventing freemartinism by interrupting the common blood circulation surgically at a stage after XX/XY chimerism had been established. Perhaps some male hormone, synthesized by the male twin at a later stage, is responsible for the inhibition of ovarian development characteristic of freemartinism; the formation of seminiferous tubules that occurs secondarily might then lead to an increased level of SDM antigen.

Primary masculinization of the indifferent gonad of an XX individual does not occur in freemartinism, nor has it been convincingly demonstrated by coculture or cotransplantation of male and female indifferent gonads. To study disturbances in the initial stages of sexual differentiation of the mammalian gonad, we must return to primary chimeras.

C. Hermaphrodites

As we have seen, about 3% of all overt chimeras made by embryo aggregation in balanced strain combinations develop as hermaphrodites. The exact proportion is likely to vary from one strain combination to another, though the few published series with a sample size of more than 20 are rather consistent: 4 hermaphrodites in 123 overt chimeras (Mintz, 1969), 2 in 103 (Mullen and Whitten 1971), 1 in 24 (McLaren and Bowman, cited in McLaren, 1976). Since 50% on average of all chimeras will be XX↔XY and there is no reason to expect any hermaphrodites among XX↔XX or XY↔XY chimeras, it is not surprising to find about 7% of hermaphrodites among known XX↔XY individuals (Table I). In unbalanced strain combinations, where one component greatly predominates, the incidence of hermaphroditism like the distortion in the sex ratio will be substantially lower. Whitten (1975) mentions a figure of 1% (12 hermaphrodites in 1200) for all overt chimeras made by himself and his colleagues; many of these were presumably from unbalanced combinations.

The number of hermaphrodite chimeras for which detailed descriptions of the gonads or any reproductive organs exist in the literature is small (for summaries, see Whitten, 1975; McLaren, 1976). Fortunately a careful study has been carried out (Whitten, et al., 1979); Eicher et al., 1980) on the fetal gonads of more than 100 hermaphrodite mosaics encountered in the BALB/cWt strain, and the chimera data such as they are agree well with the picture seen in these mosaics, as well as in other spontaneous cases of chimerism and mosaicism (see Tarkowski, 1964; Whitten, 1975). In BALB/cWt mice, the Y chromosome seems peculiarly susceptible to nondisjunction, leading to a high incidence of XO/XY and XO/XY/XYY mosaics (Eicher et al., 1980). Presumably the hermaphroditism comes about as in chimeras, whenever an appropriate ratio or distribution of female-determining (XO,XX) and male-determining (XY,XYY) cells exists within the genital ridges. In the BALB/cWt mosaics, the XO cell line seems to be at a selective disadvantage (P. S. Burgoyne and W. G. Beamer, personal communication), so that by the time of sexual differentiation XO cells are in the minority, and few or no XO/XY mosaic embryos develop as females. Other mosaic situations have been described in which a female-determining and a male-determining population of cells coexist, and hermaphrodites are found. In T(X;16)16H/X Sxr mice (McLaren and Monk, 1982; Cattanach et al., 1982), the testis-determining part of the Y chromosome (Tdy) is located on the normal X chromosome, where it is presumed to be subject to X inactivation in a substantial proportion of cells because of the simultaneous presence of the X autosome translocation T(X;16)16H; this produces a mosaic individual with female- and male-determining cells both present, according to whether Tdy has or has not been inactivated, and gonads may develop as ovaries, testes, or ovotestes. Similarly, the $\dot{X}$XY wood lemmings described by Fredga (1983) may develop as males, females, or hermaphrodites, depending presumably on the proportion of cells that express the normal X chromosome (XY, male-determining) rather than the aberrant $\dot{X}$ ($\dot{X}$Y, female-determining).

Of the BALB/cWt fetal hermaphrodites studied by Whitten et al. (1979), 76 had two ovotestes, 16 an ovotestis and an ovary, and 15 an ovotestis and a testis. No fetus was found with an ovary on one side and a testis on the other. No tendency was seen, either in these mosaics or in mouse chimeras, for the right gonad to be more masculinized than the left, although an asymmetry of this sort has been reported for hermaphrodites in man and a few other species (Polani, 1970; Van Niekirk, 1974). On the other hand, a very striking regularity was found in the distribution of ovarian and testicular tissue within the hermaphrodite mouse gonads. Of 174 ovotestes in which the position of the ovarian and testicular components could be classified, testis cords were located in the equatorial part of the gonad in 81, and at the caudal (posterior) pole in 79; in only 8 were they located at the cranial (anterior) pole, and in only 6 at the two poles, with ovarian tissue in the equatorial region. Too few chimeric ovotestes have been adequately described to know whether these would show a similar tendency for testicular tissue to be concentrated in the central and posterior regions of the gonad; nor has any convincing developmental explanation for this nonrandom organization of the ovotestis yet been proposed.

Some chimeric and mosaic ovotestes have the ovarian and testicular domains well defined, sometimes even separated by a layer of mesenchymal tissue. As Tarkowski (1964) points out, this is most likely to occur when the boundary between the domains more or less coincides with the normal ingrowth of mesenchymal tissue into the gonad. Often, however, no clear boundary exists. In a fetal ovotestis, germ cells enter meiotic prophase throughout the ovarian tissue, as in a normal ovary, and sometimes also in the testis cords, especially in those adjacent to ovarian tissue. Growing oocytes within such testis cords have been observed in neonatal hermaphrodite chimeras (Tarkowski, 1964). (For a discussion of factors affecting entry of germ cells into meiosis, see Chapter V.) In adult chimeras, hyalinized oocytes have been described in testis cords lying near ovarian tissue (P. S. Burgoyne, cited by McLaren, 1980). In adult T(X;16)16H/X *Sxr* mosaics, ovotestes with an apparently normal primary structure of testis tubules throughout the gonad may be secondarily disrupted by large local accumulations of growing oocytes and follicles within the tubules (McLaren, 1983). The fetal testes of XX↔XY chimeras in which a few germ cells in meiotic prophase are seen (Mystkowska and Tarkowski, 1970; McLaren *et al.*, 1972) are not normally classified as ovotestes; if however these aberrant germ cells are acting as indicators for some as yet undetected ovarian feature of the somatic tissue of the gonad, it might be considered that the proportion of hermaphrodites listed in Table I greatly underestimates the true incidence of intersexual development of the gonad, at least in the initial stages.

Whether a minor degree of hermaphroditism in the fetus can be corrected during the course of development, so that it is no longer detectable in the adult gonad, is not yet clear. Tarkowski (1964) suggests that the growth of oocytes and their follicles might secondarily obliterate primary testis cords, so that a fetal ovotestis with only a small amount of testicular tissue might in the adult be indistinguishable from a normal ovary. Whitten *et al.* (1979) speculate that a more common occurrence might be the hypertrophy of testis cords within an ovotestis, obliterating an initially ovarian region and giving rise to an apparently normal adult testis. Certainly the incidence of hermaphrodites detected among weanling BALB/cWt mice (0.4%) was much lower than among 15-day-old fetuses (3%); but in the fetal population all gonads were examined, while at later ages only those with abnormal external genitalia or subsequent sterility were subjected to laparatomy (Whitten *et al.*, 1979). Ovotestes may be less common, or at least less often detected, in adult than in fetal or newborn chimeric hermaphrodites: Whitten (1975) in his Table 4 lists 7 adult chimeric hermaphrodites, with a total of only 3 ovotestes (3/14), while the 3 newborn chimeric hermaphrodites of Tarkowski (1964) contained 4 ovotestes, and only 2 normal gonads (4/6).

There is also doubt as to what degree of fetal hermaphroditism would be consistent with adult fertility. Abnormalities of the reproductive ducts have often been observed in chimeric hermaphrodites (Tarkowski, 1964; Mintz, 1965; Whitten, 1975), with Müllerian and Wolffian duct derivatives running in paralled, a condition that is unlikely to be compatible with fertility. On the other hand, a small patch of gonadal tissue of inappropriate sex would not necessarily lead to sterility. A subfertile XX↔XY male hermaphrodite was reported by McLaren (1975).

What determines a gonad to develop as an ovotestis rather than as a "pure" gonad? Probably both the relative proportion of male- and female-determining cells and their arrangement is important. If 20–25% of XX↔XY chimeras develop as females (Tables I and II), and if (as suggested by Falconer and Avery, 1978) the varying proportions of the two chimera components follow a flat rather than a normal distribution (see Fig. 1), then some 0–25% of XY cells in the relevant cell population of the early gonad is presumably consistent with development in the ovarian direction. If the arrangement of male- and female-determining cells within the gonad is irrelevant, a proportion of say 25–30% of XY cells might be necessary and sufficient to induce the development of an ovotestis. Whitten *et al.* (1979) show that in BALB/cWt mosaic hermaphrodites, significantly more "pure" gonads, as compared with ovotestes, develop on the left side (21/107) than on the right (10/107). Among ovotestes, the distribution on both sides is skewed toward gonads with a low ovarian content, though "pure" ovaries were as common as "pure" testes. They suggest that the excess of "pure" gonads on the left side, also observed by Van Niekerk (1974) for human hermaphrodites, may indicate that the left gonad develops from a smaller cell pool. Certainly in the adult mouse, on average the left ovary ovulates fewer eggs (McLaren, 1963) than the right, though paradoxically the right testis weighs more than the left (Billingham, 1965). Mittwoch and Buehr (1973) found no consistent difference between the volume of fetal mouse gonads on the left and right sides, but Mittwoch (1976) reported that gonads from human fetuses were smaller on the left side than on the right in both sexes.

As to the arrangement of male-determining and female-determining cells within the gonadal rudiment, if testicular differentiation is induced by a substance of limited diffusibility emanating from XY cells (whether H-Y or SDM antigen, or some other substance), or even by contact with XY cells, then the occurrence of cohesive groups of cells of a single type should favour the development of an

XV. Chimeras and Sexual Differentiation

ovotestis, while intimate intermingling of the two types should lead to testicular development. On the other hand, the nonrandom arrangement of ovarian and testicular territories seen in mosaic and perhaps also chimeric ovotestes suggests a higher level control of gonadal differentiation that is unlikely to reflect the arrangement of individual male- and female-determining cells.

V. CONCLUSION

In mosaic insects of mixed chromosomal sex, development tends to be gynandromorphic, with each cell or group of cells expressing the phenotypic sex appropriate to its own genotype. The studies on chimeras summarized in this chapter show that this is probably not true of nemertine worms, and certainly not of mammals. The evidence is less direct for birds and amphibia, but there is no reason to think that they would differ from mammals in this respect. Sexual differentiation of the gonad appears to follow a "majority vote" system rather than proportional representation, so that the outcome is usually a normal testis or ovary, rather than an intersexual gonad. The voting system, however, is biased in the direction of the heterogametic sex, at least in mammals and birds, and in the two species of nemertine worms studied. This may reflect evolutionary conservatism, with the Y chromosome or its homologue inducing precocious gonadal differentiation and exerting a dominant influence in mixed cell populations, perhaps even using the same molecular mechanism, throughout a wide spectrum of the animal kingdom. Beguiling though this view is, it does not account for the very unconservative fact that the pathway of gonadal differentiation thus induced is male in some groups (mammals, *Lineus ruber*) and female in others (birds, *Lineus sanguineus*).

REFERENCES

Bierne, J. (1967). *C. R. Hebd. Seances Acad. Sci.* **265**, 450–477.
Bierne, J. (1970). *Am. Sci. Nat., Zool. Biol. Anim.* [12] **12**, 181–288.
Billington, W. D. (1965). *J. Reprod. Fertil.* **10**, 343–352.
Blackler, A. W. (1965). *J. Embryol. Exp. Morphol.* **13**, 51–61.
Brien, P. (1963). *Bull. Biol. Fr. Belg.* **97**, 213–283.
Byskov, A. G., and Grinsted, J. (1981). *Science* **212**, 817–818.
Cattanach, B. M., Evans, E. P., Burtenshaw, M., and Barlow, J. (1982). *Nature (London)* **300**, 445–446.
Eicher, E. M. (1982). *In* "Prospects for Sexing Mammalian Sperm" (R. P. Amann and G. E. Seidel, eds.), pp. 121–135. Colorado Assoc. Univ. Press, Boulder.
Eicher, E. M., Beamer, W. G., Washburn, L. L., and Whitten, W. K. (1980). *Cytogenet. Cell Genet.* **28**, 104–115.
Falconer, D. S., and Avery, P. J. (1978). *J. Embryol. Exp. Morphol.* **43**, 195–219.

Ford, C. E., Evans, E. P., Burtenshaw, M. D., Clegg, H., Barnes, R. D., and Tuffrey, M. (1974). *Differentiation* **2**, 321–333.
Fredga, K. (1983). *Differentiation* (Suppl.) **23**, S523–S530.
Gardner, R. L. (1977). *Int. Congr. Ser.—Excerpta Med.* **432**, 154–166.
Gardner, R. L., and Lyon, M. F. (1971). *Nature (London)* **231**, 385–386.
Gearhart, J., and Oster-Granite, M. L. (1981). *Biol. Reprod.* **24**, 713–722.
Haffen, K. (1968). *C. R. Hebd. Seances Acad. Sci.* **267**, 511–513.
Haffen, K. (1975). *Am. Zool.* **15**, 257–272.
Hajji, K., Martin, C., Perramon, A., and Dieterlen-Lièvre, F. (1984). Personal communication.
Jost, A. (1970). *Philos. Trans. R. Soc. London, Ser. B* **259**, 119–130.
Littlefield, C. L. (1984). *Dev. Biol.* **102**, 426–432.
McLaren, A. (1963). *J. Endocrinol.* **27**, 157–181.
McLaren, A. (1975). *J. Embryol. Exp. Morphol.* **33**, 205–216.
McLaren, A. (1976). "Mammalian Chimaeras." Cambridge Univ. Press, London and New York.
McLaren, A. (1978). *In* "Genetic Mosaics and Chimeras in Mammals" (L. B. Russell, ed.), pp. 125–134. Plenum, New York.
McLaren, A. (1980). *Nature (London)* **283**, 688–689.
McLaren, A. (1983). *Differentiation* (Suppl.) **23**, S93–S98.
McLaren, A., Chandley, A. C., and Kofman-Alfaro, S. (1972). *J. Embryol. Exp. Morphol.* **27**, 515–524.
McLaren, A., and Monk, M. (1982). *Nature (London)* **300**, 446–448.
Melvold, R. W., Kohn, H. I., Yerganian, G., and Fawcett, D. W. (1977). *Immunogenetics* **5**, 33–41.
Merchant-Larios, H. (1979). *Ann. Biol. Anim., Biochim., Biophys.* **19**, (4B), 1219.
Milet, R. G., Mukherjee, B. B., and Whitten, W. K. (1972). *Can J. Genet. Cytol.* **14**, 933–941.
Mintz, B. (1965). *In* "Preimplantation Stages of Pregnancy" (G. E. W. Wolstenholme and M. O'Connor, eds.), pp. 194–207. Churchill, London.
Mintz, B. (1969). *Birth Defects, Orig. Artic. Ser.* **5**, 11–22.
Mintz, B., Domon, M., Hungerford, D. A., and Morrow, J. (1972). *Science* **175**, 657–659.
Mittwoch, U. (1976). *Ann. Hum. Genet.* **40**, 133–138.
Mittwoch, U., and Buehr, M. L. (1973). *Differentiation* **1**, 219–224.
Mullen, R. J., and Whitten, W. K. (1971). *J. Exp. Zool.* **178**, 165–176.
Mystkowska, E. T., and Tarkowski, A. K. (1968). *J. Embryol. Exp. Morphol.* **20**, 33–52.
Mystkowska, E. T., and Tarkowski, A. K. (1970). *J. Embryol. Exp. Morphol.* **23**, 395–405.
Ohno, S., Christian, L. C., Wachtel, S. S., and Koo, G. C. (1976). *Nature (London)* **261**, 597–599.
Ohno, S., Ciccarese, S., Nagai, Y., and Wachtel, S. S. (1978). *Arch. Androl.* **1**, 103–109.
Polani, P. (1970). *Philos. Trans. R. Soc. London, Ser. B* **259**, 187–204.
Rossant, J., and Frels, W. I. (1980). *Science* **208**, 419–421.
Short, R. V. (1970). *Philos. Trans. R. Soc. London, Ser. B* **259**, 141–147.
Silvers, W. K., Gasser, D. L., and Eicher, E. M. (1982). *Cell* **28**, 439–440.
Simpson, E., McLaren, A., and Chandler, P. (1982). *Immunogenetics* **15**, 609–614.
Simpson, E., Chandler, P., Washburn, L. L., Bunker, H. P., and Eicher, E. M. (1983). *Differentiation* (Suppl.) **23**, S116–S120.
Simpson, E., McLaren, A., Chandler, P., and Tomonari, K. (1984). *Transplantation* **37**, 17–21.
Sivaradjam, S., and Bierne, J. (1981). *J. Embryol. Exp. Morphol.* **65**, 173–184.
Taketo, T., and Koide, S. S. (1981). *Dev. Biol.* **84**, 61–66.
Tarkowski, A. K. (1961). *Nature (London)* **190**, 857–860.
Tarkowski, A. K. (1964). *J. Embryol. Exp. Morphol.* **12**, 735–757.
Tucker, E. M., Moor, R. M., and Rowson, L. E. A. (1974). *Immunology* **26**, 613–621.

Upadhyay, S., Luciani, J. M., and Zamboni, L. (1979). *Ann. Biol. Anim., Biochim., Biophys.* **19**(4B), 1179–1196.
Upadhyay, S., Luciani, J. M., and Zamboni, L. (1981). *Int. Congr. Ser.—Excerpta Med.* **559**, 18–27.
Van Niekerk, W. A. (1974). "True Hermaphroditism, Clinical, Morphologic and Cytogenetic Aspects." Harper & Row, Haggerstown, Maryland.
Vigier, B., Locatelli, A., Prepin, J., Buisson, F. M., and Jost, A. (1976). *C.R. Hebd. Seances Acad. Sci., Ser. D* **282**, 1355–1358.
Wachtel, S. S., and Koo, G. C. (1982). *In* "Mechanisms of Sex Differentiation in Animals and Man" (C. R. Austin and R. G. Edwards, eds.), pp. 255–299. Academic Press, London.
Wartenberg, H. (1981). *Int. Congr. Ser.—Excerpta Med.* **559**, 3–12.
Whitten, W. K. (1975). *Symp. Soc. Dev. Biol.* **33**, 189–205.
Whitten, W. K., Beamer, W. G., and Byskov. A. G. (1979). *J. Embryol. Exp. Morphol.* **52**, 63–78.

CHAPTER XVI

Chimeric Tissue Combinations in the Analysis of Developmental Mechanisms in the Embryonic Kidney

LAURI SAXÉN

Department of Pathology
University of Helsinki
Helsinki, Finland

I.	The Problems	401
II.	Transmission and Spread of the Inductive Signal	402
III.	Exclusion of an Assimilatory Mode of Induction	403
IV.	Demonstration of the Migratory Capacity of Induced Cells	404
V.	Origin of the Glomerular Endothelium	405
VI.	Origin of the Glomerular Basement Membrane	407
VII.	Comment	407
	References	408

I. THE PROBLEMS

Three cell lineages can be distinguished in the permanent kidney of the higher vertebrates, the metanephros: (1) the epithelium of the Wolffian duct that gives rise to the collecting system, (2) the mesenchymal cells of the blastema that will become converted into the secretory tubules, and (3) the endothelial cell lineage of the vasculature including the glomerular loop. The spatially and temporally synchronized development of these three components is guided by a series of complex *morphogenetic interactions,* dialogues between the cell populations. The conclusive demonstration of such interactions, came from the isolation and recombination experiments of Grobstein (1953, 1955), who managed to separate the epithelial and mesenchymal components of a 11-day kidney anlage of the mouse. When the two were cultivated in isolation after separation, both failed to differentiate further; when they were brought together and subcultured, both branching of the epithelium and tubule formation in the mesenchyme were observed. Grobstein also devised a most useful experimental technique by which the two components could be separated by a filter membrane, allowing various types of experiments as described below (Grobstein, 1956).

Grobstein's (1956) demonstration of an inductive tissue interaction between the components in an embryonic kidney as well as various descriptive findings of the development of these components have led to many questions, only some of which have so far been answered. Interest has been especially focused on the mode and molecular basis of the inductive interactions. As long as the signal substances apparently carrying the message remain fully unknown, other approaches and experimental manipulations become important. We will deal here with two: experiments devised to study the mode of transmission and spread of the induction, and those dealing with the origin and development of the glomerular endothelium. The present chapter mainly discusses experiments that make use of chimeric tissue combinations; for more comprehensive reviews on the topic, the reader is referred to Grobstein (1967), Saxén *et al.* (1968, 1980), Lehtonen (1976), and Ekblom (1981).

II. TRANSMISSION AND SPREAD OF THE INDUCTIVE SIGNAL

Experimental results obtained with the filter technique of Grobstein (1956) have suggested that the inductive interaction leading to the determination and differentiation of the secretory tubules requires actual contact between the interacting cells; its prevention by interposed filters excluding cytoplasmic penetration through the narrow pores resulted in inhibited induction (Wartiovaara *et al.*, 1974; Lehtonen, 1976; Saxén *et al.*, 1976). This result immediately raised the next question: since both *in vivo* and *in vitro* transformed (induced) epithelial cells and tubules are regularly observed at a distance from the inductive epithelial–mesenchymal interface, a second mechansim for the *spread* of the induc-

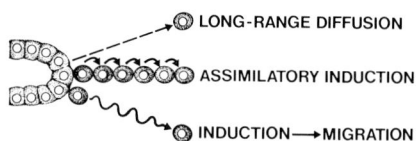

Fig. 1. Alternative mechanisms for spreading the inductive wave in the metanephrogenic mesenchyme: transmission of the signal by long-range diffusion, spreading through an assimilatory induction, or by migration of the induced cells into the deeper layers of the mesenchyme.

XVI. Developmental Mechanisms in the Embryonic Kidney

tion "wave" must be explored. Theoretically three possibilities can be outlined (Fig. 1): (1) long-range transmission of the signal, (2) an assimilatory induction where one induced cell would pass the message on to the next cell in the "field," and (3) spread of the induction implemented by cell migrations. Chimeric tissue combinations were used to test the last two alternatives after the filter experiments had excluded long-range transmission.

III. EXCLUSION OF AN ASSIMILATORY MODE OF INDUCTION

In order to test the alternative hypothesis of an assimilatory induction, the following experiment was devised (Saxén and Saksela, 1971) (Fig. 2). Isolated mesenchymal blastemas of mouse embryos with a normal karyotype were induced by the transfilter technique. After 24 hours, when induction was completed (Grobstein, 1967), the inductor was removed and the still undifferentiated mesenchyme was cut into halves; new, freshly isolated, and hence uninduced mesenchymes were then added to the culture. The new mesenchymes were obtained from embryos of the CBA/T_6T_6 strain that carries a chromosomal abnormality serving as a morphological nuclear marker. When tubules were formed in these chimeric mesenchymes, they were isolated and purified enzymatically, and monolayer cultures were then prepared from these cleaned epithelial elements. After a short-term colcemid treatment, the cells in monolayers were analyzed for their karyotype. If an assimilatory induction had taken place, one would expect either hybrid tubules or two types of tubules consisting of T_6T_6 karyotypes (induced at the second step of the experiment) as well as normal karyotypes (primarily induced). This proved not to be the case, and it was

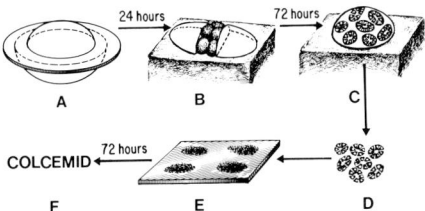

Fig. 2. Scheme of the experiment to test the possibility of an assimilatory mechanism spreading the induction in the target mesenchyme. (A) Isolated metanephric mesenchymes from 11-day mouse embryos are induced transfilter by the spinal cord. (B) After completion of induction the inductor is removed, the still undifferentiated mesenchymal explant is halved, and uninduced T_6T_6 strain mesenchymes are added. (C) After 72 hours of subcultivation tubules develop, which are then separated and enzymatically cleaned (D). (E) Monolayer cultures are prepared from the tubules, and the karyotype of their cells is determined after colcemid treatment (F). (After Saxén and Saksela, 1971.)

concluded that only cells induced by the original inductor were transformed into epithelial elements and that these were not able to pass the message on to the subsequently added, fresh mesenchymal cells. The experiments seemed, therefore, to exclude an assimilatory induction (Saxén and Saksela, 1971). The third mechanism schematized in Fig. 1 still remained to be explored.

IV. DEMONSTRATION OF THE MIGRATORY CAPACITY OF INDUCED CELLS

In order to test the last alternative mode of spreading of the inductive wave, the experiment illustrated in Fig. 3 was performed (Saxén and Karkinen-Jääskeläinen, 1975). The normal inducer, the Wolffian duct-derived ureteric bud, was routinely separated from its mesenchymal component in quail embryos (carrying the nuclear marker). Instead of the usual thorough cleaning of epithelium from the mesenchymal cells, a thin layer of "contaminant" cells was left on the epithelium. Subsequently, this bud with the few attached mesenchymal cells was combined with a piece of the metanephric mesenchyme from a chick embryo. After subculture the recombinants were histologically analyzed for the nuclear marker. Both tubules and single cells carrying the quail marker were regularly found deep in the chick mesenchyme, separated from the ureter by several layers of chick cells (Fig. 4). This shows conclusively how the induced quail cells, initially in contact with the ureter, had migrated deep into the chick mesenchyme. The authors conclude that this is the most plausible "spreading" mechansim *in vivo* also (Saxén and Karkinen-Jääskeläinen, 1975).

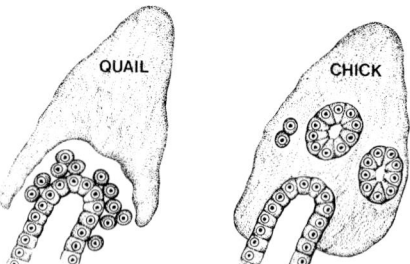

Fig. 3. Scheme of the experiment to test the migratory capacities of induced mesenchymal cells in the kidney anlage. Left: the inducer, the ureter bud, is isolated from quail embryos but a layer of "contaminant" mesenchymal cells is left attached to the ureter. Right: after combination of the contaminated quail ureter to chick mesenchyme, quail cells carrying the nuclear marker are detected deep in the chick mesenchyme, where they form tubules. (After Saxén and Karkinen-Jääskeläinen, 1975.)

XVI. Developmental Mechanisms in the Embryonic Kidney

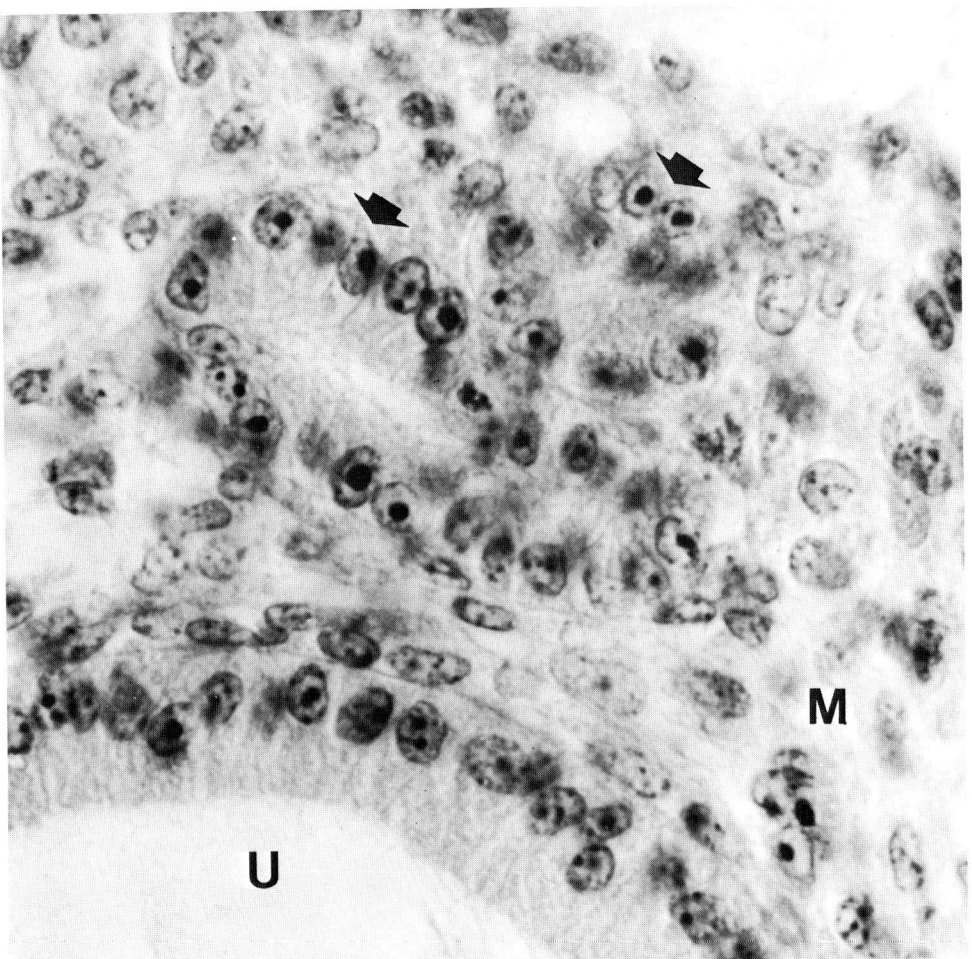

Fig. 4. Micrograph of the experimental situation schematized in the right-hand side of Fig. 3. Quail cells with the nuclear marker are seen in the chick mesenchyme (arrows) at a distance from their original location at the surface of the ureter (U). Feulgen–light green stain.

V. ORIGIN OF THE GLOMERULAR ENDOTHELIUM

In addition to the collecting epithelium from the Wolffian duct and the secretory epithelium derived from the mesenchyme (above), the nephron consists of the endothelial cells of the glomerular loop. The origin and differentiation of this third cell lineage has been the subject of somewhat controversial opinions, and

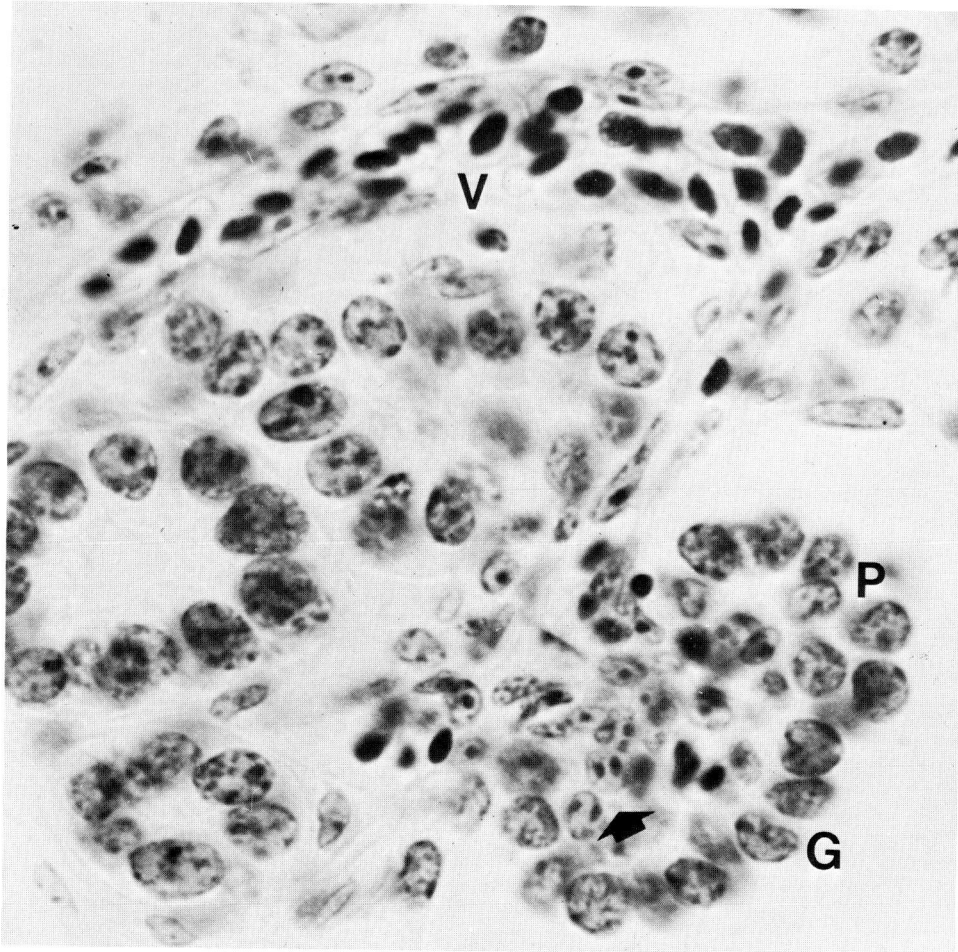

Fig. 5. Micrograph of a graft of mouse metanephric blastema on quail chorioallantoic membrane. A capillary showing endothelial cells with the quail marker (V) is invading the graft where hybrid glomeruli are seen (G). These consist of mousederived podocytes (P) and endothelial cells carrying the quail marker (arrow). Feulgen stain. Courtesy of Dr. Hannu Sariola.

three basic theories have been postulated (cf. Sariola *et al.*, 1983a): (1) the endothelial cells might originate from mesenchymal cells trapped into the lower crevice of the S-shaped body, the presumptive glomerulus; (2) they might be derivatives of mesenchymal cells migrating into this crevice, the future glomerular space; (3) they might be of exogenous origin, migrating into the mesenchymal blastema from outside capillaries, as first suggested by Osathanondh and Potter (1966). In order to test the third hypothesis [already suggested

XVI. Developmental Mechanisms in the Embryonic Kidney

by some immunofluorescence findings, (see Ekblom, 1981)], the quail marker was again used. Mouse metanephric blastemas were dissected from 11-day embryos before any vascular elements could be detected in the blastema; they were either cultured *in vitro* in organotypic cultures or grafted on quail chorioallantoic membrane (CAM). *In vitro* no endothelial differentiation within the blastema could be shown, while the CAM grafts regularly showed abundant capillary invasion. The origin of these vessels in the nephric blastema could be determined by their nuclear structure in light and electron microscopy; they proved to be almost exclusively of quail origin with rare mouse↔quail chimeric vessels. Correspondingly, the glomerular endothelium regularly expressed the quail marker, conclusively demonstrating its exogenous origin and the formation of chimeric glomeruli (Sariola *et al.*, 1983a) (Fig. 5). The chorioallantoic grafts also showed that endothelial cells of outside origin can be directed to migrate into the glomerular space and to form there chimeric structures with epithelial podocytes.

VI. ORIGIN OF THE GLOMERULAR BASEMENT MEMBRANE

Indirect evidence of the origin of the glomerular basement membrane has suggested that it might be of dual origin, as both the epithelial podocytes and the endothelial cells might contribute to its formation (cf. Sariola *et al.*, 1983b). The above technique to obtain hybrid glomeruli allowed this view to be directly tested: avascular kidney anlagen of mice were cultured on chick chorioallantoic membrane, and the constituents of the basement membrane in the chimeric glomeruli were immunohistochemically examined by means of monoclonal, strain-specific antibodies (mouse and chick antibodies against collagen type IV and mouse-specific anti-laminin antibodies). The basement membrane of the chimeric mouse↔chick glomeruli was shown to contain collagen type IV of both mouse (podocyte) and chick (endothelial) origin, thus confirming the hypothesis of a dual origin of the glomerular basement membrane (Sariola *et al.*, 1983b).

VII. COMMENT

These four experiments employing chimeric tissue combinations have led to two conclusions related to the complex development of the metanephric kidney: (1) Induction of the mesenchyme by the ureter is not spread in the target cell population through an assimilatory mechanism but by active cell migration. (2) The third cell lineage of the kidney, the glomerular endothelium, is derived from outside vasculature, and the resulting glomerular basement membrane receives contributions from both this endothelial component and from the epithelial podocytes. Techniques based on interstrain or interspecies tissue combinations have been decisive in obtaining these results and conclusions.

REFERENCES

Ekblom, P. (1981). *Med. Biol.* **59,** 139–160.
Grobstein, C. (1953). *Nature (London)* **172,** 869–871.
Grobstein, C. (1955). *J. Exp. Zool.* **130,** 319–340.
Grobstein, C. (1956). *Exp. Cell Res.* **10,** 242–440.
Grobstein, C. (1967). *Natl. Cancer Inst. Monogr.* **26,** 279–299.
Lehtonen, E. (1976). *Med. Biol.* **54,** 108–128.
Osathanondh, V., and Potter, E. (1966). *Arch. Pathol.* **82,** 403–411.
Sariola, H., Ekblom, P., Lehtonen, E., and Saxén, L.. (1983a). *Dev. Biol.* **96,** 427–435.
Sariola, H., Timpl, R., von der Mark, K., Mayne, R., Fitch, J. M., Linsenmayer, T. R., and Ekblom, P. (1983b). *Dev. Biol.* **101,** 86–96.
Saxén, L., and Karkinen-Jääskeläinen, M. (1975). *In* "The Early Development of Mammals" (M. Balls and A. Wild, eds.), pp. 319–333. Cambridge Univ. Press, London and New York.
Saxén, L., and Saksela, E. (1971). *Exp. Cell Res.* **66,** 369–377.
Saxén, L., Koskimies, O., Lahti, A., Miettinen, H., Rapola, J., and Wartiovaara, J. (1968). *Adv. Morphog.* **7,** 251–293.
Saxén, L., Lehtonen, E., Karkinen-Jääskeläinen, M., Nordling, S., and Wartiovaara, J. (1976). *Nature (London)* **259,** 662–663.
Saxén, L., Ekblom, P., and Thesleff, I. (1980). *In* "Development in Mammals" Vol. 4 (M. H. Johnson, ed.), pp. 161–202. Elsevier, Amsterdam.
Wartiovaara, J., Nordling, S., Lehtonen, E., and Saxén, L. (1974). *J. Embryol. Exp. Morphol.* **31,** 667–682.

CHAPTER **XVII**

Teratocarcinoma Chimeras and Gene Expression

C. L. STEWART

Heinrich-Pette-Institut für Experimentelle Virologie
und Immunologie an der Universität Hamburg
Hamburg, Federal Republic of Germany

I.	Introduction	409
II.	The Isolation of Teratocarcinomas and the Formation of Teratocarcinoma Chimeras	411
III.	*In Vitro* Derived Mutants of EC Cells	412
	A. EC Cells with Defects in Nucleotide Metabolism	412
	B. EC Cell Mutants Defective in Metabolic Cooperation	413
	C. Isolation of Chloramphenicol-Resistant EC Cell Lines	414
IV.	Chromosome-Mediated Gene Transfer into EC Cells	414
V.	Viral- and DNA-Mediated Transformation of EC Cells	415
	A. Viral Infection of Teratocarcinomas	415
	B. DNA-Mediated Transformation of EC Cells	420
VI.	Summary and Conclusions	422
	References	424

I. INTRODUCTION

Much of our understanding of the processes involved in the embryogenesis of eukaryotes has come from a genetic analysis of their development. Probably the best example of using such an approach has been in the study of *Drosophila*, where the identification and characterization of a wide variety of mutants has contributed extensively to unravelling the embryology of this organism (Ashburner and Novitski, 1976; Ashburner and Wright, 1978).

Ultimately the aim of developmental genetics is to study the role of individual genes in development. To this end the recent improvements in molecular biology techniques have helped greatly, especially when they have been used in conjunction with a classical genetic analysis. In *Drosophila* this combination has led to the cloning and characterization of genes affecting morphogenesis (Bender *et al.*, 1983), as well as the development of new techniques where genes can be efficiently introduced into embryos with correct tissue-specific expression occurring (Spradling and Rubin, 1982; Rubin and Spradling, 1982).

Students of mammalian embryology have not been so fortunate. Those seeking to understand this field of development have had to resort to the few phenotypically characterized developmental mouse mutants available (McLaren, 1976; Green, 1981). This problem is further compounded, since in only a few of these mutants is the identity of the affected gene known, permitting further analysis of the lesion (Gluecksohn-Waelsch, 1979; Russell, 1979; Schnieke *et al.*, 1983).

As a consequence of this, mammalian embryologists have approached the study of control of gene expression during mouse development using two methods.

The most recently developed technique has been the introduction of cloned genes into the embryo by microinjection into the zygote. It has been shown that the injected genes could be incorporated into the mouse's chromosomes and that these genes were often stably transmitted to the offspring (Gordon *et al.*, 1980; Costantini and Lacy, 1981; Wagner *et al.*, 1981; Brinster *et al.*, 1981; Palmiter *et al.*, 1982a,b; T. Stewart *et al.*, 1982; Gordon and Ruddle, 1981; C. Stewart *et al.*, 1983). The advantage of this approach over cultured cells is that the injected gene is carried in all cells of the mouse. Thus the gene will have been exposed to all possible stimuli that would occur in the cells during development necessary for turning it on or off. This should therefore lead to the proper identification of sequences necessary for tissue-specific expression. There are, however, many problems associated with this technique. The frequency of the injected sequences integrating into the embryo's genome is often low, and, although expression of the injected sequences has occurred in some instances, reproducible expression has been a problem. Furthermore, the genes are often present as multiple copies, thus complicating a molecular analysis (Gordon and Ruddle, 1981; Costantini and Lacy, 1981; Palmiter *et al.*, 1982b; T. Stewart *et al.*, 1982).

The second technique used has been the genetic manipulation of teratocarcinomas. The potential of using these tumours to alter the genome of mice was shown when it was discovered that the stem cells [embryonal carcinoma (EC) cells] could form chimeras with mouse embryos and in some individuals form functional gametes (Brinster, 1974; Illmensee and Mintz, 1976; Papaioannou *et al.*, 1975; Stewart and Mintz, 1981).

Teratocarcinomas have several advantages as vectors for introducing new genetic information into embryos. These are the following: (a) Mutations in both chromosomal and mitochondrial genomes can be selected for *in vitro* (b) Whole chromosomes can be transferred to these cells, which permits the study of genes in their correct chromosomal configuration. Furthermore, potentially interesting

XVII. Teratocarcinoma Chimeras and Gene Expression

chromosomal alterations such as translocations can also be investigated. (c) Genes incorporated into these cells can be studied both before and after the cells have been subjected to differentiation.

This chapter will therefore concentrate on the results of manipulating the genome of EC cells, with the view to introducing the altered genome into mice. Three methods have been used to change the genome: (1) the selection of mutants *in vitro*, (2) chromosome mediated gene transfer, and (3) viral- and DNA-mediated gene transfer. Although progress in this field has not necessarily fulfilled initial expectations, results obtained have suggested that the control of gene expression in EC cells may differ qualitatively from that of their differentiated derivatives.

The implications of this result for future experiments and for gene expression during embryogenesis are discussed.

II. THE ISOLATION OF TERATOCARCINOMAS AND THE FORMATION OF TERATOCARCINOMA CHIMERAS

Teratocarcinomas are malignant tumours composed of many different cell types haphazardly arranged (Damjanov and Solter, 1974). They contain a malignant stem cell population called embryonal carcinoma (EC) cells, from which differentiated cell types present in the tumours are derived (Kleinsmith and Pierce, 1964; Jami and Ritz, 1974). Teratocarcinomas have been mostly studied in mice since it has not yet been convincingly shown that cells analogous to murine EC cells are present in tumours obtained from other species (Damjanov *et al.*, 1977).

Teratocarcinomas were first discovered as rare, spontaneous testicular tumours in 129/J strain mice (Stevens and Little, 1954). By introducing genes known to affect germ cell survival, Stevens (1973, 1983) was able to increase their incidence to approximately 30%. In addition, another strain of mice was isolated that showed a high incidence of spontaneous ovarian teratocarcinomas (Stevens and Varnum, 1974). Similarities between the histology of these tumours and embryos led to the demonstration that teratocarcinomas could be readily induced in a wide variety of other strains of mice by surgical transfer of embryos to extrauterine sites (Stevens, 1964, 1970). From these tumours the first cell lines were established *in vitro* with the demonstration that the EC cells could also produce a wide variety of differentiated cell types if appropriately cultured (Rosenthal *et al.*, 1970; Kahan and Ephrussi, 1970; Bernstine *et al.*, 1973; Martin and Evans, 1975).

The potential for using teratocarcinomas as a means to manipulate the genome of mice only became apparent when it was shown that EC cells, when microinjected into blastocysts, would participate in the formation of adult chimeras

(Brinster, 1974; Mintz and Illmensee, 1975; Papaioannou *et al.*, 1975; Illmensee and Mintz, 1976). In a few chimeras, derived from EC cells passaged in mice, the gonads were colonized by derivatives of the injected EC cells. When these individuals were mated, some of the offspring produced were derived from gametes with the EC cell genotype, thus showing that some EC cells are genetically totipotent (Mintz *et al.*, 1975; Cronmiller and Mintz, 1978). This experiment has been recently repeated with germ line chimeras being produced from an EC cell line maintained in tissue culture (Stewart and Mintz, 1981).

Teratocarcinoma chimeras can now be produced by the technically simpler aggregation technique as an alternative to the blastocyst injection method (Stewart, 1982; Fujii and Martin, 1983). Probably the most significant improvement has been the demonstration that cells, which by all criteria are identical to EC cells, can be derived directly from blastocysts explanted and grown *in vitro* (Evans and Kaufman, 1981, 1983; Martin, 1981). This technique has not only reduced the time taken to isolate new lines, but it has also increased the variety of embryos from which EC cells may be isolated, e.g., early acting homozygous lethals (Magnuson *et al.*, 1982). It also appears that the majority of the blastocyst-derived lines possess a normal karyotype and are capable of forming overt chimeras at a quite respectable frequency (45 and 35% in two separate series of experiments) (Evans and Kaufman, 1983). It has recently been shown that these cells are also capable of colonizing the germ line.

Having briefly outlined the origins of EC cells and the formation of chimeras with these cells [more detailed reviews on teratocarcinomas are Graham (1977) and Martin (1980)], the next three sections will concentrate on the approaches used to select for changes in the genome of EC cells and the results of reintroducing these manipulated cells into embryos.

III. *IN VITRO* DERIVED MUTANTS OF EC CELLS

One advantage of using teratocarcinomas to alter the germ line of mice is that the experimenter is allowed to predetermine the phenotype of the EC cells by selection *in vitro*. A wide variety of mutants defective in some feature of cellular physiology can now be isolated from cells grown *in vitro*, and some of these selective techniques have been applied to EC cells (Hochstadt *et al.*, 1981; MacDonald *et al.*, 1980; Gupta and Hodgson, 1981). However, only three types of EC cell mutant have been tested for their potential to form chimeras.

A. EC Cells with Defects in Nucleotide Metabolism

The first mutant EC cells selected *in vitro* were deficient in the activity of the purine salvage enzyme hypoxanthine phosphoribosyltransferase (HPRT). This was partly because of the ease with which these mutants could be selected, but

XVII. Teratocarcinoma Chimeras and Gene Expression

also because babies lacking this enzyme have excess levels of purine, which results in the bizarre neurological disease called Lesch–Nyhan syndrome (Kelley and Wyngaarden, 1978). The study of this disease is restricted, as no equivalent animal model is known. It was therefore hoped that the production of germ line chimeras, from EC cells deficient in this enzyme, would result in mice carrying this defect.

Three attempts have been reported at producing chimeras with EC cells lacking HPRT activity. A number of mice have been produced in which there was extensive colonization of almost all organs tested, with the notable exception of the blood. However, none of the adult chimeras that were mated produced offspring carrying the EC cell genotype (Dewey et al., 1977; Stewart, 1982; Fujii and Martin, 1983).

One other mutant EC cell line, deficient in thymidine kinase activity, has been reported to produce adult chimeras. Data were presented showing that the EC cells could colonize a numher of tissues in the adult, but the frequency of chimeras was reduced compared to the wild-type parental line (Illmensee et al., 1978).

B. EC Cell Mutants Defective in Metabolic Cooperation

Most cells in an intact organism undergo metabolic cooperation, which is the transfer of small molecules between neighbouring cells via specialized cell surface structures called gap junctions (Hooper and Subak-Sharpe, 1981). The requirement for this remains obscure, although it is of interest that in developing organisms metabolic cooperation may occur and then cease between cells as they undergo differentiation (Gaunt and Papaioannou, 1979; Lo and Gilula, 1979; de Laat et al., 1980). In order to gain a greater understanding of the significance of this phenomenon, Hooper and his colleagues have isolated from EC cells a series of mutants and their revertants that are defective in undergoing metabolic cooperation with other cells (Hooper, 1982). Unfortunately, experiments designed to study the differentiation of these cells in chimeras were inconclusive, since only one malformed chimera derived from the injection of the original parental cells was reported, and no chimeras were derived after injection of the mutant cells into blastocysts (cited in Papaioannou and Rossant, 1983). The ability of these lines to aggregate with eight-cell morulae has also been studied. Only the parental cells and revertants were able to aggregate with the morulae, the mutant cells could not (Stewart and Kimber, 1984). It was, however, not clear if the inability to aggregate with the blastomeres was due to the failure to form gap junctions between the two cell types or, alternatively, to some overall pleiotropic effect of the mutation on membrane structure. Recently, M. L. Hooper (personal communication) has isolated another mutant defective in metabolic cooperation from a cell line with a known ability to form chimeras.

C. Isolation of Chloramphenicol-Resistant EC Cell Lines

Exposure of cells to chloramphenicol results in cell death due to impaired protein synthesis in the mitochondria. By selecting for EC cells resistant to chloramphenicol, the intention was to derive germ line chimeras from these cells that would be useful for studying cytoplasmic inheritance in mice (Watanabe *et al.*, 1978).

Chloramphenicol-resistant EC cells were isolated by fusing them to cytoblasts (cybrid fusion) from a chloramphenicol-resistant melanoma line. Resistance of these EC cell hybrids to chloramphenicol was maintained for 16 weeks in the absence of selective medium.

These cells were injected into blastocysts to produce chimeras. A number of mice were born in which it was claimed that the derivatives of the EC cells had colonized a variety of tissues. However, no data were presented indicating that the chimeric tissues contained chloramphenicol-resistant cells (Watanabe *et al.*, 1978).

The attempts to introduce mutant genes into the germ line of mice by selecting for the mutation in EC cells grown *in vitro* have been unsuccessful. For one reason, this failure has been due to using EC cells that even before selection were unable to form chimeras (Hooper, 1982; Papaioannou and Rossant, 1983). Some mutant EC cells were, however, able extensively to colonize tissues in the adult chimeras, thus at least allowing a study of the action of the mutation in these tissues (Mintz, 1979; Stewart, 1982). However, there has been no report of these chimeras producing offspring carrying the mutation, probably because the EC cells possessed an abnormal karyotype. An abnormal karyotype is known to affect the formation of fertile gametes (Evans, 1976). For EC cells to act as vectors for mutant genes, it will be necessary to perform the selection in a cell line known to produce functional gametes (Stewart and Mintz, 1981).

IV. CHROMOSOME-MEDIATED GENE TRANSFER INTO EC CELLS

Chromosome-mediated gene transfer by cell fusion, microcell transfer, or precipitation of isolated chromosomes directly onto cells has been used to study the genetic and epigenetic control of differentiation (Davidson, 1974; Ringertz and Savage, 1976; Fougère and Weiss, 1978; Klobutcher and Ruddle, 1981; Gmür *et al.*, 1981).

Illmensee and his colleagues have used cell fusion to introduce chromosomes from another species into EC cells in order to study the expression of genes located on these chromosomes during embryogenesis.

Initially the experiments were performed with human × mouse hybrids with an EC cell phenotype that contained a single human chromosome (17). These cells were injected into mouse blastocysts. Three chimeras were derived, and in two, weak expression of the human galactokinase locus on that chromosome was claimed (Illmensee et al., 1978).

Subsequently, a more successful series of experiments were performed using rat hepatoma × mouse EC hybrids. These retained an EC phenotype and contained a greater number of rat chromosomes and hence enzymatic and protein markers. Seven chimeras were produced, and rat-specific enzymes were detected in a number of tissues. There was some evidence that the lactase dehydrogenase pattern present in the EC cell hybrid had changed to a different pattern in the differentiated tissues. In extracts of liver, rat albumin and glycerolphosphate dehydrogenase were also expressed. Neither of these proteins were expressed in the hybrid stem cells (Illmensee and Croce, 1979; Duboule et al., 1982a,b). This demonstrated that exogenous genetic information can be introduced into EC cells and that, under appropriate conditions, permitting the differentiation of these cells, expression of the exogenous genes can occur. It was, however, not clear from these experiments if the expression was tissue specific. Only the chimeric liver tissues were described. Whether other chimeric tissues expressed albumin or glycerolphosphate dehydrogenase was not tested.

However, the use of such a system for studying gene expression is limited by the complexity of the chromosomal karyotype and uncertainty regarding the stability of the karyotype within these hybrids. Thus it would be an advantage to introduce a chromosome or chromosomes of interest into the EC cell. Recently, it has been demonstrated that individual human chromosomes can be transferred to EC cells, so that the expression of specific human genes could be studied during differentiation (Goodfellow et al., 1982). This approach has been limited to those chromosomes that carry a selectable marker, e.g., HPRT. However, the use of DNA-mediated transfer to incorporate dominant selectable markers onto human chromosomes (Tunnacliffe et al., 1983) will make it possible to construct novel EC cell populations, carrying one or a number of specified additional chromosomes.

V. VIRAL- AND DNA-MEDIATED TRANSFORMATION OF EC CELLS

A. Viral Infection of Teratocarcinomas

Like their counterparts such as λ and T-even phages, which were used in the study of bacterial gene structure, organization, and function, mammalian viruses have been used to investigate the control of gene expression in embryogenesis. These viruses exploit the cells' own machinery for their replication and expression, and their study has revealed functions necessary for their correct expression. Furthermore, because their interactions with mammalian cells are relatively well understood, they have been redesigned to act as vectors for introducing new genes into cells (Rigby, 1983).

Three viruses have been exploited to study the control of gene expression during mouse embryogenesis and in teratocarcinomas, simian virus 40 (SV40), polyoma virus, and Moloney murine leukaemia virus (M-MuLV). A detailed coverage of studies performed with other viruses in embryogenesis can be found in Kelly and Condamine (1982).

1. DNA VIRUSES

Polyoma and SV40 are both DNA viruses belonging to the papovavirus group. They both contain a relatively small double-stranded circular DNA genome and are very similar in the organization of their genomes, but they differ in their host range. Simian virus 40 can only undergo complete virus synthesis (lytic cycle) in African green monkey cells, and polyoma virus is restricted to mouse cells (Tooze, 1980). In most mouse cells, SV40 can transiently express part of its genome, but DNA replication does not occur (Tooze, 1980).

The genome of these viruses is divided functionally into two halves, the early and the late regions, which reflect the cycle of events that occurs when these viruses infect their host cells. The infectious cycle starts with transcription of the early half of the genome from a point near the origin of DNA replication. The origin of replication region does not code for proteins and is an important transcriptional control region (see below). In SV40, early region transcription results in the synthesis of two proteins, big T and little t, referred to as early gene products. In polyoma, three early proteins, big T, middle t, and little t, are produced. The synthesis of these proteins from the viruses is necessary for the replication of the DNA genome and for transcription of the late half of their genomes, which code for proteins used in the assembly of new virus particles (Tooze, 1980).

Embryonal carcinoma cells are nonpermissive for expression of both these viruses, resulting in the absence of any of the T antigens, even though it has been shown that the virus can penetrate and uncoat in the nucleus. In differentiated derivatives of these cells early gene products of both SV40 and polyoma can be detected after infection (Swartzendruber and Lehman, 1975; Segal and Khoury, 1979).

Similar results have been obtained from infected mouse embryos. The inner cell mass of the blastocyst was refractory to the expression of both viruses. However, the infection of the trophectoderm resulted in expression of the viral genomes. Postimplantation embryos only became permissive for T antigen expression at around 8 to 9 days of gestation (Abramczuk et al., 1978, cited in Kelly and Condamine, 1982). Consequently, the expression of these viral genes has attracted much attention since they appear to be developmentally regulated.

XVII. Teratocarcinoma Chimeras and Gene Expression

The inhibition of expression of both these viruses has been investigated in EC cells. Only the salient points of these studies will be presented here since a more detailed discussion on this topic can be found in Levine (1982). The overall consensus is that for both viruses there is a greatly reduced level of stable viral mRNA present in EC cells compared to levels found in their differentiated derivatives. In polyoma-infected EC cells a very low level of correctly processed RNA was found, but this seemed insufficient to produce detectable levels of T antigen (Dandolo et al., 1983). In SV40-infected EC cells no clear result could be derived. Simian virus 40 mRNA could be found only at very low levels, and this RNA was incorrectly processed (Segal et al., 1978). Whether the low levels of mRNA reflect a low level of transcription or increased mRNA degradation was not resolved (Levine, 1982). The situation, however, was complicated with the derivation of an EC cell clone containing a transcriptionally active and integrated SV40 genome linked to a herpesvirus thymidine kinase gene (Linnenbach et al., 1980). In this clone, low levels of correctly spliced SV40 mRNA were present; however, no T antigen was detectable, suggesting that additional levels of control, e.g., posttranscriptional or posttranslational, could also be operating in EC cells (Linnenbach et al., 1981). Posttranscriptional and posttranslational controls appear also to be involved in the replication of other DNA viruses in infected EC cells (Cheng and Praskier, 1982).

2. POLYOMA MUTANTS

A number of laboratories have isolated mutant polyoma viruses that express early gene products and undergo replication in EC cells (Vasseur et al., 1980; Fujimura et al., 1981a; Sekikawa and Levine, 1981). Polyoma mutants have been isolated from both F9 and PCC4 EC cells. These mutants, isolated from each cell line, are incapable of replicating in the other EC cell line. In both classes of mutant and in all isolates, the mutations have been mapped to the origin of replication of the polyoma genome (Katinka et al., 1980; Herbomel et al., 1981; Fujimura et al., 1981b). Comparison of the various isolates from the different laboratories has revealed that the majority of mutations occurred within 200 bases to the late side of the origin of replication. No clear picture has emerged concerning how these mutations are responsible for promoting transcription and replication of the viral genome, since there is extensive variation between mutants. The mutations have been found to be either simple single base substitutions or relatively complicated changes involving deletions, duplications, and translocations of sequences present in this region (Levine, 1982).

The block to expression of the polyoma virus seems, however, to be due simply to the inability of the wild-type virus to synthesize T antigen, which is necessary for replication of genomes (Tooze, 1980). Fujimura and Linney (1982) have shown that the mutations at the origin of replication are not only necessary for the transcription of the early gene products but they also permit replication of the viral genome to occur. Thus the block to expression of polyoma in EC cells could be due to failure in transcription of the early gene products, to failure in DNA replication, or both (Levine, 1982).

3. MOLONEY MURINE LEUKAEMIA VIRUS

The third virus that has been used extensively as a model gene to study gene expression in embryogenesis and teratocarcinomas has been the retrovirus, Moloney murine leukaemia virus (M-MuLV). The most important difference in this context between M-MuLV and the DNA viruses discussed above is that integration of M-MuLV into the infected cell's chromosomal DNA is a necessary stop in the life cycle of the virus. The infectious cycle involves the virus entering the cell, uncoating and reverse-transcribing the viral RNA into a proviral DNA copy that integrates into the host's DNA. This proviral copy acts as a template for synthesis of new viral RNAs (Weiss *et al.*, 1982).

The potential for using such a virus for studying gene expression in mouse embryogenesis has been demonstrated primarily by Jaenisch and his co-workers. Jaenisch (1976) showed that early mouse embryos could be infected with M-MuLV. The virus is stably integrated and in some cases is transmitted to the offspring as a single Mendelian locus. In all, 14 substrains of mice have been produced, each substrain containing a virus at a different chromosomal locus. Thus the substrain Mov-1 has a single provirus integrated on chromosome 6 (Jaenisch *et al.*, 1981), whereas Mov-14 contains tandem copies of virus integrated on the X chromosome (Stewart *et al.*, 1983).

The main points that have emerged from the study of these Mov substrains have been the following: (1) The onset of virus expression differs between substrains. It has been suggested that the site at which the virus had integrated may influence its expression (Jaenisch *et al.*, 1981). (2) Retroviruses can act as insertional mutagens disrupting embryonic development (Schnieke *et al.*, 1983). (3) Infected preimplantation embryos inhibit virus expression, and this inhibition is maintained throughout development. However, this inhibition disappears in postimplantation embryos. Embryos infected on the eighth day of gestation were able to replicate virus efficiently (Jähner *et al.*, 1982). This has defined a developmentally regulated switch in mouse embryonic cells with respect to their ability to support retroviral expression.

Retroviruses have also been used to study the control of gene expression in teratocarcinomas. It has been known for some time that embryonal carcinoma cells were nonpermissive for retrovirus infection (Teich *et al.*, 1977; Speers *et al.*, 1980), i.e., EC cells infected with M-MuLV did not express virus, whereas their differentiated derivatives could be readily infected and produced high titers of virus. The block to expression was not due to failure of the virus to enter or integrate into the EC cells' DNA. Rather, it was found that the virus integrated

XVII. Teratocarcinoma Chimeras and Gene Expression

into the chromosomal DNA and that viral transcription was suppressed (C. L. Stewart et al., 1982; Gautsch and Wilson, 1983). This block was also maintained even after infected EC cells were induced to differentiate (C. L. Stewart et al., 1982; Gautsch and Wilson, 1983).

The biological activity of the integrated proviral copies in the infected EC cells and in mice derived from infected preimplantation embryos was tested by transfection of their DNA onto cells permissive for virus expression. No expression was found; however, virus could be activated by treatment of the transfected cells with 5-azacytidine, and thus the proviral copies were not irreversibly inactivated (C. L. Stewart et al., 1982; Jähner et al., 1982). These results suggested that the inability of the EC cells to support virus expression was not due to the presence of some inhibitory factor or absence of a factor necessary for virus expression (Gautsch, 1980). The nature of this block is not clear. One of the proposals has been that methylation of the proviral genomes is important (Harbers et al., 1981), since high levels of methylcytosine are generally associated with gene inactivation (Razin and Riggs, 1980; Felsenfeld and McGhee, 1982). In infected EC cells and in mice derived from infection of preimplantation embryos, the proviral copies were all found to be highly methylated. In differentiated derivatives of EC cells and in mice derived from post-implantation infections, the proviral copies were unmethylated (Jähner et al., 1982; C. L. Stewart et al., 1982). Whether methylation is directly involved in the suppression of proviral transcription or is a consequence of the provirus being inactivated by some unknown means is unclear (Gautsch and Wilson, 1983), for an EC cell clone has been recently derived in which a single retroviral genome is present and expressed but is also highly methylated (C. Stewart, unpublished results).

In conclusion, the expression of all three viruses in teratocarcinomas and embryos appears to be developmentally regulated. Each virus has revealed a different level at which its expression can be regulated. The results from SV40-infected EC cells suggested that correct processing of the transcribed RNA must occur (Segal et al., 1978; Segal and Khoury, 1979), whereas polyoma virus and its mutants require the appropriate DNA sequences for transcription to occur (Vasseur et al., 1980; Katinka et al., 1981; Fujimura et al., 1981a; Sekikawa and Levine, 1981; Dandolo et al., 1983). Neither virus integrates into the infected cells' chromosomal DNA. The retrovirus M-MuLV may act as a more useful model for studying genes introduced into embryonic cells. Infection of EC cells and early embryos revealed that an efficient mechanism exists for inhibiting the expression of DNA that integrates into their chromosomes (C. L. Stewart et al., 1982; Jähner et al., 1982; Gautsch and Wilson, 1983). Once this block is established, it appears to be frequently maintained even when the cells differentiate, although in some instances *in vivo* and rarely *in vitro* virus expression can occur in the differentiated cells (Jaenisch *et al.*, 1982; C. L. Stewart *et al.*, 1982). A similar observation has been reported for SV40-infected EC cells (Friedrich and Lehman, 1981).

The mechanisms that affect the expression of viruses may also operate on cloned eukaryotic genes that are introduced into these cells.

B. DNA-Mediated Transformation of EC Cells

Viruses, as described in the preceding section, have been useful for uncovering some of the mechanisms that may regulate gene expression during mouse embryogenesis. It is, however, the intention of most workers in this field to discover how the animal's own genes are expressed and regulated during its development. Currently, the method most widely used is to clone the gene of interest and then reintroduce it back into the embryo or cell. This approach permits the alteration of the gene *in vitro* so that the effect of the modification, e.g., changes in its sequence, methylation, etc., can then be studied (Pellicer *et al.*, 1980a; McKnight and Kingsbury, 1982). At present methods of introducing genes into cells are inefficient, and selective techniques are required to isolate clones containing the gene of interest.

The first experiments performed showed that EC cells, deficient in thymidine kinase activity, could be transformed with the herpes simplex virus thymidine kinase (HSV *tk*) gene to a tk-positive phenotype (Pellicer *et al.*, 1980b). These experiments were extended to show that additional genes could also be introduced into EC cells either by coprecipitation (e.g., rabbit β-globin) with the *tk* gene or by covalent linkage, e.g., to the *tk* gene SV40 (Pellicer *et al.*, 1980b; Linnenbach *et al.*, 1980; Wagner and Mintz, 1982). With the SV40 transfection experiments it was shown that correct transcription and expression of the SV40 genome could occur, but only in the differentiated derivatives of the EC cells carrying the SV40 genome (Linnenbach *et al.*, 1980). In cells transformed with β-globin, the presence of the integrated globin genes was stable over a long period of time, but only weak expression of the globin mRNA was detected (Pellicer *et al.*, 1980b; Wagner and Mintz, 1982).

Subsequent experiments have been concerned with showing that EC cells can be transformed using dominant selection markers, thus removing the need for using previously mutated EC cell lines deficient in thymidine kinase or HPRT activity.

Two dominant selectable markers have been used, both based on SV40 virus as an expression vector. The vectors are pSV-2Gpt, which carries the bacterial enzyme xanthine phosphoribosyltransferase (XPRT or EcoGpt) (Mulligan and Berg, 1980, 1981), and pSV-2Neo, which carries the neomycin resistance gene (Tn5 phosphotransferase) (Jiminez and Davies, 1980; Colbere-Garapin *et al.*, 1981; Southern and Berg, 1982).

A number of EC cell lines have been reported as being stably transformed by either of these vectors. However, some results have been contradictory. It has been reported that two EC cell lines could be readily transformed with pSV-2Gpt (Wagner and Mintz, 1982; Bucchini *et al.*, 1983), although the stability of the transformed phenotype was not clearly demonstrated. However, another laboratory reported that one of these cell lines could not be transformed with the same vectors (Nicolas and Berg, 1983). Transformation and isolation of EC cell clones could only occur if the vectors were modified to include the promoter sequences from the HSV *tk* gene. The reasons underlying this discrepancy were not clear, although they may have been due to differences existing between the different cell lines, as has been found with the polyoma mutants (Levine, 1982).

The main points to emerge from these attempts to introduce exogenous genes are the following: (1) Embryonal carcinoma cell transformation is less efficient than with fibroblasts (L cells) in terms of the number of clones isolated per microgram of DNA transfected. The transformation frequency can in some instances be increased by using a different promoter (Linney *et al.*, 1983). (2) Most EC clones contain stably integrated vector sequences. (3) Transfer of EC cell clones into nonselective medium often results in complete loss of the resistant phenotype, with consequent loss of the EC cell clones when they are returned to selective medium. It was, however, intriguing that loss of the phenotype did not involve loss or drastic changes of the integrated sequences (Nicolas and Berg, 1983). This result would be of significance for experiments designed to investigate the expression of exogenous genes in EC↔embryo chimeras. Thus the phenotype of the transfected EC cells in both selective and nonselective medium would have to be clearly established, before any conclusions could be drawn concerning the effect that differentiation of the EC cells might have on the expression of the exogenous genes.

Finally, a new class of viral vectors has recently been described (Mann *et al.*, 1983; Mulligan, 1983). These vectors, based on retroviruses, may be one of the most promising to use in that they avoid a number of the problems that are associated with the introduction of cloned genes by DNA precipitation or injection. These problems have been the frequent instability of the integrated sequences and the fact that these sequences are often present as multiple copies, making the analysis of their expression difficult (Pellicer *et al.*, 1980b). The integration of retroviruses is extremely efficient as well as stable, in that all EC cells exposed to this virus were found to carry newly integrated sequences (C. L. Stewart *et al.*, 1982). Under appropriate circumstances, cells can also be isolated that carry a single viral copy (Jaenisch *et al.*, 1981; C. L. Stewart, unpublished results).

Although the expression of retroviruses has been shown to be inhibited in EC cells (Teich *et al.*, 1977; C. L. Stewart *et al.*, 1982; Gautsch and Wilson, 1983), this can be overcome by using a retroviral vector that contains a selectable marker. Embryonal Carcinoma cell clones expressing the viral vector have been derived by using a retrovirus carrying the *EcoGpt* gene (C. L. Stewart, unpublished results). Thus these vectors will not only be useful for efficiently introducing genes into EC cells, where their expression can be studied under appropriate circumstances, but they will also be useful for studying the block to retrovirus expression that operates in EC cells as well as in the early mouse embryo.

VI. SUMMARY AND CONCLUSIONS

This chapter has reviewed the various methods that have been used to alter the genome of embryonal carcinoma cells with the intention of (a) using these cells as vectors for introducing new or mutant genes into the germ line of mice and (b) studying gene expression within these cells.

Embryonal Carcinoma cells, as a system, offer a number of advantages for the study of gene expression compared with the alternative method of manipulating the genome of mice by microinjection of genes into the zygote. The variety of techniques for introducing genetic information into these cells and studying their expression during differentiation is greater. Thus selection for mutants in the mitochondrial and chromosomal genomes, chromosome-mediated gene transfer, or cell fusion as well as DNA-mediated gene transfer can all be used. By selecting for particular mutant phenotypes *in vitro* three different attempts have been made at inserting a particular mutant into the germ line of mice. These mutants were selected for deficiency in HPRT or TK enzyme activity (Dewey *et al.*, 1977; Illmensee *et al.*, 1978; Stewart, 1982; Fujii and Martin, 1983), chloramphenicol-resistant mitochondria (Watanabe *et al.*, 1978), and metabolic cooperation-defective mutants (Hooper, 1982; Papaioannou and Rossant, 1983). In none of these attempts were these mutations incorporated into the germ line, thus their effect on development could not be studied. The reasons underlying this failure have varied. An EC cell line was used that was subsequently found not to form chimeras (Hooper, 1982; Papaioannou and Rossant, 1983). Furthermore, all of the cell lines used carried chromosomal abnormalities that would inhibit the formation of functional gametes. For this approach to succeed at manipulating the genome of mice it will at least be necessary to obtain an EC cell line that, after selection for the mutant phenotype, still retains a normal karyotype. This would not, however, guarantee that such a cell could form germ line chimeras, since a number of euploid EC cell lines have been isolated that form chimeras at a very low frequency (Papaioannou and Rossant, 1983). In the long run, alternative methods may be more useful at producing mouse strains carrying the desired mutations. With the increasing availability of tests for a wide variety of enzymes and the ability to screen complex mixtures of proteins for changes

XVII. Teratocarcinoma Chimeras and Gene Expression

(Garrels, 1979), a more suitable method for isolating mutant mice may be by chemical- or radiation-induced mutation of the germ cells with subsequent screening of the offspring carrying the desired mutation (Peters, 1983). A more random, but extremely effective method, in that it permits an identification of the affected gene, would be to use retroviruses as insertional mutagens (Jenkins et al., 1981; Schnieke et al., 1983). Furthermore, it appears that the injection of DNA into zygotes may also act as an efficient method of generating developmental mutants (E. F. Wagner, personal communication).

The introduction of exogenous genes into EC cells has, however, produced substantial information concerning the control of gene expression in these cells and in early mouse embryos. Infection of EC cells with SV40, polyoma, and M-MuLV viruses revealed that EC cells cannot express these viruses, whereas their differentiated derivatives can (Levine, 1982; C. L. Stewart et al., 1982; Gautsch and Wilson, 1983). A similar situation also occurs in cells of the mouse embryo in that preimplantation embryos are nonpermissive for virus expression, but postimplantation embryos at around the eighth to ninth day of gestation can support virus expression (Kelly and Condamine, 1982; Jähner et al., 1982). Thus these viruses have apparently defined a switch in terms of mechanisms controlling gene expression that is developmentally regulated. Not all viruses tested are affected by this regulation; however, these three viruses contain in their genome promoter regions (enhancers) that regulate their expression, and similarities exist between these regions in these viruses (Khoury and Gruss, 1983).

The reasons underlying the inability of early mouse embryonic cells to express these viruses are not understood. Two possibilities can be proposed: (1) Gene expression in early embryonic cells may be under very tight control in that mechanisms exist to inhibit the "accidental" expression of genes that could be potentially disruptive to embryonic development. Such genes may include viruses, since their expression could have potentially deleterious consequences. (2) The mechanisms regulating gene expression in these cells may be different from their differentiated derivatives. Thus the failure to express the viruses or the inefficiency of transforming EC cells with cloned genes (Pellicer et al., 1980b; Nicolas and Berg, 1983) may be due to the genes lacking the appropriate signals necessary for their expression (Linney et al., 1983).

The introduction of exogenous genetic information into EC cells has been shown to be useful for studying gene expression both in vitro and in vivo. It has been shown that EC cells selected to carry an integrated SV40 genome did not synthesize SV4O proteins; when these cells were induced to differentiate in tissue culture, SV40 proteins were expressed as the cells underwent differentiation (Knowles et al., 1980; Linnenbach et al., 1981). Furthermore, developmentally regulated expression of exogenous genes carried in EC cells has been shown to occur in EC cell↔embryo chimeras (Illmensee and Croce, 1979; Duboule et al., 1982a,b). Future experiments should therefore centre on the introduction of cloned eukaryotic genes into EC cells so that their expression can be studied *in vivo* in chimeras. Such an approach should lead to the identification of factors regulating gene expression in mammalian embryogenesis.

ACKNOWLEDGMENTS

I would like to thank D. Jähner for many fruitful discussions; A. Stacey, M. Kuehn, and R. Jaenisch for critically reading the manuscript; and E. Linney, J. F. Nicolas, R. Lovell-Badge, and K. Huebner for making available to me their unpublished manuscripts. Finally, I am very grateful for the enormous efficiency of E. Danckers in preparing and typing the manuscript. The author was supported by a grant from the Stiftung Volkswagenwerk to R. Jaenisch. The Heinrich Pette-Institut is financially supported by Freie und Hansestadt Hamburg and Bundesministerium für Jugend, Familie und Gesundheit.

REFERENCES

Abramczuk, J., Vorbrodt, A., Solter, D., and Koprowski, H. (1978). *Proc. Natl. Acad. Sci. U.S.A.* **75,** 999–1003.
Ashburner, M., and Novitski, E., eds. (1976). "The Genetics and Biology of Drosophila," Vols. 1A–C. Academic Press, New York.
Ashburner, M., and Wright, T. R. F., eds. (1978). "The Genetics and Biology of Drosophila," Vol. 2C. Academic Press, New York.
Bender, W., Akam, M., Karch, F., Beachy, P. A., Peifer, M., Spierer, P., Lewis, E. B., and Hogness, D. S. (1983). *Science* **221,** 23–29.
Bernstine, E. G., Hooper, M. L., Grandchamp, S., and Ephrussi, B. (1973). *Proc. Natl. Acad. Sci. U.S.A.* **70,** 3899–3903.
Brinster, R. L. (1974). *J. Exp. Med.* **140,** 1049–1056.
Brinster, R. L., Chen, H. Y., Trumbauer, M., Senear, A. W., Warren, R., and Palmiter, R. D. (1981). *Cell* **27,** 223–231.
Bucchini, D., Lasserre, C., Kunst, F., Lovell-Badge, R., Pictet, R., and Jami, J. (1983). *EMBO J.* **2,** 229–232.
Cheng, C., and Praskier, J. (1982). *Virology* **123,** 45–59.
Colbere-Garapin, F., Horodniceanu, F., Kourilsky, P., and Garapin, A. C. (1981). *J. Mol. Biol.* **150,** 1–14.
Costantini, F., and Lacy, E. (1981). *Nature (London)* **294,** 92–94.
Cronmiller, C., and Mintz, B. (1978). *Dev. Biol.* **67,** 465–477.
Damjanov, I., and Solter, D. (1974). *Curr. Top. Pathol.* **59,** 69–130.
Damjanov, I., Skreb, N., and Sell, S. (1977). *Int. J. Cancer* **19,** 526–530.
Dandolo, L., Blangy, D., and Kamen, R. (1983). *J. Virol.* **47,** 55–64.
Davidson, R. L. (1974). *Annu. Rev. Genet.* **8,** 195–218.
de Laat, S. W., Tertoolen, L. G. J., Dorresteijn, A. W. C., and van den Biggelaar, J. A. M. (1980). *Nature (London)* **287,** 546–548.
Dewey, M. J., Martin, D. W., Martin, G. R., and Mintz, B. (1977). *Proc. Natl. Acad. Sci. U.S.A.* **74,** 5564–5568.
Duboule, D., Croce, C. M., and Illmensee, K. (1982a). *EMBO J.* **1,** 1595–1603.
Duboule, D., Petzoldt, U., Illmensee, G. R., Croce, C. M., and Illmensee, K. (1982b). *Differentiation* **23,** 145–152.

XVII. Teratocarcinoma Chimeras and Gene Expression

Evans, E. P. (1976). *Chromosomes Today* **5**, 75–81.
Evans, M. J., and Kaufman, M. H. (1981). *Nature (London)* **292**, 154–156.
Evans, M. J., and Kaufman, M. H. (1983). *Cancer Surv.* **2**, 185–207.
Felsenfeld, G., and McGhee, J. (1982). *Nature (London)* **296**, 602–603.
Fougère, C., and Weiss, M. C. (1978). *Cell* **15**, 843–854.
Friedrich, T. D., and Lehman, J. M. (1981). *Virology* **110**, 159–166.
Fujii, J. T., and Martin, G. R. (1983). *J. Embryol. Exp. Morphol.* **74**, 79–86.
Fujimura, F. K. and Linney, E. (1982). *Proc. Natl. Acad. Sci. U.S.A.* **79**, 1479–1483.
Fujimura, F. K., Deininger, P. L., Friedmann, T., and Linney, E. (1981a). *Cell* **23**, 809–814.
Fujimura, F. K., Silbert, P., Eckart, W., and Linney, E. (1981b). *J. Virol.* **39**, 306–312.
Garrels, J. (1979). *J. Biol. Chem.* **254**, 7961–7977.
Gaunt, S. J., and Papaioannou, V. E. (1979). *J. Embryol. Exp. Morphol.* **54**, 263–275.
Gautsch, J. W. (1980). *Nature (London)* **285**, 110–112.
Gautsch, J. W., and Wilson, M. C. (1983). *Nature (London)* **301**, 32–37.
Gluecksohn-Waelsch, S. (1979). *Cell* **16**, 225–237.
Gmür, R., Knowles, B., and Solter, D. (1981). *Dev. Biol.* **81**, 245–254.
Goodfellow, P. N., Banting, G., Trowsdale, J., Chambers, S., and Solomon, E. (1982). *Proc. Natl. Acad. Sci. U.S.A.* **79**, 1190–1194.
Gordon, J. W., and Ruddle, F. H. (1981). *Science* **214**, 1244–1246.
Gordon, J. W., Scangos, G. A., Plotkin, D. J., Barbosa, S. A., and Ruddle, F. H. (1980). *Proc. Natl. Acad. Sci. U.S.A.* **77**, 7380–7384.
Graham, C. F. (1977). *In* "Concepts in Mammalian Embryogenesis" (M. I. Sherman, ed.), pp. 315–394. MIT Press, Cambridge, Massachusetts.
Green, M. C. (1981). "Genetic Variants and Strains of the Laboratory Mouse." Fischer, Stuttgart.
Gupta, R. S., and Hodgson, M. (1981). *Exp. Cell Res.* **132**, 496–500.
Harbers, K., Schnieke, A., Stuhlmann, H., Jähner, D., and Jaenisch, R. (1981). *Proc. Natl. Acad. Sci. U.S.A.* **78**, 7609–7613.
Herbomel, P., Saragosti, S., Blangy, D., and Yaniv, M. (1981). *Cell* **25**, 651–658.
Hochstadt, J., Ozer, H. L., and Shopsis, C. (1981). *Curr. Top. Microbiol. Immunol.* **94/95**, 243–308.
Hooper, M. L. (1982). *Br. Soc. Cell Biol. Symp.* **5**, 195–207.
Hooper, M. L., and Subak-Sharpe, J. H. (1981). *Int. Rev. Cytol.* **69**, 45–104.
Illmensee, K., and Croce, C. M. (1979). *Proc. Natl. Acad. Sci. U.S.A.* **76**, 879–883.
Illmensee, K., and Mintz, B. (1976). *Proc. Natl. Acad. Sci. U.S.A.* **73**, 549–553.
Illmensee, K., Hoppe, P. C., and Croce, C. M. (1978). *Proc. Natl. Acad. Sci. U.S.A.* **75**, 1914–1918.
Jähner, D., Stuhlmann, H., Stewart, C. L., Harbers, K., Löhler, J., Simon, I., and Jaenisch, R. (1982). *Nature (London)* **298**, 623–628.
Jaenisch, R. (1976). *Proc. Natl. Acad. Sci. U.S.A.* **73**, 1260–1264.
Jaenisch, R., Jähner, D., Nobis, P., Simon, I., Löhler, J., Harbers, K., and Grotkopp, D. (1981). *Cell* **24**, 519–529.
Jami, J., and Ritz, E. (1974). *J. Natl. Cancer Inst. (U.S.)* **52**, 1547–1552.
Jenkins, N., Copeland, N., Taylor, B., and Lee, B. (1981). *Nature (London)* **293**, 370–374.
Jiminez, A., and Davies, J. (1980). *Nature (London)* **287**, 869–871.
Kahan, B., and Ephrussi, B. (1970). *J. Natl. Cancer Inst. (U.S.)* **44**, 1015–1036.
Katinka, M., Yaniv, M., Vasseur, M., and Blangy, D. (1980). *Cell* **20**, 393–399.
Katinka, M., Vasseur, M., Montreau, N., Yaniv, M., and Blangy, D. (1981). *Nature (London)* **290**, 720–722.
Kelley, W. N., and Wyngaarden, J. B. (1978). *In* "The Metabolic Basis of Inherited Disease" (J. B. Stanbury, J. B. Wyngaarden, and D. S. Fredrickson, eds.), 4th ed., pp. 141–166. McGraw-Hill, New York.

Kelly, F., and Condamine, H. (1982). *Biochim. Biophys. Acta* **651,** 105–141.
Khoury, G., and Gruss, P. (1983). *Cell* **33,** 313–314.
Kleinsmith, L. J., and Pierce, G. B. (1964). *Cancer Res.* **24,** 1544–1552.
Klobutcher, L. A., and Ruddle, F. H. (1981). *Annu. Rev. Biochem.* **50,** 533–554.
Knowles, B. B., Pan, S., Solter, D., Linnenbach, A., Croce, C., and Huebner, K. (1980). *Nature (London)* **288,** 615–618.
Levine, A. J. (1982). *Curr. Top. Microbiol. Immunol.* **101,** 1–30.
Linnenbach, A., Huebner, K., and Croce, C. M. (1980). *Proc. Natl. Acad. Sci. U.S.A.* **77,** 4875–4879.
Linnenbach, A., Huebner, K., and Croce, C. M. (1981). *Proc. Natl. Acad. Sci. U.S.A.* **78,** 6386–6390.
Linney, E., Donerly, S., Olinger, B., Bender, M., and Fujimura, F. (1983). *Cold Spring Harbor Conf. Cell Proliferation* **10,** 271–284.
Lo, C. W., and Gilula, N. B. (1979). *Cell* **18,** 411–422.
MacDonald, C., Hooper, M. L., Buultjens, T. E. J., and Carritt, B. (1980). *Exp. Cell Res.* **127,** 277–284.
McKnight, S. L., and Kingsbury, R. (1982). *Science* **217,** 316–324.
McLaren, A. (1976). *Annu. Rev. Genet.* **10,** 361–388.
Magnuson, T., Epstein, C. J., Silver, L. M., and Martin, G. R. (1982). *Nature (London)* **298,** 750–752.
Mann, R., Mulligan, R. C., and Baltimore, D. (1983). *Cell* **33,** 153–159.
Martin, G. R. (1980). *Science* **209,** 768–776.
Martin, G. R. (1981). *Proc. Natl. Acad. Sci. U.S.A.* **78,** 7634–7636.
Martin, G. R., and Evans, M. J. (1975). *Proc. Natl. Acad. Sci. U.S.A.* **72,** 1441–1445.
Mintz, B. (1979). In "Models for the Study of Inborn Errors of Metabolism" (F. A. Hommes, ed.), pp. 343–354. Elsevier/North-Holland, Amsterdam.
Mintz, B., and Illmensee, K. (1975). *Proc. Natl. Acad. Sci. U.S.A.* **72,** 3585–3589.
Mintz, B., Illmensee, K., and Gearhart, D. D. (1975). In "Teratomas and Differentiation" (M. I. Sherman and D. Solter, eds.), pp. 59–82. Academic Press, New York.
Mulligan, R. C. (1983). In "Experimental Manipulation of Gene Expression" (M. Inouye, ed.). Academic Press, New York (In press).
Mulligan, R. C., and Berg, P. (1980). *Science* **209,** 1422–1427.
Mulligan, R. C., and Berg, P. (1981). *Proc. Natl. Acad. Sci. U.S.A.* **78,** 2072–2076.
Nicolas, J. F., and Berg, P. (1983). *Cold Spring Harbor Conf. Cell Proliferation* **10,** 469–485.
Palmiter, R. D., Brinster, R. L., Hammer, R., Trumbauer, M., Rosenfeld, M., Birnberg, N., and Evans, R. (1982a). *Nature (London)* **300,** 611–615.
Palmiter, R. D., Chen, H. Y., and Brinster, R. L. (1982b). *Cell* **29,** 701–710.
Papaioannou, V. E., and Rossant, J. (1983). *Cancer Surv.* **2,** 165–183.
Papaioannou, V. E., McBurney, M. W., Gardner, R. L., and Evans, M. J. (1975). *Nature (London)* **258,** 70–73.
Pellicer, A., Robins, D., Wold, B., Sweet, R., Jackson, J., Lowy, I., Roberts, J. I., Sim, G. S., Silverstein, S., and Axel, R. (1980a). *Science* **209,** 1414–1422.
Pellicer, A., Wagner, E. F., El Kareh, A., Dewey, M. J., Reuser, A. J., Silverstein, S., Axel, R., and Mintz, B. (1980b). *Proc. Natl. Acad. Sci. U.S.A.* **77,** 2098–2102.
Peters, J. (1983). *Hereditas* **98,** 158.
Razin, A., and Riggs, A. (1980). *Science* **210,** 604–609.
Rigby, P. W. J. (1983). *J. Gen. Virol.* **64,** 255–266.
Ringertz, N. R., and Savage, R. E. (1976). "Cell Hybrids." Academic Press, New York.
Rosenthal, M. D., Wishnow, R. M., and Sato, G. H. (1970). *J. Natl. Cancer Inst. (U.S.)* **44,** 1001–1009.

XVII. Teratocarcinoma Chimeras and Gene Expression

Rubin, G. M., and Spradling, A. C. (1982). *Science* **218**, 348–353.
Russell, E. S. (1979). *Adv. Genet.* **20**, 357–459.
Schnieke, A., Harbers, K., and Jaenisch, R. (1983). *Nature (London)* **304**, 315–320.
Segal, S., and Khoury, G. (1979). *Proc. Natl. Acad. Sci. U.S.A.* **76**, 5611–5615.
Segal, S., Levine, A. J., and Khoury, G. (1978). *Nature (London)* **280**, 335–337.
Seikiawa, K., and Levine, A. J. (1981). *Proc. Natl. Acad. Sci. U.S.A.* **78**, 1100–1104.
Southern, P. J., and Berg, P. (1982). *J. Mol. Appl. Genet.* **1**, 327–341.
Speers, W. C., Gautsch, J. W., and Dixon, F. J. (1980). *Virology* **105**, 241–244.
Spradling, A. C., and Rubin, G. M. (1982). *Science* **218**, 341–347.
Stevens, L. C. (1964). *Proc. Natl. Acad. Sci. U.S.A.* **52**, 654–661.
Stevens, L. C. (1970). *Dev. Biol.* **21**, 364–382.
Stevens, L. C. (1973). *J. Natl. Cancer Inst. (U.S.)* **50**, 235–242.
Stevens, L. C. (1983). *Cancer Surv.* **2**, 75–91.
Stevens, L. C., and Little, C. C. (1954). *Proc. Natl. Acad. Sci. U.S.A.* **40**, 1080–1087.
Stevens, L. C., and Varnum, D. S. (1974). *Dev. Biol.* **37**, 369–380.
Stewart, C. L. (1982). *J. Embryol. Exp. Morphol.* **67**, 167–179.
Stewart, C. L., Stuhlmann, H., Jähner, D., and Jaenisch, R. (1982). *Proc. Natl. Acad. Sci. U.S.A.* **79**, 4098–4102.
Stewart, C. L., Harbers, K., Jähner, D., and Jaenisch, R. (1983). *Science* **221**, 760–762.
Stewart, C. L., and Kimber, J. (1984). In preparation.
Stewart, T. A., and Mintz, B. (1981). *Proc. Natl. Acad. Sci. U.S.A.* **78**, 6314–6317.
Stewart, T. A., Wagner, E. F., and Mintz, B. (1982). *Science* **217**, 1046–1048.
Swartzendruber, E. C., and Lehman, J. M. (1975). *J. Cell. Physiol.* **85**, 179–180.
Teich, N. M., Weiss, R. A., Martin, G. R., and Lowy, D. R. (1977). *Cell* **12**, 973–982.
Tooze, J. (1980). *Cold Spring Harbor Monogr. Ser.* **10B**, 1–1073.
Tunnacliffe, A., Parkar, M., Povey, S., Bengtsson, B., Stanley, K., Solomon, E., and Goodfellow, P. (1983). *EMBO Jour.* **2**, 1577–1584.
Vasseur, M., Kress, C., Montreau, N., and Blangy, D. (1980). *Proc. Natl. Acad. Sci. U.S.A.* **77**, 1068–1072.
Wagner, E. F., and Mintz, B. (1982). *Mol. Cell. Biol.* **2**, 190–198.
Wagner, E. F., Stewart, T., and Mintz, B. (1981). *Proc. Natl. Acad. Sci. U.S.A.* **78**, 5016–5020.
Watanabe, T., Dewey, M. J., and Mintz, B. (1978). *Proc. Natl. Acad. Sci. U.S.A.* **75**, 5113–5117.
Weiss, R., Teich, N., Varmus, H., and Coffin, J. (1982). *Cold Spring Harbor Monogr. Ser.* **10C**, 1–1396.

7
Perspectives

CHAPTER XVIII

Mammalian Chimeras—Future Perspectives

R. L. GARDNER

Sir William Dunn School of Pathology
University of Oxford
Oxford, England

I.	Introduction	431
II.	Size Regulation in Chimeric Embryos	432
III.	Refinements in Chimera Production	433
IV.	Cell Markers	436
V.	Cell Lineage	438
VI.	Tissue Growth	439
VII.	Mutant Genes	440
VIII.	Concluding Remarks	441
	References	441

I. INTRODUCTION

It is now over two decades since Tarkowski first reported the experimental production of mammalian chimeras following the aggregation of pairs of cleaving embryos in the mouse (Tarkowski, 1961). A prior claim to have obtained development to term of a chimeric embryo was made by Nicholas and Hall (1942) in experiments in which denuded rat zygotes were aggregated in pairs. In the light of subsequent experience, two points are remarkable about these rat experiments. First, some embryos evidently survived despite their being transferred directly to the uterus at the one-cell stage. Second, one of the three resulting full–term foetuses was approximately twice the normal size. Unfortunately, since a genetic marker was not used in this study, it is impossible to decide whether the giant embryo was indeed a chimera that had maintained its growth advantage, as the authors presumed (Nicholas and Hall, 1942), or whether the explanation for its massive size lay elsewhere. Unequivocal rat chimeras have been produced more recently by embryo aggregation (Mayer and Fritz, 1974), but these, like the enormous numbers of mouse chimeras that have been obtained in this way, did not seem to enjoy any size advantage over standard individuals at birth. This appears to be true also of mouse chimeras originating from aggregates of 3 or 4 cleavage stage embryos (Markert and Petters, 1978; Petters and Markert, 1980): even mouse conceptuses obtained following aggregation of as many as 15 embryos appeared to be within the normal size range early in the second half of gestation (Hillman *et al.*, 1972). Hence, down-regulation for size is evidently completed well before term (Tarkowski, 1961; Bowman and McLaren, 1970), even after considerable enhancement of cell number in the preimplantation embryo. Nevertheless, both postnatal growth and fecundity in chimeras can evidently exceed that of their component strains (Moustafa and Brinster, 1972a). However, since the occurrence of such "chimeric vigour" does not seem to have been reported by other workers, it is not clear how common it is.

II. SIZE REGULATION IN CHIMERIC EMBRYOS

The phenomenon of size regulation merits closer investigation because of its bearing on the validity of the assumption that findings based on chimeras can be extrapolated to the development of standard embryos. Aggregation of entire cleaving embryos has been the most widely adopted procedure for producing chimeras because it is technically simpler than injecting cells into blastocysts or recombining dissociated blastomeres. Nevertheless, it almost certainly entails more radical perturbation of early development than the other methods. The immediate effect of aggregating pairs of embryos is to double the number of cells that are present in morulae. Obviously, if 3, 4, or even more embryos are united, the departure from normal cell number at this stage is correspondingly greater. However, aggregation not only affects size but apparently also alters the proportions of the embryo prior to implantation. Thus, according to the limited data that are available at present, giant blastocysts contain relatively more inner cell mass (ICM) and less trophectoderm cells than standard ones (Buehr and McLaren, 1974). Therefore, aggregation appears to have disturbed the ratio as well as the absolute number of cells that are allocated to the two primary tissues of the early embryo and, consequently, to have reduced the surface available for exchange with the maternal environment relative to the mass of internal cells.

Since the process of down-regulation does not begin until after implantation (Buehr and McLaren, 1974; Lewis and Rossant, 1982) more primitive endoderm (hypoblast) and primitive ectoderm (epiblast) cells will be formed in mature ICMs of giant than standard blastocysts. Furthermore, the ratio of these two types of cell may be biassed toward primitive ectoderm if, as has been suggested,

XVIII. Mammalian Chimeras—Future Perspectives 433

similar positional factors are operating in this case as in the earlier allocation of cells to the ICM versus trophectoderm (Gardner, 1983). The extent to which later allocation events will be affected by embryo aggregation is harder to predict because of persisting uncertainty as to how size regulation is achieved. The process appears to be initiated shortly after implantation and to be completed relatively rapidly because both double and quadruple embryos are within the standard size range prior to formation of the primitive streak (Buehr and McLaren, 1974; Lewis and Rossant, 1982; G Porter-Goff, personal communication of unpublished observations). Regarding the mechanism of down-regulation, data of Lewis and Rossant (1982) suggest that it may be mediated by a transient generalized decrease in rate of cell proliferation rather than by death or withdrawal from cycle of excess cells. If this is indeed the case, it is likely that the number of cells allocated to the embryo proper is also greater in aggregation chimeras than in standard embryos (Buehr and McLaren, 1974; Petters and Markert, 1980).

Obviously, introduction of single cells or daughter cell pairs into the preimplantation embryo will have a less dramatic effect on total cell number than embryo aggregation, regardless of the stage at which it is done (Kelly, 1977; Gardner, 1978). Indeed, the effect of introducing single ICM cells into blastocysts will be relatively trivial, since host ICMs contain at least a dozen cells by the stage at which blastocysts can be injected (R. L. Gardner, unpublished observations). Unfortunately, no comparative study has been undertaken on aggregation versus injection chimeras of the same genotypic constitution. Clearly, the most appropriate way of assessing any effects of size per se would be to compare chimeras made by aggregating pairs of whole versus pairs of half ($=4/8$ blastomeres) 8-cell embryos of the same two genotypes. It would also be of interest to ascertain whether phytohaemagglutinin treatment, which is often used to assist aggregation (Mintz et al., 1973; Petters and Markert, 1980), affects cell deployment in the early embryo.

III. REFINEMENTS IN CHIMERA PRODUCTION

There are two other features of primary (or whole-body) chimeras which, regardless of whether they are produced by aggregation or injection, limits their usefulness in certain types of investigation. The first of these is the extraordinary extent to which representation of cells of the same two constituent genotypes varies between members of a group of chimeras produced by the same procedure. Most of this variability is presumably of epigenetic origin, since it is very marked even among chimeras made between inbred strains. However, if there is indeed an inverse relationship between developmental stability and degree of homozygosity, as suggested by Lerner (1954; also see Leary et al., 1983), there might be a case for using embryos from hybrid or out-bred matings in order to reduce variability between chimeras. This strategy would be expected to reduce variability only if the supposed beneficial effects of heterozygosity depended strictly on allelic differences within rather than between cells.

The second limiting feature is the extent to which the level of mosaicism tends to be equilibrated throughout individual chimeras. Strong correlations in genotypic composition between such diverse components as coat, blood, and germ line have been reported consistently (see McLaren, 1976; Falconer et al., 1978), so that attempts to associate a specific phenotype with the cell proportions in individual tissues or organs are seldom likely to succeed. This greatly diminishes the value of chimeras for elucidating the tissues in which particular genes are expressed. Clearly, there is no difficulty with mutant genes, such as dominant spotting (Green, 1981), which adversely affect the proliferation of certain stem cells since the cell population carrying such lesions will be under-represented in or altogether absent from the relevant tissue(s) in chimeras. However, there is no reason to expect that a mutation that alters or impairs the differentiated function of a particular type of cell will necessarily limit its efficacy as a stem cell. One solution to this problem is to make chimeras using donor cells from more advanced stages of development, thereby exploiting restrictions in developmental potential or degree of clonal expansion of such cells to obtain more localized mosaicism. At present, this approach can only be used to obtain chimerism that is restricted to one of the major components of the conceptus (Table I), rather than to different tissues within the definitive embryo itself. One of the limiting factors is that the early implanting blastocyst remains the most advanced host embryo that will develop normally *in utero* following manipulation *in vitro*. Furthermore, unlike those of trophoblastic or extraembryonic endodermal origin, primitive ectoderm cells from postimplantation embryos have failed to yield chimerism following blastocyst injection (unpublished observations of R. Beddington, of J. Rossant, and of R. L. Gardner). According to available data, further restrictions in potency of primitive ectoderm cells may not occur until approximately 3 days after implantation in the mouse (Beddington, 1981, 1982). Moustafa and Brinster (1972b) reported ocular chimerism in 4 foetuses (1 living and 3 dead) among a total of 117 that had developed from albino blastocysts injected with unspecified cells from 8-day wild-type embryos. However, since pigmentation constituted the only marker in these experiments, it is not clear how widely the donor cells were disseminated. This work does not seem to have been extended to see whether viable chimeric off-spring can be obtained in this way.

Use of postimplantation embryos rather than blastocysts as hosts for producing chimeras poses problems, regardless of whether donor cells are injected into them *in utero* or *in vitro*. In the former situation, the thick mass of decidual tissue enveloping the conceptus makes accurate localization of injection very difficult during the critical stages of axiation and early organogenesis (Weissman et al.,

TABLE I.
Distribution of Chimerism Obtained with Different Types of Developmentally Restricted Donor Cells in the Mouse

Donor tissue	Method of initiating chimerism	Placenta	Trophoblast giant cells	Ectoplacental cone	Extraembryonic ectoderm	Parietal endoderm	Visceral endoderm	Visceral mesoderm	Amnion	Foetus	References
Fourth day trophectoderm	Blastocyst reconstitution	NA[a]	+	+	+	−	−	−	−	−	Gardner et al. (1973); Papaioannou (1982)
Fourth day ICM	Blastocyst reconstitution	NA	−	−	−	+	+	+	+	+	Gardner et al. (1973); Papaioannou (1982)
Fifth day primitive ectoderm	Single or multiple cell injection into blastocyst	+	−	NA	NA	−	−	+	+	+	Gardner and Rossant (1979); Gardner (1982)
Fifth day primitive endoderm	Single or multiple cell injection into blastocyst	+	−	NA	NA	+	+	−	−	−	Gardner and Rossant (1979); Gardner (1982)
Sixth or seventh day visceral endoderm	Multiple cell injection into blastocyst	−	−	NA	NA	+	+	−	−	−	Rossant et al. (1978); Gardner (1982)
Seventh day parietal endoderm	Multiple cell injection into blastocyst	−	−	NA	NA	+	−	−	−	−	Gardner (1982)
Sixth or seventh day extraembryonic ectoderm	Multiple cell injection into blastocyst	NA	+	+	NA	−	−	−	−	−	Rossant et al. (1978)

[a] NA, not present at stage of analysis.

1977). Sufficient thinning of the decidua has occurred by day 11 or 12 to enable donor cells to be transplanted with greater precision. Though haemopoietic chimerism has been initiated at this stage (Fleischman *et al.*, 1982), it is perhaps rather late for producing chimerism in solid tissues of the foetus. Obviously, accurate localization of donor cells is much simpler if injections are done *in vitro*. However, the period for which normal development of postimplantation embryos can be maintained in culture is limited (New, 1978; Beddington, 1981). Hence, only short-term injection experiments can be undertaken in these conditions so that clonal analysis employing single cell transplantation is impractical. Nevertheless, valuable information concerning the normal fate and potency of cells from different regions of late primitive-streak stage embryos has been obtained in this way (Beddington, 1981, 1982). The prospect of returning these more advanced conceptuses to the uterus following manipulation is being explored because of the enormous range of experimental possibilities that could be realized if they can be persuaded to form viable young (R. Beddington, personal communication).

IV. CELL MARKERS

In addition to refinements in the production of chimeras, improvements are needed in procedures used to analyse their mosaic composition. Since cell marking techniques are discussed in detail in another contribution to this volume (West, Chapter II), only certain general remarks will be made about them here. Considerable effort has been devoted in recent years to the quest for an *in situ* genetic marker for the mouse that would enable mammalian chimeras to be exploited as fully as their avian counterparts have been since discovery of the nucleolar difference between chick and quail cells. The value of some *in situ* markers in the mouse is limited because they are applicable to only a few types of mature cells. The use of certain others is questionable because their efficacy in chimeras has yet to be established. One cannot assume that clear-cut distinctions between cells in specimens of different genotype will necessarily be maintained in chimeras. Nevertheless, this assumption seems to underly use of genetically determined differences in β-galactosidase activity as a histochemical marker (Dewey *et al.*, 1976; Dewey and Mintz, 1978), despite evidence that lysosomal enzymes can be transferred between cells (Mullen, 1977, 1978; Herrup and Mullen, 1979). In fact, Dewey *et al.* (1976) found marked discrepancies in chimeric composition between pairs of adjacent sections of cerebellum when staining for β-galactosidase was applied to the one and electrophoretic separation of allozymes of glucosephosphate isomerase (GPI) to the other. On the basis of histochemical appraisal of its activity in this region of the brain, they attributed the disparity to restricted distribution of GPI. However, this explanation is difficult to reconcile with the finding of Oster-Granite and Gearhart (1981) who claim that all cells in the cerebellum stain strongly with specific anti-GPI sera. Dewey et al. (1976) used the histochemical procedure of Orchardson and McGadey (1970) to visualize the distribution of GPI activity. Since this procedure depends on endogenous glucose-6-phosphate dehydrogenase, they may well have been detecting cellular differences in the activity of this enzyme rather than GPI. Regardless of the correct explanation, these disparities serve to underline the importance of determining the reliability of a prospective *in situ* marker in chimeric tissue in which all cells can be assigned unequivocally to one or other genotype by an independent means.

The most promising methods for detecting chimerism *in situ* in the mouse include allozyme-specific anti-sera against GPI (Oster-Granite and Gearhart, 1981), haplotype-specific monoclonal antibodies against H-2 antigens (Ponder et al., 1983), and species-specific DNA probes for use in viable chimeras between *Mus musculus* and *M. caroli* (Rossant et al., 1983). Nevertheless, even these suffer from the limitation that they depend on use of sectioned material. Hence, patterns of chimerism can only be visualized by means of the difficult and time-consuming exercise of reconstruction from serial sections. This poses even greater problems if fresh frozen rather than wax-embedded material has to be employed. It is therefore perhaps not surprising that assessment of cell mixing in chimeric specimens has been based mainly on appraisal of selected sections rather than extensive reconstruction. This sampling approach could be very misleading in tissues that are markedly anisotropic in growth characteristics unless sections from specimens cut in several different planes are compared.

An alternative strategy that has been employed for clonal analysis in *Drosophila* is to adapt staining procedures for visualizing mosaicism for use with tissue fragments rather than sections (Lawrence, 1981). This has been applied successfully to whole mount specimens of the extraembryonic membranes of the mouse conceptus using presence versus absence of cytoplasmic malic enzyme activity as a genetic marker (Gardner, 1984). Hence, overall patterns of chimerism have been visualized in permanent preparations of parietal endoderm, entire visceral yolk sac, and amnion in which normal cellular relationships are conserved. Pieces can be readily excised from such preparations for embedding and sectioning if finer details need to be resolved. This approach is obviously more likely to prove feasible in cases in which the reagents used to reveal mosaicism are of relatively low molecular weight. Also, there is clearly a limit to the thickness of specimen that can be treated in this way. Nevertheless, because of the considerable advantages it offers, its applicability to other tissues of the conceptus and adult mouse is currently being explored. Regardless of the precise method employed for detecting chimerism *in situ*, only relatively small samples can be handled for this purpose. Hence, particularly in the case of postnatal chimeras, electrophoretic analysis of GPI allozymes in tissue homogenates is likely to remain the method of choice for establishing the general distribution of chimerism. Simultaneous use of both GPI allozymes and an *in situ* marker would therefore seem to be a sensible practice.

These advances in *in situ* cell marking techniques are so recent that the task of exploiting them to tackle specific problems has hardly begun. There are, indeed, many aspects of both normal and abnormal development on which they are likely to have a major impact. Prominent among these are studies on cell lineage, tissue growth, and the effects of various mutant genes.

V. CELL LINEAGE

The basic lineages in the early mouse embryo have been established by injecting individual cells from different tissues into blastocysts and by reconstituting blastocysts from isolated trophectoderm and ICM. Allozymes of GPI were used as markers in all these studies and were resolved by electrophoresis of homogenates of dissected components of the resulting conceptuses. The lineage scheme that emerged from this work has been reviewed in detail recently (Gardner, 1983) and is clearly deficient in two major respects. The first is that, because the chorioallantoic placenta cannot be separated into its constituent tissues, the specific contributions of trophectoderm, primitive endoderm and primitive ectoderm to this organ have yet to be elucidated (Table I). It is generally assumed that all trophoblast cells in the placenta, including those of the three cell layers that comprise the exchange surface, are derived from the outer trophectoderm of the blastocyst. Likewise, all connective tissue and blood vessels are presumably of primitive ectodermal origin, the primitive endodermal contribution being confined to the endodermal sinuses of Duval (1892). However, until *in situ* marking techniques are applied to this complex structure, the lineage relationships between its components will remain a matter for speculation. The second major deficiency in the scheme is the lack of information concerning lineages within the primitive ectoderm and, most notably, that part of it which will eventually form the foetus. As noted earlier, the solution to this problem lies in finding an effective way of initiating chimerism during postimplantation development, using donor cells from specific layers and regions of the emerging definitive embryo. Again, *in situ* markers would be essential in such studies because of the intimacy with which different tissues become intermixed during the genesis of most organs. It would be of great interest to know whether the relationship between cell lineage and germ layers in the embryo is as close as has generally been supposed.

XVIII. Mammalian Chimeras—Future Perspectives

VI. TISSUE GROWTH

In general, very little is known about normal patterns of tissue growth in mammals. The ability to visualize the arrangement of cells of two different genotypes *in situ* could furnish important clues, especially if specimens were examined at various stages of development. It would be particularly informative to concentrate on specimens of the tissue or organs under study in which mosaicism was very unbalanced since, providing the bias was not consistently in one direction (i.e., neither genotype displayed an obvious selective advantage), such specimens may include cases in which all except one progenitor cell were of the same genotype. Obviously, in these particular circumstances, all cells of the minority genotype would constitute a progenitor cell clone. The sort of information one might hope to obtain from such studies would be whether growth is coherent and whether it exhibits either regional or directional preferences that might be of morphogenetic significance.

According to available data, the early mouse embryo seems to grow in a coherent manner at least until implantation is completed (McLaren, 1976; Kelly, 1979). However, there is evidence attesting to the existence of fine-grained mosaicism in a variety of tissues in both pre- and postnatal chimeras and X inactivation mosaics (e.g., McLaren, 1976). Presumably, therefore, coherent growth cannot be sustained throughout development in the definitive embryo, though nothing is known at present about when and why it breaks down. A number of factors are likely to contribute to cell mixing including morphogenetic movements, particularly those involved in the shift from two- to three-dimensional arrangements of cells following gastrulation. In addition, cell death occurs throughout embryogenesis in mammals from the blastocyst stage (Glücksmann, 1951; El Shershaby and Hinchliffe, 1974; Copp, 1978) though, in most tissues, neither the extent nor distribution of cell loss is known. Changes in the proportion of cells that are cycling may also play a significant role. Early in ontogeny, most if not all cells in a tissue may contribute to its growth. The growth fraction later declines and may eventually attain a very low value when the phase of net growth is superceded by that of maintenance or stasis. Depending on the proportion and distribution of the cells that continue to cycle and the rate of cell death, the initial spatial arrangements of progenitor cell clones might be markedly distorted by such changes. Active migration of individual cells may also contribute to cell mixing, especially in the central nervous system and the neural crest.

It may be informative to study the growth of chimeric extraembryonic membranes before proceeding to the embryo proper, on the grounds that it should be easier to tackle this problem in two dimensions than in three. In this context it is interesting to note that the distribution of the progeny of donor primitive endoderm cells is entirely different in the two extraembryonic endodermal layers of host conceptuses following blastocyst injection. Donor cells occur in one or more discrete patches in the endoderm layer of the visceral yolk sac, whereas they are intermixed with host cells in a pepper-and-salt fashion in the parietal endoderm. Furthermore, these differences in pattern of distribution are also seen in cases in which the clonal descendants of single transplanted primitive endoderm cells span both extraembryonic endodermal layers. Hence, they must reflect differences in growth characteristics of the host tissues rather than the donor cell clones (Gardner, 1984). Detailed comparison of the patterns of chimerism in the two tissues at various stages of ontogeny should help to elucidate the factors to which these differences are attributable.

VII. MUTANT GENES

The study of mutations that impair development or function has long been recognized as a valuable way of elucidating the normal role of genes in such processes. Genetically mosaic organisms consisting of mixtures of mutant and wild-type cells can be of considerable use in these studies, as is evident especially from work on somatic recombination mosaics and gynandromorphs in *Drosophila* (Gehring, 1978). So far, however, investigations on mouse chimeras have, in general, proved somewhat disappointing in this respect. This is due, in part, to the fact noted earlier that, since there tends to be a close correlation in level of chimerism between different organs and tissues, it is very difficult to pinpoint the cell type(s) in which a mutation acts unless it impairs the proliferation of the cells in question. An additional factor is that simultaneous marking of wild-type versus mutant cells with an independent *in situ* marker has seldom been undertaken so far (Mullen, 1982). Thus, in investigating mutations, such as those of the complex *T* locus in chimeras, it is assumed that all cells that display a mutant phenotype are mutant and those that do not are wild type in genetic constitution (e.g., Spiegelmann, 1978). This is clearly unsatisfactory for the reason given earlier in relation to use of β-galactosidase as an *in situ* marker. In order to establish whether such mutations are behaving in a cell autonomous manner, it is essential to have a reliable independent means of distinguishing mutant from wild-type cells *in situ*.

Possible ways of obtaining chimeras in which the distribution of mosaicism is more variable or restricted have already been discussed and, if used in conjunction with an *in situ* marker, could considerably enhance knowledge of both the type(s) of cell and the nature of functions affected by particular mutant genes. Mutations that appear to affect interactions between tissues would also become amenable to analysis *in vivo* if restriction of chimerism to individual germ layers could be achieved. At present, the only way of investigating such interactions is by means of *in vitro* experiments in which mutant inducing tissue is recombined with wild-type responding tissue and vice versa. By the stage of development at which such pairs of tissues can be separated, decisive interactions may have already taken place between them. In addition, enzymatic or other procedures used in their separation and the media used for culture of recombinants might further affect the outcome.

VIII. CONCLUDING REMARKS

Chimerism can be initiated readily in mammalian embryos prior to implantation. In species such as the mouse, the embryo is still composed of relatively few cells when it attaches to the uterine epithelium. This means that single cells transferred between preimplantation stages can contribute substantially to conceptuses or offspring, thereby enabling early development to be analysed in a strictly clonal manner. However, cells have undergone only limited restriction in potency on completion of this phase of development, particularly those from which the definitive embryo takes its origin. Hence, studies on such chimeras provide little information on cell lineage and differentiation within the embryo proper as opposed to the more precociously developing extraembryonic tissues. The problem in mammals is therefore essentially the reverse of that in birds. Whereas the later stages of avian development are readily accessible to manipulation, the earlier ones are not. While Marzullo (1970) obtained a low rate of chimerism following injection of relatively large numbers of cells between unincubated blastodisk stage embryos in the chick, this work does not seem to have been extended.

In conclusion, studies on mammalian chimeras have already contributed substantially to the solution of a wide range of problems. Their exploitation is likely to be even greater in the future, especially if methods of producing and analysing them can be further refined.

ACKNOWLEDGMENTS

I wish to thank Dr. Rosa Beddington, Dr. David Cockroft, and Mrs. Jo Williamson for help in preparing the manuscript, and acknowledge the support of The Royal Society and the Imperial Cancer Research Fund.

REFERENCES

Beddington, R. S. P. (1981). *J. Embryol. Exp. Morphol.* **64**, 73–85.
Beddington, R. S. P. (1982). *J. Embryol. Exp. Morphol.* **69**, 265–285.
Bowman, P., and McLaren, A. (1970). *J. Embryol. Exp. Morphol.* **23**, 693–704.

Buehr, M., and McLaren, A. (1974). *J. Embryol. Exp. Morphol.* **31**, 229–234.
Copp, A. J. (1978). *J. Embryol. Exp. Morphol.* **48**, 109–125.
Dewey, M. J., and Mintz, B. (1978). *Dev. Biol.* **66**, 550–559.
Dewey, M. J., Gervais, A. G., and Mintz, B. (1976). *Dev. Biol.* **50**, 68–81.
Duval, M. (1892). "Le placenta des rongeurs." Ancienne Librairie Germer, Baillière et Fils, Paris.
El Shershaby, A. M., and Hinchliffe, J. R. (1974). *J. Embryol. Exp. Morphol.* **31**, 643–654.
Falconer, D. S., Gauld, I. K., and Roberts, R. C. (1978). *In* "Genetic Mosaics and Chimeras in Mammals" (L. B. Russell, ed.), pp. 39–49. Plenum, New York.
Fleischman, R. A., Custer, R. P., and Mintz, B. (1982). *Cell* **30**, 351–359.
Gardner, R. L. (1978). *In* "Methods in Mammalian Reproduction" (J. C. Daniel, ed.), pp. 137–165. Academic Press, New York.
Gardner, R. L. (1982). *J. Embryol. Exp. Morphol.* **68**, 175–198.
Gardner, R. L. (1983). *Int. Rev. Exp. Pathol.* **24**, 63–133.
Gardner, R. L. (1984). *J. Embryol. Exp. Morphol.* **80**, 251–288.
Gardner, R. L., and Rossant, J. (1979). *J. Embryol. Exp. Morphol.* **52**, 141–152.
Gardner, R. L., Papaioannou, V. E., and Barton, S. C. (1973). *J. Embryol. Exp. Morphol.* **30**, 561–572.
Gehring, W., ed. (1978). "Genetic Mosaics and Cell Differentiation." Springer-Verlag, Berlin and New York.
Glücksmann, A. (1951). *Biol. Rev. Cambridge Philos. Soc.* **26**, 59–86.
Green, M. C. (1981). *In* "Genetic Variants and Strains of the Laboratory Mouse" (M. C. Green, ed.), p. 260. Fischer, Stuttgart.
Herrup, K., and Mullen, R. J. (1979). *J. Cell Sci.* **40**, 21–31.
Hillman, N., Sherman, M. I., and Graham, C. (1972). *J. Embryol. Exp. Morphol.* **28**, 263–278.
Kelly, S. J. (1977). *J. Exp. Zool.* **200**, 365–376.
Kelly, S. J. (1979). *J. Exp. Zool.* **207**, 121–130.
Lawrence, P. A. (1981). *J. Embryol. Exp. Morphol.* **64**, 321–332.
Leary, R. F., Allendorf, F. W., and Knudsen, K. L. (1983). *Nature (London)* **301**, 71–72.
Lerner, I. M. (1954). "Genetic Homeostasis." Oliver & Boyd, Edinburgh.
Lewis, N. E., and Rossant, J. (1982). *J. Embryol. Exp. Morphol.* **72**, 169–181.
McLaren, A. (1976). "Mammalian Chimaeras." Cambridge Univ. Press, London and New York.
Markert, C. L., and Petters, R. M. (1978). *Science* **202**, 56–58.
Marzullo, G. (1970). *Nature (London)* **225**, 72–73.
Mayer, J. F., and Fritz, H. I. (1974). *J. Reprod. Fertil.* **39**, 1–9.
Mintz, B., Gearhart, J. D., and Guymont, A. O. (1973). *Dev. Biol.* **31**, 195–199.
Moustafa, L. A., and Brinster, R. L. (1972a). *J. Cell Biol.* **55**, 183a (abst.).
Moustafa, L. A., and Brinster, R. L. (1972b). *J. Exp. Zool.* **181**, 193–202.
Mullen, R. J. (1977). *Nature (London)* **270**, 245–247.
Mullen, R. J. (1978). *In* "Clonal Basis of Development" (S. Subtelny and I. M. Sussex, eds.), pp. 83–101. Academic Press, New York.
Mullen, R. J. (1982). *In* "Genetic Approaches to Developmental Neurobiology" (Y. Tsukada, ed.), pp. 183–193. Univ. of Tokyo Press, Tokyo.
New, D. A. T. (1978). *Biol. Rev. Cambridge Philos. Soc.* **53**, 81–122.
Nicholas, J. S., and Hall, B. V. (1942). *J. Exp. Zool.* **90**, 441–459.
Orchardson, R., and McGadey, J. (1970). *Histochemie* **22**, 136–139.
Oster-Granite, M. L., and Gearhart, J. (1981). *Dev. Biol.* **85**, 199–208.
Papaioannou, V. E. (1982). *J. Embryol. Exp. Morphol.* **68**, 199–209.
Petters, R. M., and Markert, C. I. (1980). *J. Hered.* **71**, 70–74.
Ponder, B. A. J., Wilkinson, M. M., and Wood, M. (1983). *J. Embryol. Exp. Morphol.* **76**, 83–93.
Rossant, J., Gardner, R. L., and Alexandre, H. L. (1978). *J. Embryol. Exp. Morphol.* **48**, 239–247.

XVIII.   Mammalian Chimeras—Future Perspectives

Rossant, J., Vijh, M., Siracusa, L. D., and Chapman, V. M. (1983). *J. Embryol. Exp. Morphol.* **73,** 179–191.
Spiegelman, M. (1978). *In* "Genetic Mosaics and Chimeras in Mammals" (L. B. Russell, ed.), pp. 59–80. Plenum, New York.
Tarkowski, A. K. (1961). *Nature (London)* **190,** 857–860.
Weissman, I. L., Baird, S., Gardner, R. L., Papaioannou, V. E., and Raschke, W. (1977). *Cold Spring Harbor Symp. Quant. Biol.* **41,** 9–21.

CHAPTER XIX

Quail-Chick Chimeras: Concluding Remarks and Perspectives

NICOLE M. LE DOUARIN

Institut d'Embryologie du CNRS
et du Collège de France
Nogent-sur-Marne, France

The first report on the use of the quail nuclear heterochromatin as a cell marker in tissue associations of quail and chick cells appeared in 1969 in the *Bulletin Biologique de la France et de la Belgique*. This marking technique resulted from the original observation of the large amount of heterochromatin regularly associated with all cell types of embryos and adults of the Japanese quail. This very peculiar feature of the interphase nucleus is rare in animal species, although I found similar dispositions of constitutive heterochromatin in the nucleus of many other species of birds (Le Douarin, 1971). Curiously, those birds are not related in taxonomy to *Coturnix coturnix japonica* but belong mostly to the Passeriforms. Closely related genera were found to have different nuclei, some like the quail, others like the chick; however, within a given genus all species were similar in this respect.

In any case, embryologists had to exploit the natural opportunity provided by the fact that chickens and quails, close species that are bred industrially and lay eggs in abundance all year, have interphase nuclei so easy to distinguish. After pioneer work on the neural crest, an embryonic structure whose cells have highly migratory properties, efforts were focused on the ontogeny of the immune system. With the work of F. Dieterlen-Lièvre, the technique was extended to more general aspects of hematopoiesis, namely, the search for the embryonic origin of the hemopoietic stem cells that provide the individual with blood cells throughout its entire life. Rapidly, the method spread to other laboratories, other countries, and other types of cell. Important results were established, for example the origin of limb musculature from the somites (see Chapter XI), of osteoclasts from hemopoietic precursors (see Chapter VI).

After 14 years, the quail-chick marker system still offers perspectives, essentially in organogenesis and histogenesis, subjects for which the avian embryo is particularly suitable. The quail-chick system is being enriched by association with more refined techniques that open new avenues to study differentiation at the molecular level.

Monoclonal antibodies directed against differentiation antigens of quail and chick cells can be prepared in mouse with high efficiency, owing to the phylogenetic distance separating birds and mammals. It is easy to prepare such reagents with specificities restricted to either quail or chick determinants, while others recognize identical epitopes in both species. These tools are really efficient in tissue combinations devised toward solving specific ontogenetic problems. One recent example has been the identification of the cells in the avian thymus which bear MHC class II gene products (Le Douarin et al., 1983).

Another interesting antibody, the so-called anti-MB1 (Péault et al., 1983), recognizes all cells belonging to the *hemangioblastic* lineage of the quail, i.e., the endothelial cells of blood vessels and all blood cells except differentiated erythrocytes. Its specificity is restricted to quail, and no chick cell is stained by this reagent. Ontogeny of the vascular tree and mechanisms of angiogenesis, many aspects of which are still poorly known, can be studied either in the quail or in quail-chick tissue combinations by using such a tool.

In the study of differentiation, the most decisive step is *determination*, i.e., restriction of the cell developmental capabilities to a single or a group of related phenotype(s). Although marked by the onset of specific control mechanisms of gene expression, the successive restrictions of the cell developmental potencies do not immediately result in biochemically or cytologically detectable changes and, therefore, can be defined only in operational terms: the analysis of the terminally differentiated phenotype(s) yielded by a given embryonic cell in definite conditions. Carrying out such an analysis requires that the fate of a given cell can be followed with certainty during development. Except in particular cases, this cannot be done *in vivo*, and one of the useful prospects for the quail-chick marker system is to implant isolated quail cells whose fate is under question into chick tissues, either in culture combinations or, better, in the embryo *in ovo*. A clonal analysis of development can then be envisaged that might provide valid information on the state of determination of embryonic cells at definite stages of their ontogeny.

REFERENCES

Le Douarin, N. M. (1971). *C. R. Hebd. Seances Acad. Sci.* **272**, 1402–1404.
Le Douarin, N. M., Guillemot, F. P., Oliver, P., and Péault, B. M. (1983). *Int. Congr. Immunol., 5th*, Kyoto, Japan, Academic Press, *1983*.
Péault, B. M., Thiéry, J. P., and Le Douarin, N. M. (1983). *Proc. Natl. Acad. Sci. U.S.A.* **80**, 2976–2980.

Index

A

ABO groups, 166, 167, 171, 172, 174
Accessory cells
　antigen-presenting cells (APC), 221, 228, 229
　dendritic cells, 186
　macrophages, 179, 186, 221, 228, 229, 230, 232
Acetylcholine (ACh), *see* Neurotransmitters
Alligators, 381
　temperature of eggs, 381
Alloaggression, *see* Immune system
Allocation of cell, 90, 96, 97, 98, 101
Altered self, *see* Immune system
Amnion, 437
Angiogenic response, 156
Anlage fields, 72, 76, 77, 78, 80, 81, 82, 84, 85
　disposition of, 78, 81
Area centralis, 80, 81
Area or zona pellucida, 19, 21, 22, 24, 27, 29, 73, 74, 81, 82
Autonomic nervous system (ANS), 313, 314, 336, 339, 346
　adrenal paraganglion, 341, 346
　adrenergic derivatives, 314, 339, 347, 348
　adrenomedullary gland, 337, 338, 342
　cholinergic neurones, 314, 339, 347, 348
　enteric ganglia, 314, 337, 338, 339, 341, 342
　ganglia and paraganglia ontogeny, 336–339
　parasympathetic, 314, 338, 339
　sympathetic, 314, 328, 337, 338, 339, 341, 342
Autonomic plexuses
　adrenal plexuses, 338
　aortic plexus, 338
　Auerbach's plexus, 338
　coeliac plexuses, 338
　Meissner's plexus, 338
　myenteric plexus, 338
　pelvic plexuses, 338, 339
Autoradiography, 71, 73, 76, 85
5-Azacytidine, 419

B

Balanced strain combinations, 388
Basophilic cells, 139, 142, 143, 144, 148, 151
Behavior, 369–379
　behavioral characteristics, 372
　of the chimeras, 364
　variation in, 374
Behavior tests, 370—379
　alcohol preference, 370, 372–376, 378, 379
　cricket attacking, 371, 377, 378, 379
　open field activity, 370, 371, 376
　open field defecation, 371, 375, 376 378, 379
　rope climbing, 371, 376, 377, 378
Biochemical, histochemical and immunological markers, 54–63, 436, 437, 438
　acetylcholinesterase (AChE), 330, 339
　acid phosphatase, 52
　aldehyde oxidase, 57, 58
　alkaline phosphatase, 57, 115
　allozyme-specific antisera, 60, 61
　β-galactosidase, 56, 57
　β-glucuronidase, 54, 56, 57, 61, 358, 359, 360, 361, 363
　choline acetyltransferase (CAT), 339
　glucose phosphate isomerase (GPI), 55, 59, 60, 61, 94, 100, 126, 357, 358, 374, 375, 436, 437

Index

Biochemical markers *(cont.)*
  glycerol 3-phosphate dehydrogenase (GPDH), 59, 364, 415
  HPRT (hypoxanthine phosphoribosyltransferase), 412, 413, 420, 422
    activity, 413, 420
    deficiency in, 420, 422
  isocitrate dehydrogenase, 55
  liver alcohol dehydrogenase, 370
  malic enzyme, 55
  ornithine transcarbamylase, 59
  pronase, 22, 25
  TK (thymidine kinase), 413, 420
    activity, 413, 420
    deficient in, 420, 422
  tyrosine hydroxylase, 339
  XPRT xanthine phosphoribosyltransferase, 420
Blastocyst, 23, 24, 27, 29, 30, 92, 102, 103, 116, 117, 411, 412
  composite, 29
  environment, 102
  formation, 22, 23
  injection techniques, 24–29, 95, 96, 97, 104, 411, 412, 413, 415
  reconstitution, 29–31, 94
  stages, 93, 98, 102, 103
Blastodisc, 6, 7, 13
Blastomeres, 90, 93, 115, 413
  polar and apolar, 102
Blastopore, dorsal lip of, 4
B lymphocytes, *see* Lymphocytes
Body wall, development of, 281–310
Bone marrow, 197, 199, 213
  hemopoietic cells, 31
Brain, 372, 377, 379
Branchial arches, pouches, 181, 186, 187, 188
Bursa of Fabricius, *see* Primary lymphoid organs

C

Caenorhabditis elegans, 90
Catecholamines (CA), *see* neurotransmitters
Cattanach's insertion, 54
Cell line segregation, 346–348
Cell markers, 39–67, 180, 182, 187, 192, 212, 318, 330, 379
Cell mixing, 40, 62

Cell potential, 90, 91, 95, 96, 101
Cell selection, 39, 40, 53, 98, 124
Central nervous system (CNS), 314, 326, 346, 349
Cerebellum, 355–359
  cerebellar mutants, 359–365
Chemokinesis, 201
  chemokinetic factors
Chemotactism, 194–195, 198, 199, 200, 201, 203
  chemoattraction, 114, 198, 213
  chemotactic factors, 202, 203
  chemotactic index, 201, 202
  chemotactic peptides, 203
  Zigmond's chemotactism chamber, 199, 200
Chimeras, 71, 72, 73, 76, 84, 166, 170, 171, 173, 174, 175, 176, 177, 384–396
  aggregation, 21, 24, 54, 55, 116, 117, 118, 120, 219, 220, 222, 223, 226, 227, 229, 230, 231, 234, 235, 373, 374, 388, 389, 390, 432, 433
  amphibian, 152–154
  avian, 6, 7
  blood, 393
  bone marrow, 228
  chick-chick, 387
  chick-quail, 15, 71, 76, 78, 80, 81, 114, 115, 313, 328, 330, 336, 387
  complementary, 14
  C3H↔C57BL male, 124, 125, 126
  definition, 165
  dispermic, 166, 169–171, 172, 173, 174, 175, 177
  germ layer, 7–8
  hemopoietic, 10–18
  heterosexual, 384
  human, 165–177, 166, 174, 175, 177
  injection, 54, 55, 60, 61, 433
  intersexual, 113
  interspecific, 19, 24, 61, 62, 118
  irradiation, 220, 223, 227, 229, 230, 231, 232
  mammalian, 18–32, 34, 89–109, 431–443
  mouse, 369, 373–379, 388
  natural, 166
  neural tube, 8
  parabiotic, 382
  phenotypic sex of, 171
  postimplantation, 20, 32
  "pure" germinal, 118
  production of, 374

Index

rabbit, 19
radiation, 20, 51, 53, 54, 55, 57, 126
rat, 19
rat↔mouse, 19
sheep, 19, 388
sheep↔goat, 19
single-sex, 123–125, 127
spinal cord, 348–349
supplementary, 14
teratocarcinoma, 409–424
twin, 166–169, 174, 175, 177
unestablished type, 175–176
wholebody, 170
yolk sac, 13, 14, 15, 16, 135, 136, 138, 139, 140, 147, 151
Chimerism
blood group, 218
dispermic
primary and/or secondary, 166, 169, 393
rescue by, 116–118
tetragametic, 169
whole body, 169
Chloramphenicol, 414
resistant, 414, 422
Chorioallantoic membrane (CAM), 6, 7, 10, 12, 182, 192, 339, 406, 407
blood vessels, 13
graft on, 406, 407
Chorioallantoic placenta, 438
Chromosomal markers, see Nuclear markers
Chromosome, 14
chromosomal genome, 422
introduction of, 414
X chromosomes, 418
Ciliary body, 44
Ciliary ganglion, 340, 344, 345
Clones, progenitor cell, 439
clonal analysis, 436
clonal elimination, 218, 222, 223–224, 230
selection, 218
Columella, 258, 270, 274
Covariance, 373, 375, 377, 378
between behaviors, 375
factor analysis (technique), 375
of behavior groups with tissues, 378
Craniofacial development, 241–280
Cricket (*Achaeta domestica*), 371, 376, 377
Culture, 22, 198, 402, 403
medium, 22, 421
systems, 22
transfilter technique, 198, 402, 409

Cutaneous sensory corpuscles, 333–336
Grandry's corpuscles, 336
Herbst corpuscles, 335, 336
Meissner corpuscles, 336
Pacinian corpuscles, 335–336
Cybrid fusion, 414
Cytological markers, 48–53
cytoplasmic inclusions, 52
mitochondrial malic enzyme, 59, 61
Cytoplasmic inheritance, 414

D

DDK strain cytoplasm, 126
DNA hybridization, 61
DNA viruses, 416, 417, 419, 420, 423
polyoma virus, 416, 417, 419, 423
simian virus 40 (SV 40), 416, 417, 419, 420, 423
Dorsal mesentery, 148, 149, 151

E

Ear
inner ear, 45
tympanic membrane, 258
Ectoderm, 94, 95, 96, 103, 104, 105, 106, 181, 187, 432, 438
distal, 95
extraembryonic, 94, 104
head, 95
primitive (epiblast), 95, 103, 104, 105, 106, 432, 438
Ectoplacental cone, 94, 104
Embryonal carcinoma cells (EC cells), 410–423
cell hybrids, 414, 415
clones, 421
embryo chimeras (in), 423
introduction of genes into, 420
by coprecipitation, 420
by covalent linkage, 420
mutant of, 412
transformation of, 415, 421
DNA mediated, 415, 420
viral mediated, 415
Endoblast, 73, 74, 76, 77, 78, 80, 81, 84
ingression of the, 81

Endoderm, 77, 81, 95, 103, 104, 105, 106, 116
  parietal, 95, 437
  primary endoderm, 116
  primitive, 95, 103, 104, 105, 106, 432, 438, 439
  visceral, 95
Endophyll, 73, 74, 84, 86
  entophyllic crescent, 82, 83, 84
Endothelial cells, 49, 140, 148, 151, 156
Endothelial lineage, 148
Enteric nervous system, 314
Epiblast, 7, 76, 77, 82, 114, 116, 124
  perinodal, 82
Extracellular matrix, 202
Eye
  choroid, 44, 45
  cornea, 260
  iris, 44, 45, 260
  retinal pigment epithelium, 44, 45, 46, 48, 62

F

Fate, of cells, 90, 91, 93, 95, 96, 101, 102
Feminization, secondary, 384
Fibronectin, 202, 328, 330
Follicle cells, 125, 126
Foster mother, 23
Freemartins, 119
Fusion of cells, 414, 422

G

Gametes, 113, 414, 422
  formation of fertile gametes, 414, 422
Ganglion of Remak, 338, 339
Gap junctions, 413
Gastrula
  amphibian, 4
Gastrulation, 7, 76
  gastrular ingression, 78
  movements of, 74
Genes and mutants, 409–423
  albino (c), 44, 45, 46, 48, 370
  agouti (A), 45, 46, 47
  beige (bg), 52, 53, 57
  blonde mutation, 46
  brown (bw), 46, 47
  brown and dilute (b/b d/d), 45
  of cell death, 47–48
  characterization of mutants, 410
  cloning of genes, 410
  dominant spotting (W), 45, 159, 160
  down less (dl), 47
  dystrophia-muscularis (dy), 48
  ebony (e), 46, 47
  expression, 409–423
  extreme dilution allele (c/c^e), 45
  forked (f), 47
  fuzzy (fz), 47
  galactokinase locus, 415
  greasy (Gs), 47
  hooded, 46
  ichthyosis (ic), 49, 357, 360, 361, 362, 363
  jimpy (Jp), 117
  lurcher (lc), 358, 359, 360–361, 362, 364
  maroon-like (mal), 46, 47, 57, 58
  morphology (genes of), 47–48
  multiple wing hairs (mwh), 47
  myelination mutants, 366–367
  myelin synthesis deficiency (Msd), 117
  nervous (nr), 365
  neomycin resistance, 420
  nude mice (nu), 224, 226
  null alleles, 59
  pigment genes, 45–47
  pink-eyed dilution (p/p), 45, 46
  purkinge cell degeneration (pcd), 358, 359, 360, 361, 364, 365
  quaking (qk), 367
  recessive yellow (e/e), 45
  reeler (rl), 359, 363–364
  retinal degeneration (rd), 365–366
  retinal dystrophy (rdy), 48, 366
  sex-linked albino mutation, 115
  shiverer (shi), 266
  short ear (se), 47, 304, 305
  sex reversed (sxr), 52, 120
  staggerer (sg), 359, 361–362, 364
  steel (Sl), 160
  tabby (Ta), 47
  tissue degeneration (genes of), 47–48
  transfer of genes, 411, 414, 422
  T/t locus, 118
  t^x/t^y genotype, 118
  vermilion (v), 46, 47
  vestigal tail (vt), 47, 304
  waved-2 (wa-2), 47
  weaver (wv), 359, 362–363
  white (w), 46, 123, 390, 391

X-linked sparse fur mutant (spf), 59
yellow (y), 46, 47
Genetic cell markers, 94, 355, 356, 357, 360, 431, 436
Genital ridges, 114, 119, 389, 390, 392, 394
Genome, 410, 411, 412, 416, 417, 422
  retroviral, 419
  viral genome, 416
  SV40 genome, 417
Germ cell, 111–129
  in amphibia, 113, 114, 115
  in chimeras, 113, 114, 126
  determinant, 113
  genetic characteristics, 113, 125
  genotype, 126
  insects, 111–115
  lineage, 111–129
  migration, 118–119
  phenotype, 126
  primordial, 31, 114, 115, 116, 119, 123, 124, 126, 386, 387, 390, 391, 395
  selection, 124
  sexual differentiation, 113, 114
  XO germ cells, 121
  XX germ cells, 120, 121, 122, 123
  XX/Sxr germ cells, 121
  XY germ cells, 120, 121, 122
Germ-line
  chimeras, 112, 116, 118, 413, 422
  determinants, 112, 116
  of mammalian, 115
Germ plasm, 114
Girdles, development of, 300–302
Glial cells, 319, 322, 325, 326, 328, 336, 341, 344
Glomerulus, see Kidney
Gonad, 122, 123, 125
  gonadal differentiation, 381, 387, 389, 390, 391, 392, 396, 397
  gonadal sex, 387, 389
  heterosexual twin pairs, 119
  sexual differentiation
Grafts, see also Transplantation
  allografts, 71, 73, 76, 77, 82
  chorioallantoic, 10
  coelomic, 10–11
  intramesenteric, 10–11
  isochronic, 76, 78, 80
  isotopic-orthotopic, 71, 76, 78, 80, 330, 338, 342
  somatopleural, 10–11

xenografts, 71, 73, 78, 81, 82, 84, 85
Graft-versus-host reaction, 180
Granulocytes, 49, 52, 53, 57, 204
Granulopoiesis, 206
Growth factor, 346
Gynandromorphs, 99

H

Haldane's rule, 118
Harderian gland, 44
H-2, 179–216, 217–238
  antigen, 60, 218, 221, 222, 228, 229, 230
  anti-H-2 antibodies (serum blocking factors), 222
  barrier, 228
  chimerism, 222, 223
  incompatible (strains, combinations), 220, 222, 226, 227
  reactivity, 227
  recognition, 228
  region, 226, 230
  restricted ion, 225, 228, 229, 230, 233
  stimulating cells, 222
  system, 42
  type, 221, 222, 231
Hemangioblasts, 148, 151
  hemangioblastic lineage, 151, 156
Hematogenous theory, 182, 186
Hemoglobin
  patterns, 136
  switch, 138
Hensen's node, 81–84
Hermaphroditism, 393–396
  asymmetry, 394, 396
  hermaphrodite mosaics, 394
  human hermaphrodites, 396
  mixed sex (XX↔XY), 387–388
HSV tk herpes simplex virus thymidine kinase, 417, 420
Human x mouse hybrids, 414
H-Y antigens, 225, 228, 230, 392
Hypoblast, 7, 73, 74, 76, 114

I

Immune response antigen (IA)
  incompatible cells, 227
  region of H-2, 226

Idiotype(s), 233
    idiotypic determinants, 224, 234
    network, 234
Immune system, *see also* Lymphocytes and Primary lymphoid organs, 179–216, 217–238
    alloaggression, 225
    altered self, 224, 228
    humoral immunity, 180
    hypersensitivity reactions, 180
    immune response (*Ir*) genes, 226, 227, 228, 229, 230, 232
    self–nonself discrimination, 218, 219, 223, 225
    self-tolerance, 218, 219, 224, 229, 233
    tolerance, 169, 221, 222, 223, 226
Immunosurgery, 25
Induction, 72, 81–87, 402, 403
    inductive signal, 402, 403
    inductive tissue interaction, 402
    inductive wave, 402
    primary, neural, 72, 82–84
    of secondary axis, 72, 82, 84–87
Inner cell mass (ICM), 24, 25, 29, 30, 95, 97, 103, 104, 116, 416, 432, 433, 438
    allocation of cells to, 92–95
    clonal development of, 96
    lineage, 96, 106
    restriction of cells to, 101–103
Interactions
    cell–cell, 205–209
    neuron–glia, 364
Intersexes (mammals), 387, 389
Interstitial cells, 382
Intramembranous bones, 244
Iontophoresis, 40

K

Kidney, 56, 60, 61
    embryonic, 401
    glomerular basement membrane, 407
    glomerular endothelium, 405, 406, 407
    glomerulus, 401, 405–407
    mesonephros, 16, 391
    mesonephric region, 122, 389
    metanephros, 401
    metanephric blastema, 406

L

Laminin, 202
Laryngeal cartilage, 270
Leech, 90
Limbs, development of, 281–310
Limpet Crepidula, 381
Lineage
    of cell, 71, 72, 89, 90, 91, 93, 95, 96, 101, 102, 104, 106, 107, 282, 307
    myogenic cell lineage, 296
    origin of cell lineage, 290–300
    purkinge cell, 357–359
    restriction of cell lineage in mammals, 91
    somatic, 112
Lymphocytes, 10, 179, 186, 206
    B lymphocytes, 180, 204, 207, 213
    cytotoxic T cells (Tc), 218, 221, 222, 224, 225, 226, 228, 230, 232, 233
    differentiation, 183
    helper T cells (Th), 218, 219, 221, 224, 226, 228, 230, 232, 234, 235
    migration, 180
    mixed lymphocyte reaction (MLR), 221, 222, 227
    origin, 180, 181–188
    precursors, 183, 186, 190, 191, 198, 199, 213
    suppressor T cells, 218, 219, 222, 223, 224, 232, 234
    T lymphocytes, 180, 199, 213
    T cell–B cell collaboration, 226, 227, 229

M

Major histocompatibility complex (MHC), 224, 226, 228, 229
    antigens, 218, 219, 221, 224, 232
    class I antigens, 218, 219, 221, 230
    class II antigens, 218, 219, 221
    restriction, 218, 219, 229–231, 232, 233
    self-MHC, 224, 225
    tolerance system, 224
Masculinizing agent, 389, 392, 393
    SDM antigen in free martinism, 393
    XY cells (mammals), 389, 392
Maternal antibodies, 167
Meiosis, 115, 119, 120, 121, 122, 123
    inducing substance, 122

Index

meiotic cells, 122
polar granules, 112, 115
posterior pole cytoplasm, 112
prophase of, 119, 122, 395
Membranous labyrinth, 254, 255
Metabolic cooperation, 413, 422
Metanephros, see Kidney
Mesoblast, 78, 80, 81, 84, 85
  ingression of the prechordal mesoblast, 81
Mesoderm, 77, 95, 148, 151, 154, 244, 250, 251, 255, 260, 261, 267, 269, 270, 276
  extraembryonic, 95
  lateral plate, 148, 151, 154
  paraxial, 244, 250, 251, 255, 260, 261, 269, 270, 276
Mesonephros, see Kidney
Microdensitometry, 51
Microenvironment, 142
Microinjection, 40, 410, 413, 422, 423
  into the blastocyst, 411, 412, 413, 415
  into the zygote, 410, 422, 423
Migratory capacity, 404
Minor histocompatibility antigens, 228, 231, 232, 233
Morphogenetic interactions, 401
Mosaics, Mosaicism, 56, 57, 58, 165, 166, 176
  ABO, 176
  BALB/cWt, 394, 395, 396
  drosophila, 46
  embryos, 16–18
  hermaphrodites, 394
  Rh, 176
  testis (XX/Sxr), 123
  tissue, 169
  X-inactivation, 52, 59
  XX/XY, 176
Mouse strains, 370
  A/J, 370, 371, 373
  CBA/T$_6$T$_6$, 403
  C57BL/6J, 370, 371, 373
  129/J, 411
  DBA/2J, 371
  T (X; 16) 16 H/X Sxr mice, 394–395
Muscles, 250, 251, 281–310, 379
  branchiomeric, 250, 251
  development of, 281–310
  laryngeal, 250, 251
  ocular, 250, 251
Mutant, see Gene and mutants; Mutation

Mutation, 410, 413, 417, 422, 423, 434, see also Genes and mutants
  chemical-induced, 423
  in chromosomal genome, 410
  in mitochondrial genome, 410
  mutagens, 418, 423
  mutants, 409, 410, 411, 412
  mutant genes, 440
  mutant cells, 440
  polyoma viruses, 417, 421
  radiation-induced, 423
  selection of mutants, 411, 412
Myelin basic protein (MBP), 366
Myelopoiesis, 204, 210

N

Nemertine worms (Lineus), 383–386
Nervous system (mammalian), 353–368
Neural crest, 8, 9, 186, 187, 188, 241–280, 313–352, 378, 379
  migration pathways of, 325, 328, 330, 337, 340, 344, 347, 348
Neural plate, 84, 85
Neural tube
  excision, 8
  rudiment, 9
Neurectoblast, 76
Neuropeptides, 314–346
  enkephalins (EK), 341, 342
  neurotensine, 342
  peptidergic neurons, 341, 342, 347
  somatostatin (S), 341, 342
  substance P (SP), 330, 331, 332, 333, 341, 342
  vasoactive intestinal peptide (VIP), 341, 342
Neurotransmitters
  acetylcholine (ACh), 314, 346, 348
  catecholamines (CA), 314, 339, 342, 346, 348
Non- or low responder, 226, 229, 230, 232
Notocord, 8
Nuclear markers
  centromeric heterochromatin, 54
  chromatin marker, 48–51
  chromosomal markers, 48, 53–54
  DNA marker, 62–63
  Pelger anomaly, 51

Nuclear markers *(cont.)*
  quail-chick marker system, 6, 18, 48, 49, 50, 51, 62, 71, 73, 74, 76, 78, 81, 84, 114, 183, 187, 194, 318, 320, 321, 325, 326, 330, 337, 346, 403, 404, 405, 407
  nuclear marker based on ploidy, 51
  Robertsonian translocations, 54
  sex chromosomes, 54, 114, 182, 192
  T6 chromosome, 54
Nucleotide metabolism, 412
  defects in, 412
Nude mice, *see* Genes and mutants

O

Oocytes, 112, 117, 119, 121, 122, 384, 395
  genetically determined characteristics of, 126
  growing, 122
Oogenesis, 112, 115, 120, 121
  X chromosomes, 121, 122, 123
Optic cup, 4
Organizer, 4
Osteoclasts, 49, 53, 140
Otic capsule, 245, 255, 256, 273
  placode, 254
Ovary, 115, 119, 120, 121, 122, 125, 126, 127, 386, 389–394
  correlation between right and left ovaries, 391
  follicle cells, 391
  ovarian follicles, 119
  thecal cells, 391
Ovotestis, 120, 384, 389, 392, 394–397
  distribution of ovarian and testicular tissue (mouse), 394
  nonrandom organization of, 395

P

Pancreas, 16
Para-aortic foci, 151
Parabionts, 12, 14, 182, 192
  parabiosis, 12, 13
  parabiotic chimeras, 382
Parthenogenetic embryos, 116, 117
  diploid, 116

Patterning, 262–269, 276
  definition, 263
  muscles, 267
  skeletal tissues, 263
Peripheral nervous system (PNS) of Birds, 313–352
Pharyngeal pouches (III and IV), 181, 183
Phytohaemagglutinin, 23
Pigment, 42–48, 61
  melanin granules, 45, 52
  melanocytes, 45, 375, 376, 378
  pigmentation, 61
  pigmented feathers, 348
  skin colour, 170, 171, 172
Placode, 319, 320
  otic, 254, 322
  placodal cells, 330, 333, 345
Plasminogen activator, 209–212
Polarized cell division, 92, 93
Postimplantation embryos, 95, 96, 97, 102, 105, 418, 423, 434, 436
Preimplantation embryos, 18, 20, 96, 418, 419, 423
Primary gynandromorphous state, 384
Primary lymphoid organs
  Bursa of Fabricius, 10, 140, 144, 146, 180, 182, 183
    ontogeny, 203–213
  functions, 180
  ontogeny, 179–216
  thymus, 133–163, 179–216, 217–238
    colonization, 188–203
    histogenesis, 180, 181–203
Primitive streak, 76–78, 81–82, 84–86, 391
  final elongation, 81
  induction of, 85, 86
  ingression, 77
Progeny testing, 119
Purkinje cells (PC), 56, 60
  cerebellar, 100, 105
  degeneration (pcd), 358, 359–360, 361, 364, 365
  lineage, 357–359

R

Regulation (size)
  in chimeric embryo, 432, 433
  in hydra, 382

Index

Reproductive ducts, 396
  Müllerian duct, 396
  Wolffian duct, 396, 401, 404, 405
Retroviruses, 418, 419, 422, 423
  expression, 422
  Moloney murine leukemia virus (M-MuLV), 418, 419, 423
  substrain Mov14, 418
Ribs, development of, 300–302

S

Satellite cells, 344, 345
Schwann cells, 326, 333, 334, 336, 338, 344
Secondary sex glands, 391
  epididymis, 391
  seminal vesicles, 391
Self–nonself discrimination, *see* Immune system
Self-tolerance, *see* Immune system
Sensory ganglion, 313, 319, 328, 330, 336, 346, 347
  acoustic, 322
  cranial sensory, ontogeny, 319–326
  geniculate, 320, 322
  jugular and superior complex, 320, 330
  nodose, 320, 325, 326, 332, 333, 345, 347
  petrosal, 320, 325, 326, 330
  superior cervical (SCG), 319, 330, 337, 338
  trigeminal, 319, 320, 321–322, 330
  vestibular, 322
Serologically detected male (SDM) antigen, 392, 393, 396
  female XO mice, 392, 393
Sex-determining mechanism, 381–391
  chromosomal sex, 115, 120, 127, 386, 387, 391
  heterogametic (ZW) sex, 384, 386
  homogametic (ZZ) sex, 386
Sex-linked albino marker, 387
Sex ratio, 118, 387, 394
Sex reversal, 112, 113, 120, 127, 384, 386
Sex vesicle, 122
Sexual differentiation, 382, 384, 386, 387, 389, 390, 393
  induction of testicular development, 393
  influence from the brain region, 386
  XX↔XY sexual differentiation, 389

Skeleton, development of, 281–310
  birds, 282–302
  limbs and body wall, 281–310
  mammals, 302–307
Somatic environment, 120, 125, 126, 127
Somatopleure, 6
Somitomeres, 244, 251, 268, 274
  somitomeric mesoderm, 260, 276
Spermatogenesis, 117, 121, 126
  primary spermatocytes, 119, 121
  prospermatogonia, 121
  spermatogonia, 119
  spermatozoa, 117, 121, 125, 126, 127
Spinal ganglia, 326–330
  dorsal root ganglion (DRG), 326, 328, 330, 333, 341, 344, 345, 346
S-100 protein, 336
Stem cells, 134, 135, 138, 142, 144, 146, 147, 148, 152, 153, 154, 155, 156, 159, 160, 180, 182, 189, 207, 212
  bone marrow, 134, 151
  hemopoietic, 139, 148, 154, 161
  intraembryonic, 136, 137, 138, 139, 144, 147, 148
  site of formation, 142
  thymic, 151, 206
  yolk sac, 135, 136, 137, 138, 142, 147, 148, 151
Sternum, development of, 300–302
Syndrome, 187, 188, 413
  Di George's, 187–188
  Lesch–Nyhan, 413

T

T antigens, 416, 417
Teeth, 263
Teratocarcinoma, 117, 160, *see also* Tumours
  chimeras, 409–424
  ovarian, 411
Testicular feminization, 52, 117
Testis, 115, 117, 118, 119, 121, 122, 125, 126, 127, 384–397
  determining region, 122
  Leydig cells, 391, 392
  seminiferous tubules, 393
  Sertoli cells, 117, 125, 391, 392

Testis *(cont.)*
    testicular differentiation, 396, 397
    testis cords, 392, 395
Testosterone, 53
    testosterone treatment, 206, 207, 209
Thoracic aorta, 148, 151
Thymus, *see* Primary lymphoid organs
Tissue growth, 439
    cell death, 439
    coherent growth, 439
T lymphocytes, *see* Lymphocytes
Tolerance, *see* Immune system
Transfer
    embryo, 23
Transmission ratio distortion, 118
Transplantation
    artifacts, 271–273
    back-transplantation experiments, 344–346, 347
    heterotopic, 338–342
    interspecific tissue, 4
    orthotopic, 321–325
Tritiated thymidine (^{3}H-TdR), 76, 77, 82, 95, 96, 318, 320, 326, 345
Trophectoderm, 24, 25, 27, 29, 30, 104, 105, 106, 416, 432, 433, 438
    allocation of cells to, 92–95
    restriction of cells to, 101–103
    trophoblast (giant) cells, 94, 104, 438
T_6T_6 karyotypes, 403
Tumours
    malignant, 411
    melanoma line, 411
    testicular, 411
Turtles, 381
    temperature of eggs, 381
Twinning, 119

U

"Unbalanced" strain combinations, 388, 393

V

Vascular anastomosis, 13, 119
Vital markers (dyes), 40, 73, 74
    charcoal, 73, 74
    HRP horseradine peroxydase, 90, 93
    ion oxide, 73, 74

Virus, *see also* DNA viruses; Retroviruses
    expression, 418, 419, 422
    integration, 418
    transcription, 419
    vectors, 416, 420, 421, 422

W

Wood lemmings, 394

X

Xenopus, 90, 386
    HRP-marked blastomeres in, 90
XO/Sxr, 122
XX/Sxr, 122
XX↔XX chimera
    female, 121, 123
    Sxr, 120
XX↔XY chimeras, 120, 121, 122, 123, 171, 172, 388, 391, 392, 393, 395, 396
    female, 120, 121
    mixed sex, 387
    ovary, 121
    testis, 123
XXY
    chromosome constitution, 120
XY sex bivalent, 122
XY↔XY chimeras
    male, 123
    ram, 123

Y

Y chromosome, 121, 123
Yolk sac, 182, 197
    blood islands, 182
    endoderm, 116
    visceral, 437, 440

Z

Zone of polarizing activity (ZPA), 283–285, 289, 290

THE LIBRARY
ST. MARY'S COLLEGE OF MARYLAND
ST. MARY'S CITY, MARYLAND 20686